Springer Geology

More information about this series at http://www.springer.com/series/10172

Alok Krishna Gupta

Origin of Potassium-rich Silica-deficient Igneous Rocks

Springer

Alok Krishna Gupta
Mineralogy and Petrology
University of Allahabad
Allahabad
India

ISBN 978-81-322-2082-4 ISBN 978-81-322-2083-1 (eBook)
DOI 10.1007/978-81-322-2083-1

Library of Congress Control Number: 2014947649

Springer New Delhi Heidelberg New York Dordrecht London

Printed on acid-free paper

Springer is part of Springer Science+Business Media (www.springer.com)

Dedicated to
Late Professor Kenzo Yagi,
(Hokkaido University, Japan),
Professor William S. Fyfe,
(University of Western Ontario, Canada)
and Professor Peter J. Wyllie,
(California Institute of Technology, USA).

Preface

In this book the author attempts to provide up-to-date information about the geochemistry, exotic mineralogy, petrology, and experimental studies on ultrapotassic feldspathoid-bearing mafic and ultramafic rocks, which are quite distinct from the rocks of basalt family. The parental liquids for this intriguing group of rocks bear definite signature of their deep mantle source. Modern developments in earth science show that carbonates, N_2, O_2, P, H_2O, and K-rich crustal materials can be recycled into the deep earth causing mantle metasomatism. Partial melting of such a modified mantle source could be responsible for the generation of such K-rich silica-deficient magmas.

Age determination of ultrapotassic rocks show that the origin of their parental liquids are related to the last few hundred million years of mantle evolutionary history in contrast to generations of komatiitic magmas, which were partial melting products of the mantle, evolved mainly during the Proterozoic or Archaean, albeit with a few exceptions.

In the introductory chapter, the geochemical and mineralogical peculiarities of K-rich silica-poor rocks in contrast to alkaline basalts have been described. This chapter also describes the objective and scope of this book. In the next five chapters of this volume, the characteristic mineralogy (Chap. 2), classification (Chap. 3), their distribution in all the continents and oceanic islands (Chap. 4), major minor, trace and isotopic geochemistry have been discussed (Chap. 5). Physico-chemical constraints for the crystallization of leucite and melilite, their P-T stability together with the oxygen fugacity condition of their formation, are described in Chap. 6.

Experimental studies on the system nepheline-kalsilite-SiO_2 in air and under 1, 2, 3, 5, and 20 kb in presence of excess water have been summarized in Chap. 7. Genesis of pseudoleucite and the problems related to survival of leucite beyond Tertiary due to analcitization are discussed in this chapter.

Incompatible relationship between the mineral pairs, leucite and sodic plagioclase and occurrence of the former, in association with calcic plagioclase can be understood from the study of the systems leucite–albite and leucite–albite–anothite under atmospheric pressure. These results are summarized in Chap. 8.

In Chap. 9, phase relations in the system diopside–nepheline–sanidine studied in air and under 1, 2, 10, and 20 kb in presence of excess water, are summarized. The course of crystallization of leucite-bearing tephrites and basanites is also described in this chapter with reference to the system forsterite-diopside-leucite-anorthite. Association of silica-deficient feldspathoidal rocks occurring in close proximity to silica-saturated feldspar-bearing lavas can be very well understood from the study of the diopside-leucite-anorthite-SiO_2 system. Experimental results of various joins of these systems are described also in Chap. 9.

In Chap. 10, genesis of melilite- and leucite-bearing rocks are described in detail with reference to the study of the systems forsterite-diopside-leucite-akermanite and diopside-nepheline-leucite-akermanite under atmospheric pressure.

Experimental study of the system forsterite-diopside-leucite and forsterite-akermanite-leucite under 23 kb in presence of excess water at variable temperatures show that leucite and kalsilite-bearing mafuritic rocks are represented under high pressure in presence of excess water by lamproites and minettes. These results are also discussed in Chap. 10.

In certain petrographic provinces such as Coli-Albani and Somma-Vesuvius of Italy and the Bufumbira province of equatorial Africa, K-rich silica undersaturated rocks occur in close proximity to their silica-saturated analogues. This is also true in the Highwood Mountains region of the USA. Such occurrences may be understood with reference to the detailed study of the system leucite-akermanite-albite-SiO_2, studied under one atmospheric pressure by Gupta and Gupta (1985). Similar studies on the system leucite-akermanite-albite with or without anorthite have been conducted by Dwivedi et al. (2007). These results are discussed in Chap. 11.

The P-T stability of phlogopite can be understood from the study of the system forsterite-kalsilite-SiO_2 in presence of excess water at 1, 2, and 3 kb (Luth 1967). The same system has been investigated at 20 kb in presence of CO_2, H_2O and under dry condition by Wendlandt and Eggler (1980 a–c) and upto 28 kb with or without H_2O, and in presence of CO_2 by Gupta and Green (1988). These studies are discussed in Chap. 11. Foley et al. (1986) also studied the system forsterite-kalsilite-SiO_2 in presence of F up to 28 kb at variable temperatures to determine how presence of F affects stability of phlogopite. His studies are also included in this chapter.

Study of Yoder and Kushiro (1969) up to 30 kb (with or without excess water on the stability of phlogopite has been discussed in Chap. 12. Trones (2002) also investigated the stability of phlogopite up to 70 kb and variable temperatures. High pressure-temperature stability of K-ricterite by Gupta and Venkatesh (1993) and Konzett et al. (1997) has been included in Chap. 12. Experimental results of Massone (1992) on phengite as a source of potassium are also described in this chapter.

Investigation on ultrapotassic rocks under atmospheric pressure and high temperatures has been summarized first in Chap. 13. This is followed by high pressure-temperature studies on synthetic and natural rock systems by various petrologists as available till date. These studies give an insight into the nature of parental source materials of such magmas.

Structural and tectonic control in the migration of leucite-bearing mafic and ultramafic melts are discussed in Chap. 14. These studies, conducted by different structural geologists and geophysicists, are summarized in great detail in this chapter.

In Chap. 16 the genesis of K-rich feldspathoid-bearing lavas is discussed. Different hypotheses, invoked during the first half of the last century involved (1) assimilation between different magmas and rock types (Daly 1910, 1933; Shand 1933), (2) subtraction of eclogites from peridotite or picritic magmas (Holmes 1932; O'Hara and Yoder 1967), (3) gaseous transport hypothesis (Kennedy 1955), (4) zone refining process (Harris 1957), and finally, (5) partial melting model of phlogopite and/or richterite-bearing peridotites (Gupta and Yagi 1980; Gupta and Fyfe 1975). Merits and demerits of different processes have been critically analyzed in this chapter.

It has been emphasized in Chap. 15 that in the genesis of ultrapotassic magmas, partial melting of a fertile mantle source is required. Formation of such a modified mantle is related to recycling of the crustal materials into the deep earth by the process of subduction. Possible causes for the frequent occurrence of K-rich silica-poor volcanic rocks during the last few hundred million year history of the earth are also described in Chap. 15. In Chap. 16, a brief summary of the entire volume is given and petrologic conclusions are discussed.

Acknowledgments

I am grateful to the editors of the following journals for allowing me to reproduce the following figures from their journals: *Earth Science Reviews* for Figs. 1.3a, b, c, 1.5, and 3.3–3.5 and *Journal of Geophysics* for Figs. 1.4, 13.26, 14.2 and 14.11.

I am indebted to the editors of the following Journal: *Contribution to Mineralogy and Petrology* for allowing me to reprint Figs. 2.6, 2.13, 3.7, 4.11, 4.20b, 4.14, 4.30, 5.9, 5.11, 6.1, 6.12, 6.13, 7.3, 7.4, 10.7, 12.14, 12.15, 13.6a, b, c, 13.7, 13.12a, b, 13. 19 and 13.20a, b, 13.23, 13.24, and 13.25.

Thanks are due to the editor of the *Journal of Petrology* for giving me permission to reproduce Figs. 2.2, 3.1, 3.2, 3.9–3.13, 4.1, 4.12, 4.13, 4.15, 4.19, 4.22, 4.34b, c, 5.5, 5.6, 5.10, 5.12–5.17, 5.19–5.21, 5.23–5.26, 6.10, 6.11, 12.1, 12.2a–j, 12.18, 13.5, 13.17, 13.18, 13.21, 13.22, 14.10, and 14.12.

I thank the editor of the American Mineralogist for allowing me to reproduce the following Figs: 2.9a, b, 2.10a, b, 2.15, 2.16, 7.10, 9.5, 9.7, 9.8, 10.1–10.6, and 12.16. I am grateful to the editor of Mineralogy and Petrology for granting me permission to reprint the following Figs: 2.8, 2.14, 4.26, 12.4, 12.12, 12.13, 13.10–13.12, and 13.23–13.25.

Thanks are due to the editor of the American *Journal of Science* for allowing me to reproduce the following Figs. 8.2, 12.3, 12.5–12.11. I am indebted to the Director Geophysical Laboratory, Carnegie Institution of Washington for allowing me to use Figs. 6.3, 6.6, 8.1, 8.2, 8.6, and 8.7. I am thankful to the editor of Bulletin Mineralogy to allow me to print Fig. 2.11.

I thank the editor of the *Journal of Volcanology and Geothermal Research*, who has accorded permission to reproduce Fig. 2.7. The editor of the *Bulletin Geological Society of America* is thanked for permitting me to use Fig. 2.1.

Thanks are due to John Wiley and Sons for their permission to print Figs. 3.6 and 4.34a. I thank the editor of the Geological Society of America for allowing me to print Fig. 6.2, which appeared in their memoir no. 74. I thank Plenum Co of New York (Fig. 13.8a, b), Blackwell, Melbourne (Fig. 13.8a, b), Chapman and Hall Co., London (Figs. 4.31, 4.33) and Nelson and Sons (Figs. 4.8–4.10, 4.23, and 4.24) for allowing me to reproduce their figures. I thank Geological Society of London for permitting me to print Fig. 3.14.

I thank the editors of the following journals for granting me permission to reproduce the figures described below: *Indian Journal of Geology* (Fig. 4.5a, b, Calcutta University), *Journal of Geology* (Figs. 7.1, 9.1 and 10.10–10.13, Chicago University), and *Mineralogical Magazine* (Figs. 5.27, 7.2, 8.3–8.5, and 14.8).

Thanks are also due to the editors of the *Journal Fortschr Mineral* for permitting me to print Figs. 6.4 and 6.5, to *Indian Journal of Geology* for republishing Figs. 4.3 and 14.9. I thank editor of Geological Survey of America for permitting me to republish Fig. 4.5a, b. Thanks are due to editors of the following Journals for permitting me to publish the following Figs. 4.6, 4.20a, b, 5.1–5.4, 10.14, 10.15, 10. 16a, b, 10.17a, b, c, 10.18, 12.19, 13.3, 13.4, and 14.3 (Elsevier Co., Amsterdam), 4.7 (Geological Society of Australia), 4.18 (Neues Jahrbuch) 4.20a, b, 10.14, 10.15, 10.16a, b, 10.17a, b, c, 10.18, and 14.3 (Lithos, Elsevier Co)), 4.21a,b (*European Journal of Mineralogy*), 4.28 (*Canadian Journal of Earth Sciences*), 7.7–7.9 (*Canadian Mineralogists*), 6.7–6.9 (*Journal of Mineralogical and Petrological Sciences*), 5.22a–5.24a, b, c, 14.4–14.7 (*Nature*, London), 13.1 and 13.2 (*Proceedings of Japan Academy of Sciences*), 7.5 and 7.6 (*Bulletin Mineralogy*), 9.2 (*Indian Journal of Geology*), 12.17 (*Indian Journal of Earth Sciences*), 13.8c, (*Indian Academy of Sciences*), 14.1 and 14.2 (*Tectonophysics*, Elsevier Co), 15.1 (*Chemical Geology*) and 15.2 (*Science*). The Springer Verlag Company is acknowledged for allowing me to print Figs. 8.6, 9.10, 9.11, and 9.13. I thank the editor of the *Journal of Mineralogical and Petrological Sciences*, Japan for permitting me to print Figs. 11.9, 11.10, and 11.11. I thank editor of Meddeleser om Groland, Geosciences for granting me to print Fig. 4.29. I thank editor of the Japanese *Journal of Mineralogical and Petrological Sciences* for according me to print the following Figs.11.9, 11.10, and 11.11.

I thank Robert Kerrich of Saskachwan University for the measurements of nitrogen in the K-rich rocks of Damodar Valley, India. Thanks are due to Mr. Shyam Sunder Pandey of the National Centre of Mineralogy and Petrology (Allahabad University), Mr. Aninda Bose, Ms. Kamiya Khatter, and Evelyn Ebina of the Springer-Verlag Company, for their kind cooperation in typesetting and editing the entire manuscript.

I thank my two sons Dr. Neal S. Gupta (Bryant University, USA), Dr. Rick S. Gupta of Weizmann Institute, Israel, and my daughter-in-law Alokananda Rudra of Ben Gurion University, Israel for supplying me with recent publications on K-rich rocks. I thank my wife Chhaya Gupta for bearing with me, when I had been spending long hours in my study while writing this book, instead of attending on my duties related to household chores. Last but not the least, I thank Prof. M. Arima of Yokahama University, Japan for many discussions on this book and Prof. S.J. Foley of Macquarie University, Australia, for critically reviewing an earlier version of this book.

Contents

About the Author

Dr. Alok Krishna Gupta was born on December 25, 1942. He received his Ph.D. from the University of Pittsburgh (USA), and conducted his post-doctoral studies at the following Universities: Pittsburgh (USA, 6 ½ years), Western Ontario (Canada, 3 ½ years), Ruhr (Germany, 1 year), Hokkaido (Japan, 2 years), Melbourne and Tasmania Universities (Australia, 1 year each). He had been Visiting Professor at the Universities of Texas (USA) and Bristol (UK). He had been the Founder Director of the National Centre of Experimental Mineralogy and Petrology and Founder Head of the Department of Earth and Planetary Sciences, University of Allahabad. He is presently an Emeritus Professor and the Raja Ramanna Fellow of the Atomic Energy Commission. Dr. Gupta has received many awards in his distinguished career, such as Career Award (University Grants Commission), Shanti Swaroop Bhatnagar Prize (C.S.I.R), Rhode Memorial Award (Indian Science Congress Association), Best Teacher Award (Allahabad University), Jawaharlal Nehru Birth Centenary Lecture Award of the Indian National Science Academy and Birbal Sahani Gold Medal (Indian Science Congress Association). He is a fellow of all the three National Science Academies of India. He had been Vice-President of both Indian National Science Academy and National Academy of Science (India). He had been an Associate Editor of the *Proceedings Earth and Planetary Sciences* (Indian Academy of Sciences) and Managing Editor of the *Journal of National Academy Science Letters.*

About the Author

Chapter 1
Introduction

1.1 Potassium-Rich Silica-Poor Igneous Rocks: A Distinct Group, Different from Basaltic Rock Series

Basaltic magmas have been puncturing through the earth's crust and flooding its surface from time immemorial. They have been derived experimentally by partially melting of mantle peridotites (Green 1973; Takahashi and Kushiro 1983; Hirose and Kushiro 1993). Basaltic liquids thus, provide the most direct contact with the nature of the earth at depth and the processes that occur there. For this reason and because basaltic magmas are considered to be parental to other magma types, they have been subjected to a number of experimental and field studies, and consequently two types of basaltic rock series have been recognised by Yoder and Tilley (1962): (1) tholeiite-rhyolite series, and (2) alkali basalt-trachyte series. In the convergent plate boundaries, another basaltic rock series called the calc-alkaline rocks, have also been recognised. There is however, a distinct group of potassium-rich rocks, which is characterised by high K_2O content (≥3 wt%), and K_2O/Na_2O ratio (always >1). In contrast, average basaltic rocks not only contain <1 wt% K_2O (Yoder and Tilley 1962), but their K_2O/Na_2O ratio (Fig. 1.1a, b) is also <1. The SiO_2/K_2O ratio of silica-undersaturated potassic rocks is normally below 15, but in alkali basalts it is >50.

These potassic rocks sometime may have even more than 7–8 wt% K_2O, and in a few localities it may go up to 13 wt% [e.g. Leucite Hills and Highwood Mountains (U.S.A.), West Kimberley (Australia), Latera Caldera, Roccammfina (Italy) and other localities (Fig. 1.1a, b). A plot of K_2O/Na_2O ratio versus SiO_2 (Fig. 1.2a, b and c), shows that the composition of basaltic rocks once again plots at the bottom, but the same ratio in the K-rich rocks from Leucite Hills (U.S.A.), Gaussberg (Antarctica), West Kimberley and New South Wales (Australia), Vico and Roccamonfina (Italy) and other areas exhibits extreme enrichment of K_2O over Na_2O.

A.K. Gupta, *Origin of Potassium-rich Silica-deficient Igneous Rocks*, Springer Geology, DOI 10.1007/978-81-322-2083-1_1

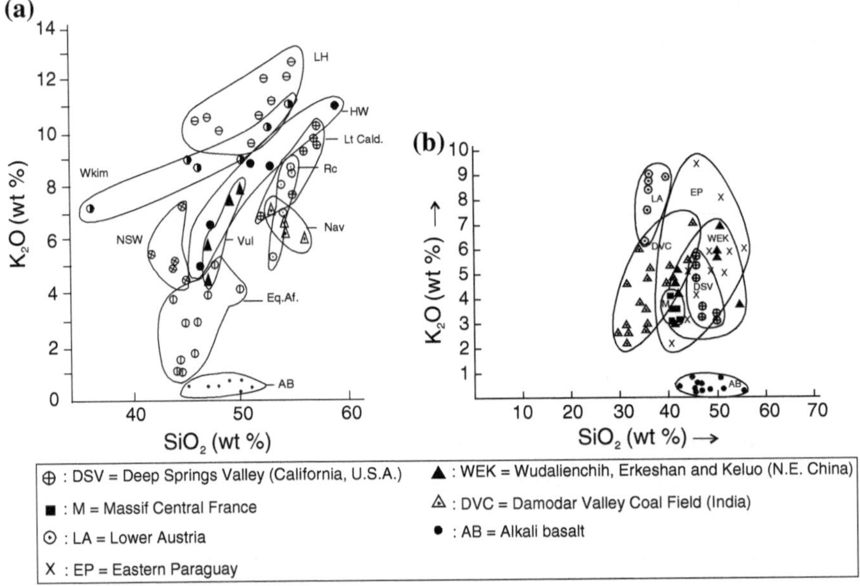

Fig. 1.1 **a** The K_2O versus SiO_2 (wt%) content of potassic rocks from different localities. *Wkim* West Kimberley, *NSW* New South Wales (both from Australia); *LH* Leucite Hills, *HW* Highwood Mountains; *Nav* Navajo Hopi (U.S.A.); *Vul* Vulsini, *Lt Cald* Latera Caldera, Roccammfina (Italy); *Eq.Af* Equatorial Africa; *AB* Alkali basalt. **b** The K_2O versus SiO_2 (wt%) content of potassic rocks from Deep Springs Valley (California, U.S.A.), Damodar valley (India), Wudalianchi, Erke-shan and Keludo (Manchuria, China), lower Austria, Massif Central (France) and Eastern Paraguay

1.2 Minor, Rare Earth and Trace Element Characteristics

The minor, REE and trace element contents of potassium-rich silica-poor igneous rocks from various localities were plotted by Foley et al. (1987) in chondrite-normalised spider-grams (Fig. 1.3a, b and c). The areas shown at the bottom of these figures depict MORB values. It may be observed from these figures that the incompatible elements in all the K-rich rocks from West Kimberley and Gaussberg (Fig. 1.3a), Toro-Ankole of S.E. Uganda (Fig. 1.3b) and most localities of Italy (Fig. 1.3c) shows that barring Ti and Y, all other elements are markedly enriched in the K-rich rocks compared to basalts of different localities.

O'Brien et al. (1991) plotted $^{87}Sr/^{86}Sr$ versus ϵ_{Nd} contents of potassic rocks (Fig. 1.4), and found that the ϵ_{Nd} concentration of MORB basalts are distinctly higher (6.5–10), than that of potassic rocks (noted always below zero). The Sr isotopic ratio of basalts (0.7020–0.7035) is much lower than that of silica-undersaturated potassic rocks from all the localities (e.g. Sunda Arc, Vesuvius, Roccamonfina, Gaussberg, southern Spain, Leucite Hills and Smoky Butte; Fig. 1.5). In the case of Spanish rocks, it is even greater than 0.716 and sometimes may be as high as 0.720.

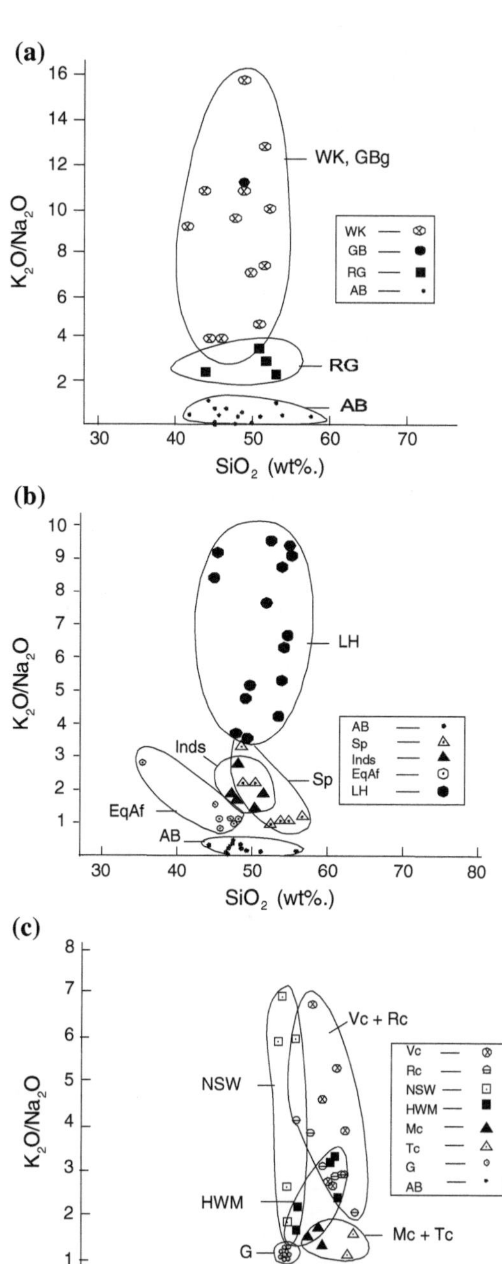

Fig. 1.2 **a** The K_2O/Na_2O ratio versus SiO_2 content of K-rich volcanic rocks of different localities. *WK* West Kimberley (Australia), *GB* Gaussberg (Antarctica), *AB* Alkali Basalt. **b** The K_2O/Na_2O ratio versus SiO_2 content of potassium-rich volcanic rocks from different localities. *LH* Lucite Hills (U. S.A.), *SP* Spain, *EqAf*: Equatorial Africa, *Inds* Indonesia, *AB* Alkali Basalt. **c** The K_2O/Na_2O ratio versus SiO_2 content of potassium-rich volcanic rocks from different localities. *Vc* Vico, *Rc* Roccamonfina (both from Italy), *NSW* New South Wales (Australia), *HWM* Highwood Mountains, *G* Gaussberg (Antarctica), *Mc* Murcia (Spain), *Tc* Tuscany, *AB* Alkali Basalt

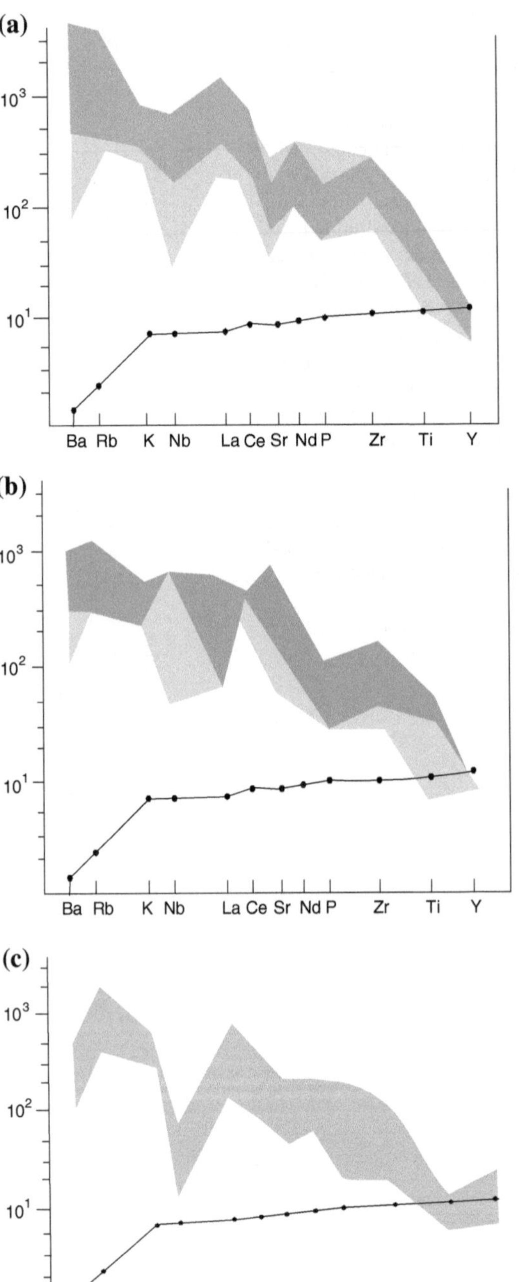

Fig. 1.3 **a** Chondrite-normalised spidergram related to mantle incompatible elements in the K-rich rocks (Group I) from West Kimberley and Gaussberg (*darker shading*). The *solid line* at the bottom depicts MORB values. **b** Chondrite-normalised spidergram related to mantle incompatible elements in the K-rich rocks (group II) from Toro-Ankole, Africa (with standard members in *dark shadings*; for further explanation see Chap. 3). The *solid line* at the bottom depicts MORB values. **c** Chondrite-normalised spidergram related to mantle incompatible elements in the K-rich rocks (group III) from Italy. The *solid line* at the bottom depicts MORB values (after Foley et al. 1987)

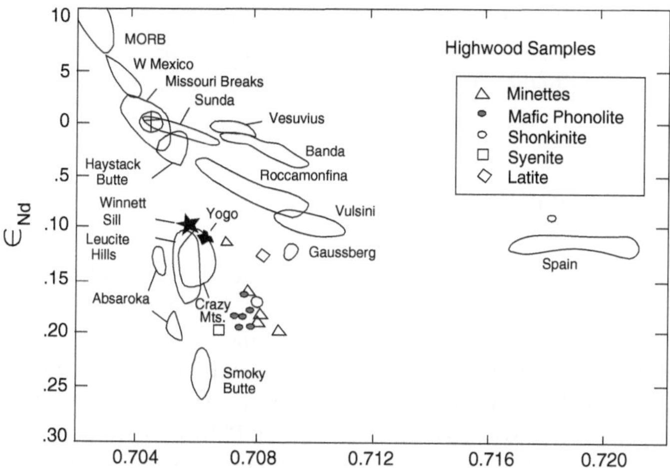

Fig. 1.4 The ε_{Nd} versus Sr^{87}/Sr^{86} ratios of ultrapotassic rocks from different localities (after O'Brien et al. 1991)

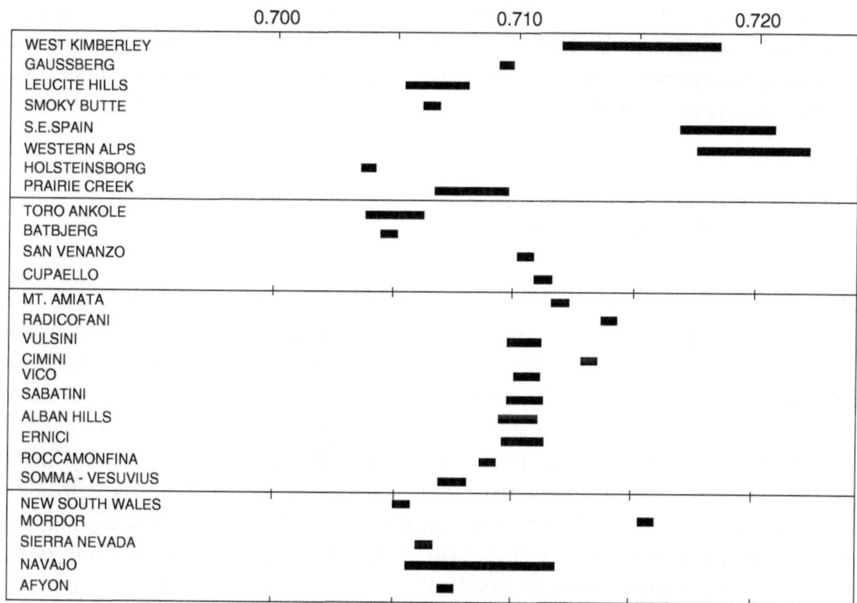

Fig. 1.5 The $^{87}Sr/^{86}Sr$ ratios of K-rich rocks from different localities (after Foley et al. 1987)

The fluorine content of the potassic rock series is particularly more abundant, where the K_2O content is much greater than the Na_2O concentration (Foley et al. 1987; Edgar and Pizalato 1995; Edgar et al. 1994). It was also pointed out by

Edgar et al. (1996) that the amount of F increases relative to K_2O in the following sequence: Alkali basalt →kamafugite→ kimberlite → lamprophyre → lamproite (lamproites having highest F content, also see Aoki et al. 1981).

1.3 Mineralogical Peculiarities

High K_2O content and silica-undersaturated nature of these rocks, are reflected in their exotic mineralogy in that they often contain phlogopite and leucite (sometimes kalsilite). In sodium-rich potassic rocks, nepheline is common, and then melilite appears as a reaction product between nepheline and clinopyroxene (Bowen 1928; Schairer et al. 1962), but plagioclase is then absent. Sodalite group of minerals are rare, but reported from some localities (Chap. 2). Phlogopite present in these rocks is usually titanium-rich. K-richterite is quite a common mineral in the lamproitic rocks. High concentration of titanium as silicate and oxide minerals is a hallmark of these rocks. Thus, sphene, ilmenite, rutile, perovskite and a roedderite-like phase often appear as accessory minerals. Such K-rich minerals like wadeite and priderite are typically found in ultrapotassic lamproites. Olivine is a common mineral in the ultramafic variety, but diopside and salite are common clinopyroxenes. K-feldspar is a ubiquitous mineral in trachytes, minettes and phonolites, whereas plagioclase feldspar is a common mineral in tephrites, basanites and tephri-phonolities. Plagioclase is however, totally absent in kamafugitic rocks and lamproites.

1.4 Scope of This Volume

These ultrapotassic silica-undersaturated melts of mantle origin intrigued such eminent earlier petrologists as von Rath (1864), Hussak (1890), Judd (1887), and Osann (1906). Washington (1906) however, was the first geochemist, who made detailed studies on these rocks form the Roman comagmatic province of Italy. Holmes and Harwood (1937) made extensive studies on such rocks from the east African rift valley in a memoir. They have been treated as a special group by Shand (1943), and Turner and Verhoogen (1960) in their books on Igneous Petrology. Geochemical and petrological reviews on potassic rocks have also been made by (Gupta and Yagi 1980), Foley et al. (1987), Bergman (1987), Mitchell and Bergman (1991) and Gupta and Fyfe (2003). Since the publication of these reviews, we have learnt more about a number of new localities (Chap. 4). Systematic age determination and study of more than forty new complex synthetic systems relevant to their genesis, are now available. Synthesis of all these data is very important.

Different types of tectonism (Chap. 14) are basically outcome of plate tectonic processes (Ninovitch and Hays 1972; Beccalova et al. 1991; Bianchini et al. 2008), which are driven by convective cooling processes in our planet. The earth cools by heat conduction at very specific sites by transport of energy and molten materials. The great many present sites include the oceanic regions dominated by mantle-derived basalts. The extensive subduction zones are characterised by rocks of the andesite family, and there are many random hot spots or mantle plumes, which cause flooding of ocean floors and continents. Highly ultramafic komatiitic lava flows flooded the surface of the Arhean crust, whereas Table 1.1 shows, highly potassic silica-undersaturated K-rich molten rocks are relatively more recent. A very intriguing question comes to our mind, how these convecting patterns have changed over the 4.5 billion-year history of the earth. Since the classic work of the great geochemists of the early part of this century, the elements have been classified according to their main sites of concentration. When one examines the classic ideas, reviewed by Mason (1966), it is clear that potassium is a typical element of the continental crust (average 1.9 %), whereas ocean ridge tholeiites average about 0.26 %. Why are these mantle-derived ultrapotassic rocks with high K_2O/Na_2O ratios so interesting? First, they come from the mantle at high temperatures. They tend to contain anomalous concentrations of many incompatible elements such as K, Rb, Sr, U,F, P, etc., along with compatible Ni, Co, and Cr indicating a mixture of crust and mantle. For key isotopic indicators like Sr^{86}/Sr^{87}, O^{16}/O^{18}, they show great variability. They thus appear to be a result of a geosphere mixing process. In addition, most of these rocks are young (for compilation of ages of K-rich silica-undersaturated igneous rocks of various localities (Table 1.1, with few ancient examples). Is this real or the result of a sampling problem? There is not much well-preserved Archaean crust in the world, and they are not easy to recognise! Over the past decades, there has been a steady debate on the extent of new surface geosphere materials being carried back to the mantle. One needs only to examine the composition of some sea floor spilites with over 4 wt% K_2O, 3 wt % CO_2, which are subducted; and modern observations leave no doubt that, as Gilluly (1971) suggested, sediments must be subducted on a large scale (see Fyfe 1992). More and more, there is evidence that the upper mantle (and perhaps even the lower mantle) is not homogeneous but rather like a fruitcake (Gupta and Fyfe 2003) and that there are thermal anomalies in the mantle resulting from deep mantle plumes. After deep subduction, the modified mantle is partially fused, releasing these unusual K-rich silica-undersaturated rocks. Do these unusual rocks formed due to flushing out from the mantle of left over subduction materials? An elegant example of this mantle complexity has recently been provided by Pilot et al. (1998, pp. 393, 678), who have reported zircons of ages 330–1600 Ma in mid-Atlantic ridge gabbros! Thus, the ultrapotassic igneous rocks perhaps provide evidence on mixing processes, thermal evolution and variation of these processes over time (cf komatiites). We need more details on the total chemistry and isotopic variation of these rocks.

Oceanic crustal samples including sediments might have been subjected to recycling into the mantle (Jia et al. 2002; Avanzinelli et al. 2009) in order to explain high Sr isotopic ratio and extreme potassium enrichment. Thus, metasomatism could develop directly and continuously from subducted potassium-bearing crust from shallow levels to a maximum depth of 300 km. The analogy between mantle processes and chromatographic fractionation in the laboratory has been used by Navon and Stolper (1987) to explain large-scale mantle metasomatism. The question is whether such a method related to mantle metasomatism by the process of

Table 1.1 Ages of potassic rocks from different localities

Sl.	Locality	Rock types	Age
1.	Leucite Hills, Wyoming, U.S.A.	Lamproite	1 Ma
2.	Navajo Province, Arizona, U.S.A.	Minette	28–19 Ma
3.	Dulce dyke swarm, New Mexico, Colorado, U.S.A.	Minette	27–26 Ma.
4.	Two Butte, Colorado, U.S.A.	Minette	35–27 Ma
5.	Central Sierra, Nevada, California, U.S.A.,	Minette	3.8–3.4 Ma
6.	ElkheadMts, U.S.A	Minette	10–8 Ma
7.	Colima Graben, Mexico	Minette,	Pliocene-present
8.	Alto-Paranaiba, Brazil	Kamafugitic rocks	57 + 0.06
9.	Matheson, Ontario, Canada	Minette	2.7 Ga
10.	Gaussberg, Antarctica	Lamproite	Holocene
11.	Virunga province, Zaire and Ruanda	Leucitetephrite	3 Ma-Present
12.	Vera, SE Spain	Lamproite	Late Miocene, 17-6 Ma
13.	Roman Province Italy	K-rich volcanics	1.3 Ma-Present
14.	Scotland and Northern England	Minette	Late Silurian-Early Devonian
15.	Devon, SW England	Lamproite	Stephanian
16.	Western Alps, Switzerland, Italy	Minette	33–29 Ma
17.	Aeolian Arc	K-rich volcanics	1 Ma
18.	East Eifel, Germany	K-rich volcanics	5.7×10^5–3×10^5 years B.C
19.	Muriah, Indonesia	K-rich volcanics	Late Pliocene
20.	Wudalianchi, Erkeshan and Keluo, N.E. China	Leucite-tephrite tephri-phonolite	9.6–7.0 Ma
21.	Lingzizong, Tibet	K-rich volcanics	50 + 1
22.	Tristan Da Cunha, Island	K-rich volcanics	Recent
23.	Utsuryo Island	K-rich volcanics	Recent
24.	Tezhser alkali complex, former U.S.S.R.	K-rich volcanics	39–37 Ma.
25.	Artem, former U.S.S.R.	K-rich volcanics	Neogene
26.	Ussuri, former U.S.S.R.	K-rich volcanics	Neogene
27.	Tokko complex, former U.S.S.R.	K-rich volcanics	Quaternary
28.	Pkhrutski massif, former U.S.S.R.	K-rich volcanics	Eocene
29.	Talyshskii zone, former U.S.S.R.	K-rich volcanics	Eocene
30.	Pyatistennyl, former U.S.S.R.	K-rich volcanics	Palaeogene
31.	Pamir, former U.S.S.R.	K-rich volcanics	Palaeogene
32.	Sandyk massif, former U.S.S.R.	K-rich volcanics	Permian
33.	Irisu complex, former U.S.S.R.	K-rich volcanics	Quaternary

(continued)

Table 1.1 (continued)

Sl.	Locality	Rock types	Age
34.	Kaindy, former U.S.S.R.	K-rich volcanics	Middle to early Permian
35.	Yakokut, former U.S.S.R.	K-rich volcanics	Quaternary
36.	Andriyanovka, former U.S.S.R.	K-rich volcanics	20 + 3 Ma.
37.	Elpinskii, former U.S.S.R.	K-rich volcanics	14.5 Ma
38.	Anomaly Pipe, former U.S.S.R.	K-rich volcanics	Upper Mesozoic
39.	Molobo, former U.S.S.R.	Lamproite	122–92 Ma
40.	Rododendron, former U.S.S.R.	Shonkinite	166 ± 5 Ma
41.	Kamchatka, former U.S.S.R.	Trachybasalt, tephrite	20 ± 3 Ma
42.	Tommot, former U.S.S.R.	Leucitetephrite	167–125 Ma
43.	Omolon, former U.S.S.R.	Leucitelamproite	Palaeocene
44.	Damodar Valley, Eastern India	Leucitelamproite and Olivine lamproite	105 Ma
45.	West Kimberley, Australia	Lamproite	24–17 Ma
46.	New South Walse, Australia	Leucitite	14–10 Ma

Navon and Stolper is feasible? Numerous field-oriented studies of Alpine massif, ophiolites and xenoliths show that a common mode of transport in the upper mantle is through the conduits formed by hydraulic fracturing. Does the method of fluid migration related to the presence of non-percolating fluid, increase pore pressure approaching the lithostatic pressure? Do the isolated pockets of fluid nucleate microcracks? With the increase in strain, do the microcracks interact to produce enechelon of crack arrays? Does the array of cracks extend and coalesce further to form fracture systems? These are the questions to be addressed, when one considers the methods of large-scale mantle metasomatism.

Phlogopite, K-richterite and apatite found in the mantle xenoliths within the K-rich lavas, are enriched in fluorine (Edgar et al. 1996). The study of Vukadinovic and Edgar (1993) on hydroxy-fluor-phlogopite, hydroxy-fluor K-richterite and hydroxy-fluor apatite however, showed that $D_f^{crystal/liquid}$ is usually >1. Fluorine therefore appears to be a compatible element in the mantle. Why then the K-rich partial melts of ultramafic mantle (containing these minerals) are enriched in fluorine, when they should have been retained in the mantle phases? The K-rich melts are also quite high in the TiO_2 contents. What is the mechanism for Ti-enrichment in such melts?

The upper atmosphere is dominated by the presence of nitrogen. It is also an essential constituent of many biological components particularly protein and their degradation products (Hall 1999), and it is therefore present in soils and sediments. Igneous and metamorphic rocks are in general poorer in nitrogen than most sediments.

The ammonium ion is very similar in its properties to the potassium ion, which is substituted readily by NH_4^+ as the ionic radius of K^+ (1.46 A°) is close to NH_4^+ (1.61 A°). Besides, their coordination number is similar. Like Rb^+ and Cs^+, NH_4^+

ion should therefore, be present in K-feldspar and different types of micas (e.g. phengite, biotite etc.) and clay minerals in the sediments.

Some peralkaline minettes are characterised by high NH_4^+ content (Hall 1999), this might suggest that parent magmas for lamproites and other K-rich rocks were generated by partial melting of secondary protoliths, subducted into the mantle during crustal recycling. Some studies conducted by Jia et al. (2002) on nitrogen content of ultrapotassic rocks should therefore be of importance in this regard.

Some interesting questions arise in case of potassic volcanism in the regions related to continental rifting, which are associated with a mantle plume or sets of plumes? In that case what are the tectonic processes involved? What kind of metasomatic reactions take place in the subcontinental lithospheric mantle boundary layer? These are the questions that have been addressed in this volume.

Chapter 2
Mineralogy

Potassium-rich silica-undersaturated mafic igneous rocks are characterised by the presence of leucite, K-feldspar, clinopyroxene and Fe–Ti oxides. Plagioclase feldspar commonly appears in tephrites, tephritic phonolites, and phonolitic tephrites. In sodium-rich potassic rocks such as basanites, forsteritic olivine appears along with nepheline. Whereas in lamproites and lamprophyres, phlogopite is a very common mineral, K-richterite may also appear along with the above-mentioned minerals (Wagner and Velde 1986). In mildly sodium-rich potassic rocks, nepheline-kalsilite solid solutions are commonly observed. Kalsilite is particularly present in mafuritic rocks. Melilite, whenever present, is formed by reaction between clinopyroxene and nepheline (Bowen 1922; Schairer et al. 1962); haüyne and sodalites may be accompanying phases along with them analcite occurs as an alteration product of leucite. Pseudo-leucite (an intergrowth of nepheline and K-feldspar) may occur instead of leucite in rocks emplaced before the Tertiary period. Priderite and wadeite are two accessory phases commonly associated with lamproitic rocks. Mg-spinel ($MgAl_2O_4$), perovskite, titanomagnetite, pseudo-brookite, ilmenite, chromite, a roedderite-like mineral and apatite are other accessory phases. Recently, Mihajlovic et al. (2002) have reported the occurence of two accessory minerals rondorfite, $Ca_8Mg[SiO_4]_4Cl_2$, and almarudite, $K\square(Na)_2(Mn, Fe, Mg)_2(Be, Al)_3[Si_{12}O_{32}]$, and a iron-rich wadalite, $Ca_{12}[(Al_8Si_4Fe_2)O_{32}]Cl_6$, from the Bellerberg (Bellberg) volcano, Eifel, Germany. Carbonates (a solid solution of $CaCO_3$, $MgCO_3$ and $FeCO_3$) occur quite often as groundmass phases. Melanite garnets are also rarely observed.

2.1 Leucite

This mineral has a cubic symmetry at temperatures higher than 630 °C, but below this temperature, it becomes tetragonal. Most leucite grains therefore, show repeated twinning on the (110) crystallographic plane, which is related to cubic-tetragonal inversion. The heat capacity measurements indicate that the transition is continuous and of second order (Lange et al. 1986). Gatta et al. (2008) studied

A.K. Gupta, *Origin of Potassium-rich Silica-deficient Igneous Rocks*,
Springer Geology, DOI 10.1007/978-81-322-2083-1_2

elastic and structural behaviour of a natural tetragonal leucite from the volcanic Lazium district (Italy). Their investigation at high pressure by an in situ single-crystal X-ray diffraction with a diamond anvil cell under hydrostatic conditions, showed a first-order phase transition, never reported in the literature before. They observed the transformation from tetragonal to triclinic symmetry at P = 2.4 ± 0.2 GPa. The phase transformation is accompanied by a drastic increase in density of about 4.7 %.

In slightly older rocks, leucite is often altered to pseudoleucite by reaction with a Na-rich fluid (Gupta and Fyfe 1975). Hovis et al. (2002) have studied two synthetic solid-solution series, namely, analcime to Rb-leucite and analcime to Cs-leucite (pollucite) to understand more fully the thermodynamic and structural behaviour of analcime-leucite and similar mineral systems. Unit-cell dimensions and volumes in these series expand with the substitution of analcime component in either Rb-leucite or pollucite, as H_2O molecules structurally replace Rb or Cs. Unit-cell volumes of the above two minerals vary linearly as a function of composition. During symmetry changes from tetragonal to isometric, as in the Rb-bearing series, the slope of the volume expansion curve changes. Once structures have reached full expansion, slopes related to volume changes flatten and are little affected by additional analcime component. Overall, the data suggest that the analcime-leucite system also can be modelled as close to thermodynamically ideal. The limited solid solution between natural analcime and leucite must be attributed to energetically favored heterogeneous equilibria involving minerals such as feldspars and other feldspathoids, and there is limited immiscibility between the end-members.

Leucite incorporates up to 28 wt% $NaAlSi_2O_6$ (Fig. 2.1) at 0.1 GPa [$P(H_2O)$ = P (Total)] and 800 °C in the join $KAlSi_2O_6$–$NaAlSi_2O_6$ (Fudali 1963). Natural leucite does not contain more than 10 wt% $NaAlSi_2O_6$. Leucites from Leucite Hills incorporate excess silica and alkali over alumina and ferric ion (Carmichael 1967; Barton 1979). It contains up to 8 wt% $KAlSi_3O_8$ at 0.1 GPa [$P(H_2O)$ = P(Total)] and 800 °C (Fudali 1963), and less than 5 wt% $NaAlSi_3O_8$ at 1 atmosphere and 800 °C (Gupta and Yagi 1980). It was observed by the latter workers that up to 6 wt% of $KFe^{3+}Si_2O_6$ is incorporated at 1 atmosphere, but it increases to 7.7 wt% under 0.2 GPa in presence of excess water. Deer et al. (1963) reported up to 1 wt% CaO in leucite from the East African Rift valley. Schairer (1948) confirmed that extensive substitution of Mg occurs in the leucite structure, where the following type of substitution takes place: Mg, Si ≈ Al, Al.

Phenocrystal leucites in mafic phonolites from Highwood Mountains are rare, and often replaced by analcite. Pseudoleucites are also observed in the rocks from this locality. Pure leucite however, occurs inside salite crystals. Such leucites contain minor Fe, Ca, Ba and Na (0.05–0.4 wt%).

Kuehner et al. (1981) observed that a leucite from Leucite Hills (U.S.A.) contain up to 2.90 wt% Fe_2O_3. The leucites incorporate 1,549–2,990 ppm Ba, 306–439 ppm Zr, 309–418 ppm Sr and 436–555 ppm Rb. Leucite is rarely preserved in the Brown Tuffaceous Pumice (Roccamonfina; Luhr and Gigannetti 1987), and analcitization of leucite is common in such rocks.

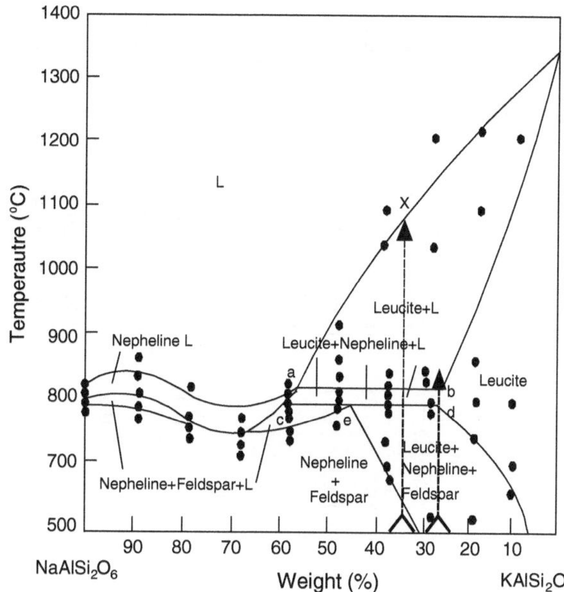

Fig. 2.1 Phase diagram of the system $KAlSi_2O_6$–$NaAlSi_2O_6$ under 0.1 GPa $P(H_2O)$ (after Fudali 1963)

The lavas of Alban Hills, central Italy, often contain fresh leucite crystals with near theoretical composition having 0.27 wt% Na_2O (Table 2.1, Francalanci et al. 1987). They are unzoned and do not have excess silica as observed in leucites from Vico and Vulsini (Holm et al. 1982).

The Gaussberg lavas have leucite crystals, which are generally twinned indicating a slow rate of cooling through the cubic-tetragonal inversion temperature. The compositional range is small and is very close to ideal leucite stoichiometry (Sheraton and Cundari 1980). They did not observe any significant difference in chemistry (Table 2.1). There is also significant amount of Ba (up to 0.3 wt% BaO) in the leucites of Gaussberg lamproites.

Vico lavas (Cundari 1975) have ubiquitous leucites, which show variable crystal development within and between flows from micro-crystalline to phenocrystal phase (1.5 cm across). Leucite often shows signs of resorption, and in rare cases is altered to analcite. Their K_2O content falls within the range of 17.6–20.0 wt%. The Na/K ratio varies significantly from 0.013 to 0.057 wt% even in leucites from the same rock specimen. A phenocrystal core has higher Na/K ratio than the rim. This observation is in agreement with the leucite data of Carmichael et al. (1974) from the similar lava type of the Roman region. The Vico leucites, however, show excess silica over the experimentally determined leucite solid solution (Fudali 1963). The extent of $NaAlSi_2O_6$ in Vico leucites varies from 1.1 to 5.0 wt% (Cundari 1975). Excess silica in leucite has also been reported by Brown and Carmichael (1969) from Korath Range and by Cundari (1973) from New South Wales.

Table 2.1 Analyses of leucite from different localities

	SiO_2	TiO_2	Al_2O_3	Fe_2O_3	FeO^a	MgO	CaO	BaO	Na_2O	K_2O	Total
1	54.54	0.08	21.13		0.36				0.05	21.33	97.49
2	55.42	0.04	23.71		0.51				0.027	21.27	101.15
3	54.16	0.05	21.85		0.76				0.11	22.59	99.27
4	55.90	0.08	20.70		1.27	0.18	0.05	0.01	0.33	21.20	99.72
5	56.00	0.14	21.50		0.58	0.06	0.03	0.01	0.22	21.20	99.74
6	54.00	0.13	22.30		0.63	0.08	0.09		0.42	21.60	99.25
7	54.47		23.42		0.40		0.01		0.12	21.12	99.54
8	54.18		23.64		0.44		0.03		0.20	20.25	99.78
9	57.96		18.38	2.90				0.04	0.04	21.13	100.45
10	59.54		18.27	2.67					0.18	19.60	100.26
11	57.71		19.62	2.03				0.09	0.05	20.77	100.27
12	58.60		18.70	2.20			0.04		0.06	20.70	100.30
13	56.30		20.30	2.00			0.03		0.03	21.50	100.20
14	55.00	0.23	23.70	0.45					0.07	20.70	100.15
15	54.70	0.17	24.20	0.52					0.21	20.14	99.94
16	55.50	0.16	22.20	0.95					0.13	21.04	99.98

(continued)

Table 2.1 (continued)

	SiO_2	TiO_2	Al_2O_3	Fe_2O_3	FeO[a]	MgO	CaO	BaO	Na_2O	K_2O	Total
17	55.30		23.00	0.63			0.03		0.39	20.90	100.20
18	55.00		23.20	0.37			0.04		0.62	20.70	100.00
19	55.00	0.21	21.40	0.96		0.28		0.09	0.10	21.80	99.84
20	54.20–55.60	0.13–0.32	20.40–22.60	0.85–1.30		0.18–0.41		0.27	0.21	20.30–22.70	

[a] FeO indicates total FeO + Fe_2O_3 contents

1–3 Leucites from Leucitites, Alban Hills, Italy (Francalanci et al. 1987)
4–6 Leucites from K-rich rocks of Sierra Nevada, van Kooten (1980)
7–8 Leucites from basanites, Laacher See Area, Germany (Duda and Schminke 1978)
9 Rim of a leucite from wyomingite, Leucite Hills, Wyoming, U.S.A. (Kuehner et al. 1981)
10 Leucite from an olivine orendite, South Table Mountain, Leucite Hills, Wyoming, U.S.A. (Kuehner et al. 1981)
11 Leucite from an orendite, North Table Mountain, Leucite Hills, Wyoming, U.S.A. (Kuehner et al. 1981)
12–13 Leucites from lamproitic rocks, Leucite Hills, Wyoming, U.S.A. (Carmicheal 1967)
14–15 Leucites from leucitites, Begargo Hills, New South Wales, Australia (Cundari 1973)
16–18 Leucites from basanites, Vesuvius (Baldridge et al. 1981)
19–20 Leucites from leucitites, Gaussberg, Antarctica (Sheraton and Cundari 1980)

The Miocene lamproites from the Fitzroy area, West Kimberley (Australia) have leucite phenocrysts, which are typically euhedral and are unaltered (Jaques and Foley 1985). They are weakly birefringent, twinned, and commonly contain inclusions of glass arranged in concentric zone. Most glassy rocks are strongly resorbed and embayed. Leucite phenocrysts containing aluminous spinel inclusions are amoeboid-shaped with strongly rounded coalesced aggregates.

Divalent cations replacing K in leucite is rare (Jaques and Foley 1985), but the CaO content in excess of 1 wt%, has been described from the K-rich volcanic rocks of the East African Rift Valley.

2.2 K-Feldspar

Sanidine is found in tephritic phonolites, phonolitic tephrites, trachytes and/or phonolites from different localities of Italy, Manchuria, Tristan da Cunha and Udsuryo Island. It is also reported from leucite-bearing minettes and lamproites as an essential mineral. For example, they occur in lamproitic rocks from Navajo, Highwood Mountains, Smoky Butte (all from U.S.A.), Spain and Damodar Valley (India). K-feldspar also occurs as minor constituents in K-rich rocks of New South Wales (Australia) and Birunga (equatorial Africa). Inclusion-free phenocrystal sanidine (up to 2.5 mm long) occurs in Brown Leucitic Tuffs from Roccamonfina (Luhr and Giannetti 1987).

K-feldspar and plagioclase coexist in leucite-bearing tephritic rocks of the Eifel region, Germany (Worner and Schminke 1984). Individual sanidine crystals in these rocks are homogeneous containing 75–90 mol% orthoclase (Table 2.2). The tie lines between plagioclase and K-feldspar (Fig. 2.2) suggest equilibrium between two types of feldspars. In this figure, composition of two types of feldspars (open and solid circles with tie lines) as determined by Worner and Schminke (1984) from Laacher See phonolites, is shown. Leucite-bearing minettes from Bearpaw Mountains, U.S.A., contain invariably sanidine ($Or_{43}Ab_{41}An_{16}$ to $Or_{93}Ab_5An_2$, Table 2.2, Macdonald et al. 1992). There is a tendency for the orthoclase content to increase with decreasing magnesia concentration in the bulk rock. These sanidines are Ba-rich containing 7 mol% celsian (Cn) molecule. The sanidines ($Or_{47}Ab_{42}An_6Cn_5$ to $Or_2An_{56}Ab_{41}Cn_1$) in Bearpaw latites have variable chemistry.

The Potassic rocks of Highwood Mountains comprise sanidines, which are by far the dominant feldspars in these rocks (O'Brien et al. 1991). They occur as euhedral phenocrysts in mafic phonolites and trachytes and in coarse-grained shonkinites and malignites. It ranges in composition from K-rich variety ($Or_{88}Ab_9Cn_2Sl_1$) to albitic sanidine ($Or_{64}Ab_{34}An_2$). Sometimes, hyalophane with up to 11.1 wt% BaO ($Or_{44}Ab_{29}Cn_{22}An_3Sl_2$), are present (Sl represents slawsonite, Sr-feldpar). In Highwood Baldy stock, perthitic and antiperthic intergrowths in feldspars ranging in composition from $Or_{88}Ab_{12}$to Ab_{100} are observed.

Table 2.2 Analyses of K-feldspar from K-rich localities

	SiO_2	TiO_2	Al_2O_3	Fe_2O_3	FeO[a]	CaO	Na_2O	K_2O	SrO	MgO	BaO	Total	An	Ab	Or	Cn	Sr-f
1	63.61		19.68		0.21	0.38	1.27	14.26	0.57		0.31	100.29	0.019	0.117	0.864		
2	63.98		19.45		0.13	0.35	2.26	13.41	0.04			99.62	0.017	0.20	0.783		
3	65.10		19.40		0.21	0.39	2.05	12.20			0.06	99.41	2.10	20.00	77.90		
4	64.80		19.00		0.19	0.53	2.14	13.40			0.12	100.18	2.60	19.00	78.40		
5	65.20		19.20		0.20	0.45	1.22	14.10			0.03	100.40	2.30	11.40	86.30		
6	64.30		19.20		0.71	1.06	1.96	12.80			0.08	100.11	5.40	17.90	76.70		
7	57.93		22.21		0.71	1.36	4.93	4.99	3.23		4.57	99.93	6.90	45.40	30.30	8.50	8.90
8	55.51		23.48		0.71	2.20	4.98	3.38	4.16		5.31	99.73	11.30	46.40	20.70	10.00	11.60
9	62.61	0.27	18.83	0.79		0.02	0.31	15.53		0.42	0.41	99.07					
10	64.23	0.41	15.16	2.48			0.66	14.88		0.54	1.92	99.42					
11	64.41	0.40	15.79	2.14			0.64	15.11		0.30	1.52	100.32					
12	63.40	0.09	16.70		1.42	0.20	2.71	13.50		0.03	0.29	98.14					
13	63.30	0.26	17.80		1.96	0.04	3.34	11.60		0.04	1.03	99.37					
14	64.80	0.12	19.20		0.35	0.83	6.74	7.56		0.83	0.05	99.67					
14	63.90	0.27	17.46	1.12			0.44	16.15		0.02	0.25	99.61					
16	63.11	0.22	14.49	3.17			0.32	16.20		0.20	0.08	97.79					

(continued)

Table 2.2 (continued)

	SiO_2	TiO_2	Al_2O_3	Fe_2O_3	FeO[a]	CaO	Na_2O	K_2O	SrO	MgO	BaO	Total	An	Ab	Or	Cn	Sr-f
17	63.61	0.33	15.76	1.93			0.55	15.83			1.21	99.22					
18	63.23	0.29	15.86	2.91			0.32	15.84		0.08	0.50	99.03					
19	64.63	0.07	18.24	0.20			0.56	15.97			0.34	100.01					

[a] Whenever FeO values are given, they indicate total FeO + Fe_2O_3 contents
1–3 K-feldspars from Leucitic Tuffs Roccamonfina, Italy (Luhr and Giannetti 1987)
3–6 K-feldspars from leucite tuffs, Latera Caldera, Italy (Turbeville 1993)
7–8 K-feldspars from Mt. Vulture complex, Italy (Melluso et al. 1996)
9–11 Sanidines from lamproites, Smoky Butte, Montana (Mitchell et al. 1987)
12–14 Sanidines from K-rich rocks of Sierra Nevada (van Kooten 1980)
15 K-feldspar from a Cancarix lamproite, Albacete, Spain (Wagner and Valde 1986)
16 K-feldspar from a Jumilla lamproite, Murcia, Spain (Wagner and Valde 1986)
17 K-feldspar from a Smoky Butte lamproite, Montana, U.S.A. (Wagner and Valde 1986)
18 K-feldspar from a Moon Canyon lamproite, Utah, U.S.A (Wagner and Valde 1986)
19 K-Feldspar from Sisco Lamproite, France (Wagner and Valde 1986)

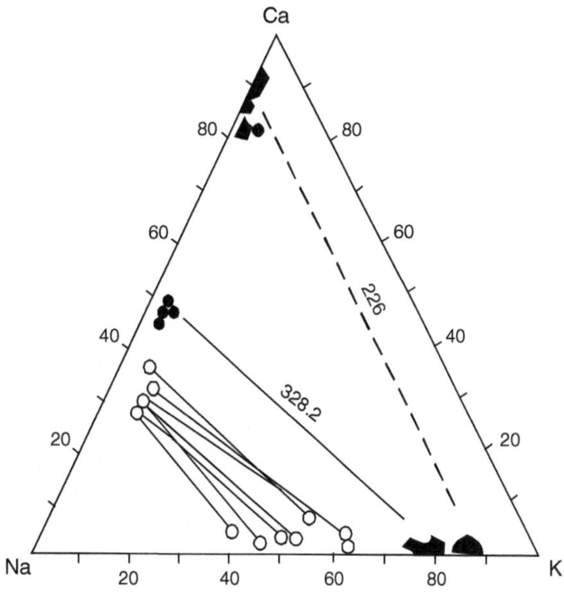

Fig. 2.2 Coexistence of calcic plagioclase in equilibrium with sanidine. There are two sets of alkali and plagioclase feldspars as shown by *solid* and *open circles*. The tie lines joining them suggest that two different types of feldspars are in equilibrium (after Worner and Schminke 1984)

The leucite-bearing rocks of Central Italy (Francalanci et al. 1987) have sanidines with compositions ranging between $Or_{88}Ab_{12}$ and $Or_{77}Ab_{23.}$ The single crystals however, do not show compositional zoning. Such elements as Ba and Sr along with other trace elements, are present in appreciable amounts in K-feldspars. Perini et al. (2003) reported the presence of K-feldspar mega crysts hosted in K-rich rocks at Monte Ciminio volcanic complex Central Italy. These megacrysts are found in leucite-free olivine latites.

The highly potassic lavas from Latera Caldera (Roccamonfina, Italy) has sanidinse phenocrysts with compositions (Table 2.2) ranging from Or_{68} to Or_{90}. More potassic and Ba-rich variants are found in the phonolitic pumices. In Trachytic and phonolitic lavas, sanidine phenocrysts range in composition from $Or_{68}Ab_{32}$ to $Or_{90}Ab_{10}$. Sanidine crystals are observed also in trachytic pumices of Latera Caldera (Turbeville 1993). They are euhedral, unzoned and untwinned, containing <0.13 wt% BaO, and they are free from any mineral and glass inclusions. In tuffs of Latera Caldera sanidine phenocrysts (Or_{69-82}), are strongly embayed and include feldspar blebs with a compositional range of $Or_{63}Ab_{37}$–$Or_{74}Ab_{26}$. Chemical composition of sanidines from K-rich lavas of Vico, are in equilibrium with plagioclase. Wyoming Orendites and jumillte from Spain contain clear crystals of sanidine of variable sizes. These feldspars include diopside, leucite and rare apatite. The sanidine crystals from Leucite Hills, contain up to 18 % $KFeSi_3O_8$ molecule with very low amount of anorthite in solid solution (0.01–0.11 wt% CaO).

Lamproites from Mount North (Australia) are constituted of 19–48 vol% of K-feldspars containing 95–98 % Or. Those from Shiprock (Montana) contain 64–72 % orthoclase molecule, whereas those from Barqueros lamproites, Spain, contain 87–89 % orthoclase (Wagner and Velde 1986). The average Fe_2O_3 content of K-feldspars is 1–2 %, but it is slightly higher in the Moon Canyon lamproite (2.5 %). Its concentration is as high as 0.62 % in Murcia lampoite and 1–3 % in Smoky Butte lamproite. In K-rich selagites from Orciatico (Italy) and Shiprock dyke (Montana), the BaO content shows large variation, 0–0.89 % and 0.3–3.32 %, respectively.

Coexistence of sanidine with anorthoclase, pseudoleucite and leucite is noted in shoshonitic and absarokitic rocks of Artem of Primorye district (north of Vladivostok), Russia. Here, K-feldspar constitutes 60–80 % of a potassic suite comprising nepheline, biotite, augite and amphibole (Borodin 1974). In many pseudoluecite tinguaites, syenites, and minette dykes at Pamir, K-feldspar occurs as a major phase (Dimitriev 1976). Sanidine is one of the major phases in the alkaline complex of the Tezhser in Caucasus, where it occurs in leucite-bearing trachytes and trachyandesites, where K-feldspar contains up to 10 % albite. Phase relations in the system kalsilite–K-feldspar at 2 GPa and variable temperatures are shown in Fig. 2.3, indicating limited solid solution of kalsilite in K-feldspar. Extent of solid solution of jadeite in K-feldspar (albite—K-feldspar) and jadeite in nepheline (nephiline—albite system) at 2 GPa and different temperatures are shown in Figs. 2.4 and 2.5, respectively.

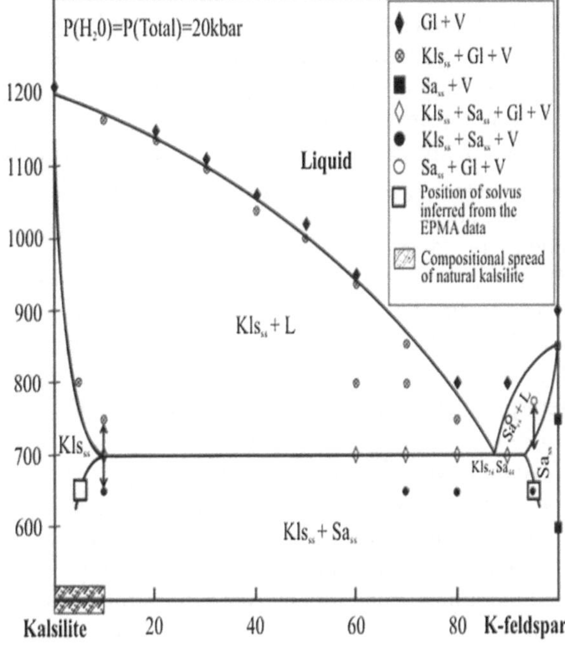

Fig. 2.3 Phase relations along the join kalsilite—K-feldspar determined at 2 GPa [$P(H_2O) = P(Total)$] and various temperatures. The *solvus* has been inferred from EPMA data for *subsolidus* runs (after Gupta et al. 2010)

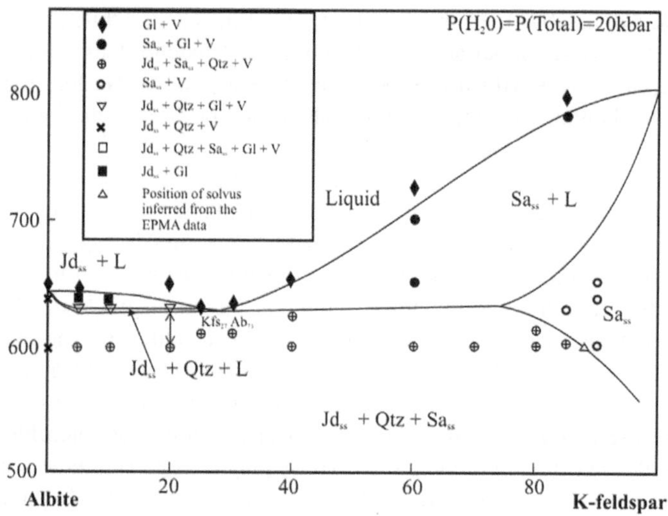

Fig. 2.4 Phase relations along the join albite—K-feldspar determined at 2 GPa [$P(H_2O)$ = P (Total)] and various temperatures (after Gupta et al. 2010)

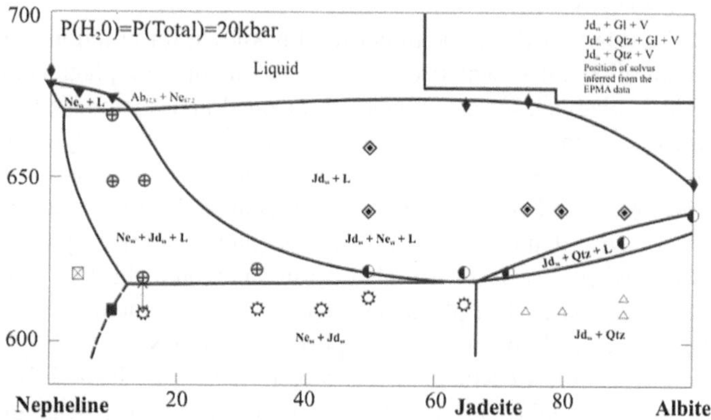

Fig. 2.5 Phase relations along the join nepheline–albite determined at 2 GPa [$P(H_2O)$ = P(Total)] and various temperatures (after Gupta et al. 2010)

2.3 Plagioclase

Plagioclase is a common mineral in leucite-bearing tephrites, basanites, tephritic phonolites and phonolitic tephrites of Vulsini, Vico, Sabatini and Somma-Vesuvius volcanic complexes, but it is absent in all lamproitic rocks. It is also found in the leucite-bearing rocks of Manchuria (China), Utsuryo islands (Korea) and Tristan da Cunha Islands. Plagioclase coexisting in equilibrium with leucite is calcium-rich

(Gupta and Edgar 1975). In their study of the join leucite-albite-anorthite under atmospheric pressure Gupta and Edgar showed that leucite does not coexist with pure albite, and reacts with it to produce nepheline$_{ss}$ and K–Na feldspar [(K, Na) $AlSi_3O_8$]. Nepheline, thus appears in the albite-rich portion of the join along with additional (K, Na) $AlSi_3O_8$.

Anorthite-rich plagioclase (up to 3 mm across) has been described from Brown Leucitic Tuffs of Roccamonfina by Luhr and Giannetti (1987). Most of the plagioclase crystals have compositions, plotting in the range of An_{85-95} with a strong mode at An_{87-89} (Fig. 2.2, Table 2.3). These plagiocases have up to 1 wt% SrO.

Mt. Vulture tephrites and phonolitic tephrites contain plagioclase, which occurs as a euhedral phase and is often enclosed within green salites together with leucite, haüyne and oxide minerals. In these rocks, the composition of plagioclase ranges (Table 2.3) continuously from An-rich (up to 81 vol%) through more sodic varieties to anorthoclase (Melluso et al. 1996). Plagioclase is found as microlites in basanites, tephrites and phonolitic tephrites. These plagioclase crystals are enriched in Sr and Ba (up to 4 wt% SrO and 7 wt% BaO). The Toppo S. Paolo phonolites are characterised by high concentration of Ba (7 wt% BaO and 2 wt% SrO). The plagioclase feldspars ($Or_5Ab_{55}An_{40}$) in minettes from Bearpaw Mountains are found to be in equilibrium with sanidine. The associated latites from the same locality contain plagioclase phenocrysts, which are idiomorphic and zoned ($Or_4Ab_{67}An_{29}$–$Or_5Ab_{68}An_{27}$).

The Latera Caldera of Roccamonfina (Italy) is known for eruption of phonolitic and tephri-phonolitic tuffs, which are constituted mainly of plagioclase crystals (An_{77-87}). These plagioclases are in general unzoned and unembayed in the wide pumices, whereas some of the black pumice fragments of overlying units have anorthitic plagioclases (An_{72-93}), which exhibit concentric zoning and more pronounced embayment and contain numerous minerals and glass inclusions. Plagioclase in Vico lavas (Cundari 1975) is occasionally glommeroporphyritic and strongly zoned and seldom resorbed (also Perini and Conticelli 2002). It is often a microphenocrystal phase. The plagioclase has anorthite content varying between 76.3 and 88.9 % in trachytic as well as tephritic phonolites.

In phonolitic tephrite of Ringgit Beser Complex of East Java (Edwards et al. 1991), the plagioclase feldspar coexisting in equilibrium with leucite, clinopyroxene and phlogopite are anorthite-rich (An_{91}).

2.4 Clinopyroxene

Diopside is the major component having variable amounts of $CaFeSi_2O_6$, $CaAl_2SiO_6$, $CaFe^{3+}AlSiO_6$, $NaCrSi_2O_6$ and $NaFeSi_2O_6$ molecules in solid solution. Orthopyroxene may be present in peridotite or pyroxenite xenoliths, but not in the silica-undersaturated potassic igneous rocks (for explanation, see Chap. 9).

Table 2.3 Analyses of plagioclase from different leucite-bearing localities

	SiO_2	Al_2O_3	FeO_t^a	BaO	SrO	MnO	CaO	Na_2O	K_2O	Total	An	Ab	Srf
1	45.47	34.39	0.69		0.76		17.41	1.60	0.29	100.61			
2	55.34	28.30	0.34		0.08		9.80	5.62	0.80	100.28			
3	55.40	28.60	0.55				9.95	5.21	0.87	100.58			
4	47.70	33.00	0.69		0.09		16.80	1.75	0.47	99.50			
5	53.50	28.40	0.60				12.30	4.10	0.83	99.73			
6	45.80	34.00	0.74			0.06	18.40	1.23	0.11	100.34			
7	46.10	33.00	0.75				18.30	1.63	0.14	99.92			
8	46.68	33.53	0.54		1.04		16.21	1.76	0.13	99.89	80.60	15.80	2.80
9	48.58	32.18	0.68		0.90		14.85	2.51	0.15	99.85	74.00	22.60	2.40

[a] FeO_t represents total FeO + Fe_2O_3 contents

1–2 Plagioclases from leucite tuffs, Roccamonfina (Luhr and Giannetti 1987)

3–5 Plagioclases from leucite tuffs, Latera Caldera (Turbeville 1993)

6–7 Pl agioclases from leucite tephrites Vico lavas (Cundari 1975)

8–9 Plagioclases from Mt. Vulture volcanic complex, Italy (Melluso et al. 1996)

Table 2.4 Analyses of clinopyroxenes From K-rich rocks of different localities

		SiO_2	TiO_2	Al_2O_3	Cr_2O_3	FeO_t^a	MnO	MgO	CaO	NiO	Na_2O	K_2O	Total
1	Core	50.32	0.77	3.81	0.28	6.27	0.08	14.16	24.43		0.16		100.16
2	Mantle	49.38	0.73	4.64	0.08	7.79		13.31	24.53		0.14		100.60
3	Rim	45.36	1.82	8.20	0.10	9.84	0.12	10.82	24.18		0.24		100.68
4	Core	44.87	1.66	8.61		9.16	0.14	10.80	24.24		0.66		100.14
5	Mantle	48.12	0.62	5.53	0.04	6.86	0.30	12.71	24.60		0.48		99.42
6	Rim	50.44	0.62	3.36	0.13	5.40	0.03	14.27	25.07		0.14		99.46
7	Core	43.80	1.52	10.3		11.00		10.00	23.80		0.33		100.75
8	Rim	45.20	1.28	8.90		9.65		11.30	23.40		0.45		100.18
9	Core	39.20	1.97	12.60		14.60	0.37	7.11	23.10		0.33		99.28
10	Mantle	45.50	1.52	8.71		10.80	0.27	9.67	23.40		0.53		100.40
11	Rim	43.30	1.51	7.59		16.30	0.87	6.93	22.70		0.80		100.00
12	Core	52.51	0.33	1.57		3.04	0.06	17.45	24.15		0.13		99.24
13	Mantle	47.83	1.02	5.67		10.49	0.33	11.26	23.70		0.47		100.77
14	Rim	46.73	1.04	6.33		10.87	0.28	11.15	23.82		0.44		100.66
15	Core	47.80	0.84	4.75		10.90	0.77	10.70	23.60		0.53		99.89
16	Mantle	43.00	2.00	9.62		10.80	0.09	10.50	23.30		0.23		99.54
17	Rim	46.10	1.36	5.57		10.80	0.54	10.20	24.90		0.44		99.91
18		54.55	0.79	0.38	0.01	2.31	0.14	16.61	25.45		0.16	0.09	100.49
19		54.20	0.51	0.05		2.08		17.92	24.96		0.20	0.01	99.93
20		51.84	0.19	1.59	0.03	10.89	0.42	12.05	22.50		0.47	0.02	100.00

(continued)

Table 2.4 (continued)

		SiO_2	TiO_2	Al_2O_3	Cr_2O_3	FeO_t^a	MnO	MgO	CaO	NiO	Na_2O	K_2O	Total
21	Core	54.28	1.69	0.22	0.38	3.18		17.83	22.10				99.68
22	Rim	54.01	2.18	0.25		3.92		17.53	22.10				99.99
23		53.00	0.92	0.52	0.56	4.30	0.12	16.70	23.30	0.06	0.39		99.87
24	Core	54.48	0.39	0.51		2.98	0.08	18.68	23.26				100.38
25	Rim	53.75	0.58	0.70		2.83	0.06	17.64	22.98				98.54
26		52.60	1.29	2.60		6.46	0.38	15.30	21.20	0.06	0.65	0.12	100.66

[a] FeO_t indicates total FeO + Fe_2O_3 content. Analyses of 18, 19, 20 and 26 are those of certain grains
1–3 Pyroxene phenocrysts (No. 1) (analyses of core, mantle and rim) from a tephritic rock of Alban Hills (Aurisicchio et al. 1988)
4–6 Pyroxene phenocrysts (No. 2) (analyses of core, mantle and rim) from a tephritic rock of Alban Hills (Aurisicchio et al. 1988)
7–8 Pyroxene phenocrysts (No. 1) (analyses of core, mantle and rim) from a leucite tephrite, Vulsini, Italy (Barton et al. 1982)
9–11 Pyroxene phenocrysts (No 2) (analyses of core, mantle and rim) from a leucite tephrite, Vulsini, Italy (Barton et al. 1982)
12–14 Clinopyrexenes from Brown Leucitic Tuffs, Roccamonfina (Luhr and Giannetti 1987)
15–17 Clinopyroxenes from a phonolitic pumice clast, Latera Caldera, Italy (Turbeville 1993)
18 Core of a pyroxene phenocryst from an orendite obtained from North Table Mountain, Leucite Hills (Kuehner et al. 1981)
19 A pyroxene grain from a wyomingite obtained from Zirkel Mesa, Leucite Hills (Kuehner et al. 1981)
20 A pyroxene grain from a wyomingite obtained from Spring Butte, Leucite Hills (Kuehner et al. 1981)
21–22 Clinopyroxenes from a lamproite, Smoky Butte, U.S.A. (Mitchell et al. 1987)
23 A clinopyroxene grain from a leucitite, Gaussberg, Greenland (Sheraton and Cundari 1980)
24–25 Diopsidic pyorxenes form an ultrapotassic rock of Spain (Lopez Ruiz, Badiola 1980)
26 Clinopyroxene from an ultrapotassic rock of Sierra Nevada (van Kooten 1980)

The lamproites from southern Spain, Smoky Butte (Montana, U.S.A.), Shiprock dyke (New Mexico, U.S.A.), Moon Canyon (Utah, U.S.A.), Sisco (Corsica), Mount North (West Kimberley, Australia) and Orciatico (Pisa, Italy) mostly plot in the endiopside field of the pyroxene quadrilateral (Wagner and Velde 1986). This mineral constitutes 6–22 wt% in a lamproite. The pyroxenes in lamproites, often occur as unzoned phenocrysts containing around 1 wt% TiO_2 and 0.1–0.3 wt% Cr_2O_3.

With the exception of Shiprock lamproite (Smoky Butte), the tetrahedral site of the pyroxenes is incompletely filled by Si and Al, average being 1.97–1.93 atom (Wagner and Velde 1986). The vacancy may be filled by remaining Fe^{3+}, when structural formula is calculated in terms of four cations and six oxygens. This deficiency in the tetrahedral site was also noticed in lamproites of Leucite Hills, Wyoming, U.S.A. (Carmichael 1967) and West Kimberley, Australia (Jaques et al. 1984). Acmite is present in many samples as a thin rim around diopside crystals. Wagner and Velde noted the presence of a green centre (enriched in Al, Fe and Na) in diopside phenocrysts of Shiprock lamproites. Similar green centre in pyroxenes of salite composition, has been described from Leucite Hills by Barton and van Bergen (1981). Such green-centred pyroxenes are considered to be disaggregation products of clinopyroxene nodules, and are common features in potassic volanics.

The K-rich lamproites of North Table Mountain at Zirkel Mesa and Spring Boat (both from Leucite Hills, Wyoming, U.S.A., Kuehner et al. 1981) contain predominantly diopside, but those from some wyomingites have salite cores, rimmed by diopsides (Table 2.4).

The clinopyroxenes from the minette dykes of Highwood Mountains are mainly diopsides ($Wo_{46}En_{46}Fs_8$) containing up to 2 wt% Cr_2O_3 (O'Brien et al. 1991), but they also occur in the salite cores of minettes from this area. Salites ($Wo_{47}En_{40}Fs_{13}$) from mafic phonolites and syenites of Highwood Mountains grade to ferrosalite and acmite. The Ti, Al and Na contents in pyroxenes increase with decreasing Cr contents. All pyroxenes have Fe^{3+}/Fe^{2+} ratios >1. The mantle and halos of salites (surrounding cavities in diopside cores) contain abundant inclusions of biotite, leucite, titanomagnetite and apatite, and all pyroxenes have a fassaitic rim, grown along different sector zones. Oscillatory zoning and diffusion halos in salites are common in potassic rocks of the Highwood Mountains region. Such halos are found as patches in Mg-rich salites. In both types of salites, the reacted margins are more Mg-rich than the [001] sectors of the unrelated crystals. This type of dissolution is commonly restricted to a single zone sector (001). In some cases the partially dissolved crystals are overgrown by more Fe-rich salite rims (O'Brien et al. 1991).

The K-rich rocks of the Alto Paranaiba Igneous Province are characterised mainly by diopsidic pyroxenes ($Ca_{48}Mg_{46}Fe_6$) containing minor amount of enstatite or hedenbergite in solid solution. The low Al and variable Ti contents of clinopyroxenes from Alto Paranaiba are similar to those of clinopyroxenes from lamproites of West Kimberley (Australia) and Kapamba (Zambia). Clinopyroxenes from Brown Leucitic Tuffs of Roccamonfina (Luhr and Giannetti 1987) are green unzoned salites, which are quite homogeneous in composition. In addition to salite,

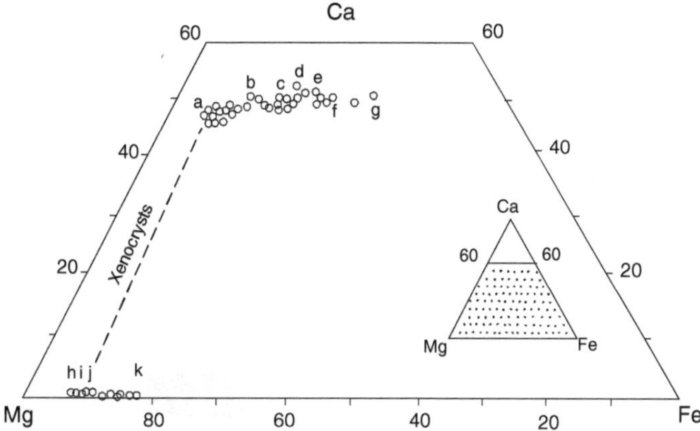

Fig. 2.6 Clinopyroxene and olivine compositions from Brown Leucitic Tuff samples plotted on to the lower portion of the pyroxene quadrilateral (mol %). Letters a–g and h–k indicate analyses of clinopyroxenes and olivine respectively (after Luhr and Giannetti 1987)

the Brown Leucite Tuffs also contain diopside aggregates (Fig. 2.6). The range of salite composition is as follows: $Ca_{49-51}Mg_{31-34}Fe_{16-19}$, whereas, diopside composition plots near $Ca_{47}Mg_{47}Fe_{6}$.

High K-series rocks (leucite basanite, leucitite, tephritic leucitite, and leucite phonolite) of Vulsini are characterised by pyroxenes, ranging in composition from diopside to salite. The compositional plot of clinopyroxenes from leucite phonolites, trachyte and tephritic leucitites is shown in Fig. 2.6a–c. These pyroxenes also contain more than enough Al to balance silica deficiency in the tetrahedral site, and substitutional relation is of the following type: $Mg^{2+} + Si^{4+} \approx Al^{3+} + Al^{3+}$. In addition to the $CaAl_2SiO_6$ molecule, the pyroxenes also contain $CaFe^{3+}AlSiO_6$ (ferri-Tschermak's component). The pyroxenes from leucite phonolites and trachytes (Fig. 2.7) do not show discontinuous zoning like the ones from tephritic leucitites.

The presence of two types of pyroxenes such as salite and diopside in some Vulsini lavas have been ascribed to variation in $f(H_2O)$ and $f(O_2)$ conditions in the same magma chamber by Luhr and Giannetti. Barton et al. (1982) however, suggested an alternate mechanism for the occurrence of diopside and salite in the same lava flow. According to them this may indicate mixing of two different types of magmas, each characterised by two different types of clinopyroxenes. High pressure of equilibration may be the reason for higher concentration of Na and Al^{vi} in pyroxenes (Thompson 1977) from Alban Hills leucitites. Pyroxenes from Vulsini tephrites and leucitites however, show that diopsides or salites are poor in Na. Clinopyroxenes from leucite-bearing rocks of Latera Caldera (Turbeville 1993) are also either diopside or salite. Dark green salite phenocrysts with similar Ti and Fe contents appear as phenocrysts (Table 2.4). Estimates based on cation deficiency shows that diopside crystals are low in acmite and Ti-bearing components, whereas the content of Tschermak's molecule is lower in the fassaites. The groundmass

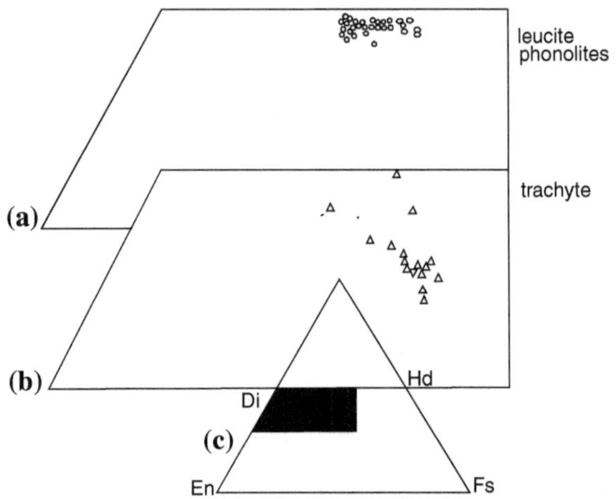

Fig. 2.7 A compositional plot of clinopyroxenes from Vulsini, **a** leucite phonolite, **b** trachyte, and **c** a plot of clinopyroxenes from Vulsini rocks in a pyroxene quadrilateral (after Barton et al. 1982)

pyroxenes in pumice and scoria fragments are salites with consistently higher Ti and Al than coexisting phenocrysts. The larger pyroxene crystals exhibit weak sector or patchy zoning and rarely concentric zoning. Many diopside crystals with fluid inclusions are deeply embayed in dark brown glassy groundmass.

Zoning in clinopyroxene is common in Vico lavas (Perini and Conticelli 2002). The pyroxenes are ubiquitous micro-phenocrystal phase in Vico lavas (Cundari 1975), and sometimes they may be 1/2 cm long along the c-axis. In a conventional plot of Ca–Mg–[Fe(total) + Mn], pyroxene compositions plot within the salite domain, and display a general Fe-enrichment trend parallel to the diopside-hedenbergite join. The distribution of Ti, relative to Al in co-existing core and rim compositions generally show contrasting trends, suggesting complex history of multiple equilibration in the pyroxene-liquid relationship. This may be influenced by temperature and/or f(H_2O) fluctuation during the crystallization of pyroxene, which contains up to 6.91 mol% $CaFeSi_2O_6$, variable amounts of $CaAl_2SiO_6$ (5.7–13.9 mol%), small amounts of acmite (1.2–3.8 mol%) and up to 2.5 mol% $CaTiAl_2O_6$ molecule in solid solution.

Clinopyroxenes from leucite tehprites, phonotephrites, phonolites and foidites from Alban Hills (Italy) are mainly phenocrysts in the post-caldera samples (Aurisicchio et al. 1988). The phenocrysts are of two distinct types: (1) normally zoned variety exhibiting green to colourless core and dark green rim, containing inclusions of opaques and (2) the reversely zoned crystals displaying a green core encircled by a light green rim. The composition of phenocrysts with light green to colourless pyroxene displays enrichment of Al, Fe_t and Ti form the core to the rim and concomitant depletion in Mg and Si. In the reversely-zoned crystals, a rim is characterised by an increase of Si and Mg and depletion of Al, Fe_t and Ti. In a conventional pyroxene quadrilateral, core compositions plot within the diopside field or above the diopside-hedenbergite join. Cores of clinopyroxenes display reverse zoning. The pyroxenes are of two different varieties (either diopside or fassite). The K-rich rocks of Hoch Eifel (Germany) contain fassites (Huckenholz 1973). Duda and Schminke (1985) thought

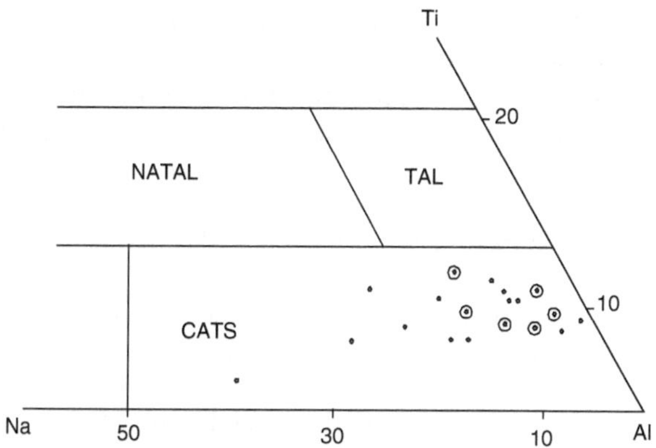

Fig. 2.8 A plot of bulk composition of pyroxenes from Alban Hills in a Na–Al–Ti diagram. The dominant core composition of clinopyroxene is plotted here. In this figure NATAL, TAL and CATAS denotes Na–Al–Ti-rich, Ti–Al-enriched and Ca-Tschermak's molecule respectively (after Aurisicchio et al. 1988)

that the green core pyroxenes from the Eifel region is related to chrome content and their crystallization was associated with polybaric differentiation of alkali basalt magma.

The dominant core composition of clinopyroxenes from Alban Hills is plotted in Fig. 2.8. In this figure NATAL, TAL and CATS denote enrichment in Na, Al (NATAL); Ti, Al (TAL); and Ca-Tschermak's molecules (CATS), respectively. In case of clinopyroxenes from the Alban Hills, Fe^{3+} is the dominant cation in the M_1 site. The ferri-Tschermak's molecules, in fassitic pyroxenes, is higher in the Al^{iv} content in comparison with diopsidic pyroxenes. Aurisicchio et al. found that the Cr content of clinopyroxenes is highly variable. It is particularly high in cores of the diopside phenocrysts ranging from 0.31 to 0.92 and sometimes reaching a value as high as 1.04 % in megacrysts. In the rims, Cr_2O_3 contents could be as high as 0.61 wt%. The TiO_2 content values in the cores of some of the pyroxenes is as high as 0.98–2.58 wt%. In the rim of the same crystal, the TiO_2 content varies between 0.84 and 2.54 wt%, respectively.

2.5 Mica

Lamproites constitute 15–30 vol% of tri-octahedral mica (Wagner and Valde 1986). Some of the micas are authigenic, whereas others are xenocrysts derived from the mantle. Large phlogopites sometimes, occur either as cores of complex micas or may serve as nuclei for crystallization of other micas. Sometimes, cores show that the TiO_2 content is around 2 wt%, surrounded by phlogopite having higher Ti, Fe and lower Al than the cores. Typical lamproitic micas have Fe/(Fe + Mg) ratio always >0.30, with concentric zoning. They show increase in Ti and Fe contents and a concomitant decrease in Al from the centre to the edge. They are characterised by deficit in the tetrahedral site [e.g. (Si + Al + Cr) < 4 atoms].

The presence of Fe^{3+} in the tetrahedral sites is well-established. Tetra-ferriphlogopite is not only found in nature, but has also been synthesised. In these micas, Si^{4+} and Mg^{2+} can substitute for 2 Al^{3+} in the tetrahedral site (Seifert and Schreyer 1971). The presence of tetrahedral Mg in micas from Smoky Butte lamproites has been demonstrated by Robert (1981). He further showed that in such micas Ti is exclusively located on the octahedral sites.

Different types of substitution in micas have been suggested by different mineralogists as summarised by Wagner and Velde (1986):

$$2\ \mathrm{Mg} = \mathrm{Ti} + \square \quad \text{Forbes and Flower (1974)} \tag{2.1}$$

$$3\ \mathrm{Mg} = \mathrm{Ti} + \mathrm{Ba} + \square \quad \text{Velde (1979)} \tag{2.2}$$

$$\mathrm{Mg} + 2\mathrm{Si} = \mathrm{Ti} + 2\mathrm{Al} \quad \text{Robert (1976).} \tag{2.3}$$

$$\mathrm{Mg} + \mathrm{Si} = \mathrm{Ti} + \mathrm{Mg} \quad \text{Robert (1976), and} \tag{2.4}$$

$$\mathrm{Mg} + 2\mathrm{OH} = \mathrm{Ti} + 2\mathrm{O}\ [\text{Arima and Edgar (1980)}] \tag{2.5}$$

The substitution of type (2.1) can be demonstrated in Fig. 2.9a and b, where Ti has been related to the total number of cations. In this substitution it has been assumed that both the tetrahedral and interlayer sites are filled, and no Fe^{3+} is present in the structure. Authigenic micas in lamproites differ from phlogopites from ultrabasic xenoliths by their higher TiO_2 contents, but xenocrystal micas are found to be similar even in their trace element contents with phlogopites from peridotites.

The micas of wyomingite and olivine orendites (Leucite Hills, U.S.A.) have cores generally enriched in Al_2O_3 and impoverished in TiO_2 than those from the micas of madupites (Kuehner et al. 1981). In comparison to phlogopites from madupites and olivine orendites, those from wyomingites have higher Cr_2O_3 contents. The phlogopite cores from the olivine orendites are enriched in Ti, Ba, Al and Fe and depleted in Mg and Si in contrast to the microphenocrystal phlogopites of the same rock type (Table 2.5).

The Highwood Mountains minettes have phlogopite phenocrysts (up to 10 cm long) characterised by high Fe/(Mg + Fe) ratio (0.1–0.2) with Cr_2O_3 content ranging up to 2.2 wt%. Biotite micro-phenocrysts in mafic phonolites from the same locality have Fe/(Mg + Fe) ratio <0.3, but are more enriched in Ti, Fe, Mg, Ca, Ba, Na and F relative to the phlogopites. As a matter of fact, some of the biotites have the BaO content as high as 7.5 wt% with (Al + Si) <0.8 per 22 oxygen.

The biotites (up to 4 mm long) of the Brown Leucitic Tuffs of Roccamonfina (Italy) have apatite inclusions with Fe/(Mg + Fe) ratio varying between 0.38 and 0.36. In CaO-poor but Fe-rich samples, this ratio is as low as 0.60 (Luhr and Giannetti 1987).

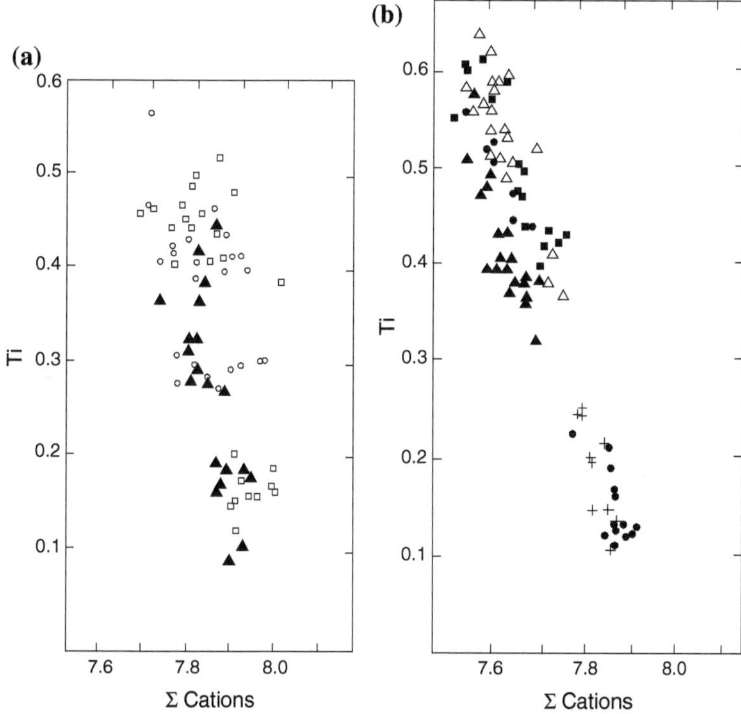

Fig. 2.9 **a** The distribution of Ti and the sum of cations in phlogopite from 3 specimens. *Open squares*: Shiprock lamproite, New Mexico; *open circles*: wolgidite, Mount North, West Kimberley, Australia; *solid stars*: Sisco lampriote, Corsica. **b** The distribution of Ti and the sum of cations in phlogopites from 5 specimens. *Solid triangles*: Concarix lamproites, Corsica; *solid circles*: Calasparra lamproite, Murcia, Spain; *solid squares*: lamproites from Jumilla, Murcia, Spain; *asterisk*: lamproites from Barqueros, Murcia, Spain; *open triangles*: Smoky Butte lamproite Montana, U.S.A (after Wagner and Velde 1986)

In an Al_2O_3–TiO_2 plot, the mica from the lamproitic rocks of Smoky Butte was found to have the highest TiO_2 content (Mitchell et al. 1987). Optically zoned micas have in general a trend of increasing TiO_2 coupled with slight decrease in the Al_2O_3 content. In a similar plot of Al_2O_3 versus TiO_2 contents, micas from K-rich rocks of Alto Paranaiba (Brazil), have chemistry similar to those of micas from West Kimberley, Smoky Butte and Leucite Hills (Gibson et al. 1994). Micas occurring in the K-rich dikes of Trace Ranchos, Limeira I, Indaia I and Pantano, are TiO_2-poor tetra-ferriphlogopites. In other intrusives (e.g. Mata do Lenco) phlogopite shows enrichment in TiO_2 from core to the rim. Phlogopites from Bociana intrusives have moderate TiO_2 (4–5 wt%), Al_2O_3 (12–13 wt%) and total FeO (11 wt%) and plot in the same fields as the groundmass micas of madupites of Leucite Hills. The micas of the lavas from Olegario, Serra do Bueno and Corrego Verjao have high TiO_2 (5–7 wt%), moderate Al_2O_3 (7–11 wt%) and FeO (6–9 wt%) contents. These are quite similar to the micas from West Kimberley and Leucite Hills. The phlogopites

Table 2.5 Analyses of phlogopite

	SiO_2	TiO_2	Al_2O_3	Cr_2O_3	FeO_t^a	MnO	MgO	CaO	SrO	BaO	Na_2O	K_2O	Cl	F	-F = O	Total
1	34.70	6.15	15.90		7.99	0.15	18.20	0.10		3.27	0.12	9.00				95.58
2	39.70	2.83	13.60	0.37	5.71	0.03	22.10	0.04		0.45	0.60	9.52	0.01	3.76	1.58	100.3
3	40.20	2.06	13.50	0.85	4.42	0.02	22.60	0.05		0.41	0.61	9.77		3.56	1.50	96.55
4	39.30	2.40	14.50	0.43	4.06	0.06	22.40	0.06		1.25	0.55	9.71	0.01	3.69	1.55	96.87
5	35.27	4.14	15.07		15.24	0.17	14.60	0.13	0.15	0.49	0.29	9.55		0.70	0.29	96.09
6	35.61	3.97	15.30		15.42	0.18	15.10	0.04		0.09	0.33	9.68		0.50	0.21	96.43
7	35.27	5.39	14.87		16.38	0.25	13.96	0.31			0.31	9.81			0.21	96.48
8	36.01	6.16	16.11		12.51	0.18	15.82	0.05			0.68	9.14	0.01	0.33		97.00
9	36.70	5.22	15.99		8.14	0.05	19.20				0.72	9.39	0.03	0.17		95.61
10	39.87	4.45	13.57	0.06	11.62	0.20	12.96	0.79			2.57	9.93	0.03	0.08		96.13
11	39.90	1.78	12.39		6.62	0.10	22.43				0.43	10.30				93.95
12	37.94	8.41	10.11		8.07	0.08	18.41				0.15	9.87	0.08	0.25		93.37
13	36.44	3.86	14.51		15.99	0.48	15.44				0.49	9.42				96.63
14	39.62	8.12	10.07		7.07		23.10					8.92				96.90
15	39.40	5.69	9.44		6.15	0.08	25.79					9.36				95.91
16	40.18	2.10	12.76		3.02		27.58					9.15				94.79
17	38.99	8.43	8.20		7.65	0.03	19.31	0.60		1.85	0.86	8.75				96.01
18	39.62	8.12	10.07		7.07		23.10					8.92				96.90
19	39.40	5.69	9.44		6.15	0.08	25.79					9.36				95.91
20	40.18	2.10	12.76		3.02		27.58					9.15				94.79

(continued)

Table 2.5 (continued)

	SiO_2	TiO_2	Al_2O_3	Cr_2O_3	FeO_t^a	MnO	MgO	CaO	SrO	BaO	Na_2O	K_2O	Cl	F	-F = O	Total
21	42.14	2.06	11.38	0.08	3.00		25.84			0.33		10.41				95.24
22	41.28	2.15	11.89	0.89	2.96	0.07	24.85			0.52	0.13	10.61				95.35
23	42.46	3.70	7.72		5.64	0.02	23.56			1.47	0.62	10.16				95.35
24	39.94	3.90	12.41	1.33	3.07	0.01	24.44			0.05	0.01	10.24				95.40
25	38.68	11.19	8.79	0.03	11.49	0.08	16.24			2.11	0.46	8.30				97.37
26	39.74	7.04	11.27	0.76	3.88	0.02	22.12			1.11	0.10	9.63				95.67
27	38.60	3.14	13.13	1.32	4.79		23.26			0.34	0.60	10.74				95.92

[a] FeO_t represents total FeO + Fe_2O_3 content

1 Mica from a leucitite, Gaussberg, Antarctica (Sheraton and Cundari 1980)

2–4 Phlogopites from ultrapotassic rocks of Sierra Nevada (van Kooten 1980)

5–7 Phlogopites from leucitic tuffs, Roccamonfina Volcano, Italy (Luhr and Giannetti 1987)

8–10 Phlogopites from a leucite-bearing rocks of Laacher See, Germany (Duda and Schminke 1978)

11 Phlogopite from a phonotephrite, Alban Hills, Italy (Aurisicchio et al. 1988)

12 Phlogopite from a lamproitic rock Damodar Valley coal field (Gupta et al. 1983)

13 Phlogopite from a potassic rock occurring at Vulsini volcanic field, Italy (Francalanci et al. 1987)

14–16 Phlogopites from ultrapotassic rocks of Spain (Lopez Ruiz and Badiola 1980)

17 Phlogopite from a K-rich lamproite of Spain (Borley 1967)

18–20 Phlogopites from lamproitic rocks, Southern Spain (Lopez-Ruiz and Badiola 1980)

21 Phlogopite from an olivine orendite, South Table Mountain, Leucite Hills, Wyoming (Kuehner et al. 1981)

22 Phlogopite from a wyomingite, Zirkel Mesa, core of phlogopite grain, Leucite Hills, Wyoming (Kuehner et al. 1981)

23 Phlogopite from a madupite, Pilot Butte, Leucite Hills, Wyoming (Kuehner et al. 1981)

24 Phlogopite from a Calasparra lamproite, Murcia, Spain (Wagner and Velde 1986)

25 Phlogopite from a jumillite lamproite, Murcia, Spain (Wagner and Velde 1986)

26 Phlogopite from a lampoite, Smoky Butte, Montana, U.S.A. (Wagner and Velde 1986)

27 Analysis of a phlogopite from the Shiprock dyke lamproite, New Mexico, U.S.A. (Wagner and Velde 1986)

from the Alto Paranaiba Igneous Province do not show any overall peculiarity, but have chemistry similar to those of kimberlites and lamproites in terms of their TiO_2 and Al_2O_3 contents.

The solubility of TiO_2 in phlogopites crystallizing in the simplified system, $K_2Mg_6Al_2Si_6O_{20}(OH)_4$–$K_2Mg_4TiAl_2Si_6O_{20}(OH)_4$–$K_2Mg_5TiAl_4Si_4O_{20}(OH)_4$ was determined at variable pressures between 1 and 3 GPa and 825–1300 °C, under a condition of $P(H_2O) < P(Total)$ by Tronnes et al. (1985). They found a complete substitution of the following types: Mg^{VI}, $2Si^{IV} \approx Ti^{IV}$, $2Al^{IV}$ and $2Mg^{VI} \approx Ti^{IV}$ in phlogopite. They tried four sets of experiments with Ti-phlogopite containing 7.1, 6.3, 4.8 and 2.4 wt% TiO_2 at variable temperatures. Their study on phlogopite co-existing with rutile and vapour indicates that Ti content of phlogopite increases with temperature at a given pressure. Their investigation on phlogopite in equilibrium with rutile and a vapour phase shows that it has 7.1 wt% TiO_2 at 1,275 °C or higher temperatures under 3 GPa. The same Ti-phlogopite is stable under following conditions: 1,110 °C and 2 GPa, above 1,025 °C and 1 GPa, above 1,275 °C and 3 GPa, above 1,110 °C and 2 GPa and finally at 1,025 °C and 1 GPa. The Ti-phlogopite breaks down to rutile and a vapour phase at lower temperatures.

The Orciatico lamproites (Conticelli et al. 1992) have micas, with fairly homogeneous chemistry. They have high TiO_2 (7–8 wt%), BaO (0.2–1.8 wt%) and MgO (16–17 wt%) content and low abundance of SiO_2 (37–40 %) and Cr_2O_3 (0.01–0.06 %). The (Si + Al) and (Na + K) are insufficient to fill the tetrahedral site (2.2) and interlayer (W) site, respectively. They suggested that possible substitution as $(Ti^{VI,\ 4+} + Mg^{IV,\ 2+}) \approx Mg^{IV,\ 2+} + Si^{IV,\ 4+}$ is more realistic (Robert 1976). They noted positive correlation between Al and Ti and following type of substitution involving $(Ba^{2+} + Al^{3+})$ and $(K^{+} + Si^{4+})$. The micas occurring in lamproites from Monticatini Valdi Cecina (Italy) are strongly zoned and have wide compositional range of TiO_2 (2–7 %), MgO (15–24 %) and FeO (4–13 %). In these micas, they noted following type of substitution: $2Mg^{IV,\ 2+} \approx Ti^{IV,\ 4+}$ (Forbes and Flowers 1974). Such a substitution as $Mg^{IV,\ 2+} + 2Si^{IV,\ 4+} \approx Ti^{VI,\ 4+} + 2Al^{VI,\ 4+}$ (Arima and Edgar 1981) is also noted. The variable K/Al ratio in these micas probably reflect f (O_2) variation (Foley 1989). Biotite occurs rarely as phenocrysts in pumice clasts and is intergrown within plagioclase and pyroxene. Isolated crystals are replaced by magnetite. Altered biotite phenocrysts have Fe/(Mg + Fe) ratio varying between 0.42 and 0.21 and F content is as high as 1.14 wt%.

The leucite tephrites and tephritic leucite phonolites from Vico lavas (Cundari 1975) have micas with Fe/Mg ratio >1:2, suggesting that they are phlogopitic. Mica in the groundmass of basic phonolitic leucite tephrite is also within the phlogopite range [(Fe/Mg) ratio = 0.39−0.33]. The pleochroic scheme of mica is as follows: α (pale yellow) $< \beta = \gamma$ (dark brown to greenish brown).

Phenocrystal phlogopites in leucite-bearing lamproitic rocks of Leucite Hills are very common minerals, but they are absent in the groundmass (Carmichael 1967). It occurs as large polysynthetically twinned crystal with evidence of resorption in wyomingites and orendites. Micas often are weakly pleochroic from yellow brown to nearly colourless, but sometimes the core shows more intense pleochroism than the rim and may include spinel. The phlogopites from Wyomingite contain up to

2.46 % F. The high F (2.16 wt%) content has also been found in high K-rich Spanish lamproites. The concentration of Ti is particularly high in phlogopites from leucitic rocks of Australia and Spain in contrast to those from Leucite Hills (Carmichael 1967, Table 2.5).

The Gaussberg leucitites have micas, which are late crystallizing phases (Sheraton and Cundari 1980). They are also titaniferous but high in BaO (1.0 wt%) and similar to those in Smoky Butte (Velde 1975) and West Kimberley (Carmichael 1967), but poorer in Al. There is more Al atom for 22 oxygens, and much of the Ti is assumed to be in the tetrahedral site.

2.6 Amphibole

Cross (1897) reported for the first time unusual pleochroism in amphiboles (lemon yellow to pink or reddish pink) from the lamproitic rocks of Leucite Hills. Similar amphiboles have also been described from the potassic rocks (jumillite) of south-eastern Spain (Wagner and Velde 1986) and West Kimberley, Australia (Jaques et al. 1984). They were originally described as magnophorite, but are essentially a variety of K-richterites [(Na, K) Na,Ca $(Mg,Fe^{2+})_5Si_8O_{22}(OH)_2$]. K-richterites belong to the calc-alkaline amphibole group, and have almost one Fe^{2+} atom in the C-site. Richterites found in Shiprock and Sisco lamproites however, are richer in iron varying between Mg_2Fe_3 and Mg_3Fe_2 (Wagner and Velde 1986). The A site (Na + K) is overfilled and the tetrahedral sites are underfilled [(Si + Al) < 8)]. The remaining tetrahedral site presumably contains Fe^{3+}, but in Calasparra lamproites, even after assigning Si, Al and Fe^{3+}, the tetrahedral site remains incomplete.

The Ti content of richterites in lamproitic rocks are usually high in kaersutites (Yagi et al. 1975), and the average is around 2 wt%, but up to 9 wt% is noted in Moon Canyon lamproites (Wagner and Valde 1986). In amphiboles with high Ti concentration (Cancarix, Calasparra and Moon Canyon), the presence of titanium creates a vacancy in the structure as can be seen in the diagram representing the variation of Ti versus the sum of cations (Fig. 2.10a and b). The P–T stability of kaersutite has been studied by Merill and Wyllie (1975).

Ti-deficient amphiboles display a negative correlation between the Ti content and number of (Si + Al) ions in the tetrahedral site. There is no relationship however, between Ti contents and sum of cations. There are some amphiboles, which are zoned. For example, the Cancarix (Spain) and Australian lamproites show darker colour towards the rim of the grains, which corresponds to an increase in Ti and Fe^{2+} and decrease in Mg. The observation may be related to the following types of substitution: $(\square + Ti + Fe^{2+}) \approx 3$ Mg. Slightly Ti-rich K-richterite coexists with phlogopite, clino-pyroxene and leucite in the leucitites of New South Wales (Cundari 1973). The formula of these amphiboles is as follows: $(Na,K)_{2.0}Ca_{0.8-0.9}(Mg, Fe^{2+}, Mn, Fe^{3+}Ti)_5(Si, Al, Ti)_8O_{22}(OH)_2$. In comparison to the richterites occurring in West Kimberley rocks, those present in the leucitites from New South Wales are more sodium rich.

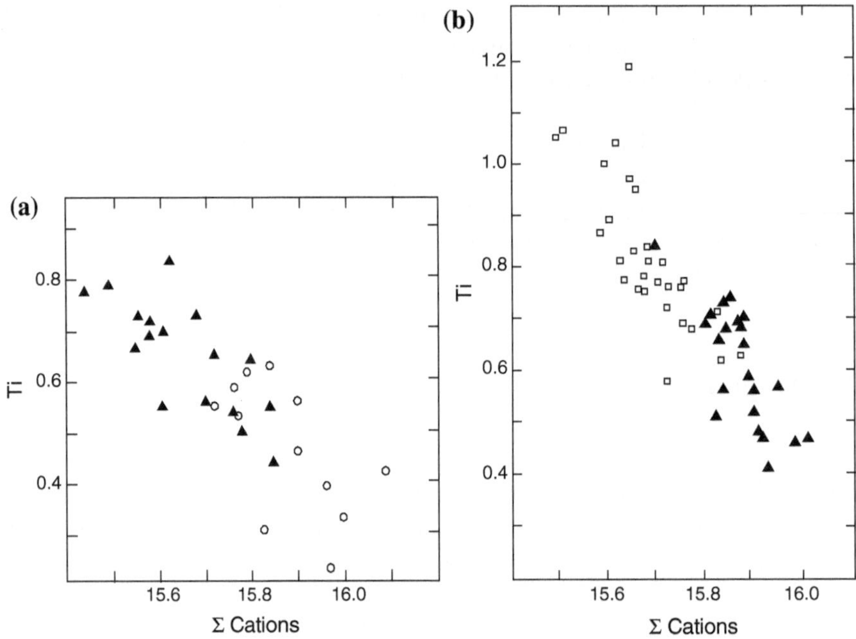

Fig. 2.10 **a** The distribution of Ti and the sum of cations in richterites from Cancalite, Cancarix, Albacete, Spain (*solid triangles*); and richterites, Calasparra lamproites from Murcia, Spain (*open circles*). **b** The distribution of Ti and sum of cations in richterite from Moon Canyon, lamproite Utah (*open squares*) and richterite from jumillite, Smoky Butte, Montana (*solid triangles*) (after Wagner and Velde 1986)

Leucite-bearing tephrites and basanites of West Eifel (Duda and Schminke 1978, Table 2.6) also contain brown yellow amphiboles. The Ti-content of these amphiboles is <5 wt% and they are kaersutites. They have higher F content in the rim compared to that in the core.

Leucite-bearing tephrites and basanites of West Eifel (Duda and Schminke 1978, Table 2.6) also contain brown yellow amphiboles. The Ti-content of these amphiboles is <5 wt% and they are kaersutites. They have higher F content in the rim compared to that in the core.

Tipolo et al. (2003) studied Solid/liquid partition coefficients for large ion lithophile elements (Ba, Rb, Sr), high field strength elements (Zr, Hf, Nb, Ta, Ti), rare earth elements (La-Yb), Pb, Th, U and selected transition elements (Sc, V) were determined by means of Secondary Ion Mass Spectrometry on potassic-richterites synthesised at upper mantle conditions (P = 1.4 GPa and T = 850–1020 °C) from silica-rich lamproites. They observed that most trace elements display an incompatible behaviour in potassic-richterites; only Sr, Ti, Sc and V show strong positive anomalies in the partitioning pattern. When (S/L)-D for potassic-richterites are compared with those for calcic amphiboles (pargasites and kaersutites) several differences become evident. In general, D-S/L are lower in

Table 2.6 Analyses of K-richterite from ultrapotassic rocks

	SiO_2	TiO_2	Al_2O_3	FeO[a]	Fe_2O_3	MgO	MnO	CaO	Na_2O	K_2O	Cr_2O_3	BaO	F	Cl	SrO	Total
1	53.76	5.48	0.61	4.33		18.58		6.16	3.92	4.26						97.10
2	54.20	5.95	1.76	5.82		17.01		4.91	4.37	4.67						98.69
3	54.70	5.90	0.59	5.40		15.55		4.67	4.56	4.54						95.91
4	53.20	5.20	1.10	5.70		17.3	0.13	5.20	5.10	4.40		0.20			0.38	97.91
5	52.70	3.72	1.21	3.23	0.94	18.76	0.09	8.48	4.05	4.00		0.80				97.98
6	38.33	4.45	14.17		10.87	12.38	0.22	11.64	1.94	2.16	0.06					96.22
7	39.87	4.45	13.57		11.62	12.96	0.2	11.79	2.53	1.93	0.06		0.08	0.03		99.09
8	38.84	4.69	13.93		12.23	12.54	0.16	11.73	2.34	1.99			0.24	0.04		98.73
9	52.89	5.14	0.41	4.51		18.65				3.64						96.16
10	54.40	3.81	0.33	7.08		18.58	0.09			2.51						97.65
11	48.85	3.23	1.62	17.85		12.36	0.21			1.52						95.89
12	51.24	6.34	0.53	5.45		18.12	0.05			5.55						97.60
13	54.50	1.64	0.93	8.05		18.13	0.11			1.32						96.42

[a] If only FeO values are given, then they indicate total FeO + Fe_2O_3 contents
1–3 K-richterite from lamproitic rocks of Smoky Bute (Mitchell et al. 1987)
4–5 K-richterite form ultrapotassic rocks of Spain (Lopez-Ruiz and Badiola 1980)
7–8 K-richterite from leucite tephrite, Laacher see (Duda and Schminke 1978)
9 Analyses of richterite from Cancarix lamproite, Spain (Wagner and Velde 1986)
10 Analyses of richterite from Barquerox lamproite, Murcia, Spain (Wagner and Velde 1986)
11 Analyses of richterite from Shiprock dyke lamproite, New Mexico, U.S.A. (Wagner and Velde 1986)
12 Analyses of richterite from Mount North lamproite, West Kimberley, Australia (Wagner and Velde 1986)
13 Analyses of richterite from Orciatico lamproite Pisa, Italy (Wagner and Velde 1986)

potassic-richterites; also, different partitioning patterns are apparent for REE and LIL elements, where S, L and D refers to solid, liquid and distribution coefficient. These differences discussed by Tipolo et al. (2003) are described in terms of the distinct crystal-chemical behaviour of the involved amphibole-end-members, with particular emphasis to the available charge balance mechanisms and to the site dimensional constraints resulting in incorporation of trace elements in various sites.

The distinct partitioning behaviours of trace elements in potassic-richterites and pargasites and kaersutites imply that melts produced from amphiboles-bearing sources may differ markedly depending on the type of amphiboles crystallised. The new partitioning data may be used to determine the role of potassic-richterite in its principal modes of occurrence, namely in lamproites, in peralkaline ultramafic veins in the lithospheric mantle, and in the deeper parts of subduction zones.

2.7 Olivine

Olivine is an important mineral in leucite-bearing basanites, lamproites, minettes (with high mg-numbers), katungite and mafurites, where it occurs as phenocrysts ($Fo_{84–92}$) and megacrysts ($Fo_{88–92}$, Table 2.7). In the potassic rocks of Alto Paranaiba Igneous Province (Gibson et al. 1995), olivine exhibits slight compositional variation, whereas olivine inclusions show compositional zonation ($Fo_{84–87}$).

Both phenocrystal and groundmass olivines are present in olivine orendites (Kuehner et al. 1981) from Spring Boat, North Table Mountain and South Table Mountain (Leucite Hills, Wyoming, U.S.A., Table 2.7). Leucitites occurring in New South Wales, Australia (Cundari 1973), are characterised by olivine ($Fo_{93–79}$) phenocrysts (0.3–3.0 mm along c-axis, Table 2.7). Iddingsites (pseudomorph after olivine) are common.

Orciatico lamproites (Italy) have olivines having wide range of compositions (Fo_{74}–Fo_{92}) with no chemical zonation. The CaO and MnO content of the olivines increase systematically with decrease in the forsterite content. Large rounded crystals with kink banding are very much common in the Orciatico lamproite.

Brown Leucitic Tuffs from Latera Caldera (Luhr and Giannetti 1987; Table 2.7) are characterised by the presence of forsteritic olivines ($Fo_{82–91}$). The rims and fractures within olivines are filled by iddingsites. Diopside crystals often form clusters with olivines, which are probably xenocrysts. Olivine occurs in the basanites of Mt. Vulture complexes. It is usually found as a phenocryst, intergrown with diopside. This phase however, does not show any reaction rim. The olivine grains have narrow compositional range ($Fo_{89–88}$, Table 2.7).

The leucite-bearing lavas of EK series from Ringgit–Beser complex, East Java (Edwards et al. 1994) contain phenocrystal forsteritic olivine ($Fo_{91–92}$). They have quenched skeletal textures and show alteration to iddingsite. Olivine phenocrysts (Table 2.7) in the minettes from Bearpaw Mountains, Montana (Macdonald et al. 1992) are usually idiomorphic and 3 mm long. Some crystals show resorption. Normal zoning towards Fe, Mn and depletion in Ni is quite common. Olivine

Table 2.7 Analyses of olivine from K-rich localities

	SiO_2	FeO	MnO	MgO	CaO	NiO	Total
1	41.25	8.00	0.09	49.83	0.10	0.62	99.89
2	40.20	9.58	0.09	49.86	0.16	0.43	100.32
3	41.07	11.41	0.14	47.25	0.16	0.33	100.36
4	40.80	8.02	0.05	50.10	0.10	0.52	99.59
5	40.30	15.40	0.18	44.00	0.16	0.18	100.22
6	40.70	10.20	0.05	49.00	0.12	0.12	100.19
7	40.17	9.35	0.18	49.30	0.42		99.42
8	40.53	10.93	0.21	47.75	0.43		99.85
9	39.46	16.17	0.27	43.55	0.23		99.68
10	41.84	7.14		50.82	0.03	0.08	99.91
11	40.41	9.73	0.33	48.77	0.15	0.18	99.57
12	40.71	9.85	0.38	49.18	0.11	0.21	100.44
13	38.55	12.27	0.61	47.56	0.62	0.25	99.86
14	39.00	18.70	0.43	41.10	0.29	0.15	99.670
15	39.30	18.60	0.38	40.90	0.42	0.10	99.70
16	40.00	11.60	0.15	47.20	0.24	0.19	99.38
17	37.80	25.30	0.53	35.80	0.57		100.00

(continued)

Table 2.7 (continued)

	SiO_2	FeO	MnO	MgO	CaO	NiO	Total
18	37.39	32.60	0.92	28.50	0.49		99.90
19	37.29	31.50	0.76	29.90	0.55		100.00
20	40.30	10.50	0.19	48.20	0.33	0.38	99.93
21	39.00–40.80	9.50–11.20	0.14–0.26	46.90–49.90	0.18–1.00	0.16–0.62	
22	40.45	11.41	0.12	47.31	0.41		99.70
23	40.84	10.88	0.22	48.31	0.43		100.68

1–3 Olivines from lamproites of Smoky Butte, U.S.A. (Mitchell et al. 1987)
4–6 Olivines from ultrapotassic rocks of Sierra Nevada, U.S.A. (van Kooten 1980)
7–9 Olivines from leucitic tuffs, Roccamonfina, Italy (Luhr and Giannetti 1987)
10 Olivine from an orendite (SK 36), South Table Mt., Olivine rimmed by phlogopite, average of four analyses (Kuehner et al. 1981)
11 Olivine from an orendite (SK 35), North Table Mt., groundmass olivine, average of two analyses (Kuehner et al. 1981)
12 Olivine from an orendite (SK 36), groundmass olivine, average of two analyses (Kuehner et al. 1981)
13 Olivine from an orendite (SK-37), Spring Butte, single analysis of groundmass olvine (Kuehner et al. 1981)
14–16 Olivines from leucitites, New South Wales (Cundari 1973)
17–19 Olivines from leucite basanites, Vesuvius, Italy (Baldridge et al. 1981)
20 Mean of 12 olivine analyses, leucitites from Gaussberg, Antarctica (Sheraton and Cundari 1980); also contains 0.03 wt% Cr_2O_3*21* Range of olivine analyses of leucitites from Gaussberg, Antarctica (Sheraton and Cundari 1980)
22–23 Olivines from the Mt. Vulture volcanic complex, Italy (Melluso et al. 1996)

crystals show overall change in composition from Mg-rich core towards Fe-enriched rim, as bulk rocks become more evolved. In some crystals compositional range of olivine is found to vary from $Fo_{93.5-87.5}$ to a larger range $Fo_{93.4-69.7}$ or $Fo_{91.0-77.4}$, suggesting that olivine could not have reached equilibrium with the host magma. Macdonald et al. (1992) considered that this diverse olivine population probably resulted from magma mixing.

Gaussburg olivine leucitites have forsteritic olivines (Fo_{91-92}) (Table 2.7), and as expected is high in NiO (0.2–0.6 wt%) and CaO (0.2–1.0 wt%), consistent with crystallization from a Mg-rich liquid under low pressure volcanic regime. In the Lyzhnaya pipe of Aldan (Russia), lamproitic pipes consist of forsteritic olivine (10–20 vol%), diopside and phlogopite. The fine-grained groundmass comprises microlites of diopside, phlogopite and K-feldspar. The Ryabinovaya pipe is a olivine-diopside lamproite, where forsteritic olivine (Fo_{92-94}) with up to 0.5 wt% NiO constitutes 15–25 vol% of the rock. Chromite inclusions (Cr_2O_3 up to 61 vol%) are present in these rocks.

2.8 Nepheline

It has a cubic symmetry at or above 1,254 °C (Bowen 1912), but hexagonal symmetry at low temperatures under atmospheric pressure. The cubic phase is called carnegieite. Nepheline usually contains albite; and at 800 °C the maximum amount of albite incorporated in nepheline is about 33 wt% (Greig and Barth 1938) under one atmospheric pressure. It has also an extensive solid solution relationship with kalsilite at 0.5 GPa (Zeng and MacKenzie 1984; Fig. 2.11). The system has a truncated solvus with a maximum temperature at 800 °C and $Ne_{30}Ks_{70}$. Phase relations of the join nepheline-kalsilite have also being studied by Gupta et al. (2010). Their isobaric diagram at 2 GPa and variable temperatures is shown in Fig. 2.12.

Holmes and Harwood (1937, p. 37) described nepheline-bearing leucitites from the potassium-rich volcanic fields of following localities in Uganda: Mabunga,

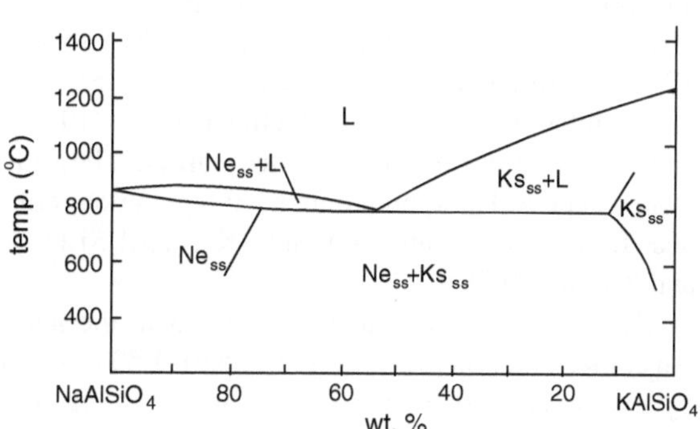

Fig. 2.11 Phase relations in the system nepheline-kalsilite at 0.5 GPa (after Zeng and MacKenzie 1984)

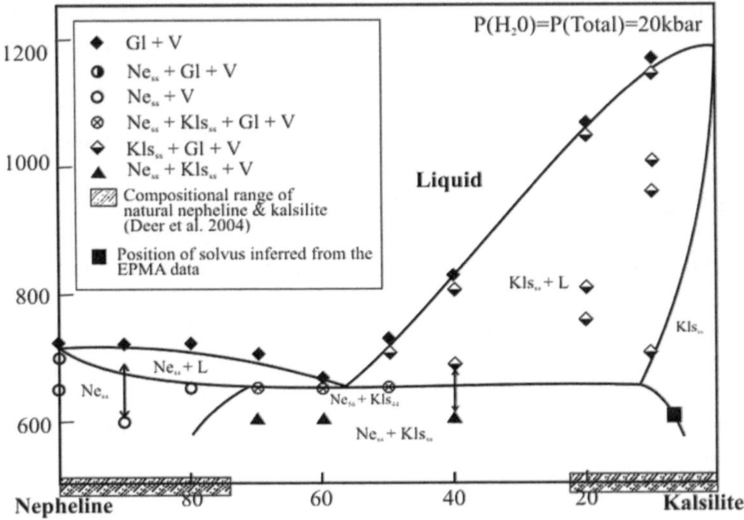

Fig. 2.12 Phase relations along the join nepheline—kalsilite determined at 2 GPa [P(H_2O) = P (Total)] and various temperatures. The *solvus* has been inferred from EPMA data for *subsolidus* runs (after Gupta et al. 2010)

Duzakara, Bunagana and Hyamachunchu. Nepheline together with melilite and leucite has also been reported from the Fort Portal area of Toro-Ankole by Holmes and Harwood (1932, p. 379). It occurs in lavas of Lake Kivu (Bowen and Ellestad 1937). Nepheline-bearing rocks are also present at Villa Senni, Italy.

Nepheline occurs as a groundmass phase in Roccamonfina leucite phonolite lavas (Melluso et al. 1996). Leucite-rich phonolitic tephrite occurs south of Lake Barciano (Sabatini district), and are characterised by nepheline in the groundmass.

Mt. Vulture volcanics (Table 2.8) contain nephelines with excess silica. For example, in Prete della Schimmia, a nepheline crystal ($Ne_{70}Ks_{25}Q_5$) with low CaO content up to 2 wt%) have been described by Melluso et al. (1996). The SrO content of these nephelines is low.

Melilite-bearing foidite occurring south of Caterira, comprises nephelines, which are more Ca-rich with values ranging from 3 to 7 wt%. One particular sample yielded a value of 8.83 wt%, corresponding to a normative anorthite content of 17 %. These nephelines are lower in K contents compared to that of the haüyne-nephelinites and melilitites. Melluso et al. (1996) thought that such high CaO contents have never been reported earlier. They further observed that nephelines from Toppo S. Paolo phonolite have high amount of silica ($Ne_{75}Ks_{13}Qz_{12}$) and plot near Barth's compositional trend: ($K_2Na_6Al_8Si_8O_{32} \rightarrow \square_2Na_6Al_6Si_{10}O_{32}$ (Dollase and Thomas 1978).

Nepheline is a rare euhedral mineral in the leucitites from New South Wales, Australia. These nephelines ($\in$ = 1.530–1.537; ω = 1.534–1.541) are characterised by excess SiO_2 (Cundari 1973; Table 2.8). It occurs as euhedral crystal in the

Table 2.8 Analyses of nephelines from K-rich mafic and ultramafic lavas

	SiO_2	TiO_2	Al_2O_3	Fe_2O_3	FeO[a]	MgO	CaO	SrO	BaO	Na_2O	K_2O	H_2O	Total
1	46.00	0.05	30.10	2.90		0.66	0.06			16.20	4.70		100.67
2	42.80	0.06	32.70	1.70		0.10	0.06			15.70	6.90		100.02
3	42.30		34.00	0.72			2.90			15.10	4.20		99.22
4	45.95		31.31	0.47			0.95	0.03	0.36	17.55	3.88		100.5
5	47.40		31.40	0.82			0.73			16.40	3.20		99.95
6	40.20	0.05	32.51	1.80		0.10	1.44			10.86	12.22	0.20	99.38
7	38.47		30.81	1.63		0.63	0.20			2.09	25.65		99.48
8	38.42	0.05	31.01	1.12			0.03			0.30	28.33	0.67	99.93
9	48.33		31.1		1.06	0.46	0.46			16.04			100.82
10	48.43		30.97		1.08	0.44	0.44			16.03			100.86
11	43.12		33.49		1.23	3.41	3.41	0.24		14.17			100.16
12	42.76		30.01		2.08	8.83	8.83	0.29		12.01			99.33
13	42.77		33.50		1.14	3.72	3.72	0.29		13.56			99.56
14	45.83		32.64	0.49			0.27			17.05	3.73		100.01

(continued)

Table 2.8 (continued)

	SiO_2	TiO_2	Al_2O_3	Fe_2O_3	FeO[a]	MgO	CaO	SrO	BaO	Na_2O	K_2O	H_2O	Total
15	49.27		29.56	0.93			Trace			16.26	3.07		99.09
16	45.59		32.46	0.61			1.22			16.25	3.31		99.44
17	48.23		31.40	0.51			0.77			16.02	3.13		100.06

[a] FeO indicates total FeO + Fe_2O_3 contents
1 Nepheline from a lava flow at Condobolin, New South Wales (Cundari 1973)
2 Nepheline from a lava flow at Cargelligo, New South Wales (Cundari 1973)
3 Nepheline from a leucite-tephrite, Sabatini, Italy (Cundari 1979)
4–5 Nephelines from a lava flow at the Roman province (Baldridge et al. 1981)
6 Nepheline from a potash ankaratrite, Zaire (Sahama and Wilk 1952)
7 Kalsilite from a venanzite, Italy (Bannister et al. 1953)
8 Kalsilite from a K-rich rock at Zaire (Sahama et al. 1956)
9–13 Nephelines from Mt. Vulture volcanic complex (Melluso et al. 1996)
14–15 Nepheline from Portobello (Wilkinson and Hensen 1994)
16–17 Nepheline from Nombi (Wilkinson and Hensen 1994)

groundmass of leucite basanites and tephrites from the Laacher See area, East Eifel, Germany.

The potassic rocks of Ishimskii, Central Kazakistan, are characterised by extrusive and subvolcanic facies rocks containing pseudoleucite phenocrysts in a groundmass comprising microlites of K-feldspar, nepheline, and altered glass. In the pseudoleucite-bearing nepheline syenites from lamproites of Central Asia (42°09′ N, 74°59′ E), nepheline occurs quite commonly.

This mineral (comprising up to 35 vol% of the rock) occurs in the alkali intrusive rocks at Changit (Maimecha Kotui Province, Russia), where it occurs in association with perovskite, aegirine-augite, apatite, K-feldspar and phlogopite.The syenites comprise pseudoleucite (2–6 cm in across) in a matrix of granular nepheline and alkali feldspar. In these rocks pseudoleucite constitutes 30–40 vol% of the rock.

The mineralogy and chemistry of nephelines crystallizing in some alkaline rocks of New South Wales were studied by Wilkinson and Hensel (1994). They noted that except for a few exceptions, the nephelines are Si-rich types, whose Qz (quartz) components exceed those defining the limits of excess SiO_2 in solid solution in the Ne–Ks–Qz–H_2O system at 700 °C and 0.1 GPa P(H_2O). They further established that the nephelines in the basanites show only limited grain-to-grain compositional variation. The nephelines in theralites and tinguaites from the differentiated Square Top intrusion of New South Wales, and in the New Zealand tinguaite however, vary significantly in Ne (nepheline), Ks (kalsilite) and Qz contents even within individual samples. These nephelines may also be strongly zoned. They found that the rims of zoned nephelines are enriched in Si and Fe^{3+}, relative to core compositions. These zoning trends contrast with the compositional trend of successive "bulk" nepheline fractions in the Square Top sequence from thearalite to tinguaite, where Qz content decreases. The nephelines coexist with high-temperature alkali feldspars. Wilkinson and Hensel noted that in the Ne–Ks–Qz system they plot on the Ne-rich side of the Barth compositional join, defined by the omission solid solution series with end-members $K_2Na_6Al_8Si_8O_{32}$ (the Buerger ideal composition; $Ne_{75}Ks_{25}$ mol%) and $\square_2Na_6Al_6Si_{10}O_{32}$ ($\square$ denotes cavity cations vacancy'). The compositions of most natural nephelines are restricted to the field defined by Ne in the Barth join. Compositions more K-rich than the ideal composition are relatively rare, but those lying close to the Ne side of the join are controlled by a number of factors which include the physical conditions attending nepheline crystallization and the chemistry of the alkaline hosts.

2.9 Kalsilite

Hexagonal kalsilite ($KAlSiO_4$) inverts to an orthorhombic variety at 850 °C under atmospheric pressure. The solid solution relationship between kalsilite and nepheline at, 0.5 and 2 GPa are shown respectively in Figs. 2.11 and 2.12. The mineral kaliophilite (another hexagonal polymorph) has a compositional range identical to kalisilite and it may be metastable at room temperature. Kalsilite is a rare mineral in

K-rich volcanic rocks. It has been described form mafuritic rocks of Uganda and also from the Alto Para-naiba igneous province of Brazil. Kalsilite-bearing lava has been described from Mt. Cimini also.

A nepheline-kalsilite-bearing intrusion occurs at Pkhrutskii (Caucasus). This intrusive covers an approximate area of 10 $km^2$and can be classified as a nepheline-kalsilite-bearing monzonite, which locally grades to essexite and fine-grained syenite. The main rock-forming minerals are nepheline, kalsilite, K-feldspar, augite, aegirine-augite and biotite.

A pseudoleucite-bearing alkali syenite occurs at Yaksha, Baikal (Russia). The syenite comprises pseudoleucite (up to 20 vol%) and nepheline. Nepheline-kalsilite intergrowth is often seen in thin section. Locally the syenite grades to kalsilite syenite. This intrusive body constitutes 65–75 % K-feldspar, 3–10 vol% biotite and up to 2 vol% kalsilite.

Predominantly katungites and mafurites have been described from Quaternary volcanic complex at San Venanzo (Perubia Province) and Cupaellow (Rieti province) of Italy, by Gallo et al. (1984). The mafurites are constituted of kalsilite, melilite and pyroxene.

Gittings et al. (1982) described pyroxenites containing an intergrowth of nepheline and kalsilite from Batbjerg complex of Greenland. Kalsilite-bearing ultrapotassic rocks constitute a 350 km long belt trending NE–SW. It is extended from the Derby Mountains of eastern Seward Peninsula to Kobuk-Selawik.

2.10 Analcite

The analcine problem and its impact on the geochemistry of ultrapotassic rocks from Serbia has been discussed by Prelevic et al. (2002). According to them Tertiary ultrapotassic volcanic rocks from Serbia occasionally display low levels of K_2O and K_2O/Na_2O. In these rocks, analcime regularly appears as pseudomorphs after pre-existing leucite microphenocrysts. The process of leucite transformation in Serbian ultrapotassic rocks is very thorough: fresh leucite survives only in ugandites from the Koritnik lava flows as well as in rare inclusions in clinopyroxene. Their study focuses on the impact of "analcimization" on the mineralogy and geochemistry of the Serbian ultrapotassic rocks, using the samples where relecit leucite has been used as a monitor for the process.

Some minettes from Bearpaw Mountains contain as much as 30 vol% of analcite phenocrysts. They often occur as rounded clear grains (2 mm across), but in some cases the crystals are turbid, and low in K_2O content (<1 wt%, Table 2.9), and the total oxide content of analcite is 88–89 wt% with 10 wt% H_2O. The analcites in the minettes are secondary formed by reaction between leucite and a Na-rich fluid (Macdonald et al. 1992).

This mineral occurs as an interstitial phase in the Square Top intrusion. Analcite displays extensive Na, Al $\Leftrightarrow$ Si substitution (Wilkinson and Hensel 1994). Their composition extends from analcite to natrolite (slightly more Si-rich than "ideal"

Table 2.9 Analyses of analcite from various ultrapotassic volcanic complexes

	SiO_2	Al_2O_3	Fe_2O_3	FeO[a]	CaO	Na_2O	K_2O	BaO	H_2O	Total
1	55.91	21.64	1.05	0.09	0.19	12.00	0.25		8.50	99.63
2	51.39	23.48	1.55	0.21	0.20	12.78	1.92		8.20	99.73
3	58.43	21.33	0.93		0.05	9.17	0.34			90.25
4	59.05	21.28	1.00		0.05	9.54	0.29			91.21
5	57.70	19.90		0.90	0.16	12.20	0.52			91.38[b]
6	55.66	21.59		1.17	0.23	12.40	0.30			91.35[b]
7	57.57	20.00		0.81	0.20	12.00	0.49			91.07[b]
8	55.86	21.57	0.44	0.02	0.94	12.67	0.66	0.01		92.17
9	54.71	22.90	0.10		Trace	14.18	Trace			91.89[c]
10	60.20	18.61	Trace		Trace	11.54	Trace			90.35[c]
11	53.72	22.96	0.50		0.31	13.46	0.1			91.05[c]
12	55.04	22.80	0.33		Trace	12.92	Trace			91.09[c]
13	53.06	24.27	0.58		0.77	12.5	0.97			92.15[c]

[a] Where only FeO values are given, they indicate total FeO + Fe_2O_3 contents
[b] The remainder consists of water (slightly above 8 wt%) and minor amout of Rb_2O, SrO and MnO (<1 wt%)
[c] 100 minus total denotes content of H_2O.
1–2 Analcite phenocrysts from feldspathoidal rocks occurring at northeastern Azerbaijan (Comin-Chiaramonti et al. 1997)
3–4 Analcites from Smoky Butte lamproites, Montana (Mitchell et al. 1987)
5–7 Analcites from leucite and analcite-bearing minette from Collima, Mexico. Analysis 6 is from Luhr and Carmichael (1981), who also reported presence of 0.01 wt% BaO. Other analyses are from Karlsson and Calyton (1991)
8 Analcite from potassic mafic lavas of the Bearpaw Mountains (Macdonald et al. 1992)
9–10 Analcite from Portobello (Wilkinson and Hensen 1994)
11–12 Analcite from Nombi (Wilkinson and Hensen 1994)
13 Analcite from Spring Mount (Wilkinson and Hensen 1994)

$NaAlSi_2O_6$). Wilkinson and Hensel observed that groundmass analcites in the basanites, mugearite and tinguaites from New Zealand, have relatively constant composition, approaching stoichiometric $NaAlSi_2O_6$. They also noticed the presence of an unusually Si-rich deutic analcite (60.2 wt% SiO_2) in vugs of the New Zealand tinguaites.

Wilkinson and Hensel (1994, Table 2.9, Fig. 2.13b) determined compositions of nephelines and analcites from different localities of the Bearpaw Mountains. These nephelines and analcites are from theralites [ST28 and ST20 (Fig. 2.13a)] and analcite tinguaites [ST40, ST16 and ST12 (Fig. 2.13b)] from Square Top, Portobello, Nombi and Spring Mount (Fig. 2.13c) (all from New Zealand). Their compositions are plotted as mole proportions with reference to $NaAlSiO_4$ (Ne)–$KAlSiO_4$(Ks)–SiO_2 (Qz) in Fig. 2.13. In this figure, dashed line (Barth line) denotes the Dollase-Thomas (1978) compositional trend of natural nephelines. The solid line marks the maximum limit for solid solution of feldspar in nepheline at 1,068 °C and 1 atmospheric pressure. This is extrapolated from the limit in the binary Ne–Qz system (Greig and Barth 1938). The dashed-dot line marks the limit of solid solution at 700 °C, 0.1 GPa P(H_2O) (Hamilton 1961). In the Fig. 2.13, nephelines (a and b) are denoted by circles and analcites by triangles (solid symbols refer to analyses made by Wilkinson and Hensel 1994). Analyses C–R (a) denote core and rim compositions of zoned nephelines in theralites ST28 and ST20, respectively. The point (Fig. 2.13c, circles) refer to nephelines (denoted by solid circles refer to analyses in Table 2.9, Nos. 1, 2), whereas solid triangle is groundmass analcite (Table 2.9, No. 5) in the Potobello analcite tinguaite. Points C and R denote the core and rim compositions of a zoned nepheline phenocrysts. The Portobello Si-rich analcite (Table 2.9, No. 6) falls outside the plot. Solid squares denote Nombi nephelines (Table 2.9, Nos. 5 and 6) and open squares refer to Nombi analcites (Table 2.9, Nos. 7, 8) linked by the tie-lines 3–5 and 4–6. The Spring Mount analcite (Table 2.9, No. 9) is represented by the solid diamond symbol.

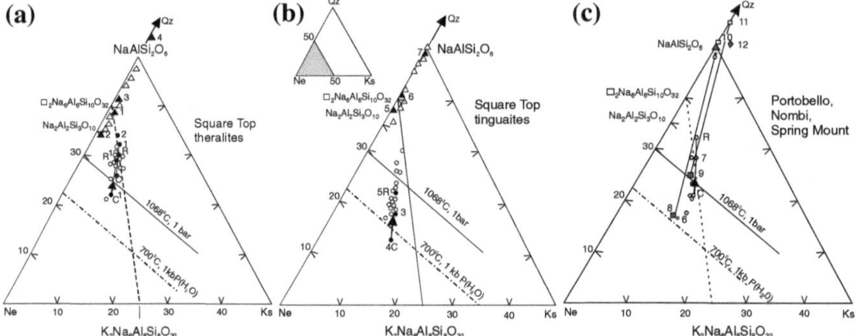

Fig. 2.13 Composition of nephelines and analcites from (**a**) Square Top theralites (ST28 and ST20), (**b**) analcite tinguaites (ST40, ST16 and ST12) and (**c**) Portobello, Nombi and Spring Mount hosts plotted as mole proportions of the components $NaAlSiO_4$ (Ne)–$KalSiO_4$ (Ks)–SiO_2 (Qz) (After Wilkinson and Hensel 1994)

Wilkinson and Hensel concluded that analcite and silica-undersaturated melt in the $NaAlSiO_4$–$KAlSiO_4$–SiO_2–H_2O system, cannot co-exist in equilibrium. Inferred solidus temperatures of the various hosts preclude a primary magmatic origin for the interstitial and groundmass analcites. These are interpreted as subsolidus phases produced by interaction of nepheline with deuteric and/or hydrothermal fluid (see also Gupta and Fyfe 1975). Analyses of nephelines and their derivative analcites indicate that the latter might have been formed from both Si-rich and more Si-poor nephelines.

Megascopic crystals of analcites are found in the Brown Leucitic Tuffs (Luhr and Giannetti 1987), where they often replace primary leucite phenocrysts. More basic leucite-bearing tuffs (CaO > 5.6 wt%) contain a lot of small analcite crystals, which are 0.02 mm across, but attain a size of 0.25 mm across in some pumices. Analcitization of leucite is a common geological phenomenon (Gupta and Fyfe 1975).

Analcites are often found in Tertiary extrusive rocks of north eastern Azerbaijan (Iran). At Harbab Khandi analcite occurs as well-developed crystals often replacing leucite in tephritic rocks. They are stoichiometric with respect to SiO_2, but are silica-oversaturated with reference to alkali. Analcites from Teic Dam and Razi are SiO_2-undersaturated and occur in three forms: (1) euhedral to subhedral or subrounded analcite in the groundmass, (2) weg-shaped subrounded analcite interfaced with large plagioclase phenocrysts in the groundmass (Table 2.9). Potassic analcite is an important phase in jumillites, fortunites and varites from southern Spain.

The presence of euhedral analcite has been described from dykes and lava flows of alkaline rocks at Pyatistennyl by Bazarova et al. (1981). Abundant micro-intrusives occur in this region, comprising leucite, analcite and pyroxene in a glassy matrix. Here, leucite and analcite coexist in the groundmass. The tuffs consist of leucite, orthoclase, plagioclase and augite.

Recently Jamtveit et al. (2009) studied replacement of leucite by analcite. They noted that a 10 % increase in volume is associated with the replacement process, and this generates stresses that eventually cause fracturing of the reacting leucite. Experimentally reacted leucite samples display characteristic fracturing patterns that include both spalling of concentric "onion-skin"-like layers near the reacting interface and the formation of cross-cutting, often hierarchically arranged, sets of fractures that divide the remaining leucite into progressively smaller domains. These structures may explain the "patchy" alteration patterns observed in natural leucite samples.

2.11 Melilite

Melilite compositions (Table 2.10) plot near the akermanite-sodamelilite join of the system akermanite-gehlenite-sodamelilite (Schairer and Yoder 1964; Schairer et al. 1965, 1967; Ferguson and Buddington 1920; Ferguson and Merwin 1919; Yoder 1973). Melilites containing small amount of $Ca_2FeSi_2O_7$ have been described from Capo di Bove Italy by Sahama (1974). He found that in the leucite-melilite

Table 2.10 Analyses of melilite from different ultrapotassic volcanic complexes (after Melluso et al. 1996)

	SiO_2	TiO_2	Al_2O_3	Fe_2O_3	FeO[a]	MgO	MnO	CaO	Na_2O	K_2O	SrO	H_2O+	Total
1	44.41	0.02	5.26		6.74	5.88	0.17	32.16	4.17	0.09			98.90
2	40.03		5.66	7.76	0.40	9.43		32.17	2.83	1.72			100.00
3	42.24	0.17	6.88	0.96	3.78	7.87	0.11	34.19	3.04	0.44			99.68
4	41.68		9.86	2.61	4.32	4.72		29.85	5.27	0.78		0.98	100.07
5	40.36		12.04	0.75	1.53	6.56		34.56	3.34	0.30		0.79	100.36
6	40.80	0.05	10.69		4.04	5.64	0.18	33.85	3.57	0.19	0.62		99.63
7	41.33		10.46		4.17	5.93	0.13	34.90	3.28	0.14	0.60		100.94
8	42.72	0.02	5.55		4.62	8.02	0.14	36.22	2.64	0.25	0.51		100.69
9	43.01		6.55		4.64	7.65	0.09	34.98	2.92	0.26	0.53		100.63
10	43.05	0.03	6.91		5.10	6.96	0.19	33.93	3.61	0.15	1.06		100.99
11	43.28		7.09		9.09	3.29	0.37	26.71	5.87	0.09	2.95		98.74
12	43.89		7.61		9.20	3.04	0.47	26.91	6.90	0.24	2.66		100.92
13	43.29	0.02	8.37		9.59	2.07	0.52	25.60	7.41	0.14	1.18		98.19

(continued)

Table 2.10 (continued)

	NaMel	Gehlen	Ak	NaFe								
6	32.80	12.50	54.70									
7	29.90	13.20	56.90									
8	24.80	2.50	72.70									
9	27.00	3.90	69.10									
10	32.00	2.10	65.90									
11	39.50		45.70	14.80								
12	41.50		36.60	21.80								
13	46.10		32.10	21.90								

[a] Where only FeO values are given, they indicate total FeO + Fe_2O_3 contents

1 Leucite-melilite-nephelinite volcano, Zaire (Sahama 1974)

2 Melilite from a K-rich olivine melilitite occurring at Capo de Bove, Italy (Buddington 1922)

3 Melilite from a K-rich rocks of Nyiragongo (Sahama and Meyer 1958)

4 Melilite from a K-rich melilite nephlinite lava from Roman province (Washington 1927)

5 Melilite from K-rich rocks of Monte Soma (Buddington 1922). Also contains 0.13 wt% H_2O

6–13 Melilites from Mt. Vulture volcanics (Melluso et al. 1996)

nephelinite lavas from Nyiragongo (Zaire), where it occurs as tiny crystal. These lavas also contain melilites, with prismatic habit, nearly perpendicular to the cavity wall. Stout prismatic melilite grains (up to 1 mm in length, averaging a few tens of a millimetre) occur in this region.

Melilite and leucite-bearing volcanic rocks occur at Mt. Vulture (Lucania, Italy). The place is located in the easternmost volcanic area of the Roman province. The most important melilite-bearing localities of this area include Prete Della Scimmia, whereas melilitite dyke is reported, and the other being Melfi hauynophyre lava flow. The flow occurs slightly outside a volcanic cone (Melluso et al. 1996).

The Prete Della Scimmia melilites are characterised by large, clear euhedral phenocrysts grading to microlites in the groundmass. In S. Caterina, melilite-melafoidites have melilites, occurring as a late groundmass phase, inter-grown with clinopyroxene, magnetite, leucite, nepheline and haüyne. At Melfi, melilite is present in haüynophyre as a light yellow-coloured groundmass phase with birefringence up to first order red and is optically negative.

The three melilite groups plot well within the field of volcanic melilites of Velde and Yoder (1977). The Pete della Scimmia melilites are very Mg-rich (up to 8 wt% MgO), alkali- and Al-poor melilites of the data set trend to become alkali-oversaturated with respect to Al (Fig. 2.14). Analysis shows that groundmass melilites contain 0.65 wt% (Ca, Na)$Fe^{3+}Si_2O_7$ (Na-ferrimelilite) having SrO contents as high as (0.5–1.5 wt%). In contrast, melilites from S. Caterina melilite-melafoidite lava flow, are gehlenite-rich (up to 13 mol%). The SrO contents are similar but slightly higher than melilites of Pete della Scimmia.

Of all the melilites those from Melfi haüynophyre are the most Na–Fe–Sr-rich ones found in Italy. Na-ferrimelilite molecule of these melilites can be as high as 22 mol % and the $CaNaAlSi_2O_7$ content is up to 46 mol%; the SrO content is high and they range from 3 to 1.18 %.

In sodium-rich leucite-bearing ultramafic rocks melilites were possibly formed by reaction between nepheline and diopside. It is observed that at 1.7 ± 0.3 GPa, melilite occurring as a reaction product of diopside and nepheline is eliminated (Singh et al. 2000). Akermanitic melilite breaks down even at atmospheric pressure and 690 °C to wollastonite and diopside (Yoder 1973), whereas at slightly higher

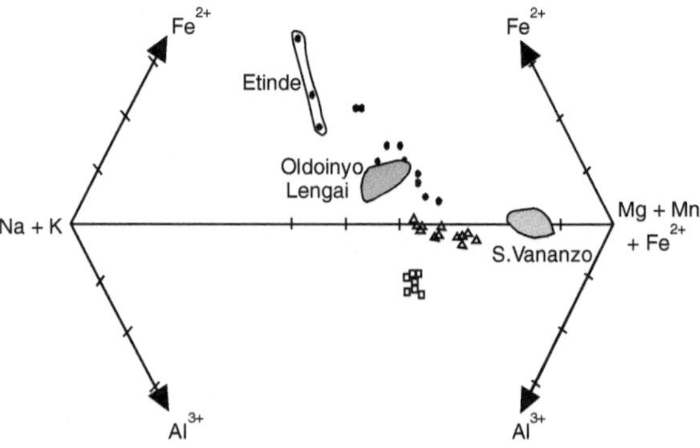

Fig. 2.14 Compositional plot of melilite in a (Na + K)–(Mg + Mn)–(Fe^{2+})–Al diagram (after Melluso et al. 1996)

pressures between 0.6 and 0.7 GPa, it breaks down to wollastonite and merwinite. Akermanite and gehlinite have a continuous solid solution relationship (Ferguson and Buddington 1920; Osborn and Schairer 1941). Sodamelilite is stable only above 0.4 GPa (Kushiro 1964), and at lower pressure it breaks down to nepheline and wollastonite.

Stopa et al. (2002) described the occurence of melilite-carbonetite rocks from the Apennines of Italy, where the rocks have paragenetic relationship with kamafugites of Grotto del Cervo, Abruzzo.

New discovery of leucite melilitites occurring as small lava flows and of kalsilite-melilite pyroclastic ejecta has been described from Montefiascone Volcanics complex of Roman Magatic Province of Central Italy by Di Battistini et al. (2001).

2.12 Haüyne

The chemical composition of Haüyne lies between nosean ($Na_8Al_6Si_6O_{24}SO_4$) and a hypothetical end member ($Ca_4Al_6Si_6O_{24}.SO_4$). It has a cubic symmetry. Some haüynes have minor replacements of Al by Fe^{3+} but the substitution of K for Na is more important in haüyne.

This mineral occurs in potassic lavas of the Roman province, central Italy (Table 2.11). At Melfi, haüynes have a relatively high content of K_2O. Haüyne often displays chemical zoning and its alumina content increases towards the core (De Fino et al. 1986). It has also been reported from phonolitic rocks at Sabatini by Cundari (1979). It occurs as "murky" phenocrysts with characteristic reddish rims.

In the Mt. Vulture volcanic rocks haüyne is a very common feldspathoid, where it occurs together with smaller amounts of leucite. Nepheline is present only as a groundmass phase in melilite-bearing rocks, and is very rare in the feldspar-bearing rocks.

Black oriented inclusions are often noted in haüyne, where it is often concentrated towards the rims. It often has a thin film at the outer rim which is inclusion-free, and has large core-to-rim and "inter-sample" variations. The SO_3 concentration varies from 11.6 to 6.78 wt%, with a clear decrease of SO_3 content from the core to the rim. Melluso et al. (1996) suggested frequent substitution of SO_3 by Cl, which is coupled with Na-enrichment. This gives evidence for increasing sodalite contents. K and Ca do not show clear trends, even though the most Ca–K, and Sr-rich haüynes are found in relatively poorly differentiated samples (basanites and tephrites). The haüynes of the phonolites are characterised by low Ca, K and SO_3 contents, decreasing towards the rims and/or to the groundmass. In melilite-bearing rocks however, the haüynes have sulphur. Caterina melilite-mela-foidite show relatively high Ca and K contents coupled with low SO_3 (7 wt%). The Melfi haüynophyre shows a higher variation in K_2O (SO_3 content varying from 6.60 to 1.8 wt%) but without coherent trends from core to the rim (De Fino et al. 1986).

Di Muro et al. (2004) studied hauynophyre bearing lava from Mount Vulture Italy. They collected a sample from a parasitic vent of the Vulture stratovolcano.

Table 2.11 Analyses of Haüyne from Monte vulture and roman province of Italy

	SiO_2	Al_2O_3	Fe_2O_3	FeO	CaO	SrO	Na_2O	K_2O	SO_3	Cl	H_2O	Total
1	33.62	28.52	0.70	0.76	5.43		17.07	3.80	9.39	0.29	1.65	101.23
2	33.50	27.40	0.32		7.90	0.22	16.30	1.20	12.50	0.46		99.80
3	33.20	26.90	0.40		7.20	0.15	14.60	5.10	11.40	0.62		99.57
4	33.10	26.50	0.34		6.50		14.50	5.90	12.00	0.52		99.36
5	39.20	31.20	0.28		0.24		23.00	1.30	6.40			101.62
6	38.00	29.80				0.10	22.80	1.10	6.90			98.70

1 Haüyne from Monte Vulture (Francalanci et al. 1987)
2–4 Haüyne from Roman Province (Balridge et al. 1981)
5–6 Sodalite from Roman Province (Balridge et al. 1981)

It is a S- and Cl-rich, leucite-melilite-bearing lava flow containing an unusually large amount of sodalite group of minerals (>23 vol%). Mineralogical and chemical study of phenocrysts has led to the identification of black haüynes, blue lazurites and of Cl-rich white or black noseans. X-ray diffraction (XRD) study confirms the occurrence of nosean having a low symmetry (P23). Raman spectra and XRD data show that S is fully oxidised to SO_4 in black haüynes and in white noseans, while it is partly reduced to form S_3^- groups in blue lazurites, which also contain H_2O molecules. Among euhedral phenocrysts, large lazurites are only faintly zoned. All other phases show variable core-rim chemical zoning and many phenocrysts are partially resorbed and/or colour-zoned. Black haüynes have highly variable S/Cl and slightly lower SiO_2/Al_2O_3 ratios, larger $Fe_{(TOT)}$ contents and more compatible trace elements than lazurites. Thin opaque nosean-sodalite rims surrounding all crystals are interpreted as a result of rapid crystallization driven by exsolution of a S-scavenging fluid phase.

2.13 Apatite

High REE content of potassic rocks is mainly due to the presence of apatite, which is a very common accessory mineral. Apatites from Leucite Hills, Wyomintg (U.S.A., Kuehner et al. 1981) contain 0.12–0.27 % La_2O_3, 0.41–0.65 % Ce_2O_3, 0.07–0.9 % Pr_2O_3 and 0.24–0.41 % Nd_2O_3 (all in wt%) (Table 2.12). Over 3 wt% of Ca in some of the apatites is replaced by the Ce-group of rare earth elements. There is often replacement of Mn, Sr and rare earth elements for Ca (Table 2.12). Apatites from the lamproitic rocks of West Kimberly, Australia are also characterised by high REE contents.

In minettes from Highwood Mountains micro-phenocrystal apatites occurs commonly (O'Brien et al. 1991). It ranges in composition from chloro-hydroxy-apatite to fluor-apatite. Rare earth element content of this phase is typically high (0.4 and 0.6 wt% La_2O_3 and Ce_2O_3, respectively). Variable F, Cl and (OH) contents in apatites imply a complex degassing history of Highwood Mountains magmatic system.

In the groundmass glass of Brown Leucitic Tuff of Roccamonfina, Italy (Lhur and Giannetti 1987), apatite (up to 0.55 mm across) is a common mineral. The F, Cl and H_2O contents of these apatites are 2.41, 0.16 and 0.58 wt%, respectively. It is enriched in SrO (Table 2.12).

2.14 Spinel

Jaques and Folley (1985) observed that aluminous spinels (pleonaste-hercynite) occur as inclusions (mostly <20 μm but rarely >40 μm) in leucite phenocrysts of the Fitzory area, Western Australia. These inclusions could be readily recognised in

Table 2.12 Analyses of apatite

	SiO_2	TiO_2	FeO_t^a	MnO	MgO	CaO	SrO	BaO	Na_2O	K_2O	P_2O_5	Cl	F	−F = 0	Total
1		0.33	0.30		0.40	54.63			0.03	0.21	36.91	0.02	5.01		98.04
2			0.42	0.03	0.27	55.10	0.22	0.04	0.10	0.62	41.70	0.04	2.78	1.17	100.15
3			0.48	0.05	0.14	54.30	0.59	0.06	0.15	0.59	41.20	0.09	3.11	1.31	99.45
4	0.99		0.19	0.03	0.10	54.98	0.58		0.23		40.37	0.16	2.41		99.60
5	0.05	0.69	0.33	–	0.00	48.05	2.07	12.20	0.15	0.11	36.50	–	0.00	–	100.01
6	0.49	–	0.30	0.03	0.31	54.10	0.29		0.15		40.10	0.75	3.30		100.40
7	0.60		0.46	0.02	0.32	53.90	0.28		0.16		40.00	0.76	1.40		100.50

[a] FeO_t indicates total FeO + Fe_2O_3 contents

1 Apatite from a lamproitic rock at Mohanpur, Damodar valley, India (Gupta et al. 1983). Also contains 0.2 wt% Al_2O_3

2–3 Apatite from an ultrapotassic basanitic lava, Sierra Nevada, U.S.A. (van Kooten 1980)

4 Apatite from a leucitic tuff, Roccamonfina, Italy (Luhr and Giannetti 1987). Also contains 0.03 wt% Al_2O_3 and 0.58 wt% H_2O

5 Barian fluorapatite, diopside-leucite lamproite, West Kimberley, Western Australia (Edgar 1989)

6 Apatite from K-rich rocks of Vesuvius, lava of1760 eruption, Ungino, Italy (Baldridge et al. 1981). Also contains 0.03 wt% SO_3, 0.12 wt% La_2O_3, 0.22 wt% Ce_2O_3, 0.14 wt% Nd_2O_3 and 0.06 wt% Al_2O_3

7 Apatite from K-rich rocks of Vesuvius, 1631 lava, La Scala, Italy (Baldridge et al. 1981). Also contains 0.11 wt% SO_3, 0.10 wt% La_2O_3, 0.20 wt% Ce_2O_3, 0.15 wt% Nd_2O_3 and 0.08 wt% Al_2O_3

fresh leucites, but could also be observed in lamproites, where the leucite is replaced and pseudomorphed by K-feldspar, zeolites etc. The alumina spinels (Table 2.13) occur as amoeboid-shaped to strongly rounded larger crystals. The spinel inclusions are found to be low in Cr content (<0.2 %, commonly <0.1 % Cr_2O_3). According to Jaques and Foley (1985) these spinels have highly contrasting compositions with high Cr content in the titanium magnesio-chromites occurring in the groundmass (Fig. 2.15).

Pleonaste, the Fe-rich variety of spinel, was found in the wyomingites from Leucite Hills (Table 2.13) by Kuehner et al. (1981). They also described the presence of magnesio-chromite (Table 2.13) having high Fe_2O_3/FeO ratio and low TiO_2 content.

Jaques and Foley (1985) have reported the occurrences of titano-magnetites (Table 2.13) as euhedral or skeletal crystals in all K-rich volcanic rocks of Western Australia, where it is found either as a late stage oxidation product of olivine, mica, K-richterite or clinopyroxenes. Ti-rich magnesio-chromite has also been described from the lamproitic rocks of Fitzroy basin, Western Australia (Jaques and Foley 1985).

Perovskite from minettes of Highwood Mountains (O'Brien et al. contains 12.8 % TiO_2 (Table 2.13), which is equivalent to 36.7 mol% of ulvospinel.

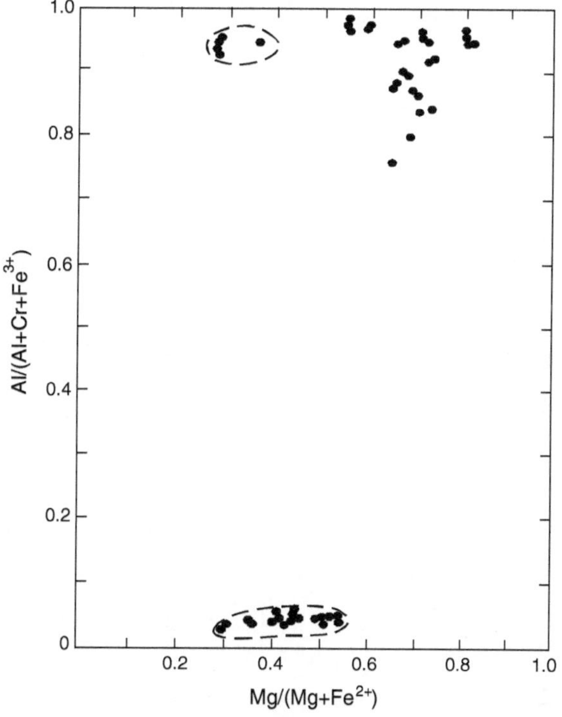

Fig. 2.15 Compositions of aluminous spinel inclusions in leucite in West Kimberley leucite lamproites contrasted with groundmass titanian magnesio-chromites in terms of Al/(Al + Cr + Fe^{3+}) versus Mg/(Mg + Fe^{2+}) (after Jaques and Foley 1985)

Table 2.13 Analyses of Spinel occurring in Different K-rich rocks from different localities

	SiO_2	FeO_t	TiO_2	Al_2O_3	Cr_2O_3	P_2O_5	MgO	MnO	CaO	Na_2O	K_2O	ZrO_2	ZnO	Total	Fe_2O_3	FeO[a]
1	0.02		0.05		0.02							0.03	0.06	78.41		78.23
2	0.09		18.95	3.36	0.11		2.27	0.44	0.29					93.48		67.97
3			1.77	2.80			1.61	0.51	0.03					91.73		85.01
4	0.36		15.03	3.19	0.15		2.01	0.99		0.18				89.51		67.60
5			12.00	0.27	0.57	0.91	1.95	0.54	0.04	0.10	0.30		0.42	95.40		78.30
6			7.02	0.68	1.00	0.56	4.44	0.59	0.10	0.04	0.24		0.21	94.58		79.70
7			13.00	5.21	0.55	0.91	2.83	0.88	0.05	0.06	0.19		0.28	95.96		72.00
8	0.12		0.21	64.87	0.23		21.77	0.05	0.02					100.50		13.23
9	0.05		0.24	63.60	0.16		14.59	0.09	0.02					100.61		21.86
10	0.10		6.79	3.56			0.67	1.49		0.02				99.33	51.36	35.34
11	0.03		5.22	4.12	0.01	0.45	2.07	0.65						93.70		81.15
12	n.d.		6.55	3.03	0.01	0.51	1.85	0.77						95.02		82.30
13	n.d.		4.28	1.96	n.d.	0.44	0.41	0.60						94.08		86.39
14			0.11	11.93	57.52		15.33	0.93						99.59		13.77
15			0.68	6.71	44.07		10.79	0.79						95.85		32.81
16			0.02	64.53	0.10		20.32	0.06						99.35		14.32
17	0.03		0.38	57.66	0.02		7.46	0.16						99.73		34.02
18	0.05		4.33	2.16	55.20		7.15	0.93						99.50		29.68
19		37.05	4.52	0.23	51.41		4.42	0.84						98.47	9.77	28.26
20		20.33	1.23	3.04	65.34		10.41	0.69						101.04	3.22	17.43

(continued)

Table 2.13 (continued)

	SiO_2	FeO_t	TiO_2	Al_2O_3	Cr_2O_3	P_2O_5	MgO	MnO	CaO	Na_2O	K_2O	ZrO_2	ZnO	Total	Fe_2O_3	FeO[a]
21		29.34	1.03	2.00	55.78		9.80	0.79						98.74	13.32	17.35
22		15.31	2.14	4.16	63.00		13.80	0.48					0.2	99.09	2.97	12.64
23		47.95	5.70	1.87	36.22		3.30	0.86						95.90	19.27	30.61

[a] Where only a few values are given, they indicate total FeO + Fe_2O_3 contents
1 Magnetite from a nepheline and leucite-bearing tephrite, Laacher See, Germany (Duda and Schminke 1978)
2 Magnetite from a leucite basanite, Laacher See, Germany (Duda and Schminke 1978)
3 Magnetite from a leucite tephrite, Laacher See (Duda and Schminke 1978)
4 Magnetite from a phonotephrite, Alban Hills, Italy (Aurisicchio et al. 1988)
5–7 Magnetite from lamproites of Sierra Nevada (van Kooten 1980)
8–9 Pleonastes from a leucite lamproite (Fitzroy, Western Australia), Jaques and Foley 1985)
10 Titanomagnetite from a leucite phonololite from Vulsini lava, Italy (Francalanci et al. 1987)
11–13 Titanomagnetites from a Leucitic tuff, Roccamonfina, Italy (Luhr and Giannetti 1987)
14–15 Magnesiochromites from an olivine orendite, South Table Mountain Leucite Hills (Kuehner et al. 1981)
16 Pleonaste from a wyomingite, Deer Butte, Leucite Hills (Kuehner et al. 1981)
17 Hercynite, aggregate in a leucite lamproites from Fitzroy, Western Australia (Jaques and Foley 1985)
18 Titanferous magnesiochromite from a lamproite, Fitzroy, Western Australia (Jaques and Foley 1985)
19–21 Representative microprobe analyses of chrome spinel from a Cancsrix lamproite, Albacete Spain (Wagner and Velde 1986)
22 Representative microprobe analyses of chromian spinel from a Sisco lamproite, Corsica. (Wagner and Velde 1986)
23 Representative microprobe analyses of chrome spinel from an Orciatico lamproite, Pisam, Italy (Wagner and Velde 1986)

Authigenic olivines often contain chromites as inclusions (Table 2.13). They are characterised by higher Cr_2O_3, lower Al_2O_3 and higher Mg/(Mg +Fe) ratio, compared to chromites found by Danchin and Boyd (1976) in harzburgites. Their low Al content is related to the peralkaline chemistry of the magma. Their high Cr content could be related to the low Al and Fe^{3+} contents of the liquid. Chromites are described from the lamproitic rocks of Cancarix (Albacete); Calasparra (Murcia), Jumilla (Murcia), Barqueros (Murcia), Smoky Butte (Montana, U.S.A.), Sisco (Corsica) and Orciatico (Italy) (Table 2.13), often contain titano-magentites ranging in composition from 16 to 0 wt% Cr_2O_3, 7 to 0.7 wt% MgO, 7 to 4 wt% Al_2O_3, and 2 to 5.4 wt% TiO_2.

In the Brown Leucite Tuffs of Roccamonfina, euhedral Ti-magnetite phenocrysts (up to 0.6 mm across) occur. In these rocks it is found as inclusions in salites. The MgO (0.41–2.70 wt%), Cr_2O_3 (trace to 0.01 wt%) and Al_2O_3 (1.96–4.25 wt%) content of titano-magnetites are variable.

Cundari (1975) observed that the Fe–Ti oxides from Vico lavas are ubiquitous, as accessory minerals. They occur as equant-shaped phenocrysts as well as micr
olites or micro-phenocrysts (Cundari 1975) in the groundmass. They noted that the TiO_2 content increases from 5.2 to 13.1 wt% and MnO up to 2.5 wt%. Melluso et al. further noted that the increase of ulvospinel and magnetite components are inversely proportional to Mg and Al contents, but are positively correlated with respect to Mn. In lamproitic rocks chromian spinels are present as a late stage product in the lamproitic rocks of Damodar Valley (India). It is found in most of the dykes of Raniganj, Jharia and Bokaro coal field. Spinel is a ubiquitous phase in the Mt. Vulture volcanic complex, where it occurs as small microphenocrysts, poikilitically enclosed by clinopyroxenes, or it may occur as a small cubic crystals included within olivines. Melluso et al. (1996) observed that these latter spinels are heterogeneous within the same olivine grains and range from Cr-rich (36 wt% Cr_2O_3) to Cr-free variety. The spinels found in the groundmass or included in clinopyroxenes contain variable amounts of Al_2O_3 (13–0.8 wt%) with highest values observed in basanites, but decreasing towards phonolites. They noted that the TiO_2 content increases from 5.2 to 13.1 wt% and MnO up to 2.5 wt%. Melluso et al. further noted that the increase of ulvospinel and magnetite components are inversely proportional to Mg and Al contents, but are positively correlated with respect to Mn. In lamproitic rocks chromian spinels are present as a late stage product in the lamproitic rocks of Damodar Valley (India). It is found in most of the dykes of Raniganj, Jharia and Bokaro coal field.

2.15 Priderite

This mineral typically occurs as an accessory phase in lamproitic rocks. Priderite has tetragonal symmetry with a structure similar to that of a cryptomelane (KMn_8O_{16}) with a general formula of ($A_{2-x}B_{8-y}O_{16}$). X stands for K and Ba, and Y represents such elements as B, Ti, Fe and Al (Carmichael 1967). He observed that

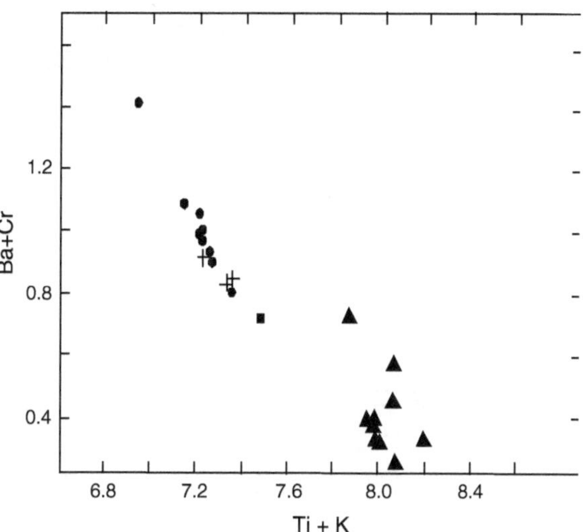

Fig. 2.16 The distribution of Ba + Cr and Ti + K in priderites. *Solid circles*: Smoky Butte lamproite, Montana; *plus symbol*, priderite from Sisco minette, Corsica, (France); *solid triangles*: priderite from wolgidite, Mount North, West Kimberley; *solid squires*: Leucite Hillis lamproite (after Wagner and velde 1986)

priderite from Leucite Hills contains more Ba than that from West Kimberley. They may have a solid solution relationship between K-priderite and Ba-priderite (Norrish 1951). Priderite in association with ilmenite is observed in Mt. North lamproite from Australia and Sisco Minette from Corsica (Wagner and Velde 1986). Analyses of priderites show that (Ba + Cr) has a negative correlation with respect to (Ti + K, Fig. 2.16).

Smoky Buttte lamproites often contain priderites (Post et al. 1982). It is enriched in Ba (up to 0.85), but K is only 0.4 per formula unit. It contains 4.3 wt% Cr_2O_3 in West Kimberley and Smoky Butte lamproites, respectively, but it can be Cr-poor also (<0.3 wt%, Table 2.14). The high Ba values suggest the existence of Ba end member of the priderite series.

This mineral has also been described as an accessory phase from lamproitic rocks of Mohanpur (West Bengal, India; Gupta et al. 1983), and Kapamba, Luangwa valley (eastern Zambia, Scott Smith et al. 1987).

2.16 Wadeite

Prider (1939) described wadeite for the first time from K-rich rocks of West Kimberley province. Henshaw (1955) determined its crystal structure, and found that it has hexagonal symmetry and high birefringence. Its ideal formula is $Zr_2K_4Si_6O_{18}$. The structure is constituted of rings of $(Si_2O_9)^{6-}$, similar to benitoite. It is also present (Table 2.15) as an accessory mineral in the ultrapotassic rocks of Leucite Hills (Carmichael 1967).

Table 2.14 Analyses of priderite

	SiO_2	TiO_2	Al_2O_3	Fe_2O_3	FeO[a]	MgO	CaO	MnO	BaO	Na_2O	K_2O	Cr_2O_3	Total
1	0.00	79.27	0.10		8.94	0.13	0.19	0.04	1.10	0.14	8.10	0.22	98.32
2	–	73.30	0.03		11.40	0.90	0.04	0.06	5.80	0.02	7.40		99.5
3	–	70.60	2.30		12.40	–	0.6	–	6.70	0.06	5.60		98.2
4	–	67.80	0.04		12.60	0.80	0.07	0.40	13.10	0.07	5.0		99.2
5	n.d.	81.56	0.02		9.51				0.26		8.65	0.42	100.42
6	n.d.	78.86	0.05		9.62				0.76		9.59	0.48	99.36
7		78.89		9.19		0.74			6.30		7.15	4.36	100.63
8		68.18		9.09		1.02			16.73		2.73	1.85	99.60
9		71.46		10.75		0.17			16.39		2.27		101.04

[a] Where only a few values are given, they indicate total FeO + Fe_2O_3 contents

1 Priderite from a Damoder valley lamproite (Gupta et al. 1983). Also includes 0.09 wt% NiO

2 Priderite from a West Kimberley lamproite (Wade and Prider 1940)

3 Priderite from a West Kimerley lamproite (Norrish 1951)

4 Priderite from a Leucite Hills lamproite (Carmicheal 1967)

5–6 Two pridirites from a Holsteinborg lamproite, West Greenland (Scott Smith, B.H. 1981)

7 Priderite from a lamproite occurring at Mount North, West Kimberley, Australia (Wagner and Velde 1986)

8 Priderite from a lamproite occurring at Smoky-Butte, Montana, U.S.A. (Wagner and Velde 1986)

9 Priderite from a lamproite at Sisco, Corsica, France (Wagner and Velde 1986)

Table 2.15 Analyses of wadeite from K-rich rocks

	SiO_2	TiO_2	ZrO_2	Al_2O_3	Fe_2O_3	FeO	MgO	CaO	SrO	Na_2O	K_2O	BaO	P_2O_5	H_2O	Total
1	48.80	2.80	27.90	0.20		0.40		0.10		0.10	19.70	0.10			100.10
2	47.40	0.80	28.50	0.80	Trace	0.60		0.10		0.10	21.50	0.10			99.90
3	39.43	1.63	21.29	5.98			0.28	5.22	0.16	2.82	18.40	1.20	3.15	1.30	100.86

1–2 Wedeites from lamproitic rocks, West Kimberly, Australia (Wade and Prider 1940)
3 Wadiate from a lamproitic rock occurring Leucite Hills (Carmichael 1967)

It is stable under following P–T conditions: 1.2 GPa (800, 1,000 °C), 1.4 GPa (1,000 °C), 1.5 GPa (1,000–1,200 °C), 1.75–1.88 GPa, (1,200 °C), 2.0–2.2 GPa (1,200 °C) and 2.5 GPa (1,200 and 1,250 °C) (Arima and Edgar 1980).

2.17 Roedderite-Like Mineral [$(Na, K)_2(Mg, Fe)_5Si_{12}O_{30}$)]

Roedderite-like minerals have been described from the lamproitic rocks of Cancarix, Albacete (Spain) and Moon Canyon, Utah (U.S.A.). This is a blue mineral (Table 2.16), belonging to roedderite-eifelite series (Abraham et al. 1983). The mineral from Cancarix is a sodium-free but iron-rich phase relative to those found in Eifel, Germany. This is a rare mineral, which is crystallised from the melt. The same phase is quite abundant in the Moon Canyon lamproite (Wagner and Velde 1986), but is less iron-rich but more enriched in Mg and Na than that from Cancarix.

Wagner and Valde (1986) suggested a substitution of the type, $Fe^{3+} \rightleftharpoons Fe^{2+} + Na^{+}$ but entry of Fe^{3+} in the structure creates a vacant site (an alkali site). Wagner and Valde suggested that the deep blue colour in the mineral may be due to the presence of Fe^{3+}. A similar phase (eifelite, Abraham et al. 1983) has been described from the K-rich volcanic ejecta of Eifel, of Germany.

2.18 Pseudo-Brookite

This mineral is a solid solution of the end members Fe_2TiO_5 (Pseudo-brookite), $FeTi_2O_5$ (ferro-pseudo-brookite), $MgTi_2O_5$ (karooite) and Al_2TiO_5. Pseudobrookite is present as an euhedral plate, hexagonal in shape. They are green or purple brown, usually opaque, but sometimes translucent (e.g. Smoky Butte lamproite, Wagner and Velde 1986). They exhibit reddish internal reflection under reflected light.

Table 2.16 Analyses of roedderite-like mineral

	SiO_2	Al_2O_3	FeO[a]	MgO	MnO	TiO2	Na_2O	K_2O	Total
1	70.29		12.60	11.44	0.23	0.02	0.04	4.48	99.10
2	69.86	0.17	9.83	13.90	0.21	0.21	0.25	5.14	99.57
3	70.60	0.50	5.80	15.70	0.21	0.07	1.80	4.20	98.88
4	71.32	0.36	0.49	17.86	0.31	0.07	5.29	4.16	99.86

[a] FeO indicates total FeO + Fe_2O_3

1 Representative electron microprobe analyses of the roedderite-like phase, from Cancarix lamproite, Albacete, Spain (Wagner and Velde 1986)

2 Representative electron microprobe analyses of the roedderite-like phase, from Moon Canyon lamproite, Utah, U.S.A. (Wagner and Velde 1986)

3–4 Roedderite in Volcanic ejecta, Eifel region Germany (Hentschel et al. 1980)

Table 2.17 Analyses of pseudobrookite from lamproites of different localities (after Wagner and Velde (1986)

	Al_2O_3	Cr_2O_3	FeO_t^a	MgO	MnO	TiO_2	Total	Fe_2O_3	FeO
1	0.02	0.82	28.97	8.65	0.33	60.33	99.12	26.38	5.24
2	0.47	0.43	19.66	9.39	0.15	69.02	99.12	8.57	11.94
3	0.30	0.99	19.86	8.92	0.14	69.04	99.25	7.71	12.92
4	0.06	0.05	23.46	7.61	0.20	66.12	97.50	11.03	13.54

[a] FeO_t indicates total FeO + Fe_2O_3 contents
1 Pseudobrookite from a Cancarix lamproites, Albacete Spain
2 Pseudobrookite from a Barqueros lamproite, Murcia, Spain
3 Pseudobrookite from a Smoky Butte lamproite, Montana, U.S.A.
4 Pseudobrookite from a Moon Canyon lamproite, Utah, U.S.A.

Around 40 % of this mineral is ferro-pseudobrookite (12–15 % pseudobrookite and 42–48 % karooite, Table 2.17). In some of the lamproites from Murcia (Spain), the pseudobrookite is very rich in MgO (up to 52 mol% karooite component), and it resembles armalcolite.

2.19 Perovskite

Wade and Prider (1940) reported the presence of perovskite as an accessory mineral in association with K-richterite in leucite-bearing rocks of West Kimberley. It is present as a small dark reddish brown mineral with high relief. The pervoskites from lamproitic rocks of Leucite Hills are enriched in Ce. Sr, Na and K. The presence of high concentration of La_2O_3 (2.03–2.05 wt%), Ce_2O_3 (4.45–4.87 wt%), Pr_2O_3 (1.47–1.64 wt%) and Nd_2O_3 (1.73–1.91 wt%) has been described by Kuehner et al. (1981, for major element chemistry, see Table 2.18).

In Prete della Scimmia, perovskite occurs as inclusions in green clinopyroxenes contained within melilitite rocks. The low totals of the microprobe analyses may be related to high REE contents. They show lower SrO (0.22–0.45 wt%) than that of the co-existing apatites (1.2 wt%).

2.20 Ilmenite

In lamproites, ilmenite occurs as a common mineral after olivine, clinopyroxene and phlogopite. They are poor in Fe_2O_3 and constitute about 64–90 vol% $FeTiO_3$ component, 11–30 % geikelite and 1–2 wt% pyrophannite. The lamproites from Barqueros (Murcia, Spain) contain Mg-ilmenite with a rutile core.

Table 2.18 Analyses of melanite and Mg-rich ankarite

	SiO_2	TiO_2	Al_2O_3	Fe_2O_3	FeO[a]	Re_2O_3	SrO	MnO	MgO	CaO	Na_2O	K_2O	CO_2	Total
1	35.70	2.64	10.30	17.10				1.31	0.19	32.80				100.04
2	32.50	5.72	5.81	22.30				0.86	0.31	32.50				100.00
3	32.82	7.19	4.84	19.44				0.85	0.21	32.18	0.08			97.61
4	27.03	17.49	1.59	19.39				0.52	1.13	31.67	0.02			98.84
5		0.03	0.02		5.45			0.02	17.07	26.44		0.01	49.24	98.32
6		48.70	0.03		7.70	11.40	5.70			23.50	2.50	0.60		100.13
7		48.20	0.05		11.40	10.60	3.80			25.10	1.00	0.30		100.45

[a] FeO indicates total FeO + Fe_2O_3 contents

1–2 Melanite garnets from a leucite-tephrites, Tavalato, Rome (Baldridge et al. 1981)

3–4 Melanite garnets from a melilitite lava occurring at Prete Della Scimmia, Mt. Vulture (Melluso et al. 1996)

5 Magnesium ankarite from a lamproite occurring at Mohanpur, Damodar Valley coal field (West Bengal, India, Gupta et al. (1983). Also contains 0.02 wt% Cl and 0.02 wt% P_2O_5

6–7 Perovskites from lamproitic rocks of Leucite-Hills, Wyoming (U.S.A.) (Carmichael 1967). Re_2O_3 refers to Rare Earth Oxides

2.21 Melanite

In leucite phonolites at Tavalato (Italy), melanite occurs both as a phenocryst as well as groundmass mineral. These phenocrysts are zoned with colourless cores having pale–yellowish–brown margins. The analyses show that they are poor in MgO, and that there is not only a significant increase in the TiO_2 and Fe_2O_3 contents but also there is a corresponding decrease in the Al_2O_3 content in the rims of the phenocrysts and groundmass phases. Melanite garnets (Table 2.18) have also been described from a melilitite lava flow at Mt. Vulture (Melluso et al. 1996).

2.22 Carbonate-Bearing Phases

Ankerite phenocysts in association with phlogopite occur in the lamproitic rocks from the Raniganj Basin, India (Table 2.18, analyses no. 5). The other accessory minerals include apatite and chromian spinal in a groundmass of the same phases along with rutile, pyrite and devitrified glass (Gupta et al. 1983). These lamproites are so much enriched in carbonates (10–15 vol%, ankarite), and phlogopite (55–60 vol%), that they termed these rocks as carbonated apatite-bearing glimmerites. Basu et al. (1997) studied the lamproitic rocks of Bokaro basin (Damodar Valley, India). They also noted the presence of $MgCO_3$-rich carbonate phases in these lamproites.

Chapter 3
Classification

Potassium-rich silica-undersaturated igneous rocks may be classified primarily on the basis of their chemistry and mineralogy, but on petrological stand point the mineralogical classification is more preferred here particularly in view of plethora of rock names given on the basis of localities from which these rocks have been described. Their nomenclature is therefore, not only confusing but also complex at times. Different types of classification schemes are described below.

3.1 Classification Based on Chemistry

3.1.1 Potassium Content as a Basis of Classification

The K-rich rock types have been described into three series by Appleton (1972): (1) low-K, (2) medium-K and (3) high-K series on the basis of a simple plot of K_2O versus SiO_2 (Fig. 3.1). The distinction of three different series has been made on the basis of three statistically discriminating lines defining different degrees of potassium-enrichment.

This scheme has been extensively used to classify rocks from Italy by Rogers et al. (1985), from Highwood Mountains (Montana) by O'Brien et al. (1991) and from Ringgit–Beser complex, Indonesia by Edwards et al. (1994).

3.1.2 Total Alkali Versus Silica Classification

Both sodium- and potassium-rich rocks have been classified by the IUGS subcommission on the systematics of igneous rocks into phonolite, tephri-phonolite, phonotephrite, tephri-basanite and foidite. This has been done on the basis of a $(Na_2O + K_2O)$ versus SiO_2 plot (Fig. 3.2). It is also known as the total alkali versus SiO_2 (TAS) diagram. Le Bas et al. (1986) used straight lines to define different

A.K. Gupta, *Origin of Potassium-rich Silica-deficient Igneous Rocks*,
Springer Geology, DOI 10.1007/978-81-322-2083-1_3

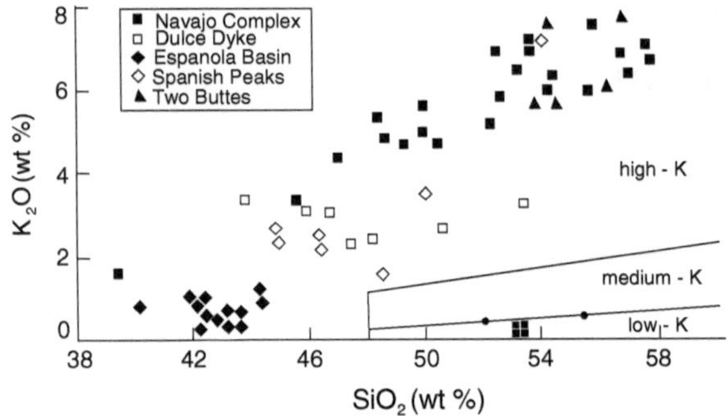

Fig. 3.1 SiO_2 versus K_2O diagram showing classification of potassium-rich rocks into low-K, medium-K and high-K series (after Gibson et al. 1993)

fields. They however, point out some limitations for this type of classification. The diagram is based on 24,000 analyses of different rock types. Regarding this classification, they suggested that there are several constraints:

1. It is purely a descriptive classification.
2. The plot has been made independent of any field location or field association.
3. No generic relationships are employed.
4. In these analyses, the CO_2 and H_2O contents are not taken into consideration, and their concentration has been subtracted first and then, all analyses were recalculated to 100.
5. The name basanite is used if olivine content exceeds 10 % in the CIPW norm.

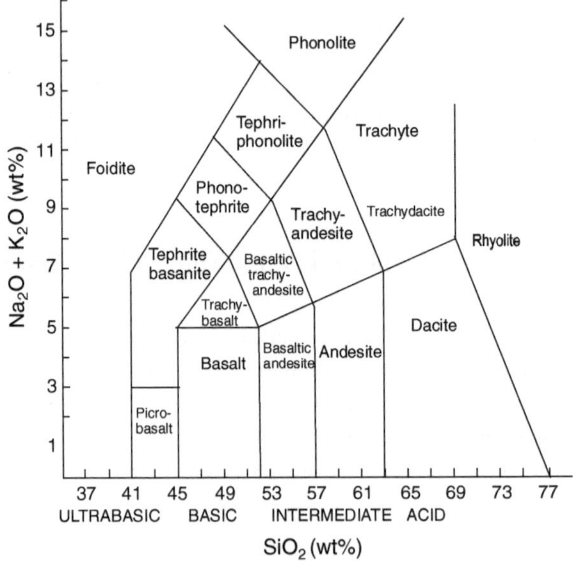

Fig. 3.2 Total alkali versus SiO_2 diagram classifying potassic rocks into different types (after Le Bas et al. 1986)

6. In the field of trachyte and trachyandesite, the distinction depends on, having more than 20 % normative quartz, and in trachyte having less. The limit of 20 vol% normative quartz (in the quartz–alkali feldspar–plagioclase–feldspathoid diagram) approximates (Na_2O + K_2O) = 9 wt%.

The rock is called trachybasalt if the K_2O/Na_2O ratio is greater than one. In case of tephrites or basanites, if the K_2O content is higher than the Na_2O content, then the prefix potassic is used. In the field of trachytes or trachyandesites, if the (Na_2O + K_2O)/Al_2O_3 molecular ratio is greater than 1, then the prefix "peralkaline trachyte" or "peralkaline trachyandesite" should be used.

In TAS classification one of the major problems is that it does not take into account other important oxides. Thus, bulk compositions of many lamproites from Leucite Hills plot in the tephriphonolite or phonotephrite field. Likewise, composition of many leucite lamproites from West Kimberley plot in the trachyandesite and phonotephrite field and some of the olivine lamproites lie in the basanite region. Basanites, trachyandesites or phonotephrites have plagioclase, but lamproites are devoid of plagioclase. This shows the limitation of such a classification scheme.

3.1.3 Chemical Classifications Based on Chemistry and Mineralogy of Type Localities

The K-series rock types have been classified into three categories by Barton (1979):

(1) Leucite Hills Type (LHT), (2) The Roman Province Type (RPT) and (3) The Toro Ankole Type (TAT). The LHT rocks (SiO_2 > 42 wt%) are characterized by the presence of Al-poor pyroxene, mica, amphibole, priderite and wadeite.

The RPT rocks (SiO_2 > 42 wt%) contain Al-rich pyroxene, Fe-poor leucite, sanidine and plagioclase and TAT rocks (miaskitic, SiO_2 < 44 wt%) have melilite and perovskite. It is clear from the description of the LHT type that it has mineralogy similar to that of lamproites. The RPT rocks are essentially tephrites, tephriphonolites, phonolitic tephrites and basanites. The TAT rocks resemble kamafugites, and this classification does not therefore, add anything new to the already existing classification schemes.

The potassic rocks have been classified into kamafugite and orendite (lamproite) class by Mitchell and Bergman (1991) after plotting the bulk compositions in a CaO versus SiO_2 diagram or in a plot of [100 × K/(K + Na)] versus 100 × ΔQ/Si diagram. Here Q refers to excess or deficiency in silica in the norm calculation, and K, Na and Si denote atomic proportions of cations. Thus, kamafugites are relatively more CaO-rich than orendite lamproites and the former group of rocks are poorer in ΔQ/Si ratio.

3.1.4 Major Oxides as Basis of Classification

Foley et al. (1987) classified potassic rocks into four groups: group I, group II, group III and group IV. The group I lamproites (Leucite Hills type of Barton (1979) and Sahama's orendite class) are characterized by low CaO, Al_2O_3 and Na_2O contents and high K_2O/Al_2O ratio usually above 0.6 but frequently >1, and mg-number as high as 87 (Figs. 3.3, 3.4 and 3.5). These rocks have high incompatible trace element contents and include peridotitic mantle xenoliths. The lherzolitic types are less predominant than the garnet and clinopyroxene-poor varieties, which have lost the basaltic melt component. Two type localities of group I rocks include West Kimberley and Gaussberg. According to Foley et al. group I rocks usually have no associated non-potassic rocks, but non standard members have calc-alkaline and shoshonitic associates. Group I rocks may occur in orogenic areas, but are usually reported from the stable continental areas.

The kamafugitic rocks, included as group II rocks, have low SiO_2 (<46 wt%) and high CaO. They have a low Al_2O_3, less extreme K_2O and have lower K_2O/Al_2O_3 ratio. Mg-number is usually above 60. Incompatible elements are less-enriched than the rocks of group I, but have a positive Sr spike. The typical localities include Toro Ankole (southeastern part of Uganda), Cupaello and San Vananzo (Italy). They are supposed to be associated with rift valleys. These rocks are characterized by the presence of melilite, perovskite and kalsilite. The mantle nodules included within group II lavas, are rocks of olivine–biotite–pyroxene (O.B.P.) series.

The Roman Province type rocks, included within group III, are characterized by high Al_2O_3 content (Fig. 3.3), low K_2O/Al_2O_3 ratio (generally <0.5, Fig. 3.4), and the K_2O concentration is higher than the group II rocks. Mg-number is lower than the other two groups of rocks (Fig. 3.4, rarely above 70), and rocks with extremely low silica contents (<42 wt%) do not occur, but those having less than 50 wt% are common. The type locality is Roman Province of Italy and Indonesian province. Because of high Al_2O_3 content, plagioclase and alumina-rich clinopyroxenes are

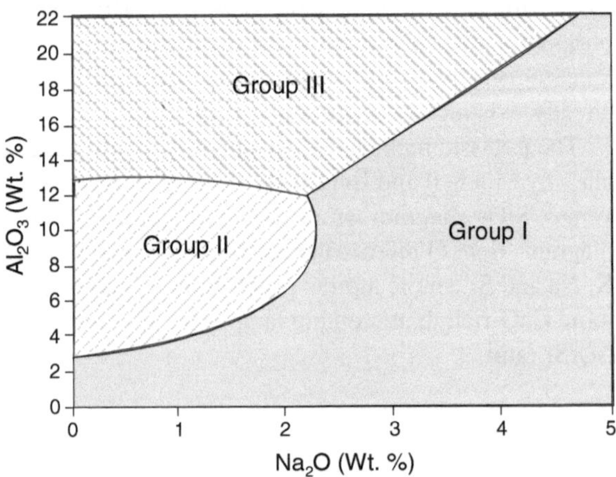

Fig. 3.3 Na_2O versus Al_2O_3 variation in ultrapotassic rocks (after Foley et al. 1987)

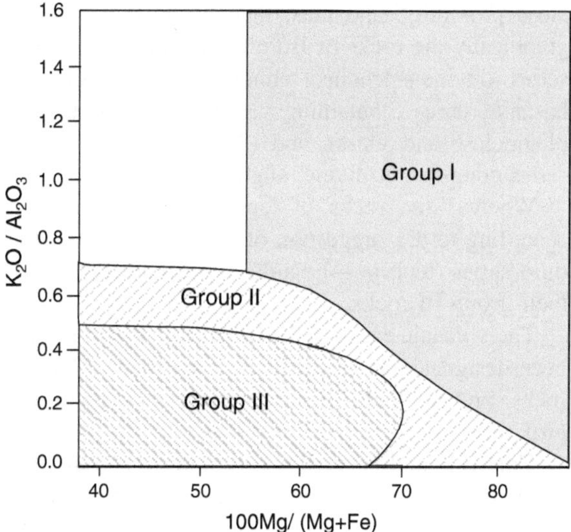

Fig. 3.4 K_2O/Al_2O_3 versus mg-number variation in ultrapotassic rocks (after Foley et al. 1987)

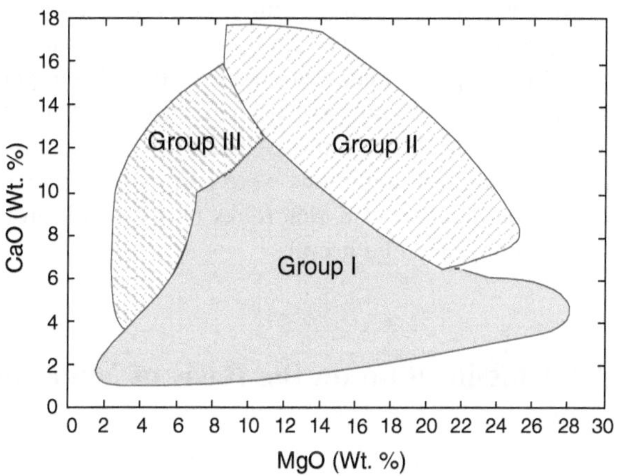

Fig. 3.5 CaO versus MgO variation in ultrapotassic rocks (after Foley et al. 1987)

common. High SiO_2 content does not allow crystallization of such silica-undersaturated accessory minerals, as kalsilite, melilite, perovskite, priderite and wadeite. The nodules are clinopyroxene-rich rocks, and have been described from Alban Hills, Roccamonfina and Somma-Vesuvius.

The group IV rocks are those, which cannot be included in either of the three groups. The type examples of group IV rocks are the potassic rocks of Navajo volcanic field, which lack mantle-derived xenoliths.

The major problem related to this type of grouping of K-rich rocks such as group 1, group 2 (K-rich rocks), group 3 (K-rich rocks) and group 4 (K-rich rocks) (I, II, III and IV), lies in the fact that the rocks of all four groups have been described from close proximity, and thus, there is no genetic significance. For example, petrographically the rocks of Bufumbira can be classified into three series: (1) leucitite series (olivine + leucite + clinopyroxene and glass), (2) leucite tephrite and leucite basanite series (containing variable proportion of leucite, olivine, augite, calcic plagioclase and glass), and (3) the potash trachybasalt (absarokite, shoshonite) series comprising olivine, augite, K-feldspar and glass.

Whereas the rocks of leucitite series can be considered as group I rocks according to the suggestion of Foley et al. (1986), those belonging to second and third series (tephrite—basanite series and trachybasalt series) should be similar to their group III rocks.

The volcanic rocks of Toro-Ankole, which are located only 100 km away, have been designated as group II rocks by them. Thus, group I, group II and group III rocks, are no doubt the transitional rock types (group IV), all located in close proximity.

Conticelli and Peccerillo (1992) described the presence of lamproites (orendites, group I), kamafugitic rocks (group II), high alumina and relatively silica-rich rocks of K-series (tephrites and phonolites, group IV) and transitional types (group-IV rocks) in central Italy.

According to Foley et al. (1987) the group I rocks might have been formed from a depleted mantle source with some rocks enriched in F, H_2O, and CH_4, but impoverished in CO_2. The group II rocks are considered to have been formed by partial melting of a mantle source, rich in CO_2 but poor in H_2O. According to them group IV rocks might have been formed under orogenic conditions. The coexistence of all the four groups of rocks in central Italy, suggest that all the diverse physico-chemical conditions were surprisingly present within such a limited area. Thus, classification of K-rich rocks into group I, group II, group III and group IV, does not appear to be useful.

3.2 Classification on the Basis of Mineralogy

There is an incompatible relationship between melilite and plagioclase in the lavas of alkalic suites (Yoder and Schairer 1969; Yoder 1973). From volcanic centres of the same petrographic province, either melilite-bearing lava is erupted, or plagioclase-bearing tephritic rocks are ejected. Petrologic study of the melilite-bearing potassic rocks and leucite-bearing basanites from the Eifel support the conclusions of Yoder and Schairer (1969). From his study of natural rocks from various areas, Yoder (1973) established that in alkaline lavas, melilite and plagioclase do not coexist but these two minerals may coexist in metamorphic rocks.

Potassic rocks therefore, have been classified into following categories in this volume:

(1) Those, which are free from plagioclase and crystallized under low $P(H_2O)$ condition (melilite may or may not be present),

(2) Those, which contain plagioclase and/or K-feldspar and also crystallized under low P(H_2O) condition (melilite-free), and
(3) Those, which contain phlogopite and/or K-richterite, and are formed under relatively higher P(H_2O) condition (lamproites).

3.2.1 Kamafugitic Rocks Without Plagioclase

The term kamafugitic rocks have been coined by Sahama (1974) to those rocks containing various combinations of olivine, augite, leucite, kalsilite and melilite. The term refers to three rock types, katungite (olivine + melilite + leucite), ugandite (olivine + augite + leucite, Fig. 3.6) and mafurite (kalsilite + leucite + melilite). The rare extreme variety of katungite without olivine is called proto-katungite. Potassic dyke rocks occurring in Navajo-Hopi has been termed as katungites. These rocks have mineralogy similar to those of katungite but lack modal kalsilite.

The potash ankaratrite–melaleucitite series have been considered by Sahama to represent a sodic and potassic analogue of kamafugitic rocks, and they are

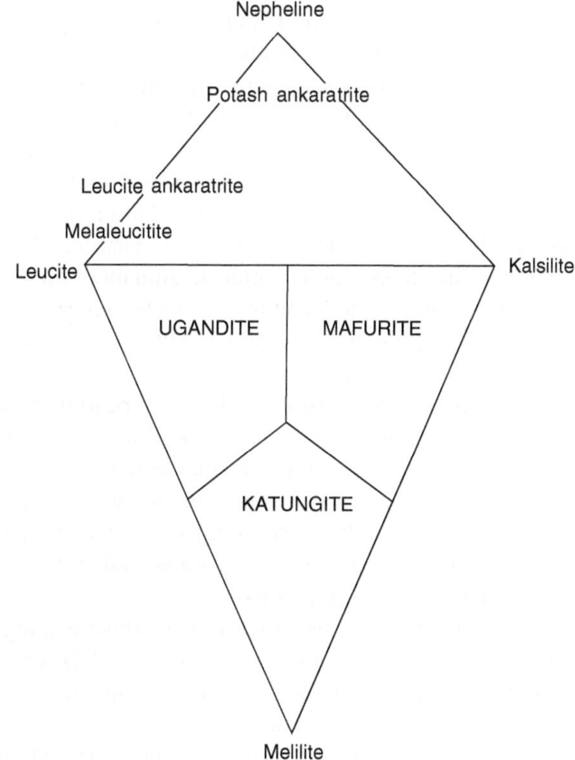

Fig. 3.6 Plot of bulk compositions of katungite, mafurite and ugandite (kamafugitic rocks) in a melilite-leucite–kalsite diagram. Bulk compositions of melaleucitite, leucite, ankaratrite and potash ankaratrite in a leucite–kalisite–nepheline diagram is also shown in the same figure (after Sahama 1974)

characterized by the predominance of augite instead of olivine. The felsic constituents are nepheline in K_2O-rich ankaratrite proper, nepheline + leucite in leucite ankaratrite and leucite in melaleucitite. A *venanzite* is constituted of olivine, melilite with some leucite and kalsilite. Diopside and phlogopite appearing as late crystallization products are also present. *Coppaelite* is a melilitite with kalsilte. Coppaelite has been described from Coppaeli di Satto (now Coppaello), Central Italy. Coppaelite and venanzite lava flows have been found to erupt in the eastern part of Roman co-magmatic province.

Washington (1927) coined the term *italite* to represent a mono-mineralic effusive rock type containing essentially entirely of leucite with minor amounts of augite, glass, and other phases. The type locality of *italite* is Villa Senni, where the rock comprises >90 % leucite, the remainder is constituted of melilite, clinopyroxene, mica, Fe–Ti oxides and apatite. The term *leucitite* (called *albanite* in old literature), refers to a rock containing leucite and augite in nearly equal proportion with a small amount of other phases. The type localities for leucitites are Lake Cargelligo, Begargo Hills, Flagstaff Hills, Bygalorie and Condoblin (all localities from New South Wales, Australia). In various localities of New South Wales, modal volume percentage of leucite (18–36 %), clinopyroxene (33–45 %), olivine (3–20 %) and Fe–Ti oxides (8–20 %) is variable. The other salic phases in these leucitites include nepheline, alkali feldspar and apatite; and the hydrous phases are constituted of phlogopite and amphibole. Leucitites have also been reported from Gaussberg, Antarctica by Sheraton and Cundari (1980). They are also reported from Mt. Jugo and Colli Albani volcanic fields.

Leucitites are associated with tephrites, basanites, phonolites, and nepheline-bearing rocks in many localities of Uganda and Italy. When olivine is present in addition to leucite and clinopyroxene, the rock is called *olivine leucitite*. In earlier literature, the term leucite basalt had been used to designate *olivine leucitites*. As a basalt always contains plagioclase in addition to clinopyroxene, the term "leucite basalt" should be avoided. Holmes and Harwood (1937) preferred to use the term *ugandite* when there was a significant amount of olivine. If the proportion of leucite is greater in an olivine leucitite, it is called *mikenite* (Finckh 1912). Ferguson and Cundari (1975) preferred to call ugandite to those rocks with a colour index in the range of 60–75 vol%.

A melilite, clinopyroxene and leucite-bearing rock is called *melilite leucitite* (earlier called *cecilite* or *vesbite*, depending on different proportion of these three phases). In Villa Senni some *vesbites* contain 50–60 % leucite, 18–23 % melilite and 16–20 % (all in vol%) clinopyroxene with such accessory minerals as titanomagnetite, apatite, calcite, etc. Some of the melilite leucitite from Capo di Bove is constituted of 42 % leucite, 23 % melilite and 16 % clinopyroxene and remainder is constituted of accessory phases.

The term *katungite* (named after the extinct Katunga volcano, of which it is the sole product) has been used by Holmes (1950) for a rock consisting of olivine, melilite and leucite with minor amounts of apatite, phillipsite, natrolite, fluorite, and rare nepheline. The rock contains 16–25 % leucite, 40–50 % melilite and 22–25 % olivine (all in vol%). The accessory phases include haüyne, zeolite, apatite, and calcite in minor amounts. In San Venanzo (Italy), a similar rock called *venanzite*, consists of leucite (30 %), melilite (47 %), olivine (17 %) (all in vol%) and other accessory phases (6 %). The name venanzite may be dropped as the term katungite has been widely used. A rock consisting of kalsilite, augite, and olivine has been called *mafurite* by Holmes (1950), the type locality of which is Birunga, Uganda.

A leucite-bearing ijolite has been called *niligongoite* by Holmes and Harwood (1937). The rock is essentially composed of leucite, nepheline, clinopyroxene and melilite with accessory olivine. Instead of the term niligongoite a more generalized term such as *melilite–nepheline leucitite* may be used. If olivine is present as an essential mineral, it should be called an *olivine–melilite–nepheline leucitite*. Such rocks are found in some localities of East Eifel, Germany and in the Nyiragongo area of Birunga Province, Uganda.

A *potassium-rich ankaratrite* has been described by Sahama and Wilk (1952) from the west of Nyamunuka crater in the Katwe-Kikorongo volcanic field of southwestern Uganda. These rocks consist of clinopyroxene, olivine, nepheline, leucite, biotite, apatite, perovskite, analcime, sodalite, and some opaque minerals as accessories.

3.2.2 Leucitic Rocks with Feldspars

The term *leucite tephrite* refers to a rock comprising leucite, augite and plagioclase. When the amount of leucite is significant it is called *vesuvite* (Holmes and Harwood 1937). Savelli (1967), Cundari and Mattias (1974) and Ferguson and Cundari (1975) used the modal components, A (alkali feldspar)–P (plagioclase)–F (feldspathoid) to classify leucite and feldspar-bearing rock types into leucite tephrite, tephritic phonolite, phonolitic tephrite, phonolitic leucitite, tephritic leucitite, etc. (Fig. 3.7). Various combinations of these rocks have been described by Ferguson and Cundari from the Bufumbira region of Uganda. Such rock types have been noted by Savelli (1967) from the Somma-Vesuvius region and by Cundari (1975) from the Vico area (both in Italy). Bulk composition of these rocks would lie in the simplified system diopside–leucite–anorthite–SiO_2 (Sect. 9.2).

The essential minerals of leucite basanites include leucite, olivine, calcic plagioclase, and clinopyroxene. The term *kivite* was used by Holmes and Harwood (1937) to describe a leucite basanite with higher K_2O/Na_2O ratio. Bulk composition of these rocks lies in the system forsterite–diopside–leucite–anorthite. Apart from the Bufumbira region, the intimate association of leucite tephrite and leucite basanite is found in the East Eifel region of Germany and different localities of Italy (Chap. 4).

Rocks consisting of augite, potash feldspar, biotite and rare plagioclase, have been termed as shonkinite. Although it is free from leucite, shonkinite often contains pseudoleucite, e.g. from Shonkin Sag, Montana (Weed and Pirsson 1896; Nash and Wilkinson 1970, 1971) and Gentungen, Indonesia (Edwards et al. 1994). Porphyritic rocks comprising olivine and augite in a groundmass of leucite,

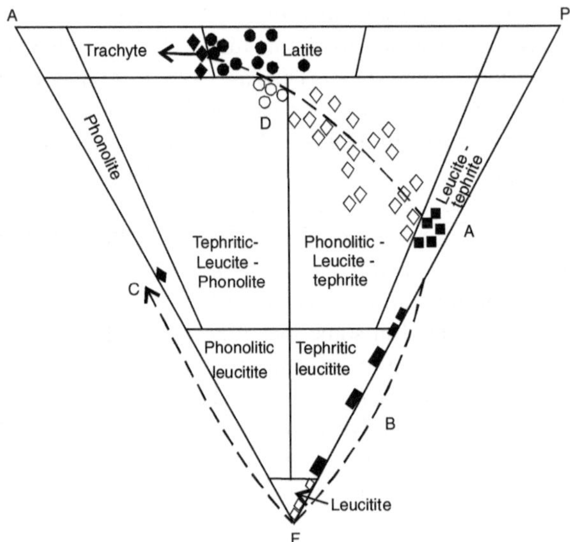

Fig. 3.7 A (alkali feldspar)–P (plagioclase)–F (feldspathoid) diagram of Birunga rocks (after Ferguson and Cundari 1975)

labradorite with orthoclase rims have been termed as *absarokite*. The rock is named after the Absaroka Range, near Yellowstone Park, USA, where it was first described.

A leucocratic rock consisting of aegirine augite and leucite or pseudoleucite has been termed as *fergusite*. *Yakutites* are rocks characterized by the presence of orthoclase, aegirine and kalsilite (up to 40 vol%). These rocks have been described from the Munrun complex in the eastern part of Siberia (Orlova et al. 1986).

When a rock is similar in mineral composition to absarokite, but contains smaller amount of augite and olivine, then it is called banakite, which may contain quartz, in that case olivine is absent.

The name *shoshonite* is coined after the type locality near the Shoshone River in Yellowstone Park. It is basically a trachy-andesite composed of olivine, augite and plagioclase phenocrysts in a groundmass of K-feldspar, plagioclase and dark-colour glass.

3.3 Classification Based on Niggli Values

Lamproitic rocks are those, which have Niggli k [molar $K_2O/(K_2O + Na_2O)$] and mg [molar $MgO/(MgO + FeO + Fe_2O_3 + MnO)$] values exceeding 0.80. Lamproites have extreme composition with respect to K_2O and MgO (high) and Al_2O_3 (low) contents (Fig. 3.8). There are however, many rocks with typical lamproitic mineralogy which has Niggli 'mg' ranging from 0.49–0.84 (e.g. Noonkanbah lamproite, West Kimberly, Australia (Jaques et al. 1986). Mitchell and Bergman (1991) also showed that most olivine lamproites from the Ellandale field of Western

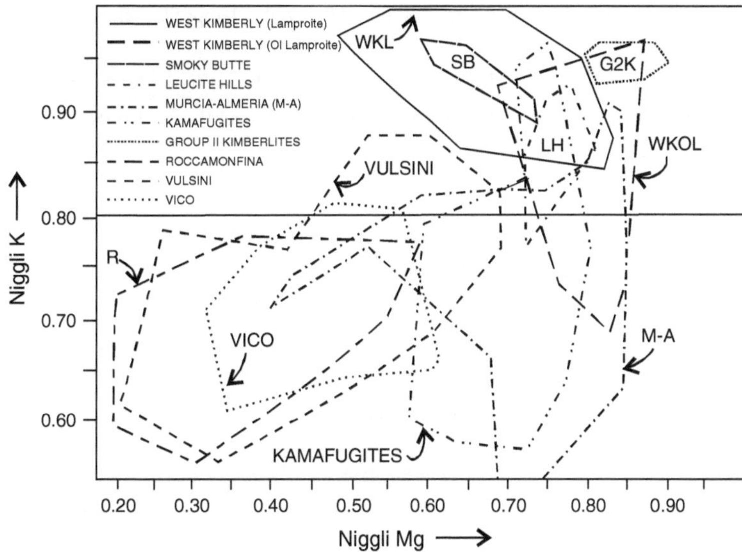

Fig. 3.8 Niggli parameter diagram for diverse potassic rocks. Lamproite data from West Kimberley (Jaques et al. 1986), Smoky Butte (Mitchell et al. 1987), Leucite Hills (Carmichael 1967; Kuehner 1980), Murcia–Almeria (Fuster et al. 1967; Venturelli et al. 1984). Data for kamafugites from Gallo et al. (1984), Peccerillo (1985), Roman province high potassium leucitites (Appleton 1972; Cundari and Mattias 1974; Holm et al. 1982, after Mitchell and Bergman 1991)

Australia have higher 'mg' and lower Niggli 'k' than the lamproites from Noonkanbah, Leucite Hills and Smoky Butte. While studying the K-rich rocks from the Aldan shield, Bogatikov et al. (1985) made a classification of these rocks on the basis of K_{mg}–SiO_2 parameters (Table 3.1).

The classification of lamproite series on the basis of simply K_{mg}–SiO_2 diagram is quite arbitrary (Mitchell and Bergman 1991). According to this classification the term lamproite is restricted to only the ultrabasic members of the series and does not follow the modern scheme of classification based on mineralogy.

According to Rock (1987) lamproites were emplaced under volcanic to subvolcanic conditions as they have leucite, otherwise they have compositions similar to those of lamprophyres. They also have a unique paragenesis in that they include K-richterite and priderite (Wagner and Velde 1986). Lamproites are also characterized by Ti-rich silicates and oxides, and do not have plagioclase. The presence of a feldspathoid (leucite or analcite or pseudoleucite) is often observed in the lamproitic rocks. Inclusion of leucite and elimination of glass in the aforesaid classification has been advocated by Mitchell (1985).

These rocks are definitely peralkaline or close to paralkalinity (Wagner and Velde 1986) as evidenced by the presence of normative acmite in the CIPW norm and are accompanied by alkali metasilicates. They are Mg-rich, and their CIPW norms show 20 % normative enstatite (or diopside) with or without forsterite. There is positive correlation between K_2O and MgO. Wegner and Velde studied lamproites from

Table 3.1 Classification and characteristics of rocks of the lamproite series (Modified after Bogatikov et al. 1991)

Rock types	Rock family	SiO_2 (wt%)	K_2O (wt%)	K_{mg}
Olivine lamproite	Ultrabasic alkaline picrite	38.8–41.0	1.6–5.1	78–89
Madupite	Alkaline ultrabasic-foidite	41.9–43.6	5.1–8.0	70–80
Wolgidite	Alkaline basic foidite	43.5–50.5	8.2–11.6	73–77
Jumillite	Alkaline basic foidite	44.4–49.0	2.8–7.4	66–85
Wyomingite, orendite and cocite	Alkaline basic phonolite	48.9–51.7	4.1–11.0	60–83
Fitzroyite, cedricite (Phlogopite leucitite)	Alkaline phonolite	51.0–59.0	8.6–12.6	55–80
Cancalite	Alkaline trachyte	51.8–57.4	7.7–10.0	67–85
Verite, fortunite	Subalkaline-trachyte	53.0–59.0	3.4–9.1	65–80

various localities, and observed that in lamproites, the K_2O content varies between 3 and 7 wt% and may be as high as 12 wt% in extreme cases. The TiO_2 content is variable and can be as high as 7 wt%. A glassy texture is observed in case of lamproites from southern Spain, Smoky Butte and Leucite Hills (Wagner and Velde 1986). In case of Australian (Prider 1982) and Damodar Valley lamproites (Gupta et al. 1983), the spinels are usually chrome-rich. In the lamproites from southern Spain, Sisco (Corsica), Moon Canyon, Shiprock and Smoky Butte (all from the U.S.A.) and West Kimberley (Australia), the F and Cl (0.2 and 0.4 wt%) contents positively correlate with high K_2O, MgO and TiO_2. This suggests the predominance of Ti-rich phlogopite present in these rocks.

Lamproitic rocks should have the following characteristics (Wooley 1987): 1. They should have molar $K_2O/Na_2O > 3$ (i.e. they are ultrapotassic) 2. The molar K_2O/Al_2O_3 ratio is usually >0.8 and more commonly >1 but below 3. The molar $(K_2O + Na_2O)/Al_2O_3$ ratio should be typically >1 (i.e. they are peralkaline). The mineralogy of different types of lamproites is discussed below:

Olivine, chrome-diopside, titaniferous phlogopite, and small amounts of leucite are the characteristic minerals of jummilites (Borley 1967). Kataphoritic amphibole (a type of K-richterite) is also present as an interstial mineral. A jumillite from Mauricia, Spain is constituted of 34 % leucite, 14 % clinopyroxene, 17 % olivine and 35 % (all in vol%) accessary minerals, which include titanomagnetite, haüyne, zeolite, apatite, etc.

Olivine and phlogopite with small amounts of leucite are the essential minerals of a *verite*. Potash feldspar may or may not be present. Calcite is often present as an accessory phase. Type localities of jumillites and verites are Mauricia and Almaria provinces of Spain.

Olivine, hypersthene, K-feldspar, phlogopite, and glass are the essential minerals of a *fortunite*. Its SiO_2 content can go up to 56.68 wt%. Pseudo-brookite is present in these lamproites as an accessory mineral (Wagner and Velde 1986). It often incorporates xenocrysts of phlogopite and Al-spinel. Orthopyroxene may be a disaggregated product of mantle xenolith in fortunite.

An olivine and phlogopite-bearing lamproite with acmite-rich pyroxene, apatite, a roedderite-like phase, rutile and calcite is called a cancalite. It contains olivine and chromian spinel xenocrysts.

Barberi and Innocenti (1967) coined the term selagite to a lamporite, containing phenocrystal olivine and phlogopite with clinopyroxene, K-feldspar, ilmenite, richterite, chromian spinel and apatite in the groundmass. The type locality is Orciatico.

Minettes are characterized by 10–40 vol% phenocrysts with a porphyritic texture (Macdonald et al. 1992). When the MgO content of the rocks exceeds 17 wt%, the phenocrystal phases are olivine, chromian spinel, clinopyroxene (diopside or salite) and phlogopite in a groundmass of phlogopite, K-feldspar and glass. There may be fresh leucite occurring as inclusions in olivine and clinopyroxene, but pseudoleucite is present as phenocrysts. In some minettes, analcime constitutes as much as 30 vol%.

Wade and Prider (1940) coined the terms *wolgidite, fitzroyite. mamilite* and *cedricite*, to various lamproitic rocks of Western Australia, the definitions of which are given below. *Wolgidite* contains diopsidic pyroxene, large grains of potash richterite, phlogopite, and somewhat turbid crystals of leucite. Relict olivine crystals are present in a serpentinous groundmass. Amphibole is more common than phlogopite and zeolite and ilmenite are present. *Fitzroyite* is a fine-grained rock, containing phenocrysts of phlogopite and leucite in a brown glassy groundmass. Rounded to sub-rounded leucite crystals contain characteristic tiny inclusions in the centers of the grains; this mineral is also present in the groundmass. Phlogopite (sometimes twinned) and abundant rutile needles are present, and the amount of leucite exceeds that of phlogopite. A typical fitzroyte is constituted of 47–50 % leucite with 5–10 % phlogopite in a serpentinous base, constituting about 40 % accessory phases including magnetite, apatite, etc. constituting less than 5 %.

The term *mamilite* was given to a fine-grained rock, consisting essentially of leucite (about 50–60 %) and potash-richterite with rutile needles and very rare phlogopite (<5 %) in a serpentinous base (<20 %). Olivine may be present in trace amount. Mica is present only in the groundmass. Leucite crystals are often altered, sometimes containing characteristic inclusions.

A lamproite, rich in leucite with small amounts of diopsidic augite embedded in a brown glassy groundmass containing small to rare amounts of potash richterite, phlogopite, has been called a cedricite. Cedricites have rutile with serpentines, pseudomorphous afer olivine. Weed and Pirsson (1996) suggested the term *missourite* for a group of rocks consisting essentially of augite (about 35–40 %), and leucite (40–50 %) with minor amounts of biotite ($\sim$5 %), analcite, zeolite, and iron ore. Washington (1927), however, suggested that olivine is not an essential mineral in missourite and listed the modal percentages of two missourites which are devoid of olivine. The compositions of these rocks lie in the field of leucitites.

Such terms as wyomingite, orendite, and madupite have been coined by Cross (1897) for leucite-bearing rocks from the Leucite Hills. Wyomingite is composed of leucite (20–25 %) with trace of olivine in a serpentinous base (~80 %) having augite, phlogopite, apatite, calcite, and magnetite in order of decreasing abundance. Phlogopite occurs as phenocrysts, whereas the other minerals are present in the groundmass. Potash richterite is sometimes present as an accessory phase. Wyomingite often shows parallel flow banding, due to the planar arrangements of phlogopite flakes with vesicular cavities, developed parallelly at the flow banding (Yagi and Matsumoto 1966).

According to Cross (1897), *Orendite* is composed of leucite (14–16 vol%), olivine (<5 vol%) in a serpentinite base. Sanidine, phlogopite, augite and calcite, are also present in the glassy groundmass. In olivine orendite, olivine may be present in small amount. The glass is filled with microlites of sanidine and augite. Sanidine forms euhedral tabular crystals, usually less than 0.1 mm in size, commonly twinned on Carlsbad law. The rocks vary in their crystallinity from a highly glassy variety to a rather crystalline type. In the glassy variety, phlogopite phenocrysts are set in a glassy groundmass, composed of pale brownish glass, filled with abundant acicular crystals of amphibole less than 0.3 mm in length, which surround the microlite-bearing cavities.

A lamproitic rock comprising small phenocrysts of diopside, enclosed by rare poikilitic crystals of phlogopite, has been termed as madupite by Carmichael (1967). The micas are usually set in a turbid groundmass of diopside and chlorite, together with other accessories. Magnetite is present and olivine is absent (Carmichael 1967). Pilot Butte at Leucite Hills, is the type locality for madupite.

Mitchell (1985) classified lamproites into two categories: (1) madupitic lamproite and (2) phlogopite–sanidine–leucite–diopside lamproite. Mitchell recommended that in case of phlogopite lamproite, phologopite should occur as phenocryst and in the case of madupitic lamproite, phlogopite should be present in the poikilitic groundmass. These two types of lamproites include additional sanidine and leucite, and their bulk compositions also plot on the diopside–phlogopite–sanidine–leucite system. Since there is no contrasting mineralogy or chemistry in case of the so-called two types of lamproites, one plotting in the limiting join diopside–phlogopite, and the other inside the tetrahedron, there is no scientific reason for such a classification. The accessory minerals in lamproites include, chromian spinel, priderite [$(K, Ba)_{2-y}(Ti, Fe^{3+})_{8-x}O_{16}$, where the value of y is close to 1 and that of x is very small], wadeite ($K_4Zr_2Si_6O_{16}$), sheherbakovite [$(Na, K)(Ba)Ti_2Si_4O_{14}$], jeppeite [$(K, Ba)_2(Ti, Fe)_{12}O_{13}$], apatite, perovskite, sphene, ilmenite (very rare) and armacolite [$(Mg, Fe)_2Ti_4O_{10}$].

3.3.1 Various Lamproitic Assemblages and Their Heteromorphic Relations to Each Other

The bulk composition of phlogopite, forsterite, akermanite, diopside, sanidine, leucite, plots in the system larnite–forsterite–kalsilite–SiO_2–H_2O (Luth 1967;

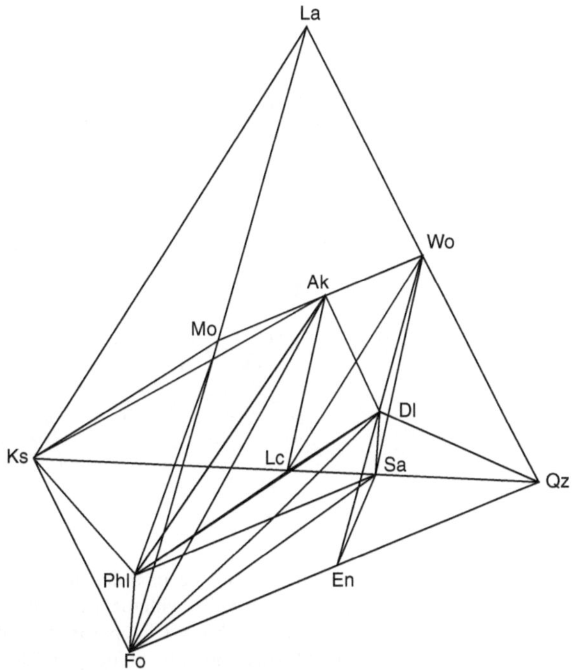

Fig. 3.9 The larnite–kalsilite–forsterite–SiO_2 tetrahedron (after Yoder 1986)

Wendlandt and Eggler 1980a, b, c; Gupta and Green 1988). If the bulk composition of an initial liquid lies in this system, then under $P(H_2O) = P(Total)$ condition, it should yield assemblages similar to many lamproites.

As such, the bulk composition of various K-rich rocks can be plotted in the larnite–forsterite–kalsilite–SiO_2 tetrahedron (Fig. 3.9). The subtetrahedra formed in Fig. 3.9, was isolated by Yoder in Fig. 3.10 for easy visualization. In Fig. 3.10, he plotted the mineralogical compositions of many lamproites.

Yoder (1986, 1989, 1990) pointed out that the assemblage, shown as a product of reaction (1), takes place under certain P–T conditions [$P(H_2O) = P(Total)$]:

$$2\ \text{Phlogopite} + \text{K-feldspar} \approx 3\ \text{olivine} + 3\ \text{leucite} + 2H_2O \qquad (3.1)$$

Two new subtetrahedra are formed because of this heteromorphic reaction, as shown in Fig. 3.11. Thus, the assemblage related to jumillite and shonkinite has heteromorphic relation to absarokite and missourite, bulk compositions of which are shown in subtetrahedra (Fig. 3.11). Kazanite (Lacroix 1926) has similar assemblage as missourite but the former has more modal phlogopite. Some of the Australian lamproites such as cedricite and manilite (leucite + clinopyroxene + olivine and phlogopite) may be included with missourite, if amphibole in each is neglected (Yoder 1986).

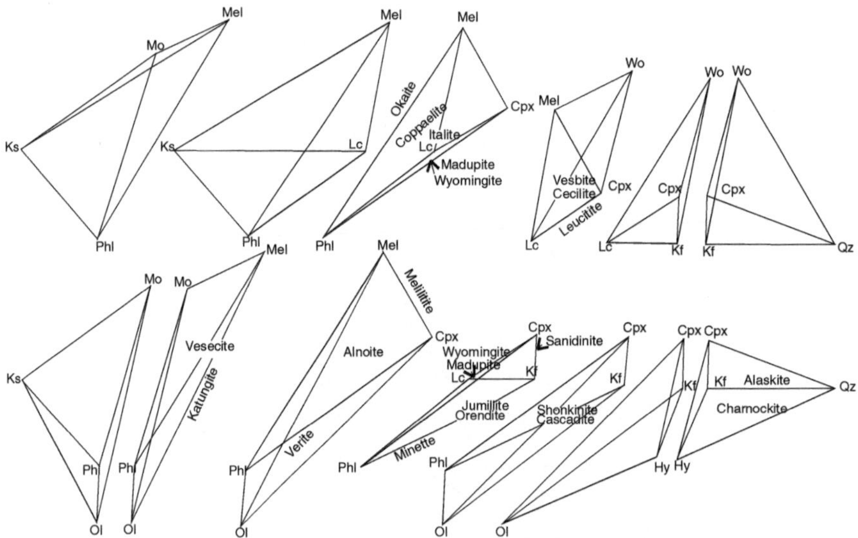

Fig. 3.10 An 'exploded' view of the subtetrahedra in the modified normative tetrahedron of Fig. 3.9. The corner of the isolated subtetrahedra are labelled with the names of the modal minerals most closely represented by the normative minerals, Cpx = Clinopyroxene, Hy = Hypersthese, Kf = Potassium feldspar, Mel = Melilite, Ol = Olivine, and Phl = phlogopite. The joins, planes and volumes (*underlined*) are named for rocks for which the assemblage is most closely representative (after Yoder 1986)

Fig. 3.11 Subtetrahedra of Fig. 3.10 from reaction of phlogopite (Phl) and potassium feldspar (Kf), to form olivine + leucite + H_2O, illustrating the heteromorphic relations of jumillite–shonkinite to missourite absarokite (after Yoder 1986)

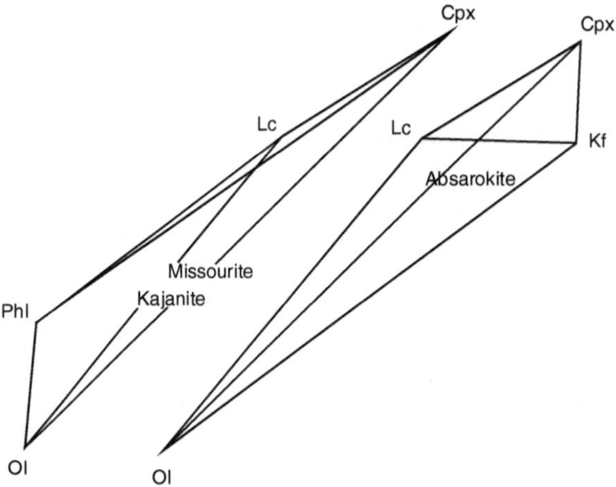

The assemblage clinopyroxene + phlogopite + leucite + olivine, similar to that found in madupite) is formed under certain P–T conditions (Barton and Hamilton 1978). The product is also similar to a missourite.

When a natural orendite and a wyomingite were heated at high temperature and low pressure (<300 bar, Barton and Hamilton 1978), an assemblage of clinopyroxene + olivine + leucite + K-feldspar was obtained. Yoder also showed that the join K-feldspar-olivine is broken in presence of H_2O at lower temperature:

$$\text{K-feldspar} + 3\ \text{forsterite} + H_2O \approx \text{phlogopite} + 3\ \text{hypersthene} \quad (3.2)$$

The assemblage, biotite + olivine + hypersthene + feldspar, is typical of a lamproite, called *fortunite*. Experimental study of Arima and Edgar (1983) showed that phlogopite clinopyroxenite has a heteromorphic relation to katungite. The assemblage is formed by the following reaction:

$$3\ \text{phlogopite} + 2\ \text{clinopyroxene} \approx 5\ \text{olivine} + 3\ \text{melilite} + 3H_2O \quad (3.3)$$

The rocks called venanzite and melilite ugandite plot in the two new subtetrahedra as depicted by Fig. 3.12. Katungite, vesbite and molibovite also plot in these two subtetrahedra. Phlogopite breaks down to olivine, leucite, kalsilite and vapour at high temperatures but low pressures (Yoder and Kushiro 1969). Note the following reaction:

$$2\ \text{phlogopite} \approx 3\ \text{forsterite} + \text{leucite} + \text{kalsilite} + H_2O \quad (3.4)$$

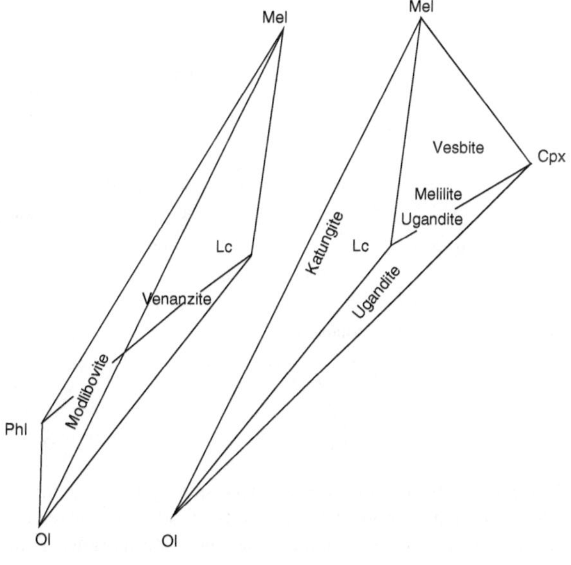

Fig. 3.12 Subtetrahedra of Fig. 3.10 resulting from reaction of phlogopite (Phl) and clinopyroxene (Cpx) and illsutrating the heteromorphic relations of alnoite–italite to venanzite–melilite ugandite (after Yoder 1986)

Because of this, all of the eight subtetrahedra in which phlogopite is a constant component, is therefore, affected and anhydrous compositions are then represented by melilite + leucite + forsterite + kalsilite and akermanite + leucite + forsterite + clinopyroxene. Melilite ugandite (melilite–olivine leucitite), phlogopite–olivine leucititewas studied at one atmospheric pressure by Gupta (1972). Yoder (1986) pointed out that these two subtetrahedra undergo a reaction as follows:

$$4\ \text{clinopyroxene}\ + 3\ \text{kalsilite}\ \approx 2\ \text{akermanite}\ +\ \text{forsterite}\ + 3\ \text{leucite} \quad (3.5)$$

Holmes (1942) recognized this heteromorphic relationship, and described a lava comprising clinopyroxene + kalsilite + forsterite and other varieties with melilite and leucite. Bulk compositions of these different types of rocks are illustrated in Fig. 3.13 as a melilite mafurite, mafurite and a leucite mafurite.

The above discussion shows that many assemblages, which have heteromorphic relation have similar chemistry but are products of different pressures and temperatures in presence of excess water. Simple classification based on only chemistry could therefore, be highly misleading. A mineralogical classification reflects not only the chemistry but the P–T condition of formation of the rocks. For example, presence of feldspathoids together with melilite and hydrous minerals in equilibrium, should suggest low P–T condition of formation. It also indicates low silica-activity condition of the magmatic fluid responsible for the generation of these rocks. A knowledge of the presence of oxide minerals (occurring as accessory) is important, as they indicate oxygen fugacity condition of formation of these rocks.

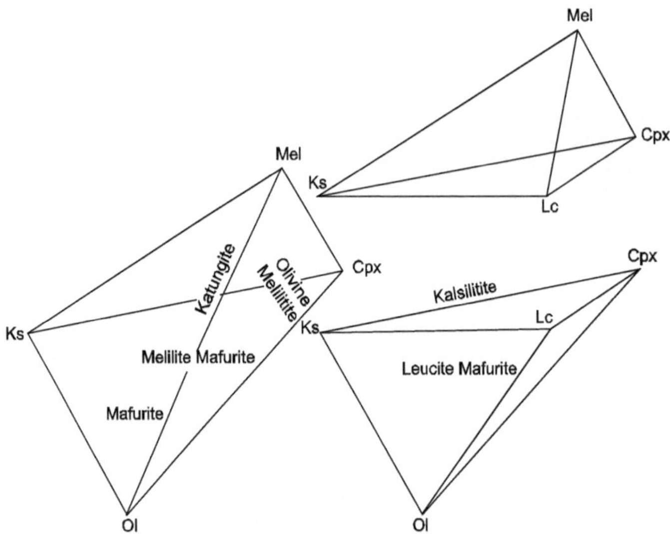

Fig. 3.13 Subtetrahedra of Fig. 3.10 resulting from reaction of melilite (Mel) and leucite (Lc) and illustrating the heteromorphic relations of anhydrous assemblages produced from the breakdown of phlogopite to melilite mafurite–leucite mafurite (after Yoder 1986)

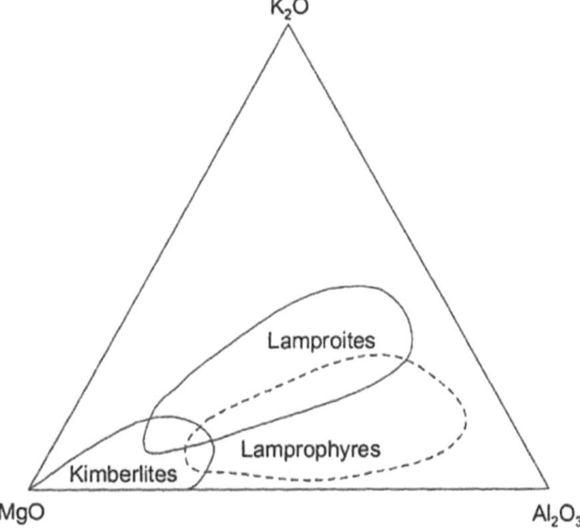

Fig. 3.14 composition plot of kimberlites, lamproites and lamprophytes in terms of K_2O–Mgo–Al_2O_3 (after bagman 1987)

3.3.2 Distinctive Criteria to Differentiate Among Kimberlites, Lamproites and Lamprophyres

Bagman (1987) plotted the compositions of Kimberlites, lamproites and lamprophytes in terms of K_2O, MgO and Al_2O_3 (Fig. 3.14) and showed that their compositions plot in three distinct fields through there are some overlaps.

Whereas kimberlites do not contain such minerals as leucite, melilite, wadeite, priderite there are magnesium rich lamproites which may contain diamond like the kimberlites suggesting their high pressure origin, whereas felspathoide-bearing lamproites were crystallized at relatively low pressure inside the crust. Although presence of K-richterite was considered as an index mineral (Kushiro and Erlank 1970) to identify lamproites, recent studies at high pressure-temperature on K-richterite suggest that these mineral is stable at a significant depth, where diamond is also stable.

Chapter 4
Different Localities of Potassium-Rich Silica-Undersaturated Igneous Rocks and Their Silica-Rich Variants

Silica-undersaturated potassium-rich igneous rocks and their derivants were earlier considered to be a rare group of rocks occurring in widely scattered localities all over the world (Turner and Verhoogen 1960). During the last three decades however, a large number of localities have been described, where these rocks are reported. They are therefore, much more abundant and widespread than previously thought of (Gupta and Fyfe 2003).

As a matter of fact they are found in all the continents. In Asia, they have been described from Wudalianchi, Erkeshan and Chi-hsing Shan of Manchuria, China and Yengbejin rift valley of the Tibetan plateau. The notable localities from Indonesia include, Mount Mouriah, Palialian, Tülering and Ringgit Beser complex of Java. In India, they have been described from Bokaro, Jharia and Raniganj sedimentary basins of the Damodar Valley Coal fields, where they occur as innumerable dykes and sills of lamprophyres and lamproites.

In Australia the two well-known localities are, West Kimberly Province of northwest Australia and New South Wales, located towards south east of that continent.

The most well known localities of potassic lavas in Africa include:

1. Ruwenzori and the localities adjoining Western Rift of Central Africa,
2. Birunga volcanic province, and
3. The Kapamba region of Luwangwa Valley, Zambia.

In Europe, they occur in the Murcia and Almaria Province of Spain, East Eifel province of Germany and many localities of Italy. Some of the important Italian localities include, Monti Vulsini, Vico, Monti Sabatini, Alban Hills, the Phlegrean Fields, Roccamanfina and the Somma-Vesuvious region. They also occur in the Massif Central region of France. In the United States of America, they have been described from Highwood Mountains (Montana), Hopi Butte (Utah), Navajo (Arizona) and Leucite Hills of Wyoming, Yoshemite National Park, Sierra Nevada and Deep Spring Valley, located east of Sierra Nevada.

A great deal of information about the occurrence of K-rich silica-undersaturated rocks of the former USSR can be obtained from Kogarko et al. (1995). Their description shows that there are at least thirty localities, where K-rich silica-undersaturated rocks are found. These localities occur in the petrographic provinces

A.K. Gupta, *Origin of Potassium-rich Silica-deficient Igneous Rocks*,
Springer Geology, DOI 10.1007/978-81-322-2083-1_4

of Kazakhstan, Central Asia, Caucasus, Baikal, Primorye, Omolon and Kamchatka. They are basically leucite-bearing but silica-rich poptassic variants associated with them are also found in these localities.

Leucite bearing rocks also have been reported from Alto Paranaiba Igneous Province of Brazil, Colima volcano of Mexico and Serra Geral Formation of eastern Paraguay. K-rich feldspathaoidal rocks have been reported also from the Aolean Arc, south of Italy, Tristan Cunha Islands in the Atlantic Ocean and Utsuryo Island, Korea. These localities are described below.

4.1 Ultrapotassic Silica-Deficient Rocks from Asia

4.1.1 Leucite-Bearing Rocks of Manchuria, China

The potassium-rich volcanic fields of Wudalianchi, Erkeshan and Keluo (WEK, NE China) are located between the Mesozoic Songliao basin and the Palaeozoic Xing'an Mountains fold belt (Zhang et al. 1995; Fig. 4.1). The Songliao basin was formed during Late Jurassic rifting behind the active Pacific margin, and the Palaeozoic fold belts of Xing'an Mountains. The potassic lavas are extended over an area of more than 1,400 km^2. Both the Keluo and Wudalianchi fields are characterized by the presence of 14 pyroclastic cones, whereas the Erkeshan field is characterized by the presence of two volcanoes. Historical records and K–Ar dating show that these potassic rocks erupted during three main episodes:

1. Miocene (9.6–7.0 Ma),
2. Middle Pleistocene (0.56–0.13 Ma) and
3. Recent (1719–1721 A.D.).

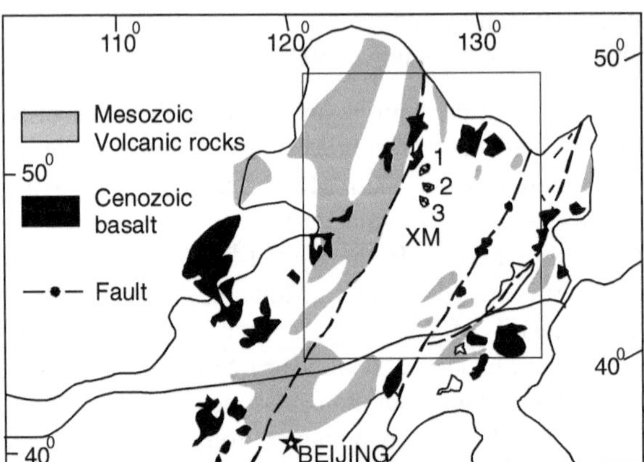

Fig. 4.1 Map of East China showing the main tectonic domains, distribution of Mesozoic and Cenozoic volcanic rocks (modified from Fan and Hooper 1991), *XM* Xing'an-Mongolian fold belt. *1* Keluo, *2* Wudalianchi and *3* Erkeshan (after Zhang et al. 1995)

According to Zhang et al. (1995), asthenospheric and crustal attenuation in Songliao basin have been established by seismic and electric conductivity measurements. They noted that the lithosphere at the centre is 80 km thick, 40 km thinner than that in the margins. The rocks from this locality are porphyritic with groundmass having vitreous to holocrystalline texture. These potassic rocks are subdivided into olivine leucitite, leucite basanite and trachybasalts. Olivine leucitites are distinguished from leucite-basanites and trachybasalts by the presence of nepheline and paucity of feldspars and high modal contents of mafic phases and leucite. The basanites contain leucite but this mineral is absent in trachybasalts. Olivine leucitites are found from three volcanoes:

(1) Dangzishan,
(2) Heishan, and
(3) Muhenanshan of the Keluo field.

The Muhenanshan volcano is the only one known so far, which is characterized by the outpouring of olivine leucitite and trachybasalts in different episodes of magmatism (Fig. 4.1).

Zhang et al. (1995) described the presence of mantle-derived spinel peridotite xenoliths at several volcanoes of each field. Orthopyroxene, clinopyroxene and sanidine megacrysts and various igneous inclusions are found from Dangzishan, Heishan and Gushan volcanoes (Keluo).

The WEK volcanic rocks are characterized by high K_2O (3.5–7.1 wt%), K_2O/Na_2O ratio (>1) and incompatible elements. They have high $^{87}Sr/^{86}Sr$ (0.7050–0.7056), low $^{143}Nd/^{144}Nd$ (0.51238–0.51250) and low $^{206}Pb/^{204}Pb$ (17.06–16.61) ratios. The major element analyses of some of the WEK rocks are given in Table 4.1.

Zhang et al. (2002) studied the Holocene volcanoes in the Jingpohu region, which are situated in the "Crater Forest" and "Frog Pool" areas along the northwest side of Jingpohu Lake, northeastern China. They dated three charcoal samples from the first and second volcanic cycles, and found that the ages of the first and second cycles are 3,430–3,490 and 2,470 years, respectively. The lavas from the Jingpohu area consist of basanites (BSN), alkali olivine basalts (AOB) and tephrites (TP). According to them Crystal fractionation models are consistent with the generation of AOB and TP from a basanitic parent. Minor fractionation of olivine, augite, magnetite and Cr-spinel is required to produce AOB compositions whereas the generation of TP requires extensive fractionation of kaersutite, phlogopite and anorthoclase with minor olivine, augite, magnetite and leucite. They suggested that the presence of kaersutite, phlogopite and anorthoclase megacrysts and mantle xenoliths were, part of a fractionation history at high pressure in the mantle. Although all basaltic rocks contain many granitic xenoliths, their geochemical characteristics show that they have not undergone any contamination of upper crust en route to the surface, but some alkali basalts might have suffered from contamination with lower crust. Relatively unradiogenic isotope ratio compared with Bulk Earth, steep chondrite-normalized REE patterns and strong incompatible element enriched patterns suggest that the magmas were derived from a mixture of an

Table 4.1 Analyses of K-Rich rocks from Tibet and North–East China

	SiO_2	TiO_2	Al_2O_3	Fe_2O_3	MnO	MgO	CaO	Na_2O	K_2O	P_2O_5	H_2O^+	LOI	Total
1	43.63	2.06	10.55	10.23	0.16	14.07	9.76	3.45	3.99	0.81		0.29	99.29
2	43.79	2.40	11.83	10.80	0.16	10.65	9.69	4.17	4.56	1.16		−0.04	99.85
3	44.83	2.21	13.30	9.92	0.16	7.95	9.29	4.35	5.34	1.17		0.86	99.82
4	52.46	3.03	14.52	8.57	0.10	3.86	5.16	3.98	6.96	1.30		−0.01	100.35
5	50.04	2.75	13.89	9.38	0.12	5.14	6.80	4.03	5.85	1.30		0.65	100.39
6	50.94	2.57	13.50	8.51	0.11	6.55	5.85	3.57	5.74	1.17		0.49	99.40
7	52.72	0.72	19.57	5.62	0.11	1.47	3.97	5.16	5.18	0.41	4.37		99.30

1–6 Major analyses of rocks from Wudalianchi, Erkeshan and Keluo, North-East China (after Zhang et al. 1995)
7 Chemical analyses of K-rich rocks fromYangbajin, Tibet (after Coulon et al. 1986)

incompatible element depleted anhydrous lherzolite a sthenospheric mantle source and an enriched, amphibole-phologite (apatite-) bearing lherzolite continental lithospheric mantle source. They proposed that the basanites are the products of very low degree partial melts (<1 %) of this source under high extension strength.

4.1.2 K-Rich Volcanics from Yangbajin Rift, Tibet

The Tibetan high plateau in the Himalayan and Karakoram Ranges are product of continuous collision between India with the southern margin of Eurasia plate during past 15 Ma (Coulon et al. 1986). Potassic and ultrapotassic lava flows and pyroclasic rocks from the volcanic fields, south of Xungba and north of Borgba (Fig. 4.2). The ages of the rocks range between 17 and 25 Ma. The K_2O content of these rocks vary from 5.98 to 6.86 wt% and the K_2O/Na_2O ratios ranges between 2.38 and 6.58, average being 3.27. The ultrapotassic rocks are characterized by phenocrysts of phlogopite, clinopyroxene and microphenocrysts of apatite and titanomagnetite set in a pale-brown glassy matrix. In some samples sanidine, clinopyroxene, Ti-phlogopite and olivine are present. Their $^{87}Sr/^{86}Sr$ ratio varies between 0.7172 and 0.7220 and $^{143}Nd/^{144}Nd$ ratio ranges between 0.51190 and 0.51200, whereas the ranges of $^{207}Pb/^{204}Pb$ and $^{208}Pb/^{204}Pb$ vary respectively between 15.68–15.72 and 39.42–39.60.

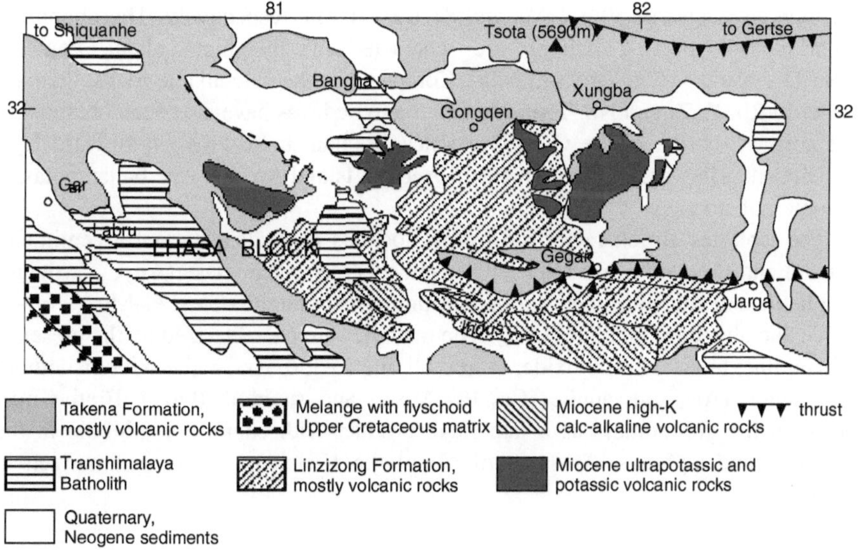

Fig. 4.2 The K-rich mafic rocks from Tibet (after Coulon et al. 1986)

Neogene volcanic rocks are also present in the southern part of Tibetan Plateau (Coulon et al. 1986). The central and southern portion of Tibet is divided into three blocks:

(1) The Lhasa block,
(2) The Quantang block and
(3) The Kunlun block.

The Yarlung-Tsangpo ophiolite suture separates the Lhasa block from the Indian Plate (Fig. 4.2). The Lhasa block is separated from the Quantang block by the Bangong Nujian suture zone, which is tectonically dismembered. The Kunlun block is separated in the south from the Quantang block, and in the north from the Eurasian plate by two suture zones. The northern part of the Lhasa block is constituted of the Precambrian gneissic basement rocks overlain by Jurassic sedimentary rocks, which in turn are overlain by rocks of lower Cretaceous series. Mesozoic volcanism took place at Gyanco, Barda, Nagqu and Anduo during Cretaceous. Large volumes of basalt, andesite and dacites are observed in these places, whereas ignimbrites are rare. East of Lhasa, volcanic rocks of supposedly Triassic age constitute a thick (1,400 m) calc-alkaline suite, composed of ignimbrites, andesite and porphyritic basalts. Towards south, there is a folded Cretaceous sedimentary formation, known as Takena formation, which is uncomfortably overlain by a volcanic sequence, known as the Lingzizong formation, which is made up of andesite and abundant rhyolitic ignimbrites.

The Lingzizong volcanics are affected by a Pleistocene rift structure towards northeast of Yangbajin. Along the eastern border of this rift the andesite and rhyolite are associated with phonolitic tephrites and trachytes. The plutonic block of Ganjdese lies southeast of the Lhasa block. Biotite and sanidine extracted from a trachyte gave ages of 50 ± 1 Ma and 81.1 ± 1.1 Ma, respectively. The phonolitic tephrite and alkali trachyte are characterized by high alkali contents $[(Na_2O + K_2O) > 1, K_2O/Na_2O > 1)]$. Compared to the calc-alkalic rocks, they are richer in Rb, Y, Zr and Nb. Tephritic and trachytic lavas have also been obtained in the northern border of the Gangdese volcanic chain indicating a northward K_2O enrichment. They are similar to rocks of Aeolian Island Arc and Azerbaijan potassic complex of Iran.

The tephrites are characterized by phenocrysts of analcite (about a centimetre across) and pseudomorphs after leucite phenocrysts, plagioclase ($An_{56–53}$), titenomagnetite and altered olivine, set in a groundmass of augite ($Ca_{47}Fe_{17}Mg_{36}$), Fe–Ti oxide and abundant sanidine (Or_{60}). The trachytes are characterized by the presence of sanidine, biotite, Fe–Ti oxide, scarce olivine (oxidized), apatite with or without plagioclase ($An_{60–40}$), augite ($Ca_{40}Fe_{22}Mg_{38}$) and sanidine (Or_{65}). Biotite may occur in the groundmass as a late stage product with edenitic amphibole in the trachytes and tephrites. Analyses of Yangbajin K-rich rocks are summarized in Table 4.1.

4.1.3 Occurence of K-Rich Silica-Deficient Rocks from Turkey

Altherr et al. (2008) have described leucite tephrites from the southern zone of Eastern Pontides. It is an isolated small outcrop (1.5 km^2) occurring in the Asutka thrust sheet. Field relations constrain the timing of volcanism between Maastrichtian and late Paleocene. The leucite tephrites consist of clinopyroxene, analcime (psudomorph after leucite), Ti-magnetite, plagioclase, sanidine, apatite and accessory biotite. The rocks are represented by three geochemical and petrographic varieties: Types I and II are mostly blocks within epiclastic debris and type III are lava flows. All the rocks are considerably altered, whereby leucite is almost totally analcimized. There are however leucite inclusions within clinopyroxene that occurs remote from the solution-bearing channels. Furthermore, calcite and chlorite/ smectite also have been formed due to low grade metamorphism. Chemical effects of alteration are primarily reflected in an unsystematic variation in the whole-rock K_2O/Na_2O ratios (0.12–1.71) and in the abundance of large-ion lithophile elements, such as Cs, K, and also Rb. Both the high modal abundance of leucite and the composition of clinopyroxene, indicate an ultrapotassic nature of the primary rocks. In primitive mantle-normalized element concentration diagrams, all samples are characterized by negative anomalies of Nb–Ta, Zr–Hf and Ti and a positive anomaly in Pb, testifying the orogenic nature of the rocks. The rocks cover restricted ranges in all initial isotopic ratios with Sr^{87}/Sr^{86} (60 Ma) ranging from 0.70537 to 0.70568, Nd^{143}/Nd^{144} from 0.512529 to 0.512585, Pb^{206}/Pb^{204} from 18.65 to 18.83, Pb^{207}/Pb^{204} from 15.65 to 15.66 and Pb^{208}/Pb^{204} from 38.64 to 38.88. There is no obvious relationship between the degree of analcimization and the isotopic composition. Chondrite-normalized rare earth element (REE) patterns show no significant Eu anomaly, but a strong enrichment of the LREE over the HREE with (La/Yb) (cn) = 11.6–14.2, whereby normalized concentration of Eu to Lu are nearly similar. The Everek Hanlan leucite tephrites represent the youngest products of Cretaceous to Paleocene arc magmatism of the Eastern Pontides, thus documenting the last stages of the Neotethyan subduction.

Coban and Flower (2006) studied the ultrapotassic rocks in the Bucak area of Isparta Angle, SW Turkey. These rocks are characterized by low SiO_2 (46.8–49.2 wt %) and relatively high MgO (10.4–11.6 wt%) contents, with lamproitic affinity (K/ Na, > 2.5; Mg#, 73–75; Al_2O_3 (9.2–11 wt%), CaO (7.4–10.6 wt%), Cr, (525,675 ppm); Ni (442–615 ppm). They are made up by phlogopite (30–40 vol%), leucite (25–30 vol%), olivine (5–20 vol%), which rarely contain Cr-spinel, clinopyroxene (5–10 vol%), sanidine (5 vol%) and richterite, with accessory apatite, magnetite and ilmenite. One sample also include negligible amount of sodalite in the groundmass, which is an unusual mineral in lamproites. Mineral phase variation and textures record discrete phases of pre-eruptive crystallization: (1) early appearance of (Cr-spinel-bearing olivine), Ti poor phlogopite ± apatite crystallizing at pressures of 1.0–2.0 GPa, at or close to the lithospheric Mechanical Boundary Layer (MBL), and (2) later appearance of Ti-rich phlogopite, clinopyroxene, richterite, leucite, sanidine, and other minor phases, at pressures of 0.1–1.0 GPa, indicating discrete, pressure-specific fractionation events. According to Coban and Flower (2006) the Bucak silica poor 'leucite' lamproites were probably generated by partial melting of phlogopite-bearing, refractory peridotite at pressures of 1.5–2 GPa, higher than those proposed for SiO_2 -saturated 'phlogopite' lamproites (1–1.5 GPa) from Afyon, located to the North of this locality. The depth (total pressure) of melt segregation probably dominates over volatile partial pressures (e.g. of CO_2, F, H_2O) in determining the SiO_2-undersaturated character of the Bucak magmas.

4.1.4 Potassium-Rich Lamprophyres and Lamproites from Bokaro, Jharia and Raniganj Basins, East India

Phlogopite-bearing mafic and ultramafic dykes and sills are often found to intrude into a sedimentary rock sequence (Upper Carboniferous to middle Triassic) of the Bokaro, Jharia and Raniganj basins located in the eastern part of India (Fig. 4.3). The sedimentary sequence consists of alternating rock types of sandstone and shale along with coal-bearing seams (Lower Gondwana formation). This sequence occupies tectonic troughs with E–W trending fault boundaries.

4.1.4.1 Bokaro Basin

Petrological and geochemical studies of the K-rich lamproitic and lamprophyric rocks from the east Bokaro Coalfield of Bihar was studied by Basu et al. (1997). They found altogether thirty dykes ranging in thickness between 0.3 and 12 m and in strike length extending from few metres to more than 2.5 km.

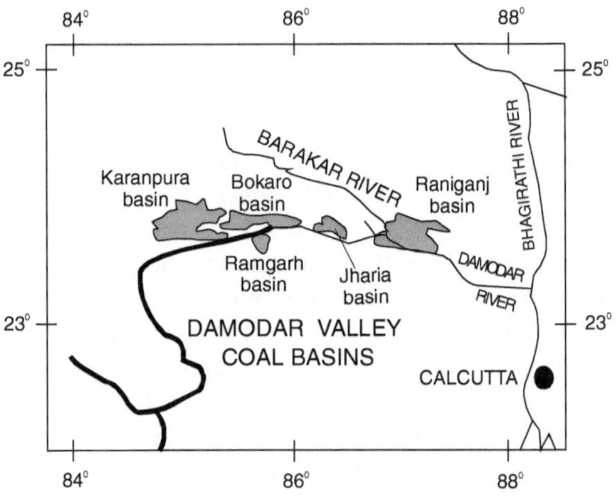

Fig. 4.3 Distribution of Gondwana basins in Damodar valley coal field (Ghosh and Mukhopadhyay 1985)

The lamprophyres intrude mostly the Barakar formation as subvertical dykes trending predominantly east–northeast to west-southwest (Raja Rao 1987). According to Basu et al. (1997), the dykes bear various contact relations with the country rocks comprising shale, sandstone and coal seams. The nature of contact may be sharp, bulbous and sometimes of tapering forms in profile sections. There may be dykes with angular offsets, branching and veined types. They incorporate xenoliths of the country rocks or ultramafic rocks of mantle origin. Rarely, they occur as sills and at times exhibit circular pipe-like cross sections. The coal seams in contact with the intrusives are indurated and transformed into coke. Some of the dykes are vesicular with amygdules of carbonate minerals, secondary silica, zeolite etc. Macroscopic bandings are common especially in the olivine lamproites with sharp or gradational inner contacts with high concentration of phenocrysts, xenocrysts or xenoliths in the central part. Basu et al. (1997) found that in some of the dykes there is a systematic disposition of fine and more glassy layers, and the lamproitic dykes indicate flow differentiation. Occasionally conjugate intersecting dyke swarms are also noticed. The dolerite dykes, whenever present cut across the entire sequence of the Gondwana formation with strikes varying between north–south and northwest–southeast. The dykes often extend into the Precambrian basement rocks.

Detailed petrological study indicates a systematic mineralogical variation representing a continuous spectrum of ultrapotassic rocks ranging from olivine lamproite through lamproite to minettes.

The minettes are reported mostly from the western part of the east Bokaro coalfield (West of Gomia, Swang, Gobindpur, etc.). They display porphyritic to panidiomorphic textures with phenocrysts of alkali feldspar, phlogopite, olivine (pseudomorphed by carbonates) and slender prisms of apatite set in a glassy matrix that contains finer crystals of alkali feldspar, olivine, zoned phlogopite, and opaque phases like magnetite, ilmenite and sulphides of Ni and Cu.

The olivine lamproites generally contain megacrysts and possible xenocrysts of phlogopite and olivine. In some samples large needles of ilmenite are set in a glassy groundmass which contains finer crystals of phlogopite, apatite and carbonates. Some of the olivines are pseudomorphed by magnesian carbonates, iron carbonates, silica and rare serpentine. In other samples they are surrounded by phlogopites exhibiting ocellar texture. The opaque minerals include magnetite, pyrite, arsenopyrite, pentlandite, millerite, ilmenite, priderite and native gold. The olivine-leucite lamproite differs from the olivine lamproite in that it contains leucite both as a phenocrystal as well as groundmass phase. Distinctive leucite twining is sometimes preserved in the usually altered leucite crystals. Serpentinized olivine xenocrysts are occasionally found to be rimmed by leucite crystals with thin intervening zone of phlogopite. The leucite lamproite contains leucite as a major phase both in the form of phenocrysts and in the groundmass. Minor phases include olivine and apatite set in a glassy matrix. The leucite phenocrysts usually contain inclusions of glass and opaque oxides. Some of the olivine lamproite (with or without leucite) dykes occurring at Kalyani, Bokaro and Jarangdih Mines display excellent flow layers along the margins. They are essentially glassy layers with small opaques and aligned phlogopite phenocrysts. The central part of some of the dykes (e.g. those in the Kalyani mines) have a coarse-grained core (about 65–75 cm thick) composed of olivine, leucite phlogopite, apatite, opaque minerals and vesicles filled with cryptocrystalline silica and zeolite. The olivine lamproites and their leucite-bearing variants contain large olivine xenocrysts (up to 1 cm × 0.5 cm). The rocks in general contain fine specks of native gold and sulphides, mostly arsenopyrite, chalcopyrite, pentlandite and millerite. Oxide phases include magnetite, ilmenite and priderite. Nodules and xenoliths of various sizes (up to about 2 cm diameter) and shapes (pear, discoid, etc.) have been noted in some samples. Olivine xenocrysts are common, but rare occurrence of xenoliths, comprising an aggregate of olivine, orthopyroxene (harzburgite) and chromian spinel is also noted.

A set of dolerite dykes of much larger dimensions (up to about 12 m thick and more than 2.5 km long along strike length) than the lamproites, traverse through the Bokaro basin.

Analyses of some of the rocks representing a large array of lamproitic rocks from the Bokaro basin are given in Table 4.2. Extensive alteration in the form of carbonation, silicification, formation of zeolites that fill the vesicles and devitrification poses some difficulty in evaluating the chemical data. Moreover, there are frequent changes in composition from one unit to the other even within a single dyke. There is also some overlapping composition among these units. The presence of carbonates is often reflected in higher concentrations of CaO and MgO, though MgO content also depends on the presence of phlogopite lamproite magma, where H_2O plays a dominant role compared to CO_2. High K_2O values usually reflect the presence of phlogopite, leucite and in some instances, alkali feldspar (viz. minette with $K_2O/Na_2O > 4$). Some of the rocks that have been identified as olivine lamproites are low in K_2O. The concentration of MgO may also be as high as 20 wt% compared to olivine lamproites sensu stricto (Rock 1987). This is attributed to the originally high modal percentage of olivine, now almost entirely replaced and pseudomorphed by magnesite and/or silica. In some extreme cases the olivine megacrysts are thoroughly carbonated along with groundmass (viz. olivine lamproites from Pichari). These rocks are now rich in magnesium carbonates (MgO: 13.10–21.32 % and CO_2: 17.64–23.56 %) with variable SiO_2 (25.30–34.10 %), CaO (4.35–9.00 %), TiO_2 (5.50–7.30 %) and K_2O contents (2.01–2.60 %).

Characteristically the lamproitic rocks are high in incompatible elements viz. Ba (up to >1,000 ppm), Sr (up to 0.44 %) and Rb (up to 500 ppm). The chondrite normalized REE patterns of different varieties of lamproites are shown in Fig. 4.4. High concentration of REE (up to 1139.4 ppm.) and significant enrichment of LREE [$(La/Lu)_n$ up to 16–5.60]. The highly fractionated rare earth pattern $(La/Lu)_n$ (up to 165) suggests that they were derived by low degree of mantle melting (Paul and Potts 1981). The significant positive correlation (r = 0.938) between (Gd/Lu)n and TiO_2 suggests that fractional crystallization of major phases such as phlogopite and perhaps to some extent the minor phases like rutile, ilmenite etc. had important effect.

The $^{87}Sr/^{86}Sr$ ratios in the samples range from 0.70394 to 0.71544. This ratio in three Raniganj lamprophyres ranges between 0.70493 and 0.71067 (Middlemost et al. 1988), but in seven Jharia ultrapotassic ultramafic rocks the $^{87}Sr/^{86}Sr$ ratio

Table 4.2 Analyses of Lamproites from Damodar Valley, India

	SiO_2	TiO_2	Al_2O_3	Fe_2O_3	FeO	MnO	MgO	CaO	Na_2O	K_2O
1	32.37	9.32	9.84	8.68[a]	–	0.12	8.09	8.26	0.13	7.09
2	35.29	4.25	5.54	11.20[a]	–	0.286	8.31	7.18	0.11	3.42
3	35.70	5.94	5.99	11.22[a]	–	0.137	10.05	5.85	0.08	4.69
4	34.11	6.56	9.08	10.44[a]	–	0.145	6.10	11.55	0.21	4.90
5	30.31	7.33	6.61	11.51[a]	–	0.17	11.20	11.76	0.12	2.82
6	38.29	3.31	7.29	2.81	5.15	0.12	10.01	12.24	1.16	5.67
7	41.20	2.50	8.48	2.38	3.85	0.09	12.30	9.68	1.11	7.38
8	39.79	2.69	9.08	3.83	9.08	0.21	10.52	12.62	2.05	1.55
9	44.50	4.81	8.32	0.10	7.62	0.13	8.18	6.85	0.33	5.52
10	33.11	4.80	7.56	0.20	10.62	0.20	12.50	5.23	0.11	2.58
11	41.10	7.20	7.34	2.36	10.08	0.16	10.65	2.89	0.13	5.32
12	36.68	6.54	8.55	5.67	6.12	0.14	8.43	10.5	0.86	2.92
13	36.86	5.94	6.40	10.29	6.30	0.20	8.87	6.74	1.00	2.64
14	42.05	4.98	7.18	0.89	17.28	0.57	8.93	2.20	0.14	1.92
15	48.72	5.23	8.62	3.15	6.57	0.15	9.60	3.40	0.12	1.47
16	49.20	3.54	7.14	0.20	7.92	0.13	6.79	4.07	0.10	5.41

(continued)

Table 4.2 (continued)

	SiO_2	TiO_2	Al_2O_3	Fe_2O_3	FeO	MnO	MgO	CaO	Na_2O	K_2O
17	42.33	5.02	8.14	9.41[a]	–	0.09	11.13	6.04	0.27	6.22
18	37.09	6.46	9.33	13.79[a]	–	0.12	10.44	8.07	0.03	2.87
	Cr_2O_3	BaO	P_2O_5	H_2O^+	H_2O^-	CO_2	SO_3	LOI	–	Total
1	0.053	0.487	2.93	–	–	–	0.17	11.78	0.86	100.18
2	0.045	0.506	6.32	–	–	–	0.31	14.77	1.50	99.03
3	0.062	0.280	2.42	–	–	–	0.48	15.61	0.82	99.33
4	0.078	0.933	5.18	–	–	–	0.45	8.75	1.17	99.66
5	0.120	0.495	3.53	–	–	–	1.47	11.59	1.07	100.11
6	–	–	1.00	–	–	–	–	11.88	–	98.93
7	–	–	0.46	–	–	–	–	9.60	–	99.03
8	–	–	0.52	–	–	–	–	7.03	–	98.97
9	–	–	1.60	1.38	0.10	9.40	–	–	–	98.90
10	–	–	2.21	1.66	0.82	18.08	–	–	–	99.70
11	–	–	0.75	1.40	0.78	n.d.	–	–	–	99.43
12	–	–	1.21	3.83	0.27	7.71	–	–	–	99.43
13	–	–	0.97	5.85	1.49	6.10	–	–	–	99.65
14	–	–	1.60	3.02	2.08	6.88	–	–	–	99.72

(continued)

Table 4.2 (continued)

	Cr_2O_3	BaO	P_2O_5	H_2O^+	H_2O^-	CO_2	SO_3	LOI	–	Total
15	–	–	0.82	3.78	0.98	n.d.	–	–	–	92.61
16	–	–	3.15	0.40	0.70	10.90	–	–	–	99.65
17	–	–	2.40	7.92[b]		n.d.	–	–	–	98.97
18	–	–	2.58	8.24[b]		n.d.	–	–	–	98.97

[a] The analyses represent total iron oxide
[b] The analyses represent total H_2O^+ and H_2O^-
1 Lamproites from Sudamdih, Damodar Valley, India (Gupta et al. 1983)
2–4 Lamproites from Mohanpur, Damodar Valley, India (Gupta et al. 1983)
5 Lamproites from Poidih, Damodar Valley, India (Gupta et al. 1983)
6–8 Lamproitic rocks from Holstinborg, East Greenland (Scott Smith B. 1981)
9–10 Lamprophyric rocks from Bokaro basin (Basu et al. 1997)
11–13 Leucite lamproite from Bodaro basin (Basu et al. 1997)
14–15 Olivine lampoites from Bokaro basin (Basu et al. 1997)
16 Leucite lamproites from Bokaro Basin (Basu et al. 1997)
17 Analysis of Mohanpur lamproite (Gupta et al. 2002)
18 Analysis of Satyanarayanpur minette (Gupta et al. 2002)

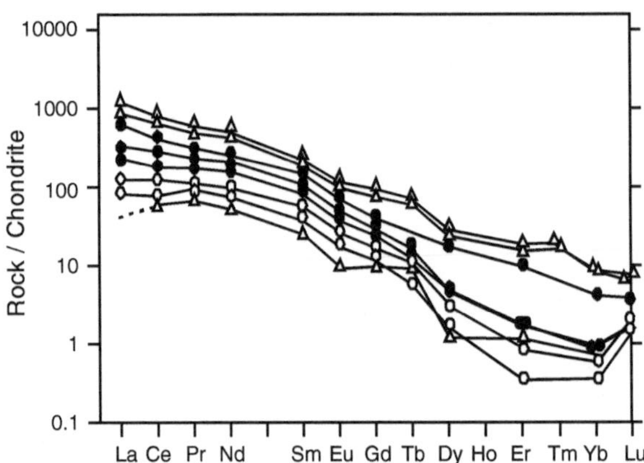

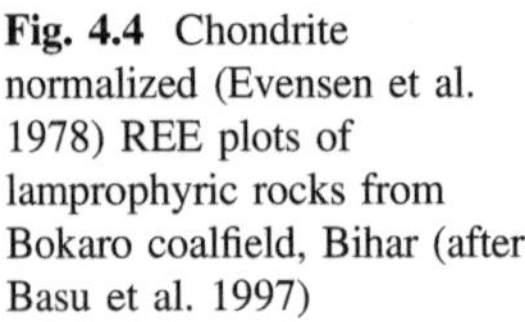

Fig. 4.4 Chondrite normalized (Evensen et al. 1978) REE plots of lamprophyric rocks from Bokaro coalfield, Bihar (after Basu et al. 1997)

ranges between 0.70526 and 0.70595 (Rock et al. 1992). The Bokaro samples encompass a much wider range of $^{87}Sr/^{86}Sr$ ratios.

In contrast to the preponderance of calc-alkaline lamproites, particularly minette, in the Jharia and Raniganj basins of the Damodar Valley, the Bokaro coalfield is characterized by widespread occurrence of lamproites (leucite lamproite, olivine-leucite lamproite and olivine lamproite). The minettes from Bokaro basin have higher amount of alkali feldspar and phlogopite with higher abundances of TiO_2 (4.75 wt%) and Al_2O_3 (7.14 wt%) but smaller amount of K_2O (4.80 wt%) and MgO (11.39 wt%) in comparison with minettes from the Jharia basin. The lamproites from Bokaro are characterized by the presence of leucite and/or olivine, priderite and show flow differentiated layers. In general, the lamproites are much less abundant in the other Gondwana basins of eastern India. Interestingly, a 2 m thick dyke in Bokaro basin shows variation from leucite olivine lamproite to olivine lamproite. The corresponding average K_2O abundances are 5.46, 3.39 and 1.87 wt%, respectively. High modal percentage of pseudomorphic olivine in the same samples indicates their ultramafic composition.

In general, however, the mineralogical assemblage in Bokaro rocks is less diverse than in the rocks of neighbouring Jharia basin (Rock et al. 1992). The phlogopites of the lamprophyric suite from Bokaro have higher TiO_2, Al_2O_3, and MgO compared to the phlogopite of Jharia (Rock et al. 1992). The mg-number varies between 0.63 and 0.87. In the Mg/(Mg + Fe)–TiO_2–Al_2O_3 diagram (Scott Smith 1989; Scott Smith and Skinner 1984), most of the Bokaro phlogopites plot close to the lamproite field and lots of them also lie in the vicinity of the minette-lamproite interface. This reflects the grading of olivine lamproite to minette through intermediate varieties.

The petrogenesis of the Gondwana lamproites from Raniganj and Jharia has been discussed by Middlemost et al. (1988), Rock et al. (1992) and Paul and Sarkar (1984). From the petrological data presented above, it is seen that the Bokaro lamproites are more diverse, and include a significant proportion of lamproites. The mantle origin of the lamproite magma of the Bokaro coalfield is evidenced by the presence of mantle-derived nodules and xenoliths of harzburgitic composition, their primitive mafic-ultramafic major element chemistry, enrichment of refractory trace elements like Ni, Cr, Co, Sc and high mg value of the phlogopite xenocrysts or phenocrysts associated with them. High concentration of REE (up to 1139.4 ppm) and considerable LREE enrichment [$(La/Lu)_n$ from 16 to 560] in the samples suggest their derivation from enriched mantle sources. The Bokaro lamproites have $^{87}Sr/^{86}Sr$ ratio in the range of 0.70394–0.71544. Available Sr isotopic data of the Jharia, Raniganj and the Bokaro rocks are similar to those of lamproites (Bergman 1987). Despite the wide range of $^{87}Sr/^{86}Sr$ ratio, it could be shown that these have not been affected by contamination (low correlation coefficient of the isotopic ratio: 0.243, with respect to SiO_2) and are believed to reflect source region characteristics. It was therefore, suggested by Basu et al. that the source region of the Bokaro lamproites was highly heterogeneous.

According to Basu et al. (1997), the K-rich rocks of Bokaro were produced by partial melting of a phlogopite–bearing harzburgitic) mantle. Extreme enrichment of incompatible elements like, K, Ti and LILE usually associated with phlogopite, apatite, rutile, priderite etc. suggests that the parent liquid is a product of partial melting of a metasomatized mantle source (Wilson 1989). The highly fractionated rare earth pattern [$(La/Lu)_n$ up to 165] suggests derivation of the potassic rocks by low degree of melting in the metasomatized mantle (Paul and Potts 1981).

4.1.4.2 Jharia Basin

According to Ghosh and Mukhopadhyay (1985), the Jharia basin is a part of an east–west-trending chain of intracratonic Gondwana basin of eastern India. They thought that at first there was initiation of a general downwarping of the basin floor, followed by its early differentiation into a number of sub-basins. Later, there was merging of the sub-basins into a single major basin because of a subsidence, which was controlled by the development of a narrow intrabasinal graben. According to them, the graben was bounded by contemporaneous normal faults. There was sinking of inselbergs along with rising of basement ridges towards the later stages. This was followed by post-depositional faulting and reactivation of old faults. Later, there was rise of basement rocks forming horst and graben-like structures and finally intrusion of dolerites, lamproites and lamprophyres took place through sedimentary country rocks.

Ghosh (1949) made petrographic studies of the intrusive rocks from the coal fields of Jharia and Dhanbad petrographic province, and from Raniganj of Burdwan district, West Bengal. He termed these rocks to be biotite lamprophyre, minette and kersantite. He also reported the occurrence of leucite from both the areas (also see Chatterjee 1974).

4.1.4.3 Raniganj Basin

Bulk rock analyses of the lamproites from Mohanpur, Poidih and Sudamdih (Table 4.2) show the following features: (i) extremely low SiO_2, Al_2O_3 and high P_2O_5, TiO_2 and K_2O contents, (ii) higher concentration of some minor and trace elements such as Ce, Y, La, Th, Nd, Rb, Sr, Ba and Zr, which are higher almost by one order of magnitude than those of the average basalts (Frey et al. 1978).

From the mineral assemblages and bulk compositions, Gupta et al. (1983) considered that the rocks studied by them cannot be termed as kersantite, minette or camptonite (Nockolds et al. 1978), but they should belong to the group 'glimmerites', as described by Holmes (1937), which consist mainly of mica with some other minor constituents. Since the rocks are also high in apatite and ankerite, they may be designated as 'carbonated apatite glimmerites' belonging to the lamproitic family. Analyses of these rocks are summarized in Table 4.2. From a study of combined Rb–Sr, Sm–Nd and U–Th–Pb isotopic systems of the potassic rocks of the Jharia and Raniganj coal fields, it was also shown by Middlemost et al. (1988) and Rock et al. (1992) that these are generated from an enriched mantle source.

Geochemical and experimental studies of rock samples from three localities of the Jharia and Raniganj coal fields were made by Gupta et al. (1983) to elucidate the genesis of these rocks. These samples were from drill cores at Sudamdih (S, Dhanbad district, Bihar), Mohanpur (M-1, M-2, M-3, Burdwan district, West Bengal) and from a dyke (P) at Poidih (also from Burdwan district). All these localities, included in Damodar Valley, are situated within 200–280 km west-northwest of Calcutta. Microscopically, all the samples are characterized by variable proportions of phenocrystal phlogopite and apatite and microphenocrystal ankerite and chromian spinel in the groundmass of phlogopite, apatite, ankerite, rutile, pyrite and devitrified glass. In addition, priderite microphenocrysts occur rarely in the Mohanpur rocks as accessory minerals. Phlogopite is always the most important constituent, whereas modal percentage of apatite is 20–30 %, ankerite 8–18 % and spinel less than 5 %.

4.1.5 Leucite-Bearing Rocks of Indonesia

4.1.5.1 K-Rich Rocks of Sunda Arc

The Banda Arc (Whiteford and Jazek 1979) and its westerly continuation, the Sunda Arc occurs in the southeast margin of the Asian lithospheric plate (Varne 1985, Fig. 4.5a). The arc has been squeezed due to northward movement of the Indo-Australian plate and the westward movement of the Pacific plate. According to Varne, northerly subduction of the Indo-Australian plate during the Neogene time led to the collision of the plate northwest of Australia with the arc. It is thus, tightly bounded to the south by the continental lithosphere, and there was back-arc thrust occurring north of the arc from Bali to Wetar during subduction following the collision.

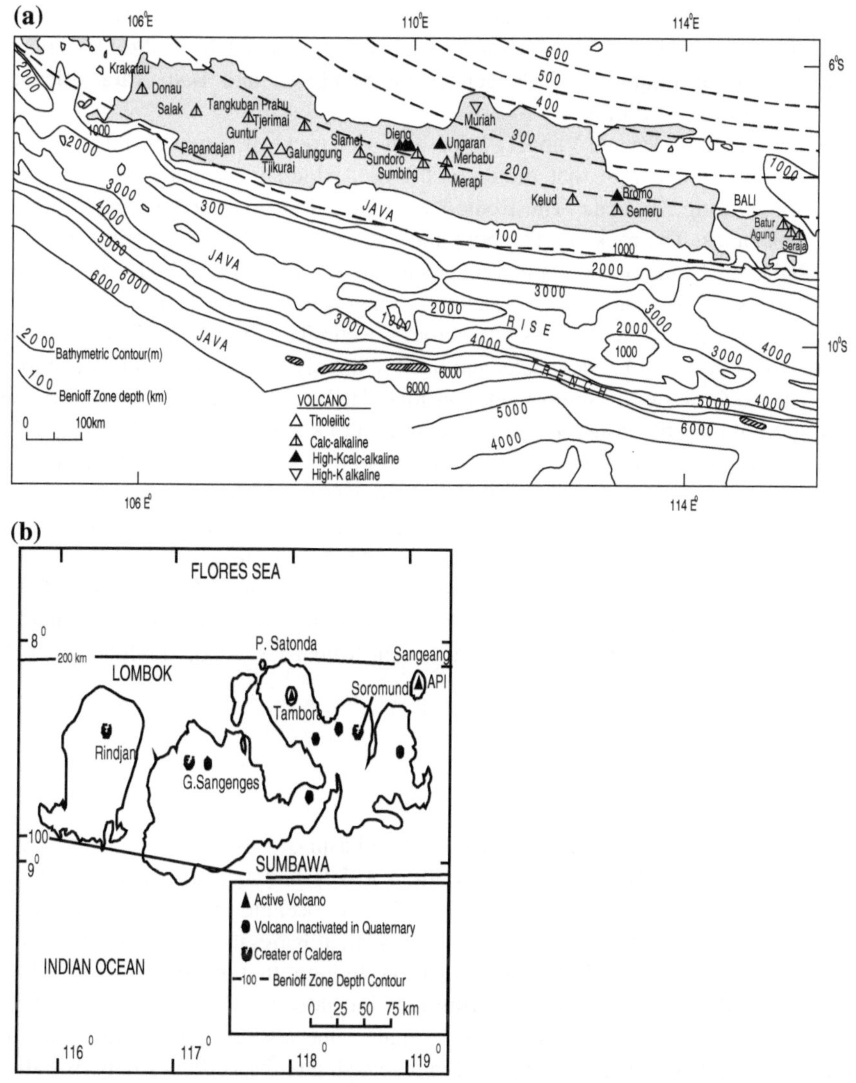

Fig. 4.5 **a** Regional tectonic setting of Java and Bali. Bathymetry (after Hamilton 1974) and Benioff zone contours (after Hamilton 1972). Volcanoes are denoted by *triangles*. **b** Locality map showing the islands of Lombok and Sumbawa and five other volcanoes (after Foden and Varne 1980). The Benioff-zone contours (after Hamilton 1974)

The active volcanoes of Bali, Batu Tara and Agung erupted melts of basalt-andesite-dacite series of rocks. Rindjani volcano in Lambok occurring towards east also erupted calc-alkalic melts. The active shoshonitic volcanoes of Tambora and Sangeang Api, located at Sumbawa, erupted however, melts of nepheline-bearing trachybasalt-trachyandesite association. Varne (1985) described nepheline-normative olivine leucitite and other related rocks, which erupted from the Quaternary volcanoes of Soromundi and Sangenges on Sumbawa and Batu Tara volcano, located northeast of Flores.

According to Foden and Varne (1980), the K-rich rocks of Tambora and Sangeang Api (Fig. 4.5b) occur in that sector of the arc, which is a relatively young arc developed during Miocene. The Benioff zone lies at an intermediate depth below these volcanoes. The rocks of Tambora and Sangeang Api are in general similar to shoshonite suite but are more silica-undersaturated. According to Foden and Varne (1980) the best known analogue of Tambora-Sangeang Api is the Roman Province of Italy, where both high and low K-series rocks occur. The Sumbawan lavas are very similar to the low-K_2O series rocks and have relatively low Ti, Zr and P concentration. The leucite series of Sumbawa is considered to be similar to the rocks of African rift region. It may be observed that concentration of such elements as K, Rb and Sr, are very similar in both cases, but the rocks from the African rift region are characterized by higher P_2O_5. The content of TiO_2 and Zr is 3–4 times more, and the Nb content of the African rocks is more than ten times than those of Indonesian leucitites. The K_2O content correlates well with the Sr and Nd isotopic ratios.

Varne (1985) established that progressive increase in K_2O/SiO_2 ratio in the lavas with SiO_2 contents less than 53 wt%, is accompanied by $^{87}Sr/^{86}Sr$ ratio ranging from 0.704 to 0.707 and decreasing Nd content from +5.3 to −3.7 (Whiteford et al. 1979).

4.1.5.2 Ringgit Beser Complex

The volcanic complex is located in the Madura basin of East Java. The Complex is 200 km above the Benioff zone (Hamilton 1979) and lies to the north of the volcanic axis. The crustal segment in East Java is oceanic and only 18 km thick. Van Bemmelen (1949) thought that the volcanism at Ringgit-Baser complex started during Plio-Pleiostocene. Edwards et al. (1994) thought that the volcano erupted for a short period and became extinct during Late Pleistocene. The southern foot of the volcano subsequently got folded and the main cone also collapsed. Shape of the volcanic ridge is somewhat maintained in the north and the vertical southern wall is also intact. The Beser ridge to the south displays anticlinal structure. The Ringgit complex towards north is divided into two parts by WSW–ENE-trending graben. There have been eruptions of three different types of volcanic rocks:

(1) Calc-alkalic,
(2) High K-calc-alkalic, and
(3) High K-alkaline series.

Edwards et al. further subdivided the high K-alkaline lavas into two subgroups based on their Nb and Ba content: the enriched potassic series (EK) and the potassic series (K). They observed that northern part of the complex comprises predominantly EK series lavas but in the southern part along with high K-alkaline lavas, calc-alkaline and high K-calc-alkaline lavas are interspersed. The volcanic eruption of Ringgit complex took place 1.2 and 1.0 Ma ago, whereas eruption of Beser complex took place 0.8 and 0.6 Ma ago. The calc-alkalic basalts and high K-calc-alkalic trachybasalts and basaltic trachyandesites are characterized by a lot of plagioclase phenocrysts (An_{89-94}) and augite; olivine may be present in more mafic varieties.

The high K-series lava is constituted of basanite, tephrite, tephritic phonolite, trachybasalt, basaltic trachyandesite and trachyandesite. The lavas of Ringgit-Beser complex are not aphyric and contain up to 25 vol% phenocrysts. Clinopyroxenes are twinned and zoned and are salites or diopsides. Pyroxenes are thus, similar in chemistry to those from K-rich lavas of Muriah and Batu Tara. More mafic varieties contain olivine phenocrysts (Fo_{88-90}) and more potassic lavas contain leucite. The trachytic lavas are devoid of leucite but are constituted of plagioclase and biotite phenocrysts. Edwards et al. (1994) found that most evolved potassic lavas contain sanidine ± hornblende. Titano-magnetite and apatite are common accessory phases, and in case of highly mafic lavas magnesio-chromite is present. The basanites and trephrites are characterized by the presence of clinopyroxene ± leucite ± nepheline ± magnetite in the groundmass. Trachytic lavas are constituted of plagioclase + clinopyroxene + magnetite ± leucite ± nepheline in the glassy matrix.

The EK volcanic series are characterized by basanite, tephrite and phono-tephrite. Zoned and twinned Al–Ti-rich salite and diopside along with leucite, are usually present as phenocrysts. Phonotephrites often contain plagioclase (An_{91}) and/or phlogopite and more mafic variety contains olivine (Fo_{91-92}). The accessory phases include titanomagnetite ± apatite ± sphene. The groundmass is constituted of clinopyroxene ± leucite ± olivine ± nepheline ± plagioclase ± magnetite.

Edwards et al. (1994) noted that the K and EK series rocks show variable K_2O content and have almost similar chemistry and cannot be separated on the basis of K–Si relationship. In fact the most magnesian EK series lavas have greater K_2O content than potassic lavas with similar MgO content. These two series of lavas can be distinguished from each other on the basis of their Ba and Nb content. The K_2O content of both series of rocks ranges between 0.98 and 6.35 wt%, and is in fact higher than that of most island arc lavas (usually <2.0 wt%) and except for calc-alkalic rocks the K_2O/Na_2O ratio is >1.

The Ringgit-Beser lavas have a wide MgO content ranging between 1.6 and 18.1 wt%, though the average is slightly greater than 4 wt%. The SiO_2 content of the lavas ranges between 42.9 and 54.2 wt%. The calc-alkalic lavas have bulk chemistry plotting near the critical plane of silica-undersaturation (Yoder and Tilley 1962). The alumina content ranges between 9.3 and 9.9 wt%. The majority of EK series lavas have Al_2O_3 content >13 wt%, a value, which overlaps with those of in-plate basalts. The Fe_2O_3 content ranges from 6.6 to 12.3 wt% and the lavas of EK series tend to be lower in Fe content. The CaO content varies between 7 and 14.8 wt %, but the majority of the samples have CaO content >10 wt%. The maximum lime content is found in lavas with 45 vol% clinopyroxene phenocrysts. The TiO_2 content is <1.12 wt%.

Ultrapotassic rocks are characterized by phenocryst of phlogopite, embedded in a groundmass containing leucite, diopside, olivine and Fe–Ti oxides, and are reported from Kajan River, Kalimantan. The phlogopites are reported to show reverse pleochroism, which are usually observed in phlogopites from lamproites. The K_2O/Al_2O_3 and K_2O/Na_2O ratios of the rocks from Kalimatan are 0.4 and 2.0, respectively. The rocks from this locality are syenitic leucitites. In the nearby locality around Karamu River of central Kalimatan, minettes have been described by Bergman (1987). Analyses of these rocks are summarized in Table 4.3.

Iddings and Morley (1915) and Rittmann (1951) studied the volcanic rocks of Indonesia. Leucite-bearing lavas were erupted in particular by Mt. Mouriah volcano (now extinct), which occurs northeast of Semarang, Java (Nelson and Whiteford 1983). Associated with this volcano are two other small volcanoes: Paliaian and Tülering. K-rich phonolitic lavas containing modal leucite were erupted from a small volcano called Lorous. Leucite-bearing rocks in association with phonolites also occur in the Bawean Island between Java and Borneo. Mt. Mouriah is largely made up of tuffs and breccia with small massive lava flows of mainly leucite tephrites and leucitites with a few olivine-bearing varieties. These rocks are of late Tertiary age. Analyses of some of the rocks given in Table 4.3 indicate that the SiO_2 content varies from 44 to about 53 %; the Al_2O_3 and CaO contents are rather high, whereas the TiO_2 contents are very low. Mineralogically these rocks are characterized by leucite, augite, K-feldspar, and calcic plagioclase in various proportions. Nepheline is usually absent in the rocks of Mt. Mouriah, whereas olivine and biotite are occasionally present. The interplay of subduction zone and interpolate processes pertaining to the genesis of K-rich silica-undersaturated rocks have been discussed by Edwards et al. (1991).

In Celebes, olivine leucitite, leucite trachyte and leucitite tuff occurs at various localities along the coast from Cape Mandar to Cape William and near the Bay of Mamudju. Leucite basanite is found in Oldeidu Kiki in the Matinang Mountains, southeast of Bowool in Northern Celebes. These lavas are associated with trachytes and trachyandesites. Potassium-rich rocks have been reported in Pic de Maros Mountain, located between Maros and Tjambo, north of Makassar in Celebes. These rocks occur as intrusive bodies of shonkinites, consisting of augite, K-feldspar, biotite, and rare plagioclase. In some localities such as Gentungen, the shonkinites contain pseudoleucite and are very similar to the shonkinitic rocks of Montana, U.S.A. These shonkinites grade into nepheline syenite, syenite porphyry, and trachyte.

4.1.5.3 Other Localities in South East Asia

Bergman (1987) described a group of rocks called, cocites near the border of Laos and North Vietnam in Indochina. They occur at Coc pia as dykes, where they cut across Mesozoic alkaline syenites and granites and are fine-graind, comprising olivine, diopside, phlogopite, magnetite, sanidine and leucite. At Sincao, north east of Laidchau, cocite group of rocks have been emplaced in Triassic sediments, and the dykes may be much younger in age. Some potassic rocks known as cocites are

Table 4.3 Analyses of Ringgit Beser, East Java (Indonesia)

	SiO_2	TiO_2	Al_2O_3	Fe_2O_3	FeO	MnO	MgO	CaO	Na_2O	K_2O	P_2O_5	LOI	Total
1	46.88	0.994	14.26	9.17[a]	–	0.157	7.67	11.80	2.84	5.00	0.929	1.27	100.97
2	45.60	0.892	12.78	9.60[a]	–	0.176	8.48	12.68	2.12	5.96	0.966	1.37	100.62
3	45.84	1.007	13.99	9.31[a]	–	0.164	8.23	11.66	2.71	5.01	0.924	0.83	99.67
4	46.44	0.671	11.12	8.64[a]	–	0.145	12.27	11.98	2.04	5.43	0.591	1.56	100.88
5	47.27	0.845	14.73	9.94[a]	–	0.202	5.57	12.03	1.95	6.35	0.825	1.10	100.81
6	45.59	0.897	15.36	11.66[a]	–	0.234	4.48	11.70	2.54	5.81	0.882	1.87	101.01
7	52.83	0.647	18.98	6.69[a]	–	0.178	2.22	7.46	3.83	6.15	0.467	1.51	100.96
8	43.79	0.961	11.66	9.75[a]	–	0.180	10.47	14.51	1.69	5.14	1.263	1.71	101.12
9	43.86	1.090	11.09	2.04	10.38	0.200	9.00	15.16	2.03	4.34	–	–	99.19
10	45.88	1.090	13.20	1.96	9.99	0.180	7.77	13.88	2.07	4.42	0.740	–	101.18
11	44.71	0.970	11.84	1.82	9.26	0.180	10.49	14.10	2.21	3.16	0.500	–	99.24
12	49.52	0.830	16.64	1.50	7.76	0.210	4.17	10.14	4.33	4.47	0.520	–	100.09

[a] The analyses represent total iron oxide

1–4 The highly potassic rocks (EK series) from Ringgit Beser complex, East Java (Edwards et al. 1994)

5–8 The potassium-rich lavas from Ringgit Beser complex, East Java (Edwards et al. 1994)

9 The K–rich rock from Sumbawaa, Sunda Arc (Foden and Varne 1980)

10–12 Leucite-bearing lavas from Sumbawa Island Indonesia (Foden and Varne 1980)

characterized by the presence of phenocrystal phlogopite and augite in a groundmass of sanidine, analcite (pseudomorphous after leucite). The cocitic rocks have high K_2O (4–5 wt%), low K_2O/Na_2O (~ 1.0) and K_2O/Al_2O_3 ratios (0.4–0.6).

4.2 Ultrapotassic Rocks of Australia

4.2.1 West Kimberley

A host of potassium-rich intrusive rocks were emplaced (Prider 1939, 1982; Wade and Prider 1940; Alsopp et al. 1985; Nelson et al. 1986) in the Fitzroy area of the Kimberley region of Western Australia (Fig. 4.6). As many of the intrusions contain diamond (Atkinson et al. 1982, 1984), these rocks have been subjected to numerous petrological, geochemical, and geochronological studies. The lamproites are confined to a broad irregular zone, which trend east–west at the south western margin of the Kimberly block (King Leopold Mobile Belt, KLMB). The belt is then extended across the Leonard shelf to the southern margin of the Fitzroy Trough.

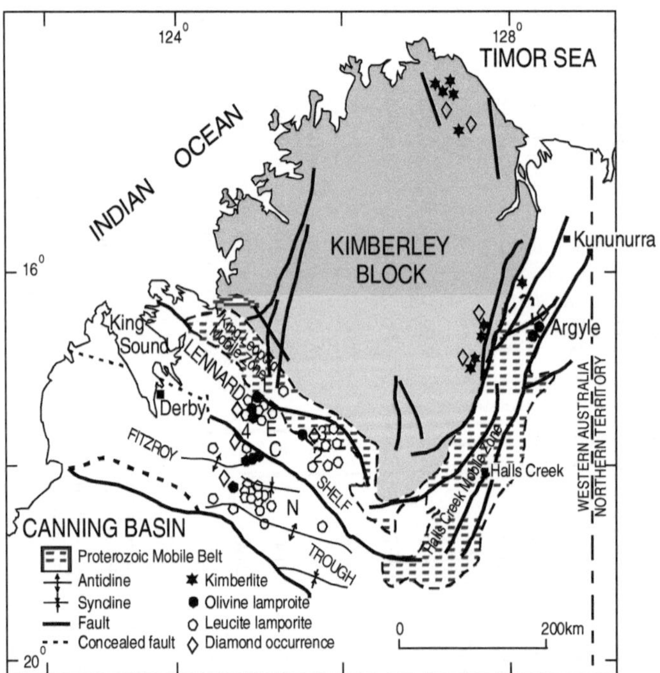

Fig. 4.6 Distribution of lamproites and diamond occurrences in relation to the tectonic feature of the Kimberley region of Western Australia (after Jaques et al. 1984) (*1* Mount North, *2* Old Leopold Hill, *3* Mount Rose, *4* Seltrust Pipe 2; Approximate positions only)

The Canning basin is situated to the southwest of the region. According to Allsopp et al. there are three main roughly east–west-trending lamproite fields occurring in this broad zone. One field is referred to as Ellendale Field, which is located within the King Leopold Mobile Belt on the northern flank of the Leonard Shelf; the other is called the Calwynyada field situated south of the Leonard Self and the last one is called the Noonkanbah field (located in the central zone of the Fitzroy Trough). There is another small cluster of lamproites occurring within the King Leopold Mobile Belt about 100 km east of the Ellendale Field and to the north of Fitzroy River. The lamproites occur as eroded craters, plugs, lava flows, lava lakes, sills, and dykes. Allsopp et al. (1985) summarized the radiometric ages of different intrusives from the West Kimberley region (Table 4.4).

Table 4.4 Isotropic ages of Lamproites from West Kimberley (Alsopp et al. 1985)

Locality	Average age	Method
Mt. North	22.2 ± 0.0[a]	K–Ar
	22.0 ± 1.6	Rb–Sr
Old Leopold Hill	20.7 ± 0.6[a]	K–Ar
	21.8 ± 0.8	Rb-Sr
Wolgidee Hills	17.8 ± 0.0[a]	K–Ar
	17.5 ± 0.2	K–Ar
	22.1	Rb–Sr
Howes Hill (west)	19.0	K–Ar
Mt. Percy	22.0 ± 0.3	K–Ar
	20.8	Rb–Sr
Camarotoechia Bore	22.1 ± 0.3[a]	K–Ar
Hoper West	21.9 ± 0.2	K–Ar
Winjana	21.9 ± 0.2	K–Ar
81 Mile Vent	21.5 ± 0.2	K–Ar
Water Reserve Sill	21.4 ± 0.2	K–Ar
Noonkanbah Hill	20.1 ± 0.2	K–Ar
"P" Hill	19.5 ± 0.7[a]	K–Ar
Dadja Hill	19.6 ± 0.2	K–Ar
Fishery Hill	19.6 ± 0.2	K–Ar
Machells Pyramid	18.4 ± 0.6	K–Ar
Mt. Gytha	17.0 ± 0.9	K–Ar
Ellendale 9	24.0 ± 1.0	Rb–Sr
Mt. Rose	23.2 ± 5.0	Rb–Sr
Seltrust Pipe 2	23.9 ± 0.8	Rb–Sr

[a] Average was calculated using a program

According to Smith and Lorenz (1986) there are more than 100 Miocene lamproites in West Kimberley, Australia (Jaques et al. 1984), covering 75,000 km^2 from Ellendale through Noonkanbah to Fitzroy crossing. There are a few dykes, which intruded into lower Proterozoic granites and metamorphic rocks of the King Leopold mobile zone; most of the lamproitic bodies intrude Phanerozoic sediments of the Leonard shelf, Fitzroy trough at the north-eastern margin of the Canning basin. According to them there are 48 lamproites in the Ellendale Field on the Lennard Shelf (Atkinson et al. 1984). Forty-three lamproites were emplaced in an elongated belt (40 km long and 10 km across, trending 305°) aligned with Oscar Fault, which is one of the major step faults bounding the Fitzroy Trough. This trough started forming during the Ordovician. Major graben development took place during the Devonian. There was sedimentation till the Triassic and around this time, the early structures were reactivated with right lateral and vertical movements on the step of the graben.

The lamproites are of two categories:

(1) Leucite lamproites comprising variable proportion of phlogopite, diopside, richterite with less than 5 % modal olivine, and
(2) Olivine lamproite without leucite but containing phlogopite, diopside, richterite and 30 vol% modal olivine.

The ranges of SiO_2 contents of leucite lamproites and olivine lamproites are 50–55 and 35–42 wt%, respectively. The H_2O^+ concentration of leucite lamproites and olivine lamproites are around 2 and 7 wt%, respectively. The CO_2 content is <0.25 wt%. Smith and Lorenz (1986) found that, of the 46 Ellendale lamproites, 14 may be classified as olivine lamproites; 28 belong to the category of leucite lamproites and four are transitional rock types with 28 % modal olivine and appreciable amount of leucite.

The Ellendale diatremes intrude sandstone of Permian formation and are usually less then 100 m in diameters and sometimes more than 1 km across. There are many diatremes which are elongated in plan view with their long axis trending west-northwest. Usually a diatreme is of the shape of champagne-glass with a narrow feeder vent, overlain by a broad shallow crater. It is often observed that during the final stage of volcanism the magma ascended through the conduit into the centre of the craters, spreading out in the form of a lava lake or dome. According to Smith and Lorenz, the near surface "lamproites" overlie the crater sediments and sometimes may overlap into the country rock. They found that the near surface "lamproites" are highly vesicular and brecciated often incorporating clastics from underlying sediments below the crater.

Smith and Lorenz concluded that volcanic activity related to nearly all Ellendale pipes started with phreatomagmatic phase. During this period the lamproitic magma interacted explosively with copious amount of groundwater from sedimentary rocks of the Permian granite formation. This was followed by a maar formation, which then collapsed due to ejection of juvenile, country rock clasts with the formation of diatrame underneath. Because of the phreatomagmatic activity at the surface, many thin pyroclastic beds of base surge origin, formed on the rim of the crater or within the maar. According to them, continued activity led to the subsidence of the diatreme portion and collapse of the crater rim. This was followed by subsidence of the interbeded pyroclastic and epiclastic floor deposit within the diatreme. They further suggested that the explosive activity resulted in the removal of groundwater as steam, subsequently a cone of depression was formed. This allowed downward extension of the diatreme. The pheratomagmatic activity stopped, when the diatreme penetrated deep enough into dry mudstone of the Fairfid formation. The magma then ascended in a non-explosive manner and intruded as diatreme and formed lava lakes of olivine lamproite and lava domes of leucite lamproite within the initial maar crater. Smith and Lorenz (1986) found that in case of many leucite lamproite pipes, early pheratomagmatic eruption was followed by ascent of the magma (small microphenocrysts of phlogopite) into the maar crater. In the next stage, the lamproite magma with large phlogopite phenocrysts ascended from a stratified magma reservoir. The predominant rock types of the West Kimberley area are: fitzroyite, cedricite, and mamilite. Wolgidite is the most predominant rock type at Moulament Hill, Mt. North, and Wolgidee Hill. Fitzroyite is the most common rock type found along the Fitzroy River basin, where it often includes quartz amygdules. Cedricite and mamilite are typical rock types in the Mt. Credric and Mamilu Hills, respectively. There are also some other varieties, which are transitional in character between cedricite and fitzroyite, and are similar to the wyomingite of the Leucite Hills. Some of the transitional rocks are found in "P" Hills. Analyses of K-rich lamproitic rocks from West Kimberley, Australia, are given in Table 4.5, but for comparison between geochemistry of West Kimberley rocks with that of Spanish rocks, see Nixon et al. 1984).

4.2.2 New South Wales, Australia

The potassic lava flows extend over a small area in fold belts of Southern Highlands (Wellman 1970). This region has a peneplaned basement complex comprising goesynclinal sediments and granitic rocks, which are overlain by a thin Cainozoic sedimentary sequence. The fault planes and major lineaments strike along north-northwest direction. According to Cundari (1973), the general outcrop patterns of the K-rich volcanic rocks are distributed along the same trends (Fig. 4.7).

The K–Ar ages of the rocks suggest that the lavas were erupted 14–10 Ma years ago (Wellman 1970). The volcanoes at Lake Cargelligo and Tullibigeal are well preserved but at Bygalorie the domal structure of the volcano has been reduced to a 'volcanic skeleton'. The other notable localities of K-rich lavas include El Capitan, Bergargo Hills, Flagstaff Hills, Byrock, Condoblin, Griffith and Harden. The K-rich rocks cover an extensive area at Tullibigeal (43 km^2), Bygalorie (46 km^2) and Lake Cargelligo (12 km^2), but in other localities such as El Capitan (6 km^2), Begargo Hills (2 km^2), Flagstaff Hills (1 km^2), Byrock (1 km^2), Condoblin (2 km^2), Griffith (1 km^2) and Harden (0.1 km^2), they cover relatively small areas.

Table 4.5 Analyses of Leucite-bearing rocks from West Kimberley and New South Wales (Australia)

	SiO_2	TiO_2	Al_2O_3	Fe_2O_3	FeO	MnO	MgO	CaO	Na_2O	K_2O
1	36.02	5.31	5.32	5.37	0.89	0.04	7.79	15.12	0.16	7.25
2	51.19	4.89	8.53	6.12	1.38	0.06	7.15	5.82	0.58	9.02
3	52.45	5.85	8.64	5.48	0.94	0.13	6.42	2.01	0.38	10.42
4	54.48	5.57	9.87	4.89	1.70	0.09	5.35	1.89	0.88	11.06
5	44.83	4.22	8.01	5.08	5.46	0.16	13.53	8.27	1.62	6.09
6	43.40	4.47	9.00	5.01	6.46	0.18	12.47	9.71	2.20	4.27
7	43.30	3.70	8.30	5.20	5.74	0.17	14.50	9.57	1.55	4.60
8	44.24	3.80	8.85	4.82	6.28	0.17	13.43	9.15	2.47	4.32
9	41.82	5.46	7.54	6.55	4.52	0.15	13.15	9.27	0.66	5.92

	P_2O_5	H_2O^+	H_2O^-	CO_2	BaO	SrO	ZrO_2	FeS_2	SO_3	ΣREE	Total
1	1.15	1.92	0.90	9.84	1.58	0.22	0.19	0.21	0.27	–	99.72
2	0.79	1.99	1.26	–	0.60	–	–	0.22	0.11	–	99.71
3	1.58	1.99	2.89	–	1.19	–	–	–	–	–	100.37
4	0.40	1.36	0.89	–	0.64	0.16	0.07	0.20	0.10	–	99.69
5	1.28	1.06	–	–	–	–	–	–	–	0.58	100.19
6	1.22	0.96	–	–	–	–	–	–	–	0.56	99.91

(continued)

Table 4.5 (continued)

	P_2O_5	H_2O^+	H_2O^-	CO_2	BaO	SrO	ZrO_2	FeS_2	SO_3	ΣREE	Total
7	0.80	1.60	–	–	–	–	–	–	–	0.61	99.64
8	1.00	0.84	–	–	–	–	–	–	–	0.48	99.85
9	1.34	2.60	–	0.26	–	–	–	–	–	1.05	100.29

1 Analyses of wolgidite, West Kimberley, Australia. It also includes 0.17 % Cr_2O_3 (Wade and Prider 1940)
2 Analyses of cedrite, West Kimberley, Australia (Wade and Prider 1940)
3 Analyses of fitzroyite, West Kimberley, Australia (Wade and Prider 1940)
4 Analyses of mamilite, West Kimberley, Australia. It also includes 0.09 % (Wade and Prider 1940)
5–6 Leucitites from Tullibigeal, New South Wales, Australia (Cundari 1973)
7–8 Leucitites from Lake Cargelligo, New South Wales, Australia (Cundari 1973)
9 Leucitites from El Capitan, New South Wales, Australia (Cundari 1973)

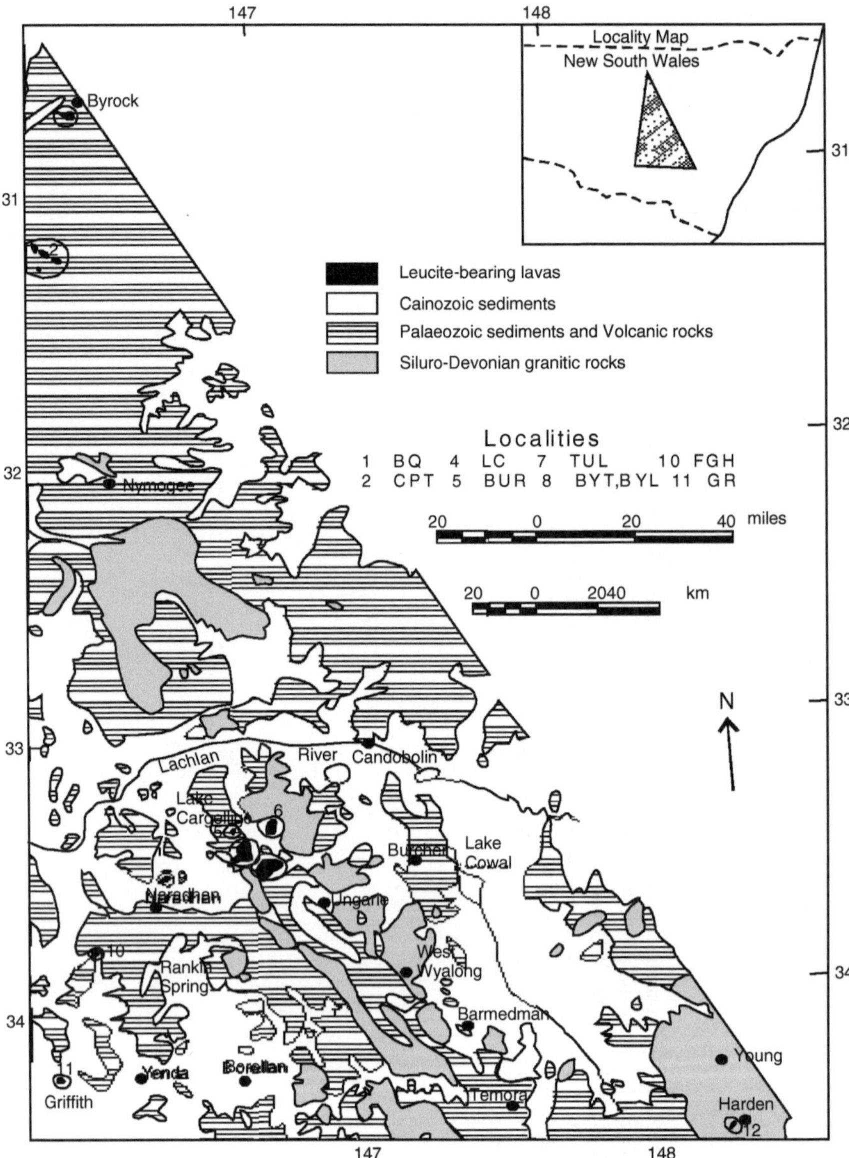

Fig. 4.7 Distribution of leucite-bearing lavas in New South Wales, Australia (after Cundari 1973)

Although Cundari termed these rocks as leucitities, on the basis of modal mineralogy, some of the rocks at Byrock with 10 vol% phlogopite and K-richterite, may be termed as lamproites (BQ-5, BQ-7). Some of the rocks occurring at Begargo Hills (BEH-A, B and C) and Condoblin (CND-4) may be called phonolites (leucite + clinopyroxene + K-feldspar + glass). Analyses of leucitites from New South Wales, Australia are given in Table 4.5.

4.3 Potassium-Rich Silica-Deficient Rocks from Africa

4.3.1 Birunga Volcanic Field

The rift system in Africa is extended from the Red Sea in the north to Malawi (Wooley and Jones 1987) in the south for a distance of 2,800 km, and can be divided broadly into three major rifts: the Ethiopian Rift, the Kenyan or Gregory Rift and the Western Rift. Holmes (1965) mentioned that these rifts are located on plateaux. The Ethiopian Rift cuts across the Ethiopian Highlands and the Kenyan and Western Rifts across the East African Plateau. According to Rogers et al. (1998) the dimensions of these circular plateaux are similar to those of uplifted areas of the oceans associated with oceanic island (Courtney and White 1986). Alkali volcanism in this region is associated with crustal warping and there is possibly a plate tectonic control (Bailey 1964). The width of the rift valleys at Lake Albert, Lake Tanganyika, Rukwa, Lake Rudolf, Lake Natron, Ruaha and Lake Nyasa, varies between 30 and 55 km. The rift valley is constituted of the following segments (Fig. 4.8):

(1) The Nyasa section and its branches towards the southern end of the central plateau (Rukwa, and Ruaha):
(2) The Western Rift from lake Tanganyika through Lake Kivu, Lake Edward and Lake Albert and Ruwenzori massif;
(3) The Gregory Rift, east of Lake Victoria and
(4) Lake Rudolf and the Ethiopian section.

The topography broadly correlates with a negative Bouger anomaly (Bullard 1936). This suggests that plateaux are dynamically supported by convective activity in the underlying asthenosphere (Ebinger and Sleep 1998; Rogers et al. 1998). There is also a general consensus that the rift system is supported probably by more than one plume (Macdonald 1994).

The Birunga province is situated on the international boundaries of Zaire, Rwanda and Uganda in the western branch of the East African Rift System (Fig. 4.9). The Birunga volcanic field has 8 volcanoes (Fig. 4.10), namely, Mikeno, Magahinga, Muhavura, Visoke, Sabinyo, Karisimbi, Nyamuragira and Nyiragongo. Six of the eight volcanoes are now extinct; only Nyiragongo and Nyamuragira are active. There are several smaller volcanic cones also. According to Rogers et al. (1998) the most easterly volcano Muhavura has a very youthful morphology and appears to be intermediate in age between the currently active volcanoes and more deeply eroded Sabinyo.

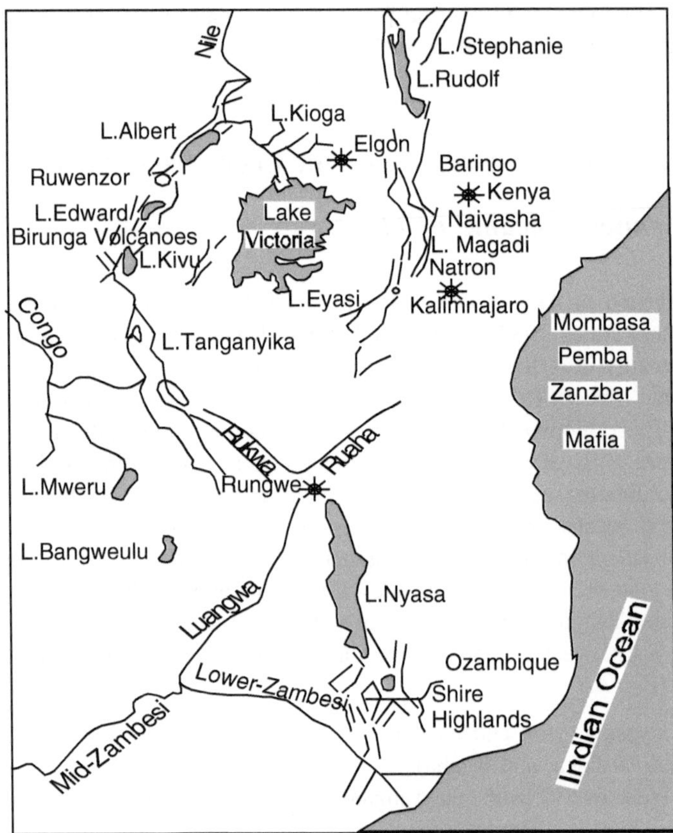

Fig. 4.8 Map of the Great Rift Valley of Africa from Zambesi to Ethiopian border (after Holmes 1965)

The two active volcanoes namely, Nyiragongo and Nyamuragira (Fig. 4.10), which are separated by less than 15 km, are characterized by the production of contrasting lava flows. The Nyamuragira is a typical Birunga volcano erupting K-basanites and more evolved derivatives such as K-hawaiites. These lavas are feldspar, olivine and clinopyroxene-phyric with leucite phenocrysts or leucite in the groundmass. The Nyiragongo volcano however, is known for erupting leucite nephelinites and melilitites comprising nepheline, leucite and melilite with olivine and clinopyroxene.

Sahama (1973) studied samples of lavas of Shaheru-Nyiragongo-Baruta, and found that they are mainly latites, leucite phonolites and leucite tephrites. Except for tephrites other rocks of Nyiragongo are nepheline-normative. In tephrites, leucite predominates over nepheline. The content of clinopyroxene is far more than that of olivine in the melanocratic rocks. There is continuous increase in the ($Na_2O + K_2O$)

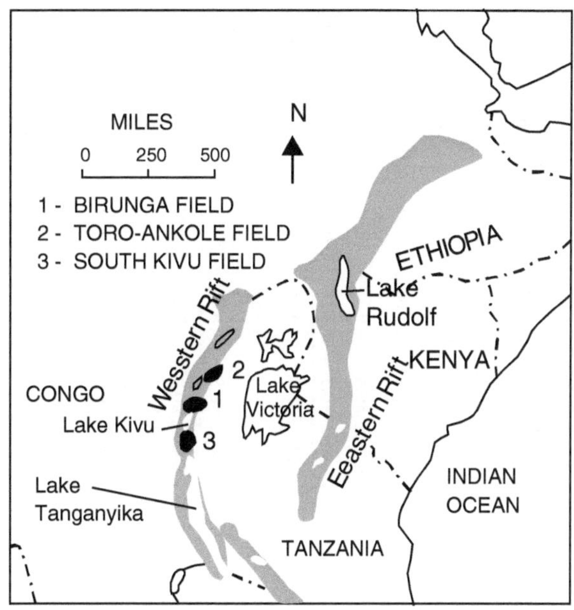

Fig. 4.9 Map of the Birunga volcanic and Toro Ankole field (after Holmes 1965)

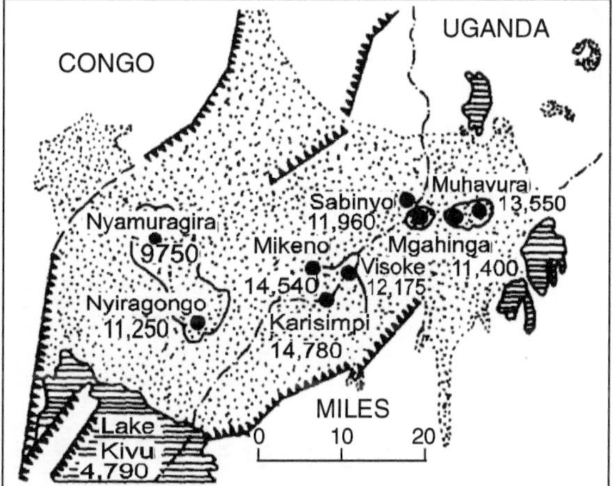

Fig. 4.10 The sketch map showing the position of heights (in ft) of the eight major volcanic complexes north of Kivu (after Holmes 1965)

and K_2O/Na_2O ratio from Lake Kivu through Nyamuragira, Nyiragango and to two different centres of the lava plane (Table 4.6).

The three most easterly volcanoes, Muhavura, Gahinga and Sabinyo lie along an east–west line suggesting their positions as fault-controlled (Rogers et al. 1998). Whereas the Muhavura and Gahinga were erupted lavas of the K-basanites series (e.g. Nyamuragira, Mikeno and Karisimbi) to the west, the lava of the Sabinyo volcano, is unique in the Birunga volcanic province in that they are silica-saturated orthopyroxene-bearing K-trachytes and latites. The majority of the lavas from

Table 4.6 Analyses of Potassic rocks from Toro-Ankole and Bufumbira, Equatorial Africa

	SiO_2	TiO_2	Al_2O_3	Fe_2O_3	FeO	MnO	MgO	CaO	SrO	BaO	Na_2O	K_2O	P_2O_5	H_2O^+	H_2O^-	Others	Total
1	40.52	5.28	8.17	6.33	6.29	0.21	11.55	12.55	0.22	0.19	2.50	3.75	0.44	1.31	0.40	–	99.71
2	36.71	5.54	9.30	9.44	4.24	0.26	6.34	14.08	0.34	0.25	2.40	6.05	1.11	1.82	0.42	1.30	99.60
3	39.18	6.29	9.86	8.03	5.13	0.23	6.63	13.31	0.25	0.23	2.87	5.00	0.85	1.08	0.55	0.35	99.84
4	43.73	1.84	9.85	3.04	8.03	0.17	20.18	8.70	–	–	1.54	1.73	0.37	0.47	0.08	0.43	100.16
5	40.47	3.52	5.38	4.03	6.47	0.13	24.84	8.06	–	–	0.68	3.46	0.29	1.11	0.57	1.07	100.08
6	44.70	3.40	11.9	2.30	9.80	0.23	10.30	11.10	–	–	1.96	2.80	0.50	0.72	–	0.24	99.95
7	43.70	3.41	10.0	4.00	7.00	0.21	11.20	13.80	–	–	1.89	2.90	0.52	0.68	–	0.19	99.50
8	44.30	2.80	11.6	2.60	8.30	0.22	11.50	10.90	–	–	2.70	3.10	0.59	0.80	–	0.36	99.77
9	38.05	3.84	7.54	8.41	2.80	0.21	13.55	13.90	–	–	1.31	3.02	0.95	2.27	3.00	0.82	99.57
10	43.73	1.84	9.85	3.04	8.03	0.17	20.18	8.70	–	–	1.54	1.73	0.37	0.47	0.08	0.05	99.78
11	40.00	4.75	7.68	5.38	4.77	0.15	15.46	9.79	–	–	0.65	7.04	0.42	1.66	0.97	0.54	99.26
12	47.10	2.54	13.36	11.42	–	0.18	9.92	10.39	–	–	2.19	2.39	0.45	–	–	–	99.94
13	39.06	4.36	8.18	4.61	4.98	0.26	17.66	10.40	–	–	0.18	6.98	0.61	1.42	0.50	0.26	99.46
14	40.47	3.52	5.38	4.03	6.47	0.23	24.84	8.06	–	–	0.68	3.46	0.29	1.11	0.57	0.15	99.26
15					–				–	–				–	–	–	

(continued)

Table 4.6 (continued)

	SiO_2	TiO_2	Al_2O_3	Fe_2O_3	FeO	MnO	MgO	CaO	SrO	BaO	Na_2O	K_2O	P_2O_5	H_2O^+	H_2O^-	Others	Total
	44.92	3.17	11.51	12.17		1.18	9.11	12.45			2.19	2.95	0.52				99.17
16	46.68	3.50	11.68	12.29	–	0.19	8.52	12.82	–	–	1.73	2.56	0.41	–	–	–	100.38
17	46.61	3.45	13.48	11.96	–	0.19	6.95	10.15	–	–	2.32	3.73	0.58	–	–	–	99.42
18	47.25	3.27	14.67	12.46	–	0.18	6.77	8.79	–	–	2.74	3.57	0.56	–	–	–	100.26

1–3 Leucite-bearing ankratrite from Toro-Ankole (Brown 1971)
4 Leucite basanite from Bufumbira (Holmes and Harwood 1937, p. 61)
5 Leucite basanite from Toro Ankole (Holmes and Harwood 1932, p. 415)
6 Phonolitic tephrite from Bufumbira (Ferguson and Cundari 1975)
7 Leucitite from Bufumbira (Ferguson and Cundari 1975)
8 Leucite basanite from Bufumbira (Ferguson and Cundari 1975)
9 Katungite from Katunga, Uganda (Arima and Edgar 1983; Anlyst K. Ramlal and S. Kuehner)
10 Olivine ugandite from Uganda (Holmes and Harwood 1937)
11 Biotite mafurite from Uganda (Holmes 1942)
12 Leucite basanite from Karsimbi, Bufumbira (Rogers 1992)
13 Kalsilite mafurite, East African Rift valley (Higazy 1954)
14 Ugandite from East Africa (Higazy 1954)
15–16 Major element analyses of K-rich rocks from Muhavura, Eastern Birunga (Rogers et al. 1998)
17–18 Major element analyses of K-rich rocks from Gahinga, Eastern Birunga (Rogers et al. 1998)

Gahinga and Muhavura are characterized by the presence of plagioclase, clinopyroxene, olivine, leucite and Fe-oxide. They are classified as K-basanites and K-hawaiites. More evolved samples from Muhavura and Gahinga include K-benmorites. The lavas of Gahinga and Muhavura are shoshonitic.

The age of Birunga volcanism ranges from <100 ka to 10 Ma. The three ages from the Sabinyo mineal separates have ages as follows: 64 ± 25, 113 ± 35 and 176 ± 30 ka. Mikeno lavas are more than 200 ka old.

All lavas are shoshonitic with $1 < K_2O/Na_2O < 2$ and strongly enriched in incompatible elements. The $^{87}Sr/^{86}Sr$ ratios vary from 0.70586 in the K-basanites to 0.70990 in the latites. The $^{143}Nd/^{144}Nd$ content ranges from 0.5124–0.51206, and Pb isotopes follow sub-vertical trends on isotope diagrams ($^{206}Pb/^{204}Pb$: 19.30–19.51, $^{207}Pb/^{204}Pb$: 15.69–15.93 and $^{208}Pb/^{204}Pb$: 40.28–41.5). The $^{40}Ar/^{39}Ar$ ages of leucite and phlogopite separates suggest that the latites have ages between 100 and 200 ka and that of K-basanite <100 ka. The latites are hybrid magmas produced by mixing between a K-basanite melt with a silicic melt from the deep crust. The low-silica K-basanites reflect interaction between a mafic K-basanitic melt ($^{143}Nd/^{144}Nd$ ratio ~0.5120 and $^{87}Sr/^{86}Sr$ ~ 0.707), and nephelinite ($^{143}Nd/^{144}Nd$ ~ 0.51267 and $^{87}Sr/^{86}Sr$ ratio ~0.7045). Both were derived from the mantle lithosphere with source ages of 1 and 0.5 Ga, respectively, and the youngest ages correspond to the deepest magma sources. The magma production rate in the Birunga is low (~0.04 km^3/year), and reflect prolonged (10–15 Ma) heating of the lithosphere by the East African mantle plume.

The interesting feature of the Western Rift is the presence of Ruwenzori horst, which is a nonvolcanic mountain in Africa (16,794 feet high). The massif narrows down towards Lake Albert with fault scraps on either side. The volcanic products are as follows. In the Western Rift a few thousand year old carbonatite lavas have been recognized. An older carbonatite intrusion is seen in the northeast of Birunga, north of Lake Kivu. The pyroclastic rocks and lavas at the Fort Portal region range from carbonatites to K-rich silica-undersaturated lavas and lapillies containing leucite, K-rich nepheline, kalsilite, pyroxene, olivine and melilite. Silicate bombs and lapilies are found in the volcanic fields around Ruwenzori. Lava flows are observed only around isolated volcanoes of Katunga in the south and volcanoes of Fort Portal in the north. Intrusive tongues and ejected blocks of carbonatites are found in the Katwe area.

The lavas erupted from these volcanoes comprise mainly such feldspathoids as leucite, kalsilite, nepheline and melilite. The lava flows in the volcanic field south of Lake Kivu are mainly constituted of olivine basalts, with a few flows of trachytes, rhyolites and pyroclastics.

Detailed petrological investigation of lavas from Mikeno, Magahinga and Muhavura (Fig. 4.10) and some smaller cones from the adjoining areas were made by Ferguson and Cundari (1975). According to them the lavas range from olivine basanite to leucite phonolite and slightly silica-saturated trachytes. The lavas include xenoliths of mica pyroxenites, and peridotites with or without mica. According to them, there are two distinct petrographic series originating from a melabasanite. Earlier, melabasanite was termed as ugandite or olivine ugandite.

These rocks have a colour index of 60–75. The first series is constituted of leucite tephrite, tephritic leucitite, leucitite, phonolitic leucitite and leucite phonolite. This series has been termed as a leucitite series by Ferguson and Cundari (1975). The second series (phonolitic tephrite series) is constituted of leucite tephrite, phonolitic leucite tephrite, tephritic leucite phonolite, latite and trachyte. Melabasanites are the matrix of olivine, titanaugite and microcrystalline glass containing plagioclase and minor amount of leucite. Leucite tephrites are the basic members grading from an olivine-rich variety with subordinate amount of leucite to leucite-rich type with rare amounts of olivine. Phenocrysts of titanaugite and ulvospinel occur with leucite and/or olivine. Plagioclase is found to be restricted to the groundmass in equilibrium with leucite, titanaugite, Ti-magnetite. With disappearance of plagioclase, the rocks become leucitites with the phenocrysts of leucite and titan-augite in a fine-grained matrix of leucite, titanaugite, Ti-magnetite, rare nepheline, olivine and microcrystalline glass.

Leucite tephrites are the most basic rocks of the phonolitic tephrite series. The tephrites are characterized by phenocrysts of titanaugite, olivine and rare ulvospinel with microphenocrysts of titanaugite, plagioclase, leucite and Ti-magnetite (±glass). In more evolved rocks, alkali feldspar is present and the leucite tephrite series grade to phonolitic leucite tephrite and tephritic leucite phonolite. This series includes xenoliths of clinopyroxene and olivine (±ulvospinel) or simply clinopyroxene, olivine and biotite (±ulvospinel) and xenocrysts of biotite and rare kaersutite.

De Mulder (1985) made geochemical investigation of rocks of Birunga volcanic field with particular emphasis on the lavas erupted by the extinct Karsimbi volcano, the largest volcano in the Birunga field. According to De Mulder most of the K-rich rocks of Karsimbi are either leucite basanite or K-trachyte. Different lava types of this region are genetically related by fractional crystallization of observed phenocrystal phases from a parental basanite. The most primitive basanitic lavas with high mg-number and Ni and Cr contents have the characteristics inherited from mantle depths.

Bell and Powell (1969) considered that evolution of Birunga volcanics are product of some crustal assimilation. Rogers (1992) however, thought that although this may be true in case of K-rich mugearites, trachytes and some more Si-rich variants (De Mulder et al. 1986), the negative high field strength element anomalies and lack of Eu anomalies in the mafic and near primary leucite basanites does not suggest crustal assimilation in the evolution of magmas from this area.

Toro-Ankole is located about 125–155 km north-northeast of the Bufumbira field (Fig. 4.9, Brown 1971). The lavas of the area belong to the olivine leucitite and olivine melilitite family and are completely devoid of feldspars. Katungites are also fairly common. The lavas often include xenoliths of phlogopite peridotite, glimmerite and partially fused granitic rocks. Chemical analyses of the leucite-bearing rocks from the Western Rift are summarized in Table 4.7.

Table 4.7 Analyses of Lamproitic rocks from Kapamba, Luangwa Valley (Scott-Smith et al. 1987)

	SiO_2	TiO_2	Al_2O_3	Fe_2O_3	MnO	MgO	CaO	Na_2O	K_2O	P_2O_5	H_2O^+	CO_2	Trace elements	Total
1	51.12	1.71	7.69	7.34	0.08	12.09	5.27	1.71	5.69	0.62	1.9	2.4	0.53	98.15
2	46.06	2.58	9.51	10.14	0.13	11.62	8.25	4.23	1.76	0.72	3.8	0.4	0.64	99.84
3	47.07	2.59	9.18	10.01	0.12	10.54	7.95	4.49	1.9	0.41	3.9	0.7	0.59	99.45
4	45.51	2.83	9.31	9.94	0.26	10.4	9.19	0.63	4.98	1.26	4.3	0.4	0.63	99.64
5	42.63	2.44	7.72	10.94	0.15	13.59	9.47	4.25	1.39	0.83	3.7	0.3	0.65	98.06
6	50.7	2.2	8.7	8.43	0.13	8.62	7.49	2.33	5.6	0.85	3.5	0.3	0.82	99.67

4.3.2 Korath Range, Ethiopia

There is a small group of volcanoes in Korath Range of southern Ethiopia (Brown and Carmichael 1969). The basement rocks of these volcanoes are constituted of basanites and tephritic lavas, which are characterized by the presence of phenocrystal olivine, phlogopite, leucite and nepheline in a groundmass of plagioclase, olivine, phlogopite, leucite and nepheline. Ti-poor kaersutite has also been found to be an accompanying phase. The analyses of plagioclases show that their compositions vary from An_{80} to An_{52}. The pyroxenes are titanaugites containing significant amount of alumina as $CaAl_2SiO_6$ or $CaTiAl_2O_6$ molecules in solid solutions. Leucite contains 1–1.5 wt% Na_2O in solid solution.

4.3.3 The Kapamba Lamproites of the Luangwa Valley, Eastern Zambia

A group of pipes and dikes of lamproitic rocks occur along the Kapamba River, a tributary of the Luangwa River, about 150 km west–northwest of Chipata in Eastern Zambia (Scott Smith et al. 1987). The Kapamba lamproites occur in the Irumide tectonic belt of the Kibaran Province. The intrusive bodies are confined to the downfaulted Luangwa Graben, trending northeast. Scott Smith et al. (1987) consider the major faulting events took place at first during the post Karoo and early Jurassic period and later during Tertiary to Recent time.

The Kapamba Lamproites exhibit a NW–SE trend, which differs from the NNE–SSW-trending fault of the Luangwa Graben. There are three other provinces of post Karoo intrusions, which are also associated with Luangawa Graben. Panela and North Lungawa fields occur respectively, within 150 km east and northeast of Kapamba; and the Isoka group lies 300 km away in the northern part of the Luangawa Valley.

The Kampaba lamproites comprise 14 pipe-like bodies (P_1–P_{14}) occupying a 25 km long NW–SE-trending zone. Based on Rb–Sr mica analyses, C.B. Smith (cited by Scott Smith et al. 1987) fixed the age of the lamproites at 220 Ma.

The pipes are mainly composed of volcanoclastic materials and the clast includes juvenile lapilli with rounded pebbles of quartz (derived presumably from the country rocks). The xenoliths include older metamorphic basement rocks. According to these investigators, the rocks are mostly lapilli tuffs differing in grain size. The tuffs include graded bedding suggesting pyroclastic flows. Presence of cross-bedding indicates pyroclastic surge deposits.

Magmatic rocks often include autolithic breccia, and in many regions, evidence of multiple intrusions is noted. The pipe rocks display concentric structure having xenocrysts of crustal rocks in the marginal tuff with signs of crater collapse. The geology of these pipes is similar to the West Kimberley pipes (Ellendale 4 and 9 and Seltrust 2). The juvenile lapillis (usually <2 cm) are characterized by porphyritic texture with phenocrystal (up to 8 mm) to microphenocrystal olivine (0.25–0.5 mm) set in a fine-grained glassy groundmass. Lapillis in some pipes comprise laths of phlogopite, while in others, are characterized by the presence of phenocrysts of leucite and clinopyroxene in a groundmass of leucite, clinopyroxene and phlogopite. Because of glassy and vesicular nature of the juvenile lapillis, they suggested that most of the volcanoclastic rocks to be pyroclastic tuffs.

The magmatic rocks occurring within the pipe-like intrusions are characterized by macrocrystic or porphyritic texture with olivine set in a finer groundmass. In case of some magmatic rocks, the groundmass is constituted of phlogopite and clinopyroxene, whereas the other variety is constituted of leucite, clinopyroxene, phlogopite, potassic richterite, sanidine, opaque minerals, perovskite, apatite and glass.

Phlogopites are TiO_2-rich (5–9 wt%) and relatively poor in Al_2O_3 (4–11.5 Al_2O_3). The Na_2O content (0.3–1.3 wt%) is similar to other lamproites, but fluorine content is significant (1–5 wt%). Olivines are forsteritic with Mg/(Mg + Fe) ratio ranging from 0.8 to 0.934.

Clinopyroxenes are diopsidic with variable amounts of TiO_2 (0.4–2.7 wt%), Al_2O_3 (0.4–2.2 wt%) and Na_2O (0.2–0.9 wt%). The amphiboles are mainly potassic richeterites (4.0–6.3 wt% TiO_2, 4.8–9.5 wt% FeO and 2.7–9.2 wt% K_2O). The groundmass spinels are titano-magnetites sometimes with chromite-rich core.

Leucites are nearly isotropic and in some instances are replaced by sanidine and analcite. Primary sanidine (FeO: 2.1 wt%, Na_2O: 2.7 wt%) occurs as a coarse-grained mineral. Ilmenite (<0.1 mm) may be a primary or xenocrystal phase. Composition of primary glass is similar to phlogopite but having lower MgO (10 wt%) and higher FeO (20–22 wt%) content.

Chemical analyses of the rocks are summarized in Table 4.7, which suggest that lamproites have variable K_2O content (sometimes higher than 5 wt%) and high K_2O/Na_2O ratio (3–5). There are however, rocks with high Na_2O content (>0.48 %). These are the rocks, where leucite is replaced by analcite and /or nepheline. Scott-Smith et al. supported the mechanism of Gupta and Fyfe (1975) regarding the alteration of leucite to analcite. The rocks have MgO content varying between 8.62 and 12.09 wt% and is higher than that of Al_2O_3 (7.69–9.51 wt%).

The olivine lamproites from northwestern part yielded a lot of diamonds and are similar in mineralogical compositions to the lamproites from Prairie Creek, Ellendale and Argyle. Diamonds in these rocks are considered to be xenocrystic. Chemical analyses of lamproitic rocks from Kapamba are given in Table 4.7.

4.3.4 Leucite Lamproites from Pniel, Post Masburg, Swartruggens, South Africa

Phlogopite and leucite-bearing dykes have been described in association with diamond-bearing kimberlites at the Hellam Mine from Swartruggens (33°54′S, 18° 58′E), South Africa by Skinner and Scott (1979). These rocks are characterized by

Table 4.8 Analyses of potassic rocks from Highwood Mountains, Montana, U.S.A. (After O'Brien et al. 1991)

	SiO_2	TiO_2	Al_2O_3	Cr_2O_3	Fe_2O_3	MnO	MgO	CaO	BaO	SrO	Na_2O	K_2O	P_2O_5	LOI	Total
1	47.24	0.67	12.54	0.06	9.15	0.16	10.72	8.97	0.50	0.16	2.80	4.85	0.78	2.28	100.90
2	49.74	0.83	13.21	0.01	9.40	0.16	6.24	9.55	0.57	0.21	3.07	6.25	0.99	0.64	100.90
3	51.40	0.60	16.90	0.01	6.80	0.10	3.30	5.36	0.60	0.22	2.89	8.85	0.57	2.11	99.67
4	52.88	0.63	17.34	–	7.95	0.20	1.30	5.08	2.37	1.36	0.30	8.61	0.18	1.82	100.00
5	46.79	0.77	10.86	0.12	10.01	0.17	14.14	7.75	0.22	0.10	2.12	4.59	0.63	1.93	100.20
6	47.33	0.59	12.77	0.04	8.59	0.15	9.28	9.22	0.45	0.14	1.80	6.48	0.92	1.37	99.13
7	46.29	0.62	8.05	0.08	8.78	0.15	16.84	9.18	0.56	0.13	2.21	5.01	0.83	2.11	100.80
8	58.94	0.21	17.83	0.01	2.83	0.08	1.51	2.76	0.82	0.16	2.83	10.89	0.18	1.24	100.30

1–2 Mafic phonolites, Highwood Mountains, Montana
3–4 Leucite phonolites, Highwood Mountains, Montana
5 Olivine Minette, Highwood Mountains, Montana
6 Shonkinite, Highwood Mountains, Montana
7 Jumillite, Highwood Mountains, Montana
8 Tinguaite, Highwood Mountains, Montana

the presence of phlogopite, olivine, diopside, leucite and spinel. There are also lamproite dykes at Pniel, Post Masburg and Pielansburg in Johannesburg.

4.3.5 K-Rich Rocks from Mt. Etinde, West Africa

Mt. Etinde is located on the Cameroon line, northwest of the Congo Craton and south of the West African Craton (Nkoumbou et al. 1995). It is essentially a volcano, which had been erupting mainly nephelinitic rocks during the Quarternary period, but it also erupted mela-nephlinites, leucitites, leucite nephlinites and leucite-melilite-bearing nephlinites. The last mentioned rock type includes phenocrysts of haüyne, pyroxene, nepheline, leucite and aenigmatite. The eruption of lava of Mt. Etinde took place essentially 0.065 $\pm$ 0.1 and 1.1 Ma ago.

4.4 The Lamproitic Rocks from Antarctica

The lamproitic rocks have been described from Gaussberg, Mount Bayliss and Priestly Peak from Antarctica by Sheraton and Cundari (1980), Sheraton and England (1980) and Collerson and McCulloch (1983). According to Stephenson (1972) and Nougier (1972) the Kerguelen-Gaussberg ridge has been the center of extensive hot-spot activity during the last 27 Ma.

Gaussberg (68°48′S, 89°19′E) is an extinct volcanic cone located on the coast of Wilhem II land, Antarctica (Fig. 4.11). The cone is 370 m high without a crater and consists of pillow lavas. The pillows are best exposed around the lower part of the mountain and are 50 cm–2 m across (Sheraton and Cundari 1980). The most of the lavas are vesicular. The pillows have a well-defined black glassy crust about 3–5 cm thick. Partially fused crustal xenoliths (up to 15 cm across) mostly of granitic composition are common.

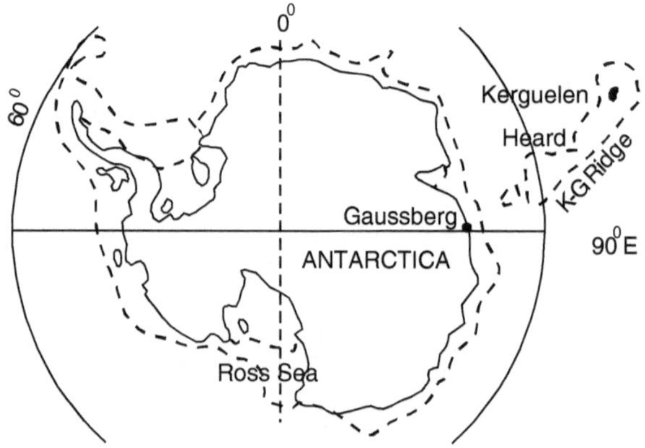

Fig. 4.11 Location of Gaussberg volcanic field in Antarctica (after Sheraton and Cundari 1980)

According to Sheraton and Cundari, the K–Ar technique yielded an age of 20 million years for the whole-rock from Gaussberg. According to Sheraton and Cundari, Gaussberg has a tectonic setting on a passive continental margin. Gaussberg lies at one end of the so-called Kerguelen Gaussberg Ridge. The rocks are characterized by the presence of subhedral crystals of olivine, clinopyroxene and leucite (up to 1 mm across) in a yellow-brown glassy matrix. Phlogopite and amphibole also occur as late crystallizing phases. The Mg/(Mg + Fe) ratio of the rocks is around 0.70.

The phlogopites are titaniferous with high BaO (1.0 wt%) and are similar to the micas of West Kimberley (in composition). The rocks are essentially leucitite with chemistry similar to those of Leucite Hills, Wyoming (U.S.A.) and West Kimberley, Australia.

According to Sheraton and Cundari (1980) the glass is characterized by the presence of quench crystals of leucite, diopside, phlogopite, amphibole and ilmenite. Chromite constitutes about 50–60 vol% of the whole rock. The Gaussberg rocks are highly potassic (K_2O content around 12 wt%) with K_2O/Na_2O ratio being close to five. The alumina content is lower than the total alkali content. The olivine leucitites from Gaussberg also contain cognate xenoliths of phlogopite and olivine-bearing leucitite. It may also include accidental xenoliths of spinel lherzolite.

There are also two isolated K-rich dykes in the Australian Antarctic territory. One of the dykes is five metre thick and located at Mt. Bayliss (73°26′S, 62°50′ E) in the southern Prince Charles Mountains, MacRobertson land. According to them the K–Ar dating of K-richterites suggests an age of 413 ± 10 million years for these dykes. Age determination of a K-arfvedsonite, occurring within the rock yielded an age of 430 ± 12 million years. The other dyke at Priestly Peak (67°11′S, 50°12′E) in Enderbyland is essentially a phlogopite-bearing syenite, termed melasyenite by Sheraton and England (1980).

According to them Mount Bayliss rocks show textural evidence indicating reaction of leucite with liquid to form K-feldspar.

The three Antarctica lamproites are enriched in TiO_2 (3–5 wt%), P_2O_5 (1.5–3.3 wt%) and lithophile elements (Ba: 400–1,500 ppm, Zr: 900–1,800 ppm; Sr: 1,300–3,000 ppm; Nb: 40–150 ppm and Ce: 270–335 ppm).

4.5 Potassium-Rich Silica-Undersaturated Igneous Rocks of the United States of America

4.5.1 Volcanic Fields of Highwood Mountains, Montana

The petrographic province of the Highwood Mountains (Fig. 4.12) was studied by Larsen and Buie (1938), Larsen et al. (1941), Marvin et al. (1980) and more recently by O'Brien et al. (1988, 1991). It is a deeply eroded volcanic and intrusive complex (52 ± 1 Ma). The volcanic field is characterized by an older series of

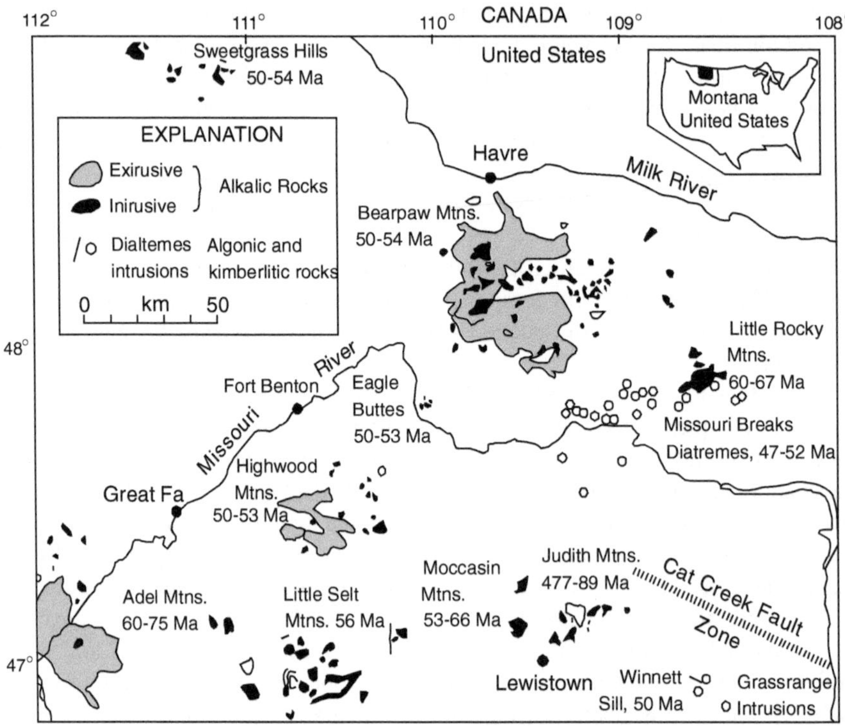

Fig. 4.12 Volcanic fields of Highwood Mountains and Bearpaw Mountains (after Macdonald et al. 1992)

quartz-normative latites, which are either overlain or intruded by a variety of potassic silica-undersaturated mafic to felsic flows, dikes, sills and stocks constituting a younger series (O'Brain et al. 1991). The Highwood Mountains igneous rocks overlie and intrude a clast-supported water-laid conglomerate unit, which was deposited on Cretaceous Eagle sandstone and Colorado Shale.

The young volcanic series is 700 m thick and is constituted mainly of mafic phonolite lava flow and breccias with tuffs having nearly 10 % crystal. There are two minette lava flows. An individual flow of mafic phonolite is less than 10 m thick.

O'Brien et al. described intrusion of several stocks (1–7.5 km^2) of shonkinite/malignite, alkali syenite/monzonite pile. These rocks are hypabyssal equivalent of mafic phonolites. The Baldy Stock alkali syenite/monzonite also outcrops in this area. The K–Ar ages of these intrusives are 52.5–51.5 ± 1.8 Ma for Highwood Baldy syenite and 50.4–51.1 ± 1.8 Ma for Shonkin Sag and Square Butte laccoliths, which were intruded into the sedimentary rocks near the Highwood Mountains.

During the emplacement of the stock, there was doming of the sedimentary sequence. High angle normal and reverse faults also were developed during the process of doming. The stocks, flows and the country rocks are cross cut by dikes and sills of mafic phonolites and minettes. There are also dikes of microsyenite, micromalignite, pseudoleucite phonolite, tinguaite and alkali trachyte. O'Brien et al. also described rare occurrence of olivine kersantite, latite and trachyte.

The last-mentioned rock type is the oldest member of the potassic series having an age of 53.1 ± 1.8 Ma (Marvin et al. 1980). Fifteen km northeast of Highwood Mountains, occur an intrusive plug called the Haystock Butte, comprising monticellite peridotite and minor alnoite.

According to Marvin et al. the Haystock Butte has an age of 48.2 ± 1.8 Ma. The latites are quartz-normative having following phenocrystal assemblage, plagioclase + hornblende ± augite ± biotite in a matrix of sanidine + quartz + biotite + Fe–Ti oxide + glass. According to O'Brien et al. mafic phonolites are constituted of phenocyrstal salite + pseudoleucite (or analcite) + olivne ($Fo_{84\text{-}65}$) ± biotite ± glass with accessory apatite and titano-magnetite. They found that more felsic leucite phonolites have less pyroxene, no olivine but more sanidine. The mafic phonolite has an assemblage similar to olivine leucitite, though pristine leucite phenocrysts are rare. Salite and olivine phenocrysts contain pristine leucite but whenever fluid had access to them along fractures they have been analcitized. Conversion of leucite to analcite was found to be consistent with the hypothesis involving reaction of leucite with a Na-bearing solution at low temperatures as suggested by Gupta and Fyfe (1975). The porphyritic texture of mafic phonolite dikes and flows indicate continuous crystallization presumably under subvolcanic conditions. According to them the near solidus crystallization of biotite as well as common occurrence of leucite implies that this magma crystallizing at shallow depths were enriched in aqueous fluid.

In this locality, there are six major stocks of which four are shonkinites and malignites. There are more primitive and more evolved mafic phonolites also. O'Brein et al. described rocks having textural continuity between shonkinite and mafic phonolite. The rocks of these intermediate varieties are termed microshonkinite. The assemblage comprising salite + sanidine + nepheline + olivine + biotite + Fe–Ti oxides + apatite ± plagioclase is predominant in shonkinites. The microshonkinites often contain spherulitic pseudoleucite phenocrysts. The coarse-grained shonkinites contain pseudoleucites, characterized by symplectic intergrowth of sanidine and nepheline. The well-known Shonkin stock contains missourites and fergusites. The missourite is hypidiomorphic and contains leucite, olivine (Fo_{81}), salite and Ba–Ti-rich biotite (5 cm across). There are dikelets composed essentially of leucite; hence they are italites. There is also a fergusite dike within the stock. The fergusite is composed entirely of pseudoleucite and salite with a small amount of biotite and olivine containing leucite inclusion. There is also a jumillite (much higher mica/leucite ratio than missourite) dike comprising olivine (Fo_{95}), salite, phlogopite, leucite, sanidine and minor amount of K-richterite.

An olivine kersantite comprises phenocrystal olivine (Fo_{90}), diopside and phologopite in a matrix of phlogopite, diopside, apatite and chromite rimmed by magnetite, analcite and plagioclase. Of the exposed portion of the minettes, 25 % are dikes and sills. Minettes are usually younger than mafic phonolites. According to O'Brien et al. (1991), the olivine minettes are characterized by porphyritic texture with phenocrystal phlogopite, diopside and olivine (Fo_{87-92}) in a matrix of sanidine, diopside, phlogopite and Fe–Ti oxides. The most evolved rocks of the area include syenite and monzonite. The Highwood Baldy Stock is characterized by the presence of three different types of alkali syenites (white, brownish grey and bluish grey) and a monzonite. Several types of trachyte dikes are also present in this area. Biotite clinopyroxinite xenoliths up to 30 cm across, occurs in the bluish grey syenite of the Highwood Baldy Stock. Xenoliths composed of phlogopite, diopside and/or olivine, are common in some minette dikes.

O'Brien et al. (1988) made systematic studies on minette, lamproite and mafic phonolitic rocks of the Highwood Mountains province of Montana. According to them the potassic mafic volcanic rocks and intrusives of the Eocene Highwood Mountains province, display mineralogical, textural and chemical evidence of several episodes of fractional crystallization and mixing of phonolitic magmas and minette at shallow depth. They concluded from the composition of the phenocrysts, bulk rocks and REE abundances of most primitive minette magmas that they were derived by partial melting of a phlogopite-bearing garnet peridotite mantle. Although the two groups of rocks are characterized by different phenocrystal assemblages, the bulk composition of both the groups overlaps significantly.

According to O'Brien et al. the mineralogical differences observed in both groups of rocks is related to difference in volatile fugacities during crystallization, which may indicate low pressure degassing. The Highwood Mountains province is characterized by an older series of quartz-normative latite flows and dykes and a younger series of K-rich silica-undersaturated mafic phonolite flows. There are also dykes and sills of rare lamproitic intrusions. Shonkinites are intruded as four major stocks. There are also relatively fine-grained intrusives of compositions similar to minettes, which occur as smaller intrusive bodies. The minettes are characterized by phenocrysts of diopsides (Al_2O_3 content 1 wt%, mg # 92) and phlogopite with or without olivine (Fo_{89-92}) in a groundmass of sanidine, biotite, salite, and Fe–Ti oxide. The mafic phonolites are constituted of salite (mg # 76, Al_2O_3 = 4 wt%), pseudoleucite and olivine (Fo_{77-60}) phenocrysts in a groundmass comprising salite, sanidine and Fe–Ti oxide. Biotite, nepheline and glass may or may not be present in the matrix. Leucite-bearing lamproitic rocks also occur rarely.

They are characterized by the presence of phenocrystal phlogopite, diopside, leucite and, olivine (at present pseudomorphed by talc and serpentine) in a matrix of phlogopite, diopside, leucite, olivine, and Fe–Ti oxides. Although cumulates and xenoliths are common in all magmatic rocks, mantle xenoliths are absent.

Although major and trace element abundance does not show any distinction between mafic phonolites and minettes, the most primitive variety of minettes are slightly enriched in MgO than the phonolites, which display more evolved composition. According to O'Brien et al. the lamproitic dyke is chemically similar to primitive minette. They considered that the syenite trend may be related to fractionation of biotite pyroxenite assemblage from a primitive mafic phonolite magma. This is in agreement with the occurrence of large bodies of pyroxenites in the area. There are two trends converging at MgO = 11.5 wt%. The samples which plot on the higher CaO branch are representative of shonkinites, rich in salite and biotite.

They further concluded that the lower CaO branch may represent an olivine control trend. They observed that K_2O remains constant within a wide range of MgO values, and concluded that Ba and Ni plots have the same trends, which demonstrate that a Ba- and K-rich phase crystallized during the entire period of crystallization. The BaO content increases from ~0.5 wt% in phlogopites from primitive minettes to 4.08 wt% in biotites from more evolved shonkinites. This may indicate that mica and clinopyroxene fractionation was responsible for maintaining a bulk D_{Ba} value ~1.0. They suggested that the presence of plagioclase phenocrysts in these latter rocks indicate that they were derived from different source regions possibly within the lower crust.

O.Brien and Thirlwall (1988) characterized the source components of the potassic mafic magmas of the High Wood Mountains province. They further suggested early crystallization of F-poor phlogopite and evidence of explosive eruption, which indicate that the parent magma was characterized by the presence of a H_2O–CO_2-rich fluid phase, and emplacement of mafic phonolites was associated with degassing. According to O'Brien et al. the majority of the evolved mafic phonolites and minettes contain mixed phenocrystal assemblages, which may be due to periodic mixing between the two magmas. The mafic phonolites sometimes contain diopside xenocrysts, occurring at the centre with salite forming the rim. The phonolites also contain phlogopite (~1 cm long) with resorbed outlines and zoned olivine (Fo_{87-90}) xenocrysts. The minettes are constituted of zoned clinopyroxene, salite and olivine ($<Fo_{77}$ or lower) with resorbed phlogopite which are often surrounded by Ti- and Ba-rich biotite.

The primitive minettes are enriched in light REE and Ba (2,000–5,000 ppm), Sr (850–1,200 ppm), Ni (250–330 ppm) and Cr (500–725 ppm). These figures are consistent with residual garnet, mica and clinopyroxene in the source. The chemistry of olivine, diopside, and phlogopite phenocrysts, in the primitive minettes is similar to those found in the phlogopite-garnet lherzolite xenolith in the minette called, "the Thum" (Ehrenberg 1982). O'Brien et al. suggested that the Highwood minette magma was produced by partial melting of a phlogopite and garnet-bearing peridotitic mantle. Chemical analyses of K-rich rocks from Highwood Mountains are summarized in Table 4.9.

4.5.2 The Bearpaw Mountains

The igneous rock complex at the Bearpaw Mountains of northcentral Montana constitute a near circular outcrop of about 60 km in diameter and has a height of 2,269 m (Macdonald et al. 1992, Fig. 4.12). The volcanic rocks are observed in a WSW–ENE-trending anticlinal arch containing Palaeozoic to upper Cretaceous sedimentary rocks and numerous intrusive bodies. Igneous activity is considered to have taken place during Eocene (54–50 Ma, Marvin et al. 1980).

Table 4.9 Potassic Mafic Lavas of The Bearpaw Mountains (after Macdonald et al. 1992)

	SiO_2	TiO_2	Al_2O_3	Fe_2O_3	MnO	MgO	CaO	Na_2O	K_2O	P_2O_5	LOI	Total
1	49.80	0.52	9.52	8.80	0.14	17.24	6.06	1.89	4.02	0.42	0.47	98.88
2	47.93	0.55	9.30	9.20	0.16	16.65	6.69	1.37	4.89	0.57	1.43	98.74
3	53.39	0.48	11.87	7.59	0.12	12.44	5.25	2.81	4.47	0.46	0.12	99.00
4	48.11	0.77	10.25	9.55	0.16	12.43	8.38	1.17	5.50	0.80	2.01	99.13
5	51.77	0.72	12.48	8.92	0.16	8.36	7.57	2.80	4.03	0.52	1.95	99.28
6	48.92	0.73	13.18	8.73	0.16	8.03	7.47	2.09	6.37	0.71	2.20	98.59
7	52.26	0.80	12.34	9.30	0.18	7.39	7.50	2.44	4.82	0.59	1.79	99.41
8	48.15	0.82	11.69	9.89	0.17	7.10	8.29	2.10	7.59	0.75	2.63	99.18

1–8 Analyses of minettes
1–4 and *7* Blocks in pyroclastic deposit
5 Flows of minette composition
6 Minette dyke
8 Minette from younger series

The Eocene igneous rocks constitute two groups: (1) a silica-saturated series ranging from minette to trachytes (their coarse-grained equivalents are shonkinites and syenite) and (2) a silica-oversaturated trachyte and trachy-dacite series.

The latites contain abundant crustal xenoliths, but the intrusive rocks of the minette-trachyte group show in-situ differentiation feature. In the Rockyboy stock, different rock types include biotite pyroxenite, various types of shonkinites (pseudoleucite, plagioclase and hypersthene), nepheline-aegirine syenite, porphyritic potassic syenite and carbonatite. Macdonald et al. found that with exception of a carbonatite, the entire range represents a single composite stock, which covers a complete compositional range of intrusions at Bearpaw Mountains.

The minettes range in composition from 20 to 6 wt% MgO. The more magnesium varity is characterized by the presence of forsterite, Cr- spinel, diopside (with or without phlogopite). Macdonald et al. noted that olivine-free rocks are characterized by following two assemblages: (1) salite + phlogopite + pseudoleucite and (2) salite + phlogopite + analcites (secondary after leucite). Primary leucite is also present within olivine as inclusions. The mineral chemistry and in particular, texture of clinopyroxenes indicate that there was mixing between different varieties of minette magmas.

Latites are characterized by the presence of phenocrysts of sanidine, plagioclase, clinopyroxene ± phlogopites ± amphibole ± magnetite ± apatite in a feldspathic groundmass. The latites contain brownish glass with mafic clots consisting of following assemblages:

(1) Biotite + clinopyroxene,
(2) Biotite + opaque + oxide, and
(3) Biotite + plagioclase + carbonate.

Macdonald et al. considered that the compositional variation is also related to accumulation of olivine, spinel and clinopyroxene phenocrysts and preservation of mantle xenoliths in the minette.

With reference to the phase equilibria relation in the system diopside-forsterite-K-feldspar (Esperanca and Holloway 1987) studied at different pressures (1.0, 1.5, 1.7 and 3.0 GPa), Macdonald et al. concluded that the minette magma was generated at depths equivalent to that of 3.0 GPa from a source having an assemblage of olivine, phlogopite and diopside. The primary magma further underwent fractionation of olivine and diopside.

The minettes are characterized by higher concentration of large ion lithophile elements (LILE) and light rare earth elements (LREE). They contain K_2O up to 6.18 wt%, Ba: 5,491 ppm, Sr: 2,291 ppm, and Ce: 99 ppm. The $^{87}Sr/^{86}Sr$ ratio lies between 0.707 and 0.710 and ϵ_{Nd} ranges between 10 and −16. According to Macdonald et al. latites were formed by fractional crystallization of minette magma in combination with assimilation of crustal rocks, which resulted in enrichment of SiO_2 and the Na_2O/K_2O, $^{87}Sr/^{86}Sr$ ratios. Analyses of Bearpaw minettes are given in Table 4.9.

Table 4.10 Analyses of K-Rich rocks from Smoky Butte and Sierra Nevada (U.S.A.)

	SiO_2	TiO_2	Al_2O_3	Fe_2O_3	FeO	MnO	MgO	CaO	Na_2O	K_2O	BaO	P_2O_5	H_2O^+	H_2O^-	CO_2	Total
1	52.94	5.14	8.55	5.65	–	0.07	8.34	4.51	0.96	8.84	–	2.21	–	–	–	98.06
2	52.19	4.96	9.04	8.91	–	0.08	7.98	4.90	1.14	7.85	0.69	0.22	0.95	0.54	–	99.45
3	53.53	5.23	9.08	5.52	–	0.08	7.78	4.43	1.93	7.85	0.77	0.22	2.69	1.72	–	100.83
4	51.89	5.17	9.05	5.32	–	0.08	7.90	4.44	1.97	7.63	0.88	0.25	4.05	1.27	–	99.90
5	50.90	5.30	9.30	5.10	–	0.10	7.00	6.10	1.40	7.00	1.10	1.60	2.90	–	2.20	100.00
6	49.80	2.80	8.50	7.20	–	0.10	11.7	5.80	1.30	6.60	0.70	1.20	2.60	–	2.70	101.00
7	51.70	1.80	12.70	3.70	2.80	0.13	5.70	6.40	2.00	8.40	–	1.50	1.80	0.89	0.05	99.57
8	53.50	1.57	12.30	3.82	2.20	0.10	7.36	4.65	2.16	8.22	–	1.56	0.26	0.26	0.05	98.01
9	51.00	1.50	11.60	4.60	2.80	0.16	9.20	7.40	2.00	7.00	–	1.60	0.67	0.24	0.05	99.82
10	47.90	1.20	11.80	6.20	2.80	0.16	11.10	8.40	2.80	5.10	–	1.60	0.33	0.30	0.05	99.74

1–6 Lamproitic rocks from Smoky Butte (Mitchell et al. 1987)
7–10 Leucite basanite from Sierra Nevada (van Kootan 1980)

4.5.3 Smoky Butte

Lamproitic rocks from Smoky Butte (47°19′N, 107°3′W) were studied by Bergman (1987) and Mitchell et al. (1987). The locality lies west of Jordan in the Garfield County, Montana. The butte is located at the centre of the series of the dikes and plug-like bodies extending for a length of 3 km. The intrusive bodies lie along the axis of a syncline. The rocks of Smoky Butte are K-rich and their chemistry is similar to lamproitic rocks of Leucite Hills, Wyoming and West Kimberley province of Australia. According to Mitchell et al. (1987), the K–Ar ages of the rocks of Smoky Butte are 27 Ma old, and in effect, are older in age compared to those of Leucite Hills, Wyoming. The dikes of northern section trend N32°E, but those at the south trend N25°E. They are separated by a system of post-intrusion faults with 200–300 m offset. The fault plane strikes east–west.

The dikes are aphyric and melanocratic. They may also occur as light grey or brown vesicular or galssy rocks with olivine and phlogopite phenocrysts. The dikes have thickness ranging from a few cms up to 2 m at Shiprock. They may exhibit flow differentiation features and have chilled margins.

In the northern section, highly vesicular lamproites and also autolithic and heterolithic lamproite breccias are found. Similar lamproites are also found in Instrument Butte in the southern section. Matison (cited by Mitchell et al. 1987) described them as plugs, but they are cone-shaped buttes. According to Mitchell et al. these buttes comprise complex breccias characterized by extrusive and hypabyssal K-rich rocks. Ropy vesicular flow tops can be observed at Top Contact Butte and Radial Dike Butte. Mitchell et al. described quite a few flows (about 20 cm thick), which are superimposed on one another at Top Contact Butte. They reported the occurrence of vesicular pumice at Radial Dike Butte. Lamproitic breccia containing sandstone xenolith and associated with ropy lava occurs at the western part of Half Sediment Butte.

Bergman (1987) described the occurrence of extrusive K-rich rocks at Smoky Butte. Mitchell et al. also found extrusive rocks occurring as bedded welded autolithic lamproite lapilli and welded agglutinate tuffs in clay-rich matrices. These extrusive rocks overlie or are marginal to the intrusives.

Complex breccias are absent in the southern section, where dikes occur as nearly parallel intrusive bodies. The largest dike occurs at Smoky Butte, where the earlier dikes are surrounded by latter batches of magma, which got devitrified to coarse-grained rocks. Mitchell et al. described following four assemblages from the Smoky Butte rocks:

(1) Olivine-phlogopite-armalcolite-diopside-glass,
(2) Olivine-phlogopite-armalcolite-leucite-glass,
(3) Phlogopite-armalcolite-diopside-leucite-sanidine ± glass, and
(4) Phlogopite-armalcolite-diopside-sanidine-K–Ti-richterite with or without glass.

The order of crystallization in the above-mentioned assemblages (1–4) is in the same order as mentioned above. Chemically the lamproitic rocks of Smoky Butte (Table 4.10) are similar, but mineralogically or texturally the rocks are different because of different cooling history. According to Mitchell et al. (1987), the rocks may be termed as hyalo-armalcolite phlogopite lamproite and sanidine-phlogopite lamproite. These rocks were formed from rapidly quenched, high temperature, uncontaminated lamproitic magma. These lamproitic rocks are highly enriched in TiO_2 and thus, the phlogopites are rich in TiO_2. Armacolites are also abundant. The minettes are characterized by minerals, which are Al-deficient and display very little compositional variation. Sanidine and nepheline intergrowth occurs as pseudoleucite. The lamproites are enriched in Ta, Hf, and light REE (La/Yb = 162–280) and/or characterized by high MgO and Cr_2O_3 contents. According to them the lamproites were derived from a metasomatized harzburgite source (2.5 Ga old) and they were emplaced at their liquidus temperatures.

4.5.4 Potassic Rocks of Navajo-Hopi Province

Magmatism associated with Rio Grande rift (Fig. 4.13) has been discussed in detail by Gibson et al. (1993) and Ehrenberg (1982). The rift is Late Paleocene to Quaternary in age and has a complex geological history. According to them a rift is a part of a much wider series of events that took place in the western U.S.A. during the Oligocene, when the area was subjected to extension and widespread magmatism in the Basin and Range province. Cordwell (1982) stated that topographically the Rio Grande rift widens from central Colorado into New Mexico and Texas. He thought that the axis of the rift is delineated by a broad uplift and an axial graben that is typically 40 km wide and 15 km deep. There are several localities which are associated with potassic volcanism: Navajo Hoppi, Dulce dike swarms in New Mexico and Colorado, Espanola Basin (New Mexico), Cieneguilla, Petaca, La Madera, Ojo Caliente, El Rito Creek Vent, Cerro Negro, Spanish Peaks and Two Buttes. Of these, Navajo Hopi, Dulce Dike swarms, Spanish Peak and Two Buttes are known for K-rich mafic rocks.

The Navajo volcanic field (Mc Dowell et al. 1986) outcrops over an area of 100,000 km^2 in the western part of the U.S.A. on the Colorado Plateau. The province (37°N, 110°W) is located about 280 km from the present day axis of the Rio Grande rift (Fig. 4.13). According to Gibson et al., the Navajo volcanic field is of Late Oligocene to Early Miocene in age (25 Ma, Roden et al. 1979; 26–24 Ma, McDouwell et al.1986). In this locality minettes and trachybasalts crop out as lava flows and dikes, sills and volcanic necks. These rocks include spinel- and garnet lherzolite. Analyses of K-rich rocks from Navajo Hopi are given in Table 4.11.

Table 4.11 Analyses of K-Rich rocks from Rio Grande Rift

	SiO_2	TiO_2	Al_2O_3	Fe_2O_3	MnO	MgO	CaO	Na_2O	K_2O	P_2O_5	Total
1	53.67	1.67	11.23	7.10	0.10	8.20	7.64	1.85	7.19	1.01	99.66
2	54.43	1.68	11.42	6.96	0.13	8.17	7.59	2.19	6.38	0.93	99.88
3	54.14	1.96	11.40	7.40	0.11	7.69	7.45	2.23	6.59	0.88	99.85
4	55.73	1.04	11.75	6.67	0.10	8.39	7.07	2.58	6.01	0.84	100.18
5	45.65	2.87	9.24	11.11	0.17	14.71	10.39	1.43	3.38	1.06	100.01
6	50.25	1.29	12.55	9.15	0.15	9.40	9.51	2.00	5.02	0.40	99.72
7	54.33	1.42	14.25	8.39	0.22	4.16	6.60	2.36	7.59	0.78	100.10
8	53.84	1.07	13.57	7.32	0.12	7.65	7.43	2.71	5.74	0.38	99.83

1–5 Lamproitic rocks from Navajo (Gibson et al. 1993)
6–8 Lamproitic rocks from Two Buttes, Arizona (Gibson et al. 1993)

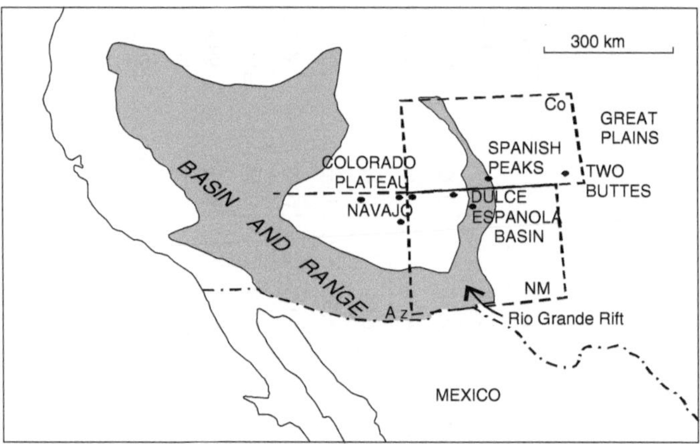

Fig. 4.13 Location of the Navajo volcanic complex and Dulce dyke (Colorado plateau), Spanish Peaks and Espanola basin in RioGrande Rift and Two Buttes (Colorado) (after Gibson et al. 1993)

4.5.5 Dulce Dike

There are dike swarms at Dulce (Fig. 4.13) in the Jicarilla Apache Indian Reservation (37°N, 107°W) about 102 km west of the present day rift axis. The dikes trend NNE–SSW parallel to Early Miocene phase of extension of the Rio Grande rift in intrusive sediments of Cretaceous and Tertiary age.

The dike swarms outcrop in an area about 50 km long and 30 km wide and include, minette, vogesite and spessartite. The K/Ar techniques yield an age of 27–24 Ma. The K_2O content of the two minettes of Dulce dike swarm from Montezuma Creek varies between 2.21 and 3.00 (wt%) and has an age of 24.3–25.0 Ma.

4.5.6 Spanish Peaks

The Spanish Peaks (Fig. 4.13) is located in the westernmost part of the Great Plains (37°30′N, 105°W), 80 km east of the present day rift axis (Gibbson et al. 1991; Gibson et al. 1993). This locality is famous for radial dike swarms. The ages of apatite and zircon for the two dikes, as determined by the Fission Track method, range between 27.5 and 20 Ma (Smith 1984, cited by Gibson et al. 1993). The radial dikes have variable chemistry, and they can be described as trachyandesite, minette, kersantite, trachyte, rhyolite and K-rich mafic dikes. The dikes vary in width between 0.3 and 30 m and may extend up to 23 km.

Table 4.12 Lamproitic rocks from Leucite Hills, Wyoming, U.S.A

	SiO_2	TiO_2	ZrO_2	Al_2O_3	Fe_2O_3	FeO	MnO	MgO	CaO	SrO	BaO	Na_2O	K_2O	P_2O_5	LOI	Total
1	46.50	2.30	0.19	11.50	5.30	0.14	0.09	8.06	7.63	0.41	0.62	2.88	10.40	1.81	2.80	100.63
2	47.50	2.35	0.19	11.80	5.42	0.14	0.09	8.24	7.80	0.42	0.63	2.94	10.60	1.85	–	99.97
3	46.50	2.30	0.19	11.50	5.30	0.14	0.09	8.06	7.63	0.41	0.62	2.88	10.40	1.81	2.80	100.6
4	53.07	2.41	0.26	8.96	3.86	0.91	0.08	11.17	3.56	0.27	0.34	1.15	10.72	1.24	1.72	99.8
5	55.14	2.58	0.27	10.35	3.27	0.62	0.06	6.41	3.45	0.26	0.52	1.21	11.77	1.40	2.24	99.59
6	47.50	2.35	0.19	11.18	5.42	0.14	0.09	8.24	7.80	0.42	0.63	2.94	10.60	1.85	–	99.97
7	53.80	4.42	0.02	10.60	3.55	1.09	0.10	7.59	5.48	0.22	0.64	1.56	11.20	1.66	–	101.93

1–2 Wyomingite from Hatcher Mesa (Barton and Van Bergen 1981)
3 Chemical analysis of Wyomingite 17472 from Hatcher Mesa (after Barton and Van Bergan 1981)
4 Olivine Orendite, South Table Mountain, Leucite Hills (Carmichael 1967). It also includes 0.08 % Cr_2O_3
5 Orendite, North Table Mountain (Carmichael 1967). It also includes 0.04 % Cr_2O_3
6 Analyses recalculated on volatile-free basis (after Barton and Van Bergan 1981)
7 Average composition (volatile-free) of 20 wyomingites and orendites from Leucite Hills, compiled from unpublished analyses and that of Carmicheal 1967 (after Barton and Van Bergan 1981)

4.5.7 Two Buttes, Colorado

The Two Buttes (Fig. 4.13) are located 310 km from the present day rift axis and lies in the Arkansas valley (37°30′N, 102°30′W), 240 km east of the Canyon city. They are intruded into Triassic shales and sandstones and Cretaceous limestones. Bergman (1987) classified them as minettes. According to Gibson et al. (1993), the K/Ar technique yielded ages ranging between 27 and 35 Ma. Chemical analyses of some of the K-rich mafic magmas are shown in Table 4.11.

Williams (1936) described potassium-rich volcanic rocks from Hopi Butte, Arizona and Navajo Mountain, Utah. In Hopi Butte, the alkaline rocks occur as pyroclastics and lava flows of limburgites, monchiquites, and nepheline-bearing trachytes. The eruption of lavas in this area took place sometimes between late Pliocene and Pleistocene. Williams described the occurrence of olivine leucitite from Todilto Park of this petrographic province.

4.5.8 Potassic Rocks from Leucite Hills, Wyoming

Leucite-bearing lavas and hypabyssal rocks were first reported by Cross (1897) and later by Kemp and Knight (1903). These rocks were more recently studied by Kuener (1980). They cover an area of 2,000–2,500 km^2 in the Sweetwater County, located in the southwestern part of Wyoming (Fig. 4.14). The locality lies 100 km

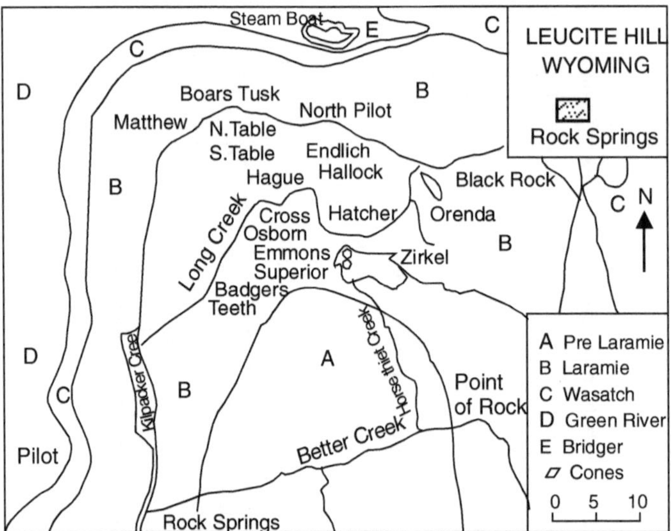

Fig. 4.14 Sketch map describing different localities known for leucite-bearing lamproites from Leucite Hills (after Barton and van Bergen 1981)

north of the border of two states of Colorado and Wyoming. These ultrapotassic lavas are found as surface flows, dike rocks and plugs. According to Barton and van Bergen (1981), these rocks were formed mainly under volcanic to subvolcanic conditions and sometimes, there are cinder or pumice cones up to 76 m high occurring on some of the lava outcrops, and were possibly formed after the outpouring of the ultrapotassic lavas. McDowell (1966, cited by Barker 1974) determined the ages of mica crystallizing in these rocks by using K–Ar technique and found them to be 1.1 ± 0.4 Ma old. The lavas and dikes punctured through sandstone, oolitic limestone and shale. Of the 22 outcrops, Pilot Butte is the only locality, where madupite is reported, otherwise in all the other 21 localities, the rocks are mainly wyomingite and orendite. According to Carmichael (1967), the type localities of wyomingites are Boars Tusk, Steam Boat Spring, South Table Mountain, North Table Mountain and Zirkel Mesa. Orendites are found particularly in Orendite Butte and South Table Mountain.

Barton and van Bergen (1981) reported the occurrence of xenoliths (<1 to ~ 5 cm^2 area) particularly from all the exposures but these are quite abundant in the volcanic rocks from Boars Tusk and Badger's teeth. Yagi and Matsumoto (1966) recorded cognate xenoliths of leucitite in wyomingite and orendite from the Emmons Cone area. The xenoliths are however, common in the lava flows, dikes and plugs. Although crustal rocks such as sandstone, limestone and granites are quite common; xenoliths of anorthosite and gabbro presumably obtained from the basement complex are also recorded within the lava flows. Yagi and Matsumoto also found such mantle xenoliths as harzburgite, pyroxenite and mica pyroxenite within the K-rich lavas. Barton and van Bergen reported the occurrence of green salitic pyroxenes as megacrysts, and also as core in microphenocrysts of wyomingite lava from Hatcher Mesa at Leucite Hills. They also reported the presence of alumina-rich phlogopite (16–21 wt% Al_2O_3), apatite, Fe–Ti oxide, Mg-rich olivine (Fo_{93}) and orthopyroxene (En_{61}) as megacrysts or as inclusions in diopside phenocrysts. They found all these phases in ultramafic xenoliths occurring within lavas. Barton and van Bergen thought that these xenoliths are disaggregation products of mantle xenoliths during the movement of the ultrapotassic lava. Some analyses of Carmichael (1967) and Barton and van Bergen (1981) on the Leucite-bearing lavas are given in Table 4.12.

Mirnejad and Bell (2006) studied the whole-rock major and trace element chemistry of the Leucite Hills rocks. They also determined the O, Sr, Nd and Pb isotopic composition of these leucite-bearing lavas, which are 3.0–0.89 Ma old. The two main groups of rocks namely, madupitic lamproites and phlogopite lamproites, are geochemically distinct and cannot be related to one another by either fractional crystallization or crustal contamination. According to them it is likely that the geochemical differences between these two rock types are related to variation in source mineralogy and depth of partial melting. The high Mg-number and large ion lithophile element abundances and negative epsilon Nd values of the lamproites indicate a mantle source, that has experienced stages of both depletion and enrichment. The negative Ta and Ti anomalies in mantle-normalized trace element diagrams and low time-integrated U/Pb, Rb/Sr and Sm/Nd

Table 4.13 Major oxide chemistry of rocks from Deep Springs Valley, Sierra Nevada, California and Eastern Paraguay

	SiO_2	TiO_2	Al_2O_3	Fe_2O_3	FeO	MnO	MgO	CaO	Na_2O	K_2O	H_2O^+	H_2O^-	P_2O_5	LOI	Total
1	46.54	1.98	10.14	11.03	–	0.17	12.73	8.97	2.14	5.41	–	–	1.37	–	100.50
2	47.32	1.82	10.36	10.26	–	0.16	12.36	8.68	2.06	5.15	–	–	1.33	–	99.50
3	46.61	1.81	10.62	10.19	–	0.16	12.14	9.13	1.99	4.75	–	–	1.30	–	98.70
4	48.20	1.92	10.35	4.98	5.42	0.20	10.91	7.93	2.00	5.81	–	–	0.56	1.61	99.89
5	47.36	1.66	14.86	1.31	6.96	0.12	6.36	8.80	1.08	9.45	–	–	0.20	1.46	99.62
6	48.33	1.91	12.92	2.56	7.24	0.17	6.62	9.08	2.54	5.81	–	–	0.86	1.25	99.29
7	44.11	2.01	12.82	4.45	5.15	0.19	9.53	10.32	3.34	5.41	–	–	0.75	1.23	99.31
8	43.21	1.65	9.89	–	10.75	0.14	12.12	11.58	3.73	4.33	0.26	0.12	1.37	–	99.15
9	45.72	1.49	11.57	–	8.86	0.16	14.00	10.58	2.61	3.90	0.31	0.09	1.22	–	100.50
10	47.56	1.54	10.18	–	6.99	0.10	11.73	8.55	2.14	6.40	1.21	0.41	1.58	–	98.39

1–3 Potassic rocks from Deep Spring Valley, California (after Beard and Glazner 1998)
4–7 Representative whole rock analyses of potassic rocks from Eastern Paraguay (after Comin-Chiramonti et al. 1997)
8–10 Analyses of K-rich rocks from Kings River region, Sierra Nevada, California (after Feldstin and Lange 1999)

ratios of both lamproite groups and other Cenozoic igneous rocks from the Wyoming Province indicate an ancient metasomatic enrichment (>1.0 Ga) of the mantle source associated with the subduction of carbonate-bearing sediments. Other chemical characteristics of the Leucite Hills lamproites, especially their high K_2O and volatile contents, are attributed to more recent metasomatism (<100 Ma) involving influx from upwelling mantle during back-arc extension or plume activity.

4.5.9 The Potassic Lava Suite from Central Sierra Nevada, California, U.S.A

Small volumes of absarokite and minette lavas (43–57 wt% SiO_2) erupted along the western slope of Sierra Nevada, California, during the late Pliocene, within the Kings River drainage system (Dodge and Moore 1981; Feldstein and Lange 1999, Table 4.13). The absarokites contain phenocrysts of olivine ± augite, whereas the minettes are constituted of K-feldspar and phlogopite ± augite ± olivine; Absarokits are distinguished by lack of K-feldspar phenocrysts. Feldestein and Lange estimated that pre-eruptive magmatic temperatures and pressures for a felsic and a mafic minette of that region to be 1138 and 1,144 (± 50) °C, and 1.2 and 1.6 (± 0.4) GPa, respectively. These magmas were characterized by extreme enrichments in the large ion lithophile elements (e.g. 1.9–8.1 wt% K_2O, 1,380–3,719 ppm Ba), depletion in high field strength elements (Ba/Na_{pm} of 7–33), and high oxygen fugacities (1–3 log units above the Ni–NiO buffer). Trace element ratios (e.g. Ba/Rb 20–100) of the rocks are distinct from those observed for the mid-oceanic ridge basalts and ocean island basalts. According to them variation in K and Ba with respect to other incompatible elements would require that phlogopite with or without potassic amphibole was an important accessory phase during the magma generation. According to Feldstein and Lange (1999) the buoyant ascent of the Kings River magmas through ~40 km thick sialic crust should require pre-eruptive volatile concentrations (H_2O and F) >2 wt%. Volcanism probably was triggered as a part of regional response to Basin and Range extension, which resulted in asthenospheric upwelling and therefore higher heat flow to the subduction-modified lithosphere.

Van Kooten (1980) described ultrapotassic lava flows from southeastern part of Yoshemite National Park (Fig. 4.15). According to him, the eruption took place through a number of isolated vents. In this region, they occur as plugs and flows, which are sometimes 5 km across. The rocks described by him as potassium-rich olivine basalt are characterized by K_2O content ranging between 3 and 4.36 wt%. This, observation together with the fact they are associated with leucite basanite

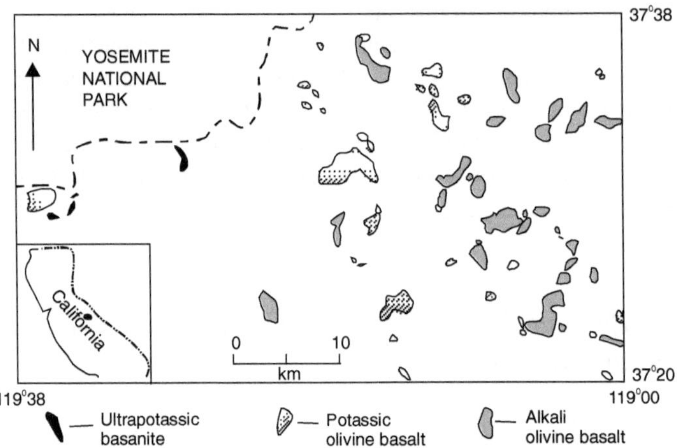

Fig. 4.15 Sketch map of different localities known for K-rich rocks south of Yosemite National Park, Sierra Nevada, California (U.S.A) (after van Kooten 1980)

suggests, that they are not alkali basalts but rocks of shoshonite-absarokite-basanite series.

On the basis of $^{40}K/^{40}Ar$ and $^{40}Ar/^{39}Ar$ isotopic analyses of phlogopite and sanidine from the basanite flows, they obtained an age of 3.4–3.6 Ma for these rocks, where as Darlymple (1963, 1964) obtained an age of 3.3–3.8 Ma for the potassium-rich mafic rocks.

The basanites are characterized by the presence of abundant phlogopite phenocrysts together with olivine and diopsidic pyroxene in a groundmass containing leucite (<0.1 mm), feldspar, olivine and pyroxene. There are also pseudo-brookite and apatite in the groundmass. The phlogopite phenocrysts are rimmed by small magnetite and olivine crystals. The groundmass phlogopites are significantly rich in TiO_2 up to 8.58 wt%, whereas the phenocrystal phlogopites contain between 2.06 and 3.56 wt% TiO_2. The Ti and Al content of the pyroxenes were found to increase with decreasing a_{SiO2} in the liquid as observed by Gupta et al. (1973) in their study related to solubility of Ti as a function of alumina and silica concentration. Sanidine in ultrapotassic basanites contains up to 5.8 wt% BaO. A basanite should contain plagioclase, feldspathoid, clinopyroxene in addition to olivine. It is not clear from the description of Van Kooten (1980), whether plagioclase is present in these rocks or not.

The so-called potassic olivine basalt is characterized by the presence of small amount of phenocrystal sanidine with or without leucite. Phlogopite also occurs in the groundmass. Phenocrysts of olivine with or without diopside occur with apatite, magnetite and pseudo-brookite. The vol% of plagioclase is less than that of sanidine.

The potassic lavas from Sierra Nevada often include spinel lherzolite nodules. According to van Kooten (1980) the temperature of equilibration of olivine-pyroxene pair in the lherzolite nodules is 1,300 °C at a depth of 100–125 km. Chemical analyses of K-rich rocks from Sierra Nevada are given in Table 4.10.

Absarokite and minette magmas (43–57 wt% SiO_2) erupted along the western slope of Sierra Nevada, California, within the Kings River Valley during the Late Pliocene (3.4–3.6 Ma, Van Kooten 1981). The absarokites contain phenocrysts of olivine ± augite in a groundmass of predominantly augite and feldspar (plagioclase ± alkali feldspar ± anorthoclase ± hyalophane) along with leucite ± nepheline ± analcite ± olivine ± apatite ± titano-magnetite ± phlogopite ± pseudobrookite. The second most abundant lava type found in this area is minette containing phenocrysts of phlogopite, augite and/or olivine ± chromite. The groundmass phases are same as those found in the absarokite. The K_2O content of absarokites ranges between 2 and 6 wt% and the MgO content varies between 9 and 15 wt%. The K_2O content for the minette varies between 3.6 and 8.1 wt%. The MgO content of the minettes is usually less (7.54 and 7.82 wt%). Absarokites are typical rock types found in the Absaroka Range near Yellowstone Park, U.S.A. (Meen and Eggler 1987)

4.5.10 Pliocene Potassic Volcanic Rocks from Deep Springs Valley, California

Potassic Rocks from Deep Springs Valley, California (Fig. 4.16) consist of lava flow remnants that lie upon Jurassic granitoids (Beard and Glazner 1998). The lava flows are essentially olivine leucitites, basanites and trachybasalts having K_2O content varying between 5.4 and 3.2 wt%. The whole-rock K–Ar method yielded an age of 2.6 ± 0.1 Ma. These lavas show positive correlations between such

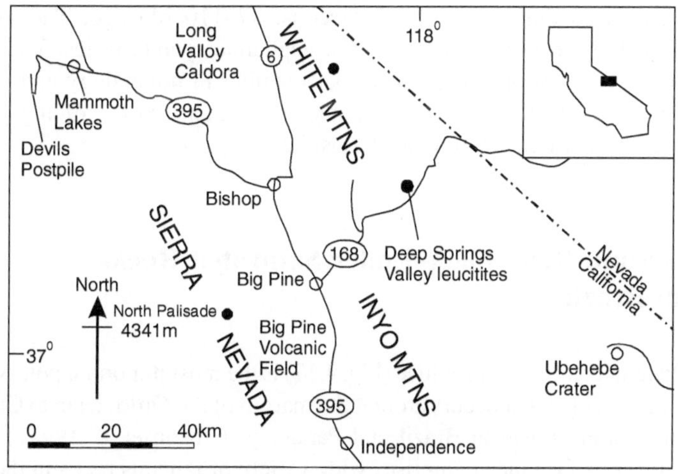

Fig. 4.16 Sketch map of Deep Spring valley leucitites (after Beard and Glazner 1998)

incompatible elements as Rb, Sr, Zr, Ba with LREE and K. The compatible element contents likewise increase with increasing MgO content. Beard and Glazner (1998) observed that the $^{87}Sr/^{86}Sr$ ratios decrease and the ϵ_{Nd} values increase with decreasing MgO. The Sr and Nd isotope compositions of these lavas are extreme, compared to most other continental and oceanic rocks. The $^{87}Sr/^{86}Sr$ ratios range from 0.7121 to 0.7105 and ϵ_{Nd} values range from 16.9 to 15.4. The Pb isotopic ratios are as follows: ($^{206}Pb/^{204}Pb$ ~ 17.2, $^{207}Pb/^{204}Pb$ ~ 15.5 and $^{208}Pb/^{204}Pb$ ~ 38.6).

Castor (2008) studied the Sulphide Queen carbonatites and associated alkaline complexes, which contain the rare earth ore bodies at Mountain Pass in California. The carbonatites are moderately dipping tabular intrusion associated with 1.40 Ga old alkaline plutons of similar sizes and orientation. The Mountain Pass carbonatite and alkaline rocks lie in a 130 km long narrow belt. Both the carbonatite and the associated alkaline rocks are chemically and mineralogically unusual. The mafic members of the Mountain Pass alkaline rock suite belong to the ultrapotassic group, and are most akin to the lamproitic subdivision. The association with significant amounts of carbonatite is also unique. Chemical similarities suggest that although the carbonatites, are slightly younger (1.38 Ga), they are related to the ultrapotassic rocks. It is not clear if the carbonatite and associated alkaline rocks came from a common magmatic parent, but they likely came from a common enriched source in the mantle. Analyses of K-rich rocks from Deep Spring Valley are given in Table 4.13.

4.5.11 Other Localities in U.S.A

Kalsilte-bearing ultrapotassic rocks have been described from a 350 km long belt, which trends northwest-southeast, and stretches from the Darby Mountains of the Eastern Seaward peninsula to Kobuk-Selawik low lands. The potassic rock complex was emplaced during middle Cretaceous, 97–110 Ma ago. The rocks predominantly belong to peraluminous K-rich nepheline syenite series. The belt also includes minor amount of ultrapotassic rocks similar in mineralogy and chemistry to potassic rocks of Highwood Mountains of Montana. Some of the rocks are characterized by high concentration of kalsilite.

4.6 Potassium-Rich Silica–Under Saturated Rocks from Brazil

The Alto Paranaiba Igneous Province (Fig. 4.17) is the most important potassium-rich petrographic province that occurs around the margin of the Ordovician to Cretaceous sedimentary Parana Basin in Brazil and Paraguay (Gibson et al. 1995). The Alto Paranaiba Igneous Province comprises wide variety of igneous rocks in the form of dikes, pipes, plugs, diatremes, lava flows, pyroclastic deposits and plutonic

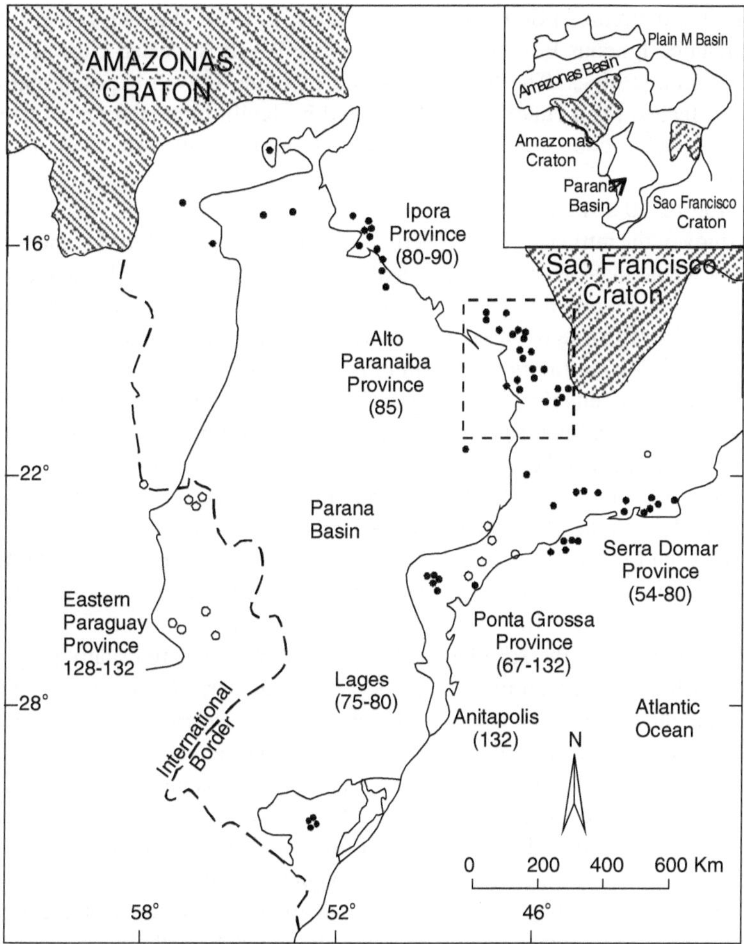

Fig. 4.17 The Alto Paranaiba Igneous Province, Brazil (after Gibson et al. 1994)

complexes. According to them the most voluminous surface expression of K-rich mafic volcanic rocks in the province are the lavas and tuffs of the Mata da Corda formation. Gibson et al. noted the occurrences of vents and crater facies rocks of these lavas at Serra do Bueno and Presidente Olegario. The lavas and tuffs outcrop throughout the petrographic province from Presidente Olegario in the north to Sao Gotardo and possibly further south to Oliveira, covering an area of more than 4,500 km^2. They overlie the lower Cretaceous Areado formation and attain a thickness of more than 100 m in the Sao Gotardo-Quintinos-Carmo do Paranaiba region.

The locality around Presidente Olegario (18°25′S, 46°26′W) was probably the site of numerous volcanic centres from which thick sequences of lavas and tuffs were erupted. The intrusion of Mata do Lenco contains xenoliths of glimmerite, harzburgite and dunite.

There is a diatreme at Serra do Bueno near the town of Bambui to the south of Alto Paranaiba Igneous Province (19°43′S, 46°03′W). It is located close to several K-rich ultramafic pipes. The Serra do Bueno diatreme is well exposed and comprises two facies, (1) a crater facies dominated by lapilli tuffs, and (2) a magmatic hypabyssal facies consisting of kamafugitic (katungite, mafurite and ugandite) intrusions. The $^{40}Ar/^{39}Ar$ analyses of olivine phenocrysts from the intrusion yielded an isochron of 90 ± 4 Ma.

The study of the intrusive rocks of Serra do Bueno (Gibson et al. 1994) and Corrego Varjao show that many rocks contain leucite, which is readily altered to analcite. The alteration of leucite possibly perturbed the isotopic system, and hence age determination of rocks of Corrego-Varjao yielded two ages, one determined from a melilite and other from a kamufugitic rocks, and the ages were found to be 77.8 ± 1.1 and 57.3 ± 0.6 Ma, respectively. According to Gibson et al. (1995) the magmatism in the entire Alto Paranaiba Igneous Province might have taken place 80–90 Ma ago.

There is now a general consensus among petrologists from Brazil that in terms of their constituent mineral assemblages and mineral chemistry the extrusive and intrusive rocks of Alto Paranaiba Igneous Province are similar to the kamafugitic family described from East Africa by Holmes (1950). However some of the K-rich ultramafic intrusives from Tres Ranchos, Indaia I and Limeria I may be termed as kimberlite.

The K-rich lavas of the Alto Paranaiba Igneous Province are usually porphyritic and have a fine-grained groundmass. They are characterized by olivine, clinopyroxene, ilmenite, perovskite and phlogopite. The accessory phases included K-richterite, magnetite, leucite, kalsilite and glass. On the basis of their phenocryst content they may be described as olivine- and clinopyroxene-phyric rocks.

The olivine-phyric lava at Presidente Olegario is composed of phenocrystal olivine (up to 2 mm) and phlogopite (<1 mm, set in a groundmass of olivine, diopside, phlogopite, richterite, perovskite, magnetite and glass).

In Carmo do Paranaiba, the K-rich lavas contain a large quantity of phenocrystal clinopyroxene, phlogopite, apatite, and leucite together with microphenocrystal ilmenite and perovskite set in a fine-grained matrix of clinopyroxene, analcitized leucite, magnetite, phlogopite, melilite and kalsilite. The pyroxene and phlogopite phenocrysts are 1 cm in length and display compositional zoning.

According to Gibson et al. (1995), the K-rich lavas and hypabyssal rocks display a continuous variation in MgO and FeO content from 11 to 16 wt%, and the mg-number varies between 39 and 87. The lavas in general have high MgO content (<20 wt%), but quite often between 18 and 20 wt%. The K_2O/Na_2O ratio varies from 1 to 16, the rocks are therefore, potassic to ultrapotassic and mafic to ultramafic. The high CaO (11–18 wt%), low SiO_2 (45–32 wt%) and low Al_2O_3 (5–8 wt%) contents resemble kamafugites from East Africa.

The Mata do Lenco intrusion is the most extreme member of the group (K_2O/Na_2O ratio = 8.1), and may be described as perpotassic (K_2O/Al_2O_3 = 0.77) and peralkaline ($K_2O + Na_2O/Al_2O_3$ = 0.86). The SiO_2 content is low (39 wt%) and MgO content (19 wt%) is high. Chemistry of kamafugitic rocks from Alto Paranaiba Igneous Province from Brazil is given in Table 4.14.

Table 4.14 Analyses of K-Rich rocks of Alto-Paranaiba Igneous Province, Brazil (Gibson et al. 1995)

	SiO_2	TiO_2	Al_2O_3	Fe_2O_3	MnO	MgO	CaO	Na_2O	K_2O	P_2O_5	LOI	Total
1	39.20	6.33	5.25	14.71	0.20	17.35	11.56	0.68	2.19	0.57	3.96	98.0
2	40.87	6.63	5.75	15.02	0.22	16.12	11.83	0.58	1.39	0.33	6.06	98.7
3	41.32	3.30	7.06	10.60	0.19	15.18	13.32	1.74	3.44	1.37	4.06	97.5
4	40.49	4.39	6.58	13.71	0.20	15.69	12.39	1.45	2.60	1.39	3.88	98.9
5	39.58	6.18	5.63	14.14	0.20	19.54	9.81	1.09	2.19	0.21	4.57	98.6
6	42.60	3.27	6.96	10.69	0.19	15.07	14.61	1.12	2.52	1.52	4.24	98.6
7	41.42	3.31	7.04	10.64	0.19	15.35	13.34	1.90	3.43	1.38	3.85	98.0
8	39.37	6.38	5.37	14.83	0.21	17.16	11.25	0.70	1.84	0.57	3.91	97.7
9	36.32	6.25	6.47	17.53	0.27	13.47	12.70	0.52	1.90	2.11	9.59	97.5

1–3 Kamafugitic lava from Presidente Olegario, Alto Paranaiba Igneous Province
4 Kamafugitic lava from Serra do Bueno, Alto Paranaiba Igneous Province
5 Kamafugitic lava from Corrego Varjao, Alto Paranaiba Igneous Province
6 Kamafugitic lava from Limeria II, Alto Paranaiba Igneous Province
7 Kamafugitic lava from Indaia II, Alto Paranaiba Igneous Province
8 Kamafugitic lava from Presidente Olegario, Alto Paranaiba Igneous Province
9 Kamafugitic lava from Camo do Paranaiba, Alto Paranaiba Igneous Province

Araujo et al. (2001) studied K-rich mefic rocks, of the Alto Paranaiba Province from southwestern Minas Gerais, Brazil. They not only studied petrography and mineralogy of the whole-rock but also determined isotope geochemistry with the objective of better understanding the Cretaceous magmatism and characteristics of the magma sources. Because of the variety and complexity of lithotypes examined in this research and the paucity of detailed studies of these Brazilian rocks in the literature, this study also attempted to establish parameters that allow for a clear distinction between kimberlite and kamafugite. They described fifty-two occurrences, and classified the rocks as kimberlites and kamafugites. Among the kamafugites, both ugandite (characterized by the presence of leucite) and mafurite (with kalsilite) end members have been characterized. Mineral compositions were found to be efficient in distinguishing between kimberlites, mafurites and ugandites occurring in the Alto Paranaiba petrographic province, primarily on the basis of phlogopite composition. The Re–Os isotope systematics permitted a better understanding related to the sublithospheric mantle source and associated magmatism in that region. Kimberlites, mafurites and ugandites have different Os^{187}/Os^{188} ratios (0.117–0.129, 0.127–0.145 and 0.142–0.147, respectively). The Rb–Sr and Sm–Nd isotope systematics failed to indicate first-order differences between kamafugites and kimberlites, whilst Pb^{206}/Pb^{204} ratios for the kimberlites are higher than those for the other rock types. Kimberlite and kamafugite isotopic compositions appear to be related, to the mixture of at least two dominant mantle components: one with an isotopic signature similar to that of lithospheric peridotites, i.e., with Os^{187}/Os^{188} ratios of the order of 0.118, similar to those observed in mantle-derived xenoliths entrained in kimberlites intruded in the Kaapvaal, Wyoming, and Siberian cratons; another with higher Os^{187}/Os^{188} ratios of the order of 0.135, within the range of ratios reported for pyroxenite veins in alpine-type peridotites and ocean island basalts. Different melting depths of heterogeneous lithospheric sources by a mantle plume are suggested to explain the isotopic characteristics of the Alto Paranaiba Alkaline Province rocks.

4.7 Silica-Undersaturated Potassic Lavas from Canada

4.7.1 Kirkland Lake, Ontario

Alkaline lavas and tuffs have been described from the Timiskaming Group of the Abitibi volcanic belt, which is one of the largest greenstone belts of the world (Cooke and Moorhouse 1969; Basu et al. 1984). Cooke and Moorhouse (1969) described four volcanic sequences, each separated by a considerable thickness of sedimentary rocks, each containing lavas and pyroclastic rocks in the Kirkland Lake area (48°10′N, 80°00′W). The rock types include andesite, basalt, trachyte and 'leucite'-bearing rocks. Fresh leucites are not reported, but pseudoleucites (0.5–2 cm across) constituting 5–9.5 vol% of the lavas, have been described. Mineralogically these rocks can be classified as leucite tephrites, through trachytes to phonolites.

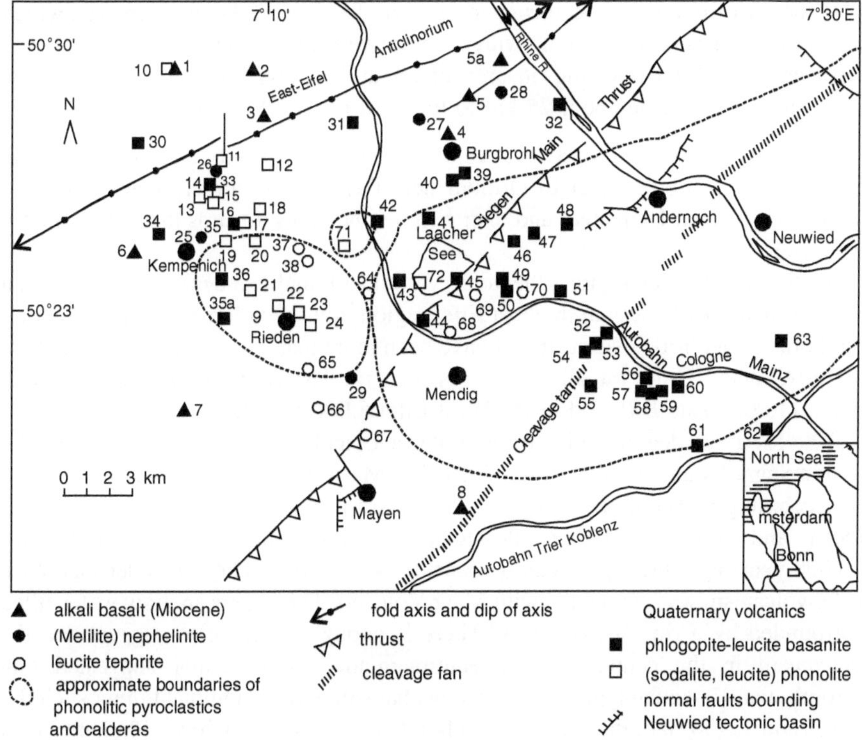

Fig. 4.18 The Quaternary volcanic fields of East Eifel (after Duda and Schminke 1978)

4.7.2 Spotted Fawn Creek, Yukon

In the west central part of Yukon pseudoleucite-bearing tinguaites have been described by Templeman-Kluit (1969) from Spotted Fawn Creek (64°22′N, 133° 40′W). These rocks are metaluminous having K_2O/Na_2O = 3.1 and K_2O/Al_2O_3 = 0.7.

4.8 The K-Rich Silica–Poor Lavas of Europe

4.8.1 Ultrapotassic Rocks of Germany

The Quaternary volcanic fields of Eifel (Duda and Schminke 1978) are located west of Rhine River and north of Mosel River (Fig. 4.18). According to Worner et al. (1987, also see Worner and Schminke 1984) and Ziegler (1992) there are two volcanic fields in this region:

(1) the East Eifel region occurring around Laacher See (30 km in diameter), which covers an area of 400 km^2 with 70 eruptive centres, and
(2) the West Eifel covering an area of about 500–600 km^2 with 200 eruptive centres, which lie about 25 km to the west of East Eifel region.

The latter is extended for a distance of 50 km along NW–SE direction from Bad Bertrich at the Mosel River to Ormont near the Belgian border. The centres are sometimes located 1 km apart and their number increases towards the centre of both the fields.

In the West Eifel region, the total volume of mafic magmas was less than 0.5 km^3 and may be less in the East Eifel region, but the volume of leucite-bearing phonolitic to tephritic magma might have been more than 2 km^3. According to them the K/Ar and ^{14}C ages available suggest that the last eruption in East Eifel took place 11,000 years B.P. and in the West Eifel 8,000 year B.P.

The K/Ar age determination of basanites and nephelinites lavas in the East Eifel region suggests an age of 0.35–0.69 and 2.7 Ma in the West Eifel region.

According to Schminke and Duda volcanism in this area started in early Pleistocene and culminated during Late Quaternary and as the last activity took place 10,000 years ago, the region can be considered as active. They consider that West Eifel volcanism was associated with Maar Volcano. The craters are now water-filled surrounded by a rim of pyroclasts. There is enough evidence to suggest that the volcanism in the centres was phreatomagmatic. They further suggested that Strombolian/Hawaiian type of eruption mechanism was predominant in most cinder cones and 10–30 % of the centres in both fields produced lava flows. Plinian type of eruption took place in Rieden, Wehr and Laacher See. Analyses of K-rich rocks from Germany are summarized in Table 4.15.

Basanites are dominant rock types in East Eifel, but nephelinites, melilite nephelinites and leucitites are common in West Eifel. In the latter area, basanitic lavas are reported from Bad Bertrich-Wartgesberg. It is observed that K_2O/Na_2O is ≈ 1 in West Eifel, but >1 in East Eifel. One interesting aspect of East Eifel volcanic field is the presence of many eruptive centres of intermediate and more evolved silica-undersaturated alkalic rocks. There is however, paucity of any such rocks in the West Eifel region. Peridotite nodules of large dimensions are common. Some of the xenoliths in the volcanoes from West Eifel might suggest that the magmas in the West Eifel region ascended relatively rapidly (Duda and Schminke 1978). Peridotitic xenoliths are small and rare in East Eifel volcanoes. In some of the East Eifel localities, they are found in such localities as Kempenich, Heider Kopf and Sattelberg (Duda and Schminke 1978). According to Lloyd and Bailey (1969, 1975) amphibole and phologopite-bearing pyroxenite xenoliths in Eifel show evidence of mantle metasomatism in the source rocks.

The volcanic centres of pumiceous tuffs originally covered an area, approximately 120 km long and 40 km wide area. Highly potassic rocks of the East Eifel region occur as massive or vesicular lava flow, associated with tuffs and scoria, and lapilli-filled layers, often containing bombs with chilled margins. Leucitic rocks occur at Steinrausch (near Hill Shein), Erntsberg (near Hiterweiln), Hardt (near Mahren),

Table 4.15 Analyses of K-Rich rocks from Laacher See (Germany)

	SiO_2	TiO_2	Al_2O_3	Fe_2O_3	FeO	MnO	MgO	CaO	Na_2O	K_2O	P_2O_5	H_2O^+	CO_2	Total
1	39.3	2.9	14.0	6.3	4.3	0.2	8.9	14.0	3.9	4.4	0.7	0.6	0.4	99.9
2	40.7	2.7	13.9	4.7	5.2	0.2	8.7	13.2	4.3	4.3	0.7	0.4	0.1	99.1
3	42.4	2.7	13.6	3.7	6.4	0.2	10.8	12.4	2.6	3.1	0.5	0.7	0.1	99.2
4	42.6	2.8	13.5	3.6	6.8	0.2	9.5	12.9	3.0	3.2	0.5	0.5	–	99.1
5	43.8	2.7	13.5	4.3	6.1	0.2	10.7	11.8	2.9	3.4	0.5	0.4	0.1	100.4
6	43.8	2.7	14.2	5.6	4.9	0.2	9.3	11.7	3.1	3.3	0.5	0.7	0.1	100.1

1–2 Leucite-bearing nephelinite from Laacher See, Germany (Duda and Schminke 1978)
3–6 Leucite basanite from Laacher See, Germany (Duda and Schminke 1978)

Dockweil, Dockweiler, Kyllerkopf (lower lava), Gaussberg (near Waldsdorf), Mosenberg, Dohnberg, and Alter Vos (near Berlinger).

4.8.2 Tertiary and Quaternary Magmatism in Massif Central France

Wilson and Downes (1991) discussed Tertiary and Quaternary volcanism in Massif Central, France (Fig. 4.19). Primitive mafic lavas of this region have SiO_2 contents ranging between 39 and 46 wt%. The other elemental abundance is as follows: MgO > 7 wt%, Ni > 150 ppm and Cr > 250 ppm. Alkali basalts and basanites predominate and are the parental magmas of two minor differentiation series [alkali basalt-trachyandesite-trachyte-rhyolite and basanite-tephrite-phonolite

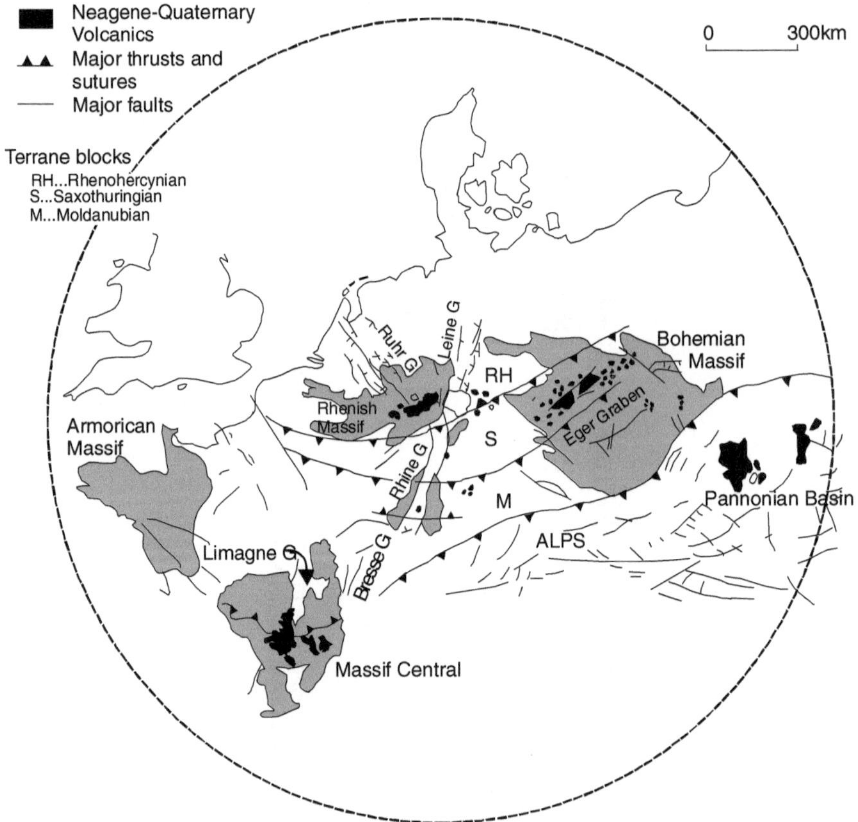

Fig. 4.19 Sketch Map of Western and Central European Volcanic Complexes (after Wilson and Downes 1991)

Table 4.16 Analyses of K-Rich rocks from Massif Central France, Austria and different localities of Italy

	SiO_2	TiO_2	ZrO_2	Al_2O_3	Fe_2O_3	FeO	MnO	MgO	CaO	SrO
1	55.00	0.59	–	16.80	2.60	3.00	0.13	3.60	6.20	–
2	53.10	0.70	–	17.90	3.60	2.80	0.13	3.30	6.10	–
3	46.32	1.16	–	19.59	8.49	–	0.16	3.56	10.02	–
4	40.50	2.00	–	13.02	11.14	–	0.25	3.61	15.6	–
5	57.50	0.56	–	18.00	2.32	1.63	0.10	0.99	3.17	–
6	52.80	0.74	–	17.70	3.51	3.28	0.14	2.23	7.08	–
7	60.16	0.70	–	20.29	3.20	–	0.31	0.38	1.54	–
8	55.21	1.04	–	19.59	5.80	–	0.12	2.13	5.13	–
9	48.24	0.76	–	15.11	3.29	4.12	0.14	7.23	11.53	–
10	48.35	0.77	–	16.35	4.35	3.40	0.15	5.49	10.11	–
11	46.82	0.74	0.05	16.21	5.54	2.34	0.15	5.35	11.19	0.21
12	46.34	0.76	0.06	18.06	3.80	4.06	0.17	3.38	9.66	0.28
13	53.88	0.70	–	17.40	3.63	4.26	0.13	2.28	5.60	–
14	54.32	1.06	–	17.66	3.07	3.88	0.14	2.05	5.18	–
15	52.93	0.57	–	18.79	3.30	1.81	0.12	1.85	5.44	–
16	53.85	0.52	–	19.22	2.30	2.16	0.11	1.78	4.77	–
17	57.79	1.51	–	11.79	2.24	3.12	0.08	8.23	3.46	–
18	58.01	1.50	–	11.70	2.29	3.07	0.09	8.27	3.53	–
19	42.71	3.60	–	9.09	11.57	–	0.15	14.33	11.23	–
20	43.61	3.62	–	8.80	11.08	–	0.15	13.44	11.45	–

(continued)

Table 4.16 (continued)

	SiO_2	TiO_2	ZrO_2	Al_2O_3	Fe_2O_3	FeO	MnO	MgO	CaO	SrO
21	38.25	2.52	–	13.64	–	7.89	0.02	19.55	1.63	–
22	39.24	2.80	–	13.69	–	8.64	0.03	19.96	0.58	–

	BaO	Na_2O	K_2O	P_2O_5	H_2O	CO_2	F	LOI	ΣREE	Total
1	–	1.50	8.40	0.45	1.48	0.12	–	–	0.51	100.40
2	–	2.40	6.60	0.48	2.10	0.02	–	–	0.39	99.62
3	–	4.06	4.70	1.07	–	–	–	1.11	–	103.20
4	–	3.96	4.86	1.26	–	–	–	4.42	–	100.60
5	–	2.67	9.51	0.23	3.13		–	–	–	99.81
6	–	4.02	5.17	0.43	2.76		–	–	–	99.86
7	–	5.74	7.01	0.10	–	–	–	0.36	–	99.79
8	–	4.63	5.37	0.43	–	–	–	0.64	–	100.10
9	–	1.07	6.99	0.53	–	–	–	1.17	–	100.20
10	–	1.20	8.53	0.49	–	–	–	0.82	–	100.00
11	0.18	1.85	7.23	0.46	–	–	–	2.25	–	100.60
12	0.22	2.40	8.44	0.50	–	–	–	1.32	–	99.45
13	–	3.11	6.70	0.50	–	–	–	1.09	–	99.31
14	–	4.38	6.02	0.49	–	–	–	1.08	–	99.33
15	–	4.47	5.26	0.21	–	–	–	4.48	–	99.23
16	–	3.81	8.07	0.20	–	–	–	3.62	–	100.40
17	–	1.31	8.06	0.85	–	–	–	1.55	–	99.99

(continued)

Table 4.16 (continued)

	BaO	Na_2O	K_2O	P_2O_5	H_2O	CO_2	F	LOI	ΣREE	Total
18	–	1.44	7.95	0.79	–	–	–	1.36	–	100.00
19	–	2.57	3.16	0.75	–	–	–	–	–	99.16
20	–	2.38	3.66	0.82	–	–	–	–	–	99.01
21	–	0.36	9.13	1.20	1.97	0.61	2.90	–	–	99.67
22	–	0.45	8.93	0.40	2.28	0.39	2.45	–	–	100.00

1–2 Analyses of leucite-bearing trachyte from Vico Volcanic area Italy (Cundari and Mattias 1974)
3 Analyses of phonotephrite from Mt. Vulture (Melluso et al. 1995)
4 Analyses of melilitite from Mt. Vulture (Melluso et al. 1995)
5–6 Phonolite (no. 5) and tephriphonolite (no. 6) of Letera Caldera, Italy (Turbeville 1993)
7–8 Ischia Island, Italy (Di Girolamo et al. 1995)
9–10 Leucite-tephrytes, Sabatini district (Conticelli 1989)
11–12 Tephritic leucitite, from Vulsini Latiun, Italy (Barton et al. 1982)
13–14 Trachyte scorias from Montagna Speccata volcano, Phlegrean Fields (Di Girolamo et al. 1984)
15–16 Brown leucitic tuff from Roccamonfina, Italy (Lhur and Giannetti 1987)
17–18 Orciatico Volcanics, South Tuscany (Conticelli et al. 1992)
19–20 Analyses of Leucite melanephelinites from Massif Central France (Wilson and Downes 1991)
21–22 Major element data for glimmerites from Austria (Becker et al. 1999)

(Downes 1984, 1987a; Briot 1988; and Briot et al. 1991)]. Rare nephelinites also occur but have markedly different geochemical characteristics from the basanites and alkali basalts and constitute a separate group, not associated with differentiated lavas. The nephelinites vary widely in their K_2O/Na_2O ratio ranging from sodic melilite nephelinites ($K_2O/Na_2O < 1$) to potassic leucite mela-nephelinites ($K_2O/Na_2O > 1$). Representative analyses of the main mafic lava types are presented in Table 4.16.

Leucite mela-nephelinites have K_2O contents higher than those of normal basanites and alkali basanites on the total alkali-silica (TAS) diagram of Le Bas et al. (1986). In contrast, the melilite nephelinites have low total alkali contents. Basanites plot on the TAS diagram between 40 and 46 wt% SiO_2, and total alkali content varies between 5.0 and 6.6 wt%, whereas alkali basalts plot between 42 and 48 wt% SiO_2 and 4 and 6 wt% total alkalis.

Concentration of TiO_2, P_2O_5, K_2O and Na_2O is generally higher in the basanites than in the alkali basalts, although Al_2O_3 and MgO are similar. The nephelinites, however, have significantly lower Al_2O_3 contents (9–10 wt%) compared with basanites and alkali basalts (12–15 wt%). The K_2O/Na_2O ratios fall in the range of 0.3–0.7 for typical basanites, alkali basalts, and melilite nephelinites, whereas the leucite nephelinites fall in the range of 1.2 and 1.7. The Massif Central is a region of uplifted and eroded Hercynian metamorphic basement, composed mainly of granitoids, schists, and amphibolites. The major phase of rifting (Limagne graben) occurred during the Oligocene, and was associated with regional uplift. Minor volcanic activity between 0.5 and 35 Ma pre-dated graben formation and involved the eruption of small volumes of under-saturated alkalic lavas (mostly nephelinites) along basement faults (Lucazeau et al. 1984). Volcanic activity decreased considerably during the main phase of graben formation and then resumed during 20 to ~0.04 Ma ago (early Miocene to late Quaternary) with major phases at 7–4 Ma (Pliocene) and 4–2 Ma (Viilafranchian). Within the Massif Central there are many individual volcanic provinces, which include, Cantal (11–3 Ma), Mt. Dore (4.3–0.4 Ma), Aubrac (8–6 Ma), Deves (5.4–0.8 Ma), Velay (10–1 Ma) and Chaine des Puys (50–4 Ka). Primitive mafic magmas are predominantly alkali basalts and basanites although nephelinites are present in a few localities (Downes 1987). The nephelinites include both sodic (melilite nephelinites) and potassic (leucite mela-nephelinites) types. There may be systematic difference in the Na_2O/K_2O ratio of primitive magmas erupted in different volcanic areas but at present there are insufficient data to substantiate this hypothesis.

4.8.3 K-Rich Rocks from Lower Austria

In the southern Bohemian massif (central Europe) veins of medium to coarse-grained glimmerites occur in peridotite massifs and at the contact with large bodies of felsic granulites (Becker et al. 1999). Crystallization and metamorphism of the glimmerites and the granulites were coeval. The glimmerites contain phlogopite and minor amounts of apatite and graphite. These veins are ultrapotassic with high concentration of large ion lithophile elements, light rare earth elements (LREE) and such mantle compatible elements as Cr and Ni. There is enough evidence to suggest reaction of metasomatic agents with peridotites. Further more, glimmerites show strongly fractionated REE patterns, negative anomalies of Nb, Ta and Ti and with except for one example, very low Zr and Hf abundances. The absence of feldspars and their low Na_2O and CaO abundance rule out the possibility of crystallization of phlogopites from a melt. The felsic granulites have low Th, U and Cs abundance and high Rb/Cs ratio plementary to the glimmerites. Both rock types overlap in initial Sr–Nd isotopic composition and show negative Eu anomalies. According to Becker et al. (1999) the chemical evidence and high F abundance in prograde biotite inclusion in granulite garnets indicate that the glimmerites may have crystallized from a fluid, liberated during the high temperature-pressure breakdown of F-rich phlogopite in the felsic granulites. The fluid must have contained HF, H_2O and carbon species as major volatile components. The minor occurrence of mica-poor dolomite veins with glimmerites suggests that liquid immiscibility might have played a significant role in the formation of these rocks. Glimmerites probably represent potential precursors to potassic magmas in collision zones.

4.8.4 Potassic Volcanism in Italy

The potassic volcanism in Italy was widespread during Quaternary along the Tyrrhenian margin from the Vulsini to Vulture volcanic complex (Beccaluva et al. 1991, Fig. 4.20 a and b). The existence of calc-alkaline shoshonite and associated ultrapotassic Oligocene volcanic rocks have been extended further in the north-western Alps by Venturelli et al. (1984). According to these petrologists, volcanism in this region was related to extensive block-faulting and rifting during the period between Upper Miocene and Present on the western side of the Apennine chain. The continent-continent collision in the Mediterranean region took place during Eocene (Patacca and Scandone 1987) and Lower Miocene, and the volcanic activity of the Roman Province took place after the continent-continent collision. According to Beccaluva et al. (1991) the entire volcanic complex is located between the Neogene compressional front of the Apennine-Maghrebian nappes and the Tyrrhenian back-arc basin followed by Pliocene oceanic-type crust. They thought that on a regional scale the volcanism of the Roman Province can be treated as the northernly expression of the Quaternary orogenic activities between the Aeolian and Southern Tuscania. They further considered that although there was block faulting and rifting, the present day extensional tectonic setting is different from major continental rift such as the East African and Rhine graben systems.

The Campanion volcanism took place in the southern most part of a complex graben system extending for about 700 km along the Tyrrhenian margin of Italy from Tuscany to Campania. According to Beccaluva et al. (1991), in contrast to the Latium and Roccamonfina complexes in the Roman Province, Italy, there was

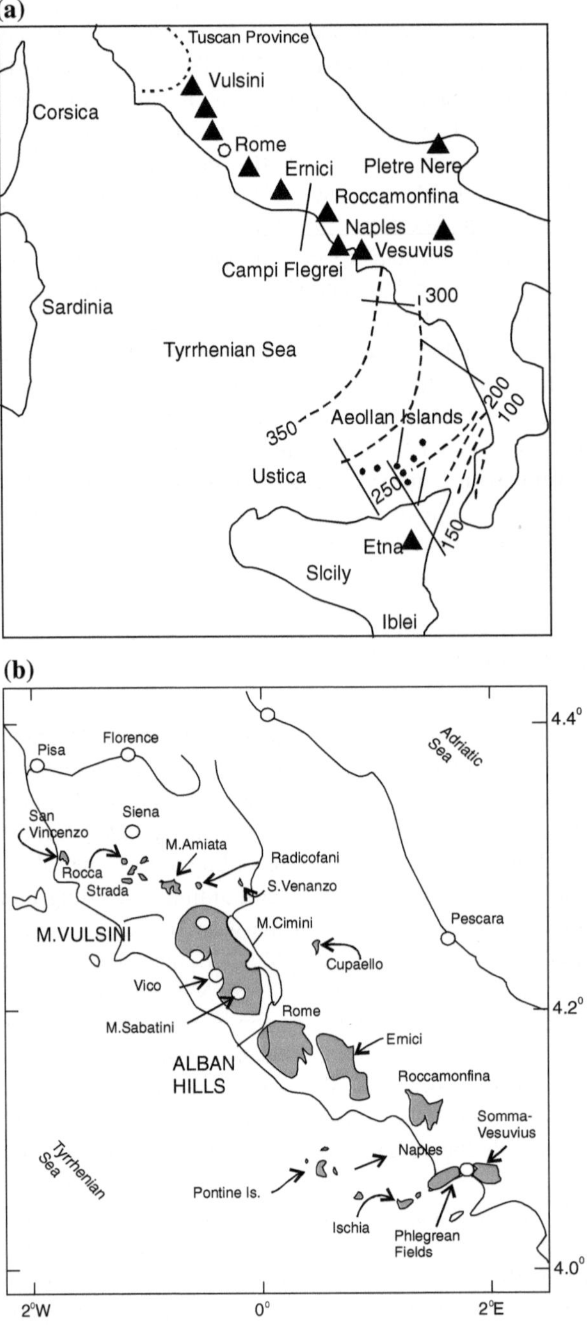

Fig. 4.20 **a** Location of different Italian volcanic complexes (*solid triangles*) known for K-rich-SiO_2 undersaturated rocks (after Beccaluva et al. 1991). **b** Distribution of volcanic complexes at San Venanzo,Vico, Cupaello, M-Amiata, Ernici, Phegrean Fields and other regions of central Italy (after Ferrara et al. 1986)

hiatus in volcanic activity 0.10–0.05 Ma ago. Volcanism around gulf of Naples (Vesuvius, the Phlegrean fields and Ischia) are still active. The Campanian volcanism has been essentially continuous from the Late Pliocene to the present. Volcanism of the Aeolian Arc region, south of Italy is also active at present.

Geochronological data show that Campanian volcanism was essentially continuous from Late Pliocene to the Present. The other rocks of the Campanian field occur at Ponza and Palmarola islands, and in the bore-holes occurring northeast of the Phlegrean Fields.

The ages of rocks of shoshonitic and leucite-bearing lavas of Ponza (1–0 Ma), Ventotene (0.8 Ma) and Roccamonfina (1.54 Ma) show that they are quite young. Lava flows of leucite tephrites are also noted in Ventotene and in submarine centre of Ischia channel (the Formiche di Vivara Volcano). Leucite-bearing mafic rocks are also observed in Phlegran Fields, Ischia, Procida, Vivara and Ventotene.

Although Foden and Varne (1980) thought that the rocks of Banda and Sunda of Indonesia and those of Aeolian Arc (Italy) are very similar in that they have shoshonitic, leucitic and calc-alkalic rocks closely associated together. Beccaluva et al. (1991) thought that the rocks of Roman Province are quite different from those of island arcs. An account of various ultrapotassic petrographic provinces of Italy is given below.

Conticelli et al. (2006) studied the petrologic geochemical and isotopic characteristics of potassic and ultrapotassic magmatism in central and southern Italy. They also made inferences on the genesis of these rocks and nature of mantle sources for these Miocene to Quaternary magmatic rocks, occurring along the Tyrrhenian border of the peninsular Italy. According to Conticelli et al. these potassic rocks can be divided into three different petrographic provinces.

They thought that the Tuscan Magmatic Province is the northernmost province, in which mantle-derived potassic and ultrapotassic, leucite-free volcanic rocks occur, prevailing over high potassium calc-alkaline rocks, and covering a time span of activity between 14.2 and 0.19 Ma. These rocks are silica-saturated to silica-oversaturated and range from high potassium calc-alkaline to ultrapotassic lamproite, through potassic and ultrapotassic shoshonitic series.

The Roman Magmatic Province extends from Northern Latium to the Umbrian and Campanian areas, arranged in a volcanic belt along the Tyrrhenian border of the Apennine chain. It is made up of rare shoshonitic and leucite-bearing rocks. Some rocks may contain melilite, and therefore, belong to the kamafugitic group. Minor high potassium calc-alkaline rocks are also found in the Pontine archipelago and in drill cores in the Campanian plain. Volcanism has been active from 0.7 to 0.1 Ma in the northern districts of the Latian area (i.e., Vulsinian, Vico, Sabatinian, Alban, Hernican and Auruncan) whereas, in the southernmost portion of the province, the Neapolitan district, shoshonitic and ultrapotassic magmatism are consistently younger than the Latian ones, ranging from 0.3 Ma to present. Historical eruptions in the Neapolitan district are indeed recorded in the Phlegrean Fields, Procida, Ischia and Vesuvius volcanoes.

The Lucanian Magmatic Province is the easternmost volcanic region characterized by rocks rich in both Na and K. Most of the rocks are hauyne- and leucite-bearing, and were erupted at Monte Vulture volcano between 0.6 and 0.1 Ma. Carbonatites have been described in the last phase of activity.

Minor amounts of K-rich rocks are also found in the Aeolian Arc, in the southern Tyrrhenian Sea. These rocks are intimately associated with calc-alkaline rocks at Vulcano,Vulcanello, and Stromboli.

According to them enrichment in K_2O and related incompatible trace elements is accompanied by strong to mildly fractionation of high field strength elements with respect to large ion lithophile elements. This can be attributed to the input of a crustal component into the mantle source of the magma prior to partial melting. Variations in trace element enrichment and isotope characteristics of the three magmatic provinces are thought to be the result of different metasomatic events and complex processes of partial melting of the mantle sources. Peculiar geochemical and isotopic characteristics of the Lucanian and Neapolitan regions are the result of different channelling of in-plate materials through lateral inflow from the foreland during the roll-back of the Ionian subduction. Metasomatism affected lithospheric mantle sources characterized by variable degrees of depletion.

They suggest that peculiar petrologic, geochemical and isotopic features of the mafic magmas are however consistent with a post-orogenic subduction-related geodynamic setting for the production of their parental magmas.

4.8.4.1 South Tuscany

The intrusion of igneous bodies of Orciatico (10°44′22″E, 43°26′10″N; Fig. 4.21a) and Montecatini Val di Cecina (10°46′08″ E, 43°23′30″N; Fig. 4.21b) are exposed a few km northwest of Volterra town (Conticelli et al. 1992; Poli et al. 1984). These two bodies are located on the western edge of vol d'Era graben. According to Conticelli et al. the emplacement of these two igneous bodies was controlled by the north–northwest–south–southeast-trending fault formed during post-tectonic extensional regime. The emplacement age of these two took place 4.1 Ma ago.

The Orciatico igneous body intruded into Pliocene-sediment in the form of a sill and laccolith fed by a narrow northwest–southeast-trending dyke (Fig. 4.21a). The outer part of the body has chilled margins. Mineralogically the rocks are characterized by the presence of abundant olivine and clinopyroxene with phlogopite, K-feldspar and oxides of microlites (An olivine-bearing minette). The inner part of the body has a holocrystalline texture with abundant olivine phenocrysts, phlogopite plates, minor amount of clinopyroxenes and rare K-richterites. Oxides and K-feldspar are the most abundant phases in the groundmass.

The igneous body at Montecatini Val di Cecina was intruded in the form of a shallow level plug. It is not only bordered but dissected by recent normal faults.

According to Conticelli et al. the Orciatico complex crystallized from mantle-derived magmas, and the most primitive rock types of this complex might not have gone through any differentiated process. These petrologists think that volcanics at Montecatini Val di Cecina were derived from the Orciatio magma by a process involving assimilation and fractional crystallization.

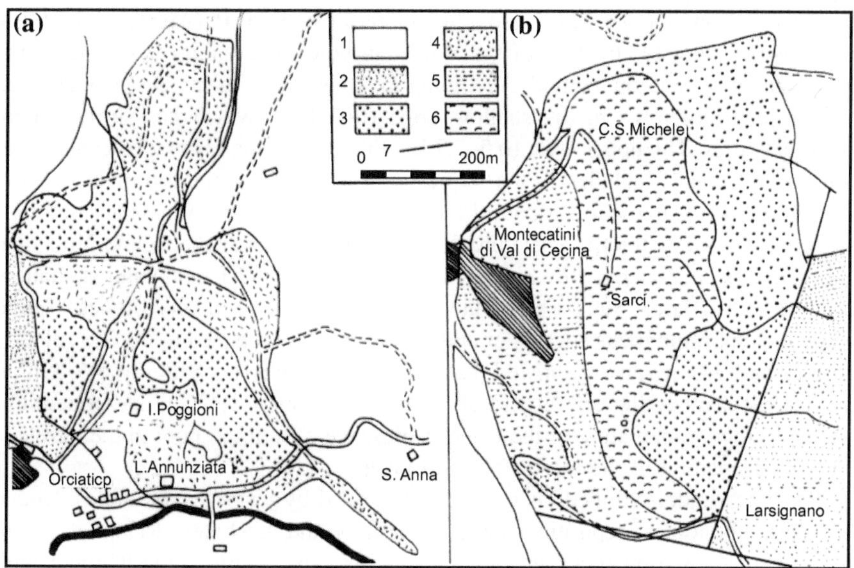

Fig. 4.21 Geological sketch map of the Orciatico (**a**) and Montecatini val di Cecina (**b**) igneous body, and surrounding rocks. *1* Pliocene and Quaternary sediments; *2* Miocene lacustrine and marine sediments; *3* thermally metamorphosed rocks ("Termantite"); *4* Orciatico orendite; *5* Montecatini val di Cecina igneous rock; *6* Montecatini val di Cecina igneous rock; *7* faults and inferred faults (after Conticelli et al. 1992)

The most primitive members of the ultrapostassic rocks are found at Tuscany. The Orciatico lavas have high $^{87}Sr/^{86}Sr$ and low $^{143}Nd/^{144}Nd$ isotopic ratios as well as very high incompatible elements. It suggests that the mantle source compositions were depleted in the basaltic components at first, but after sometimes, the enrichment of the LILE components took place. Gallo et al. (1984) described Quaternary volcanic centre at San Venanzo in the Perubia Povince and Cupaellow in the Rieti Province. These centres are located about 50–60 km east of the volcanic center of Latill Province. These are predominantly katungite and mafurite lavas. The rocks are very similar in geochemistry to the Toro-Ankole province in central and equatorial Africa. These rocks are however, slightly lower in TiO_2, Nb and Sr, often including kalsilite and melilite.

4.8.4.2 Vulsini Complex

The Vulsini complex (2280 km^2) constitutes the northernmost extension of the Roman comagmatic region (Holm et al. 1982). The volcanic rocks of the Vulsinian district include tuffs, ignimbrites, lava flows and cinder cones. Although the tuffs and ignimbrites dominate, the rocks are in general latitic, trachytic or phonolitic in composition (Sparks 1975). This petrographic province is characterized by highly potassic rocks, and includes numerous volcanic complexes most of which have been active from about 1 Ma ago to a few thousands year B.P. (Amato et al. 1991). The seismic velocity structure of the Vulsinian volcanic complex was determined using seismic refraction method. According to Amato et al. the area is characterized by complex structures in the shallow crust (0–7 km), which has strong lateral heterogeneities. The upper layer (0.5–2.7 km thick) corresponds to a volcanic cover and the upper part includes flysch sequence. The region to the top of the middle layer corresponds to the lower part of the flysch unit and to the Mesozoic carbonate sequence (2–5 km thick). The third layer extends beneath the whole Vulsinian region and is characterized by high P-wave velocity (6.7–7.1 km/s) and at shallow depth, it is 5–7 km/s. Such a high velocity is not related to the occurrence of high grade metamorphic rocks such as granulites and metamorphosed tuffs. The velocity profiles and 3-D inversion showed that there is a velocity decrease by 5 % within the deepest layer of the central part of the volcanic complex. They interpreted it to be due to high temperature of the layer or presence of a liquid pool in the intrusive body or deepening of the top of the third layer.

The silica-undersaturated lavas are characterized by leucitites, basanites, leucite tephrites, phonolites, tephritic leucite phonolites and trachytes. Relatively more silica-saturated K-rich rocks include shoshonites, shoshonitic andesites and high K-trachytes, occurring at Latera Caldera (Schneider 1965). Neither mantle xenoliths nor cumulate blocks are reported from these rocks. Blocks of well crystallized ejecta containing clinopyroxene and leucite phenocrysts (present in almost equal volume percentage) have been described. Most of the lavas are porphyritic with 1–15 vol% phenocrysts included in a matrix of clinopyroxene, leucite, plagioclase, sanidine, dark phlogopite, minor amount of ore minerals with or without olivine.

The K-rich undersaturated rocks are intermediate to metaluminous. In contrast to perpotassic rocks of East African Rift Valley, Leucite Hills and West Kimberly, these rocks have high K_2O/Na_2O ratio (2–8), Al_2O_3 (10.67–20 wt%), moderate amount of total iron oxide (3.60–8.64 wt%), and low TiO_2 (1.72–2.67 wt%) contents. The degree of undersaturation is characterized by the presence of 20–40 vol% leucite and nepheline in the most basic lavas, and 0–10 % nepheline and leucite in phonolites and trachytes. Diopside, leucite, anorthite and nepheline constitute more than 80 vol% of the norm.

Barton et al. (1982) studied clinopyroxene zoning in the lavas of Vulsini Latium in Italy. They studied two tephritic leucitites, two leucite phonolites and one trachyte. They noted three compositionally distinct types of pyroxenes:

1. Salite (forming resorbed pyroxene phenocrysts),
2. Diopside (forming anhedral cores in pyroxene phenocrysts), and
3. Clinopyroxenes of intermediate composition between (a) and (b).

According to Barton et al. (1982) the compositional zoning of pyroxenes may be related to mixing of two magmas both of tephritic leucitite in composition, one relatively Fe-rich with salite phenocrysts and evolved, and the other more primitive (Mg-rich) having diopside phenocrysts. Nappi et al. (1991) described evidence of incremental growth in the Vulsinian calderas. On the basis of volcanological field studies, drilling and gravity anomaly data, they recognized four volcanic activities in the Vulsinian volcanic district. They are as follows (oldest to youngest):

(1) Paleo-Bolsena,
(2) Bolsena,
(3) Montefiascone and
(4) Latera.

The third and fourth one were evolved contemporaneously. The volcanic activities range between 580,000 and 127,000 years B.P. (Fornaseri 1985; Santi 1990). According to Nappi et al., the volcanic complex of Vulsini district is characterized by one or more eruptive cycles, which may be summarized as follows:

(a) Initial phase involving eruption of mainly effusive leucite-bearing lava flows associated with Strombolian type activity,
(b) Explosive activity of Plinian type near the intersection of major regional faults, and
(c) Final phase, depicted by volcanic collapses associated with hydro-magmatic and/or magmatic activities.

Nappi et al. thought that not all calderas were generated by rare catastrophic and voluminous ash flow eruption, which is associated with caldera depression and involves following mechanism (Williams and McBirney 1979):

- Subsidence of coherent crustal blocks and rift fractures, and
- Collapse due to Krakatoa type of eruption resulting in the formation of caldera.

Bolsena caldera (Fig. 4.22) is older of the two calderas located south of the Siena-Radicofani graben. Nappi et al. (1991) suggested that down-sagging might have been the reason for the downwarped topography of the basin resulting in the formation of a downward sagging of the caldera. The topology of Bolsena caldera most likely have been controlled by the geometry of the Mesozoic-Cenozoic carbonate basement, which reflects a structure hinged to the southwest with large scale lowering to the northeast as shown by Bouger gravity anomaly in the eastern and northern part of the caldera. They think that the caldera may not be solely due to the collapse of magma chamber occurring during ignimbrite eruptions of Paleo Bolsena and Montefiascone volcanic complexes. This resulted due to the eruption of basal ignimbrites and Orvieto ignimbrites. These were largest explosive events during the activity of the three complexes. According to them the Bolsena caldera was produced by several phases of tectonic subsidence since the beginning of volcanic activity in the Vulsini district till Recent, resulting in the drowning of neolithic villages in the lake.

Nappi et al. consider that the Montefiascone caldera occurring within the main Bolsena caldera represents the youngest volcanic collapse due to an explosive activity of short duration and its topography is nicely preserved.

Ignimbrite eruption from northeast to southwest of caldera resulted in an incremental growth of Latera caldera, which is associated with piecemeal and

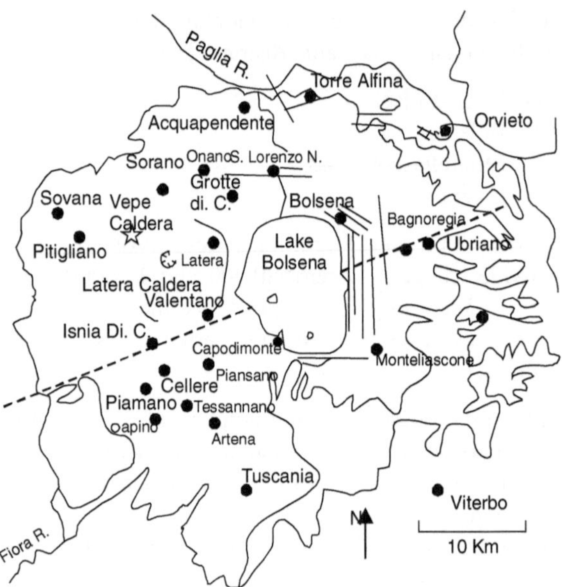

Fig. 4.22 The volcanic complex (*solid dots*) associated with Bolsena caldera showing adjacent fields of Latera caldera (after Turbeville 1993)

chaotic collapse. There was also subsidence in the northwestern sector. Deep drilling has shown the presence of volcanic breccia in the southwestern section of Latera caldera. There are four breccia layers (A, B, C and D) overlying earlier activities of Paleo Bolsena and Bolsena complexes. This area is interpreted as a source of oldest Latera ignimbrite. The Latera explosive activity shifted towards northeastern side, where ignimbrites designated as E and F were erupted. The most recent collapse of Latera Caldera affected the northwestern sector producing the Veple caldera showing subsidence due to a ring-fractured system. Analyses of K-rich rocks from Vulsini Complex are given in Table 4.16.

4.8.4.3 Vico Complex

The Vico volcanic complex (Fig. 4.20b) with six pyroclastic formations and lava flows occur south of the Vulsini complex. The lavas are characterized by low magnesia, iron and lime and relatively low SiO_2 contents, compared to most other volcanic complexes of Italy. The lavas are porphyritic comprising leucite, clinopyroxene, alkali feldspar and plagioclase with accessory mica and Fe–Ti oxides (up to a cm across). According to Cundari and Mattias (1974), the rock has glomeroporphyritic texture, and clinopyroxene ± plagioclase ± mica is widespread in the lavas. The rocks are mainly,

(1) Trachyte,
(2) Phonolite, and
(3) Tephrite and their variants.

From 1–3, there is a systematic increase in plagioclase/alkali feldspar ratio and increasing modal leucite towards tephritic types and total mafic constituent of the tephrites contain less than one third of the bulk rock.

The lavas erupted from the central vents are phonolitic leucite tephrites (e.g. upper flow of Mount Venere) and tephritic leucite phonolite (Mount Venere lower flows). Cundari and Mattias (1974) also reported the occurrence of leucite trachyte and tephritic leucite phonolite inside the caldera. All the above three lava types were also erupted through the vents at the peripheries. Analyses of some leucite-bearing rocks from the Vico complex are given in Table 4.16.

More recently, Perini et al. (2003) discussed about the evolution of magmas from the Vico region. According to them Vico volcano erupted potassic and ultrapotassic magmas, ranging from silica-saturated to silica-undersaturated types, in three distinct volcanic periods over the past 0.5 Ma. During Period I magma compositions changed from latite to trachyte and rhyolite, with minor phono-tephrites. During periods II and III, there were eruption of magmas primarily phono-tephrite to tephri-phonolite and phonolite; however, magmatic episodes involving leucite-free eruptives with latitic, trachytic and olivine latitic compositions also occurred. In Period II magmas [Sr^{87}/Sr^{86} (initial) = 0.71037–0.71115] were derived from a primitive tephrite parental magma. Modelling of phonolites with different modal plagioclase and stroncium (Sr) contents indicates that low-Sr phonolitic lavas were differentiated from tephri- phonolite by fractional crystallization of 7 % olivine + 27 % clinopyroxene + 54 % plagioclase + 10 % Fe–Ti oxides + 4 % apatite at low pressure, whereas high-Sr phonolitic lavas were generated by fractional crystallization at higher pressure. More differentiated phonolites were generated from the parental magma of the high-Sr phonolitic tephra by fractional crystallization of 10–29 % clinopyroxene + 12–15 % plagioclase + 44–67 % sanidine + 2–4 % phlogopite + 1–3 % apatite + 7–10 % Fe–Ti oxides. In contrast, leucite-bearing rocks of Period III [Sr^{87}/Sr^{86} (initial) = 0.70812–0.70948] were derived from a potassic trachybasalt by assimilation-fractional crystallization with 20–40 % of solid removed having r = 0.4–0.5 (where r is assimilation rate/crystallization rate) at different pressures. Silica-saturated magmas of Period II [Sr^{87}/Sr^{86} (initial) = 0.71044–0.71052] appear to have been generated from an olivine latite similar to some of the youngest erupted products. A primitive tephrite, a potassic trachybasalt and an olivine latite are inferred to be the parental magmas at Vico. These magmas were generated by partial melting of a veined lithospheric mantle sources with different vein-peridotite/wall-rock proportions.

4.8.4.4 Monti Sabatini

The Sabatini volcanic complex (Fig. 4.20b) includes a number of vents, which cluster around the lake-filled collapsed caldera of Bracciano located 30 km northwest of Rome. Apart from Lago de Bracciano, the other important caldera in this locality includes Lago de Martignano. The Bracciano caldera is located 30 km northwest of Rome. The volcanic history of the area started with eruption of tephritic ignimbrites and lavas, which preceded the eruption of Vico volcano. This was followed by another stage of volcanic activities, which coincided with the eruption of the Vico lavas. The third stage was associated with eruption of tephritic and leucitic lavas. The major volcanic activity of Sabatini took place after the completion of the eruption of the Vico complex, which is located adjacent to it.

Voluminous pumice-rich deposits were erupted from the Sabatini volcano, which is far more in volume than the volume of lava flows associated with the eruption of pyroclasts, which took place repeatedly, forming well stratified weakly cemented tuffaceous materials. Bulk chemistry and mineralogy of the Sabatini lavas show that they plot in the tephritic leucitite and tephritic phonolite field. According to Cundari (1979), tephritic leucitites are uncommon in Vico, but common in Sabatini. This may indicate that Sabatini lavas may constitute afterward extension of the Vico volcanic field.

The Sabatini leucitites do not contain feldspars, which are considered to be a rare feature in lavas occurring north of Rome, but are common in Alban Hills (Fornaseri et al. 1963). Another interesting feature of the Sabatini lavas is the occurrence of olivine, nepheline and haüyne. Chemical analyses of some leucitites from Sabatini are given in Table 4.16.

4.8.4.5 Alban Hills

Fornaseri et al. (1963) described the detailed geology of the rocks of Alban Hills (Fig. 4.20b). The area is characterized by different sequence of lava flows.

(1) Vallerano flow,
(2) Osa flow,
(3) Capo di Bove flow,
(4) Monte Falcone flow,
(5) Acqua Acetosa lava pile, and
(6) Santa Maria delle Mole lava flow.

There are also ejected lava blocks from Monte Cavo and Peperino and volcanic tuffs haremmana Inferiore. Aurisicchio et al. (1988) classified the lavas as leucite tephrites, phonotephrites, phonolites and foidite. According to Di Radicatti et al. (1981) the Villa Senni Tuff, which is an important pyroclastic unit of Alban hills, yielded an age of 338 ± 8 ka. However, according to Ambrosetti et al. (1972) the ages of the Alban Hills rocks vary from 43,000 to 700,000 years. The lavas from Alban Hills often include xenoliths and ejecta of biotite leucitite, magnetite leucitite, and biotite xenocrysts.

According to Aurisicchio et al. (1988) clinopyroxene crystals in the lavas of Alban Hills, occur as phenocrysts and microphenocrysts in the groundmass. The clinopyroxenes are the predominant phenocrystal phases in the post caldera samples. Aurisiccio et al. found coexistence of two core-rim evolutionary trends within the same rock:

(1) Diopside-salite, and
(2) Salite-diopside.

Gaeta et al. (2006) investigated chemical and isotopic compositions of clinopyroxene crystals from juvenile scoria clasts, lava flows, and hypabyssal magmatic ejecta representative of the whole eruptive history of the Alban Hills Volcanic District. The Alban Hills is a Quaternary ultra-potassic district that was emplaced into thick limestone units along the Tyrrhenian margin of Italy. Alban Hills volcanic products. Even the most differentiated rocks, are characterised by low SiO_2. They suggested that the low silica activity in evolving magmas can be ultimately due to a decarbonation process occurring at the magma/limestone interface. They proposed that, the differentiation process was initiated by crystallisation of clinopyroxene ± leucite ± apatite ± magnetite coupled with assimilation of a small amount of calcite and/or interaction with crustal CO_2. By combining age, chemical data, strontium and oxygen isotopic compositions, and REE content of clinopyroxene, they gave an insight into the evolution of primitive ultrapotassic magmas of the Alban Hills Volcanic District over an elapsed period of about 600 kyr.

Geochemical studies of clinopyroxene crystals, consistent with data coming from other Italian ultrapotassic magmas, indicate that Alban Hills primary magmas were generated from a metasomatized lithospheric mantle source. In addition, their study shows that the Sr^{87}/Sr^{86} and LREE/HREE of Alban Hills magmas continuously decreased during the 600 ± 35 ka time interval of the Alban Hills eruptive history, possibly reflecting the progressive depletion of the metasomatized mantle source of magmas.

According to them, there is no evidence of mixing process responsible for the coexistence of both diopside and salite in the same lava flow. Variation in the aqueous vapour pressure may be responsible for different crystallization sequences of the phenocrysts. Salitic pyroxenes either reacted with liquid or are replaced by olivine, phlogopite and Ti-magnetite. According to Aurisicchio et al. the predominant role of water in the evolution of Alban Hills magma is supported by high F content of the lavas erupted by the Alban Hills volcano. This possibly resulted in the enhancement of water solubility in the magma. There is common occurrence of mica in the groundmass of the rocks. The explosive character of the Alban Hills volcanism also indicate the presence of an aqueous vapour phase.

4.8.4.6 The Phlegrean Field

The Phlegrean Field (Fig. 4.20b, Di Girlamo et al. 1984) is constituted of a complex volcanic system that developed over the last 50,000 years (Beccaluva et al. 1990). There was first early submarine volcanic activity represented by eruption of tuffs, latititic and trachytic lavas. Subaereal volcanism is associated with pyroclastic products and lavas in different parts of the Phlegrean Fields. In the last phase there was ejection of coarse-grained brecciated material (Breccia Museo) containing a large variety of lava blocks including leucitic lavas and basement rocks, which have been hydrothermally altered. This breccia occurs on opposite sides of the 12 km wide Phlegrean depression at Torrefumo and Soccavo. These rocks have been regarded as a proximal facies and source rocks for Campanion ignimbrites, which are represented by large pyroclastic flow deposits that erupted 33,000 years ago. These pyroclastic deposits cover the entire Campanian plain. This pyroclastic unit probably erupted from a fissure system, judging from the west–northwest–east southeast mega-breccia alignment.

The Breccia Museo member (Melluso et al. 1995) was formed by explosive eruption that occurred in the southwest sector of the Phlegrean Field about 20 ka ago. The product of this eruption, range in composition from the trachytic to trachy-phonolitic lavas with K_2O content decreasing from 9.5 to 7 wt% and the Na_2O correspondingly increases from 2.6 to 7.2 wt% with differentiation (Nb from 23 to 122 ppm). The phenocrysts are mostly sanidine (Or_{88-63}) with subordinate plagioclase (An_{33-27}), clinopyroxene ($Ca_{47}Mg_{44}Fe_{09}$–$Ca_{46}Mg_{35}Fe_{19}$), biotite, titanomagnetite and apatite. The very recent post-caldera activity of the Phlegrean Field has been subdivided into three main volcanic phases. The first phase started with submarine volcanic activity (38,500–10,500 years B.P.) erupting tuffites with intercalations of trachyte-latite lavas. This was followed by caldera-filling with trachytic tuffs, which erupted from several trachytic pyroclastic vents, both in marine and subaereal environments. This is the Neopolitan yellow tuff phase. The second phase (10,500–3,000 years B.P.) was mostly subaereal with eruption of pyroclastic trachyte mainly from vents concentrated towards the west (Baia and Fondi di Baia) and in the central eastern sectors. Some shoshonitic and latitic lavas also erupted from the vents at Minopoli, Concola and Fondo Riccio. The last subaereal phase (4,500 years B.P.–1538 A.D.) resulted in the eruption of pyroclastic products and lava flows of trachytic and trachyphonolitic composition. The eruption took place from a number of vents, concentrated towards the caldera centre.

The spectrum of volcanic products in the Phlegrean Fields are petrologically distinct and part of a co-magmatic sequence ranging from trachybasalts to trachyphonolites, all appear to be related by a fractional crystallization process (Armienti et al. 1983). The REE data are in agreement with this general scheme.

The Phlegrean Fields volcanism was characterized by strong block faulting of the basement by a complex magma system at depths. According to Beccaluva et al. (1990), this magma system was fed by parental magmas fractionating independently in a partially open system. According to them, fractional crystallization process gave rise to distinct liquid lines of descent developed in a complex fracture system, where fractionating magmas are compositionally zoned upward and sometimes mixed with each other with mafic melts from depths.

4.8.4.7 Roccamonfina

Roccamonfina complex (Fig. 4.20a) lies 50 km NNW of Somma-Vesuvius along the western coast of Italy (Luhr and Giannetti 1987). The evolutionary history is long and petrologically complex. Giannetti (1979) divided the history of the volcano into two stages, which corresponds to two distinct magma series: the leucite-rich high K-series of stage 1, and the leucite-poor to leucite-free K-series of stage 2 (Appleton 1972). Luhr and Giannetti (1987) thought that stage 1 activity constitutes bulk of the volcanism associated with Roccamonfina. This is the period when the main cone was constructed from repeated lava, pyroclastic fall and pyroclastic eruption associated with mudflows. During this period the leucite-rich high K-series (mainly leucite tephrites up to 13 wt% CaO) to leucite phonolites (up to 2 wt% CaO) were erupted. The pyroclastic activity was associated with phonolitic lavas, which were predominant towards the end of stage 1 activity. Luhr and Giannetti (1987) considered that a major pyroclastic flow erupted near the end of stage 1, about 385 ka years ago. These tuffs are compositionally zoned and have been reported as Brown Leucitic Tuff. These pyroclastic flows have a minimum volume of 3–5 km^3. Luhr and Giannetti divided the Brown Tuff into three facies: white, brown and orange, on the basis of their colour, lithic content and matrix concentration. The pumices have crystal content of 9.9–0.6 vol%. In terms of vol% of minerals, they comprise, green salite > plagioclase > sanidine > biotite > titanomagnetite > analcite (after leucite) > apatite > pyrrhotite. Plagioclase is anorthite-rich (An_{85-95}) and sanidine (Or_{75-90}) has very low albite content. In the Brown Leucitic Tuff primary leucite has been completely analcitized. The rocks are usually characterized by diopside and forsterite-rich olivine (Fo_{83-92}). Diopside often grades to green salite. According to them, complex zonation of pyroxene may suggest their formation by mixing of two different types of magmas.

Except for Na_2O and K_2O the leucite-bearing tuffs have chemistry similar to the high K-series lavas of Roccamonfina. There is a nicely developed summit caldera at Roccamonfina. The diameter of the caldera is 5.5–6.5 km across. There is another smaller semi-circular caldera (gli Stagli), which touches the main caldera. The main caldera was probably formed before the eruption of Brown Leucitic Tuff. The gli Stagli caldera was formed after the eruption of the Brown Leucitic Tuff, which in turn is overlain by rocks of potassic series. There is also a 30 m thick sequence of leucite-bearing high K-series tephra above.

According to Luhr and Giannetti, the beginning of stage II activity of Roccamonfina is marked by the eruption of compositionally-zoned white trachytic tuff about 300,000 years ago. This was followed by a complex sequence of pyroclastic eruption in the northern part of the volcano resulting in the formation of "gli Stagli" caldera and a thick section comprising leucite latite, 'leucite basalt', and trachytic pyroclastic fall and flow deposit. Finally eruption of Yellow Trachyte Tuff took place in the Conca Campania and Vezzara. This was followed by further eruption of leucite basalt, leucite latite and trachybasalt. This eruption is referred to as Stage (II) eruption, the rock type of which is dominated by lavas of K-series. Except for 'leucite basalts' and leucite latites, the other rock types are free from leucite. Stage II lavas are depleted in K and related incompatible elements compared to Stage I lavas, which are basically high K-series rocks. Stage II rocks are characterized by lower $^{87}Sr/^{86}Sr$ and higher $^{143}Nd/^{144}Nd$ and $^{206}Pb^{207}Pb$ ratios. The final volcanic events at Roccamonfina involved deposition of flow units from the Campanian ignimbrites, a compositionally-zoned pyroclastic deposit, which erupted about

30,000 years ago in the Phlegrean Fields. This field lies about 50 km south–southeast of Roccamonfina.

This is one of the Plio-Pleistocene volcanic centres situated to the west of the Appenines. The rocks of this locality fall distinctly into two series (Appleton 1972): (1) a highly potassic series, consisting of nepheline– and leucite–normative lavas. (2) a potassium-poor series, consisting of nepheline or quartz–normative basalts, trachybasalts, and biotite augite latites. Series (1) is richer in K, P, Ba, Ce, Rb, Sr, Th, than Series (2). According to Appleton (1972) the highly potassic lavas of Roccamonfina can be further subdivided into the following groups:

(a) Clinopyroxene-microphyric lavas,
(b) Leucite-macrophyric lavas,
(c) Aphyric or microphyric lavas,
(d) Biotite-clinopyroxene-plagioclase(An_{72-79}) magnetite-phyric lavas, and
(e) Lavas containing sanidine as the main phenocrystal phase.

Group (c) lavas contain leucite, clinopyroxene, and plagioclase (An_{66-75}) as phenocrysts. Group (b) lavas contain phenocrysts of clinopyroxene, plagioclase and magnetite with leucite ranging up to 40 vol%. Sanidine forms only in those lavas, which contain more than 55 % SiO_2. Some of the leucite-macrophyric lavas contain glommero-aggregates of clinopyroxene, plagioclase and magnetite with or without apatite, sometimes surrounded by leucite phenocrysts indicating an early period of crystallization of the glommero-aggregates prior to that of leucite. Appleton concluded that an intermediate or high pressure process produced the parent magma with specific levels of enrichments in potassium and associated elements; and fractionation of biotite gabbro (which is found as nodules) from such a liquid at low pressure would produce a chemical variation towards salic derivatives.

Appleton suggested that the evolutionary history of Roccamonfina rocks has two parts. In the first phase there was a wide variation in the level of incompatible element enrichment on a series of rather similar primitive magmas (Cox et al. 1976). Later, due to crystal fractionation at low pressure a large variety of more salic derivatives were formed. In the later stage highly potassic magmas were evolved due to the fractionation of biotite pyroxenites followed by biotite gabbro. It was argued that K-deficient magmas were formed by the precipitation of olivine gabbro. Gianneti found that apatite and opaques are usually present along with the major phases (quoted in Cox et al. 1976). They argued that if isotopic variation in the erupted magmas is related to the early event associated with fractionation of incompatible elements in the primitive magmas, then correlation between isotopic composition and absolute concentration of incompatible elements should take into account variation related to low pressure fractionation at a later stage. If K, Rb, Sr, Ba and Zr content of the samples are correlated with $^{87}Sr/^{86}Sr$ ratios, then good positive correlation is obtained from most primitive volcanic series. They however, noted that Ti showed no correlation with respect to Sr isotopic ratio, suggesting that the most primitive lavas might have fractionated sufficient amount of titaniferous phases to eliminate any relationship that might have existed. They emphasized that although there is a good correlation between Sr isotopic ratio with other incompatible elements, evidence of modification of such element contents during low pressure fractionation has been observed.

According to Cox et al. (1976), if the incompatible element concentration is not related to low pressure fractionation then the correlation with $^{87}Sr/^{86}Sr$ ratio is possibly a primary magmatic feature, which was inherited from the parental liquids in equilibrium with mantle host rocks. They concluded that absolute incompatible element content obtained in the five most calcic samples is possibly not much different from their primary values. Even if the primary magmas were much more magnesian than those erupted (i.e. the magma lost 50 % of the ferromagnesian phases during its ascent to the surface), then the incompatible element content should have increased by a factor of two. According to Cox et al. in case of Roccamonfina, not all members were erupted and a compositional gap exists between the low-K and high-K series of lavas. It is however noted that the composition of the lavas of Monte Somma located about 50 km south of Roccamonfina fills the compositional gap between the low- and high-K series of Roccamonfina. The K, Rb, Sr, Ba and Zr contents of lavas of Roccamonfina have not been modified greatly from their primary magma values by fractionation. Thus, the Sr isotopic ratios and concentration of the above trace elements should provide ample evidence regarding the mantle source region. They thought that the Sr-isotope and trace element variations may reflect either (1) disequilibrium melting of a chemically homogeneous whole rock, or (2) melting of an inhomogeneous mantle source rock on a large scale with respect to Sr-isotopes and bulk composition. They argued that magma with highest Sr isotopic ratio could then be explained by selective incorporation of a phase by the liquid with high Rb/Sr ratio. Magma with lowest Sr isotopic ratio should be produced by a greater incorporation of a phase with low Rb/Sr ratio. High Sr isotopic ratio can be produced by disequilibrium melting of phlogopite-rich source. In this model high Sr isotopic ratio can be obtained if partial melting takes place. Melts however, formed by more advanced partial meltings have lower $^{87}Sr/^{86}Sr$ ratios and Sr content. This model however does not explain the linear correlation between Sr isotopic ratios and Sr contents. Likewise, the other linear correlation between Sr isotopic relation and K_2O, Rb, Sr and Ba are also difficult to be produced by disequilibrium melting models unless a small degree of partial fusion took place. Chemical compositions of some leucite-bearing rocks from this area are given in Table 4.16.

Contichelli et al. (2009) have studied the rocks of Roccamofina volcano The Roccamonfina volcano is characterized by two stages of volcanic activity that are separated by volcano-tectonic caldera collapses. Ultrapotassic leucite bearing rocks are confined to the pre-caldera stage and display geochemical characteristics similar to those of other volcanoes in the Roman Province. After the major sector collapse of the volcano, occurred at ca. 400 ka, shoshonitic rocks erupted from cinder cones and domes both within the caldera and on the external flanks of the pre-caldera Roccamonfina volcano. On the basis of new trace element and Sr–Nd–Pb isotope data, they show that the Roccamonfina shoshonitic rocks are distinct from shoshonites of the Northern Roman Province, but are very similar to those of the Neapolitan volcanoes. The last phases of volcanic activity erupted sub-alkaline magmas as enclaves in trachytic domes, and as lavas within the Monte Santa Croce dome.

Ultrapotassic rocks of the pre-caldera composite volcano are plagioclase-bearing leucitites characterized by high levels of incompatible trace elements with an orogenic signature having troughs at Ba, Ta, Nb, and Ti, and peaks at Cs, K, Th, U, and Pb. Initial values of $^{87}Sr/^{86}Sr$ range from 0.70926 to 0.70999, $^{143}Nd/^{144}Nd$ ranges from 0.51213 to 0.51217, while the lead isotope rations vary between 18.788 and 18.851 for $^{206}Pb/^{204}Pb$, 15.685–15.701 for $^{207}Pb/^{204}Pb$, and 39.048–39.076 for $^{208}Pb/^{204}Pb$. They observed that shoshonites show a similar pattern of trace element depletions and enrichments to the earlier ultrapotassic leucite-bearing rocks but have a larger degree of differentiation and lower concentrations of incompatible trace elements. On the other hand, shoshonitic rocks have Sr, Nd, and Pb isotopes consistently different than pre-caldera ultrapotassic leucite-bearing rocks. The $^{87}Sr/^{86}Sr$ ratio ranges from 0.70665 to 0.70745, $^{143}Nd/^{144}Nd$ varies from 0.51234 to 0.51238, $^{206}Pb/^{204}Pb$ ranges from 18.924 to 19.153, $^{207}Pb/^{204}Pb$ varies from 15.661 to 15.694, and $^{208}Pb/^{204}Pb$ ranges from 39.084 to 39.212. High-K calc-alkaline samples have intermediate isotopic values between ultrapotassic plagioclase leucitites and shoshonites, but the lowest levels of incompatible trace element contents. They have argued that ultrapotassic magmas were generated in a modified lithospheric mantle after crustal-derived metasomatism. Interaction between the metasomatic agent and lithospheric upper mantle produced a low-melting point metasomatized veined network. They think that partial melting of the veins alone produced pre-caldera leucite-bearing ultrapotassic magmas. It was possibly triggered by either post-collisional isotherms relaxation or increasing temperature due to increasing heat flow through slab tears. Shoshonitic magmas were generated by further melting, at higher temperature, of the same metasomatic assemblage with addition 10–20 % of OIB-like astenospheric mantle material. They suggest that addition of astenospheric upper mantle material from foreland mantle, flowing through slab tearing after collision was achieved.

4.8.4.8 Somma-Vesuvius

The Somma-Vesuvius complex (Cundari and Sulviulo 1987) is marked by the presence of the famous active volcano Vesuvius, which is half encircled by the breached caldera called Monte Somma (Fig. 4.23). The first eruption of the volcano took place in 79 A.D., followed by a long period of quiescence. It erupted again in 1,631, and then after a gap of 130 years it erupted in 1879. The last two eruptions were observed in 1906 and 1944 (Fig. 4.24).

The Somma-Vesuvius volcanic complex may be considered as a classic locality for leucite-bearing assemblages. Rittmann (1933) described evolution of Vesbian complex and considered that temporally the trachytes were the earliest, followed by Somma-Vesuvius suite of leucite tephrites, tephritic leucitites (earlier termed as vesbites) and their phonolitic variants.

The trachytic rocks underlie the pyroclastic deposits, which erupted from the adjoining Phlegrean complex, preceded by eruptive cycles of the Somma activity.

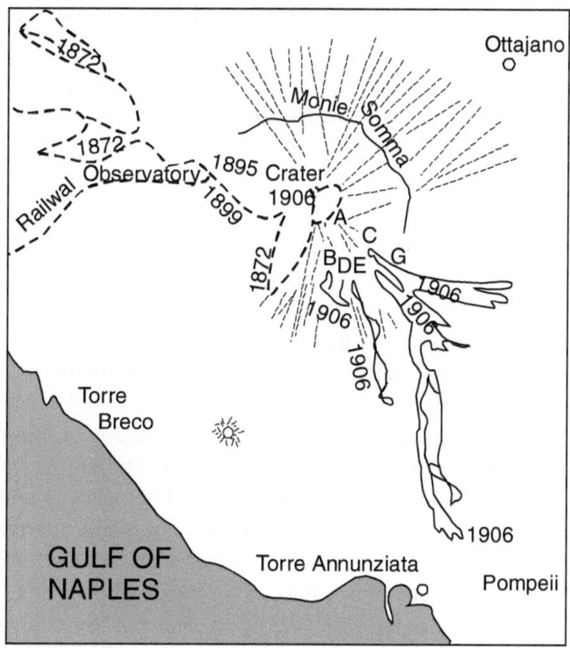

Fig. 4.23 Vesuvius Volcano, half encircled by the breached caldera known as Monte Somma (after Holmes 1965)

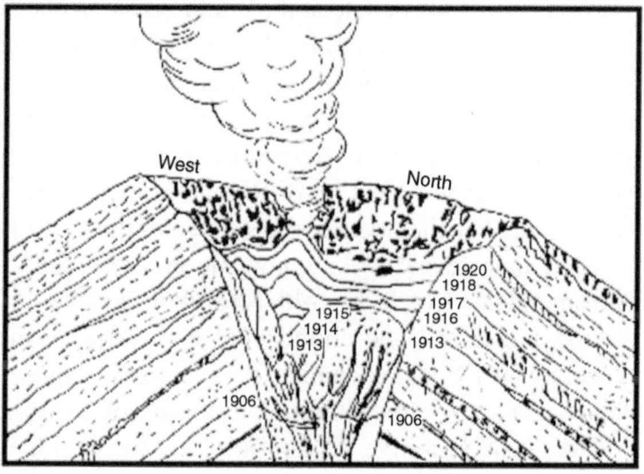

Fig. 4.24 Digramatic representation of the growth of Vesuvius volcanic cone (after Holmes 1965)

The Vesuvian eruption often involved expulsion of magma charged with gases, which got accumulated during a long period of quiescence and mild activity. Quite often discharge of explosive gases through the conduit could be found, whereas coherent lava flows escaped through fissures and vents on the flanks. During its explosive phases, there are vast clouds in the form of 'cauliflowers', which are luminous during the night. This type of eruption is popularly known as Plinian type of eruption (Fig. 4.24). During this phase, there is violent upsurge of globular masses of gas and vapour, which rise to several kilometres.

A detailed study of the pyroclastic deposits of the AD 79 "Pompai" Plinian eruption of Vesuvius was studied by Cioni et al. (1995). The pyroclastic material is characterized by K-tephritic to K-basanitic compositions. These were emplaced at temperatures >1,150 °C. They had high volatile contents (2–5 wt%) comprising H_2O, Cl, F, and S.

Chemical analyses of lavas discharged during Plinian type of eruption of the Somma-Vesuvius complex show that they form a distinct trend characterized by higher values of alkali and lower values of silica with respect to the Somma lavas (Cundari and Le Maitre 1970). The difference however, decreases towards the basic end of trends, when there is increase in mafic minerals with respect to alkali alumino silicates. According to Cundari and Le Maitre, leucite is the dominant alkaline phase in Vesuvius lavas associated with minor amounts of plagioclase and sanidine, whereas leucite is a minor constituent in Somma lavas coexisting with a large amount of plagioclase and sanidine. They found that products of the Plinian type of eruptions are mainly glassy phonolitic derivatives, which are genetically related to the lava flows erupted earlier, and they formed their natural extension towards higher silica values. The most salic types (Thornton and Tuttle differentiation index equal to 88–93) plot near the low temperature minimum of the phonolitic compositions (Qz = 27–34; Ne = 35–47; Ks = 25–27) of Hamilton and Mackenzie (1965). The magma appears to have undergone crustal differentiation.

Peccerillo (2001) made some comparative geochemical studies among the lavas of Vesuvius, Sromboli and Phlegrean Fields, and discussed about their geodynamic and volcanological implications. Data reveal close compositional affinities between these two volcanoes. The abundant 13–15 Ka old Stromboli leucite-tephritic rocks have radiogenic isotope signatures, and abundances and ratios of incompatible elements with the exception of Rb and K, which are identical to those of Vesuvius. The Phlegraean Fields also show close affinities to these volcanoes.

Their data reveal close compositional affinities between these two volcanoes. The abundant 13–15 Ka old Stromboli leucite-tephritic rocks have radiogenic isotope signatures, and abundances and ratios of incompatible elements with the exception of Rb and K, which are identical to those of Vesuvius. The Phlegraean Fields also show close affinities to these volcanoes.

The most primitive rocks from Vesuvius, Phlegraean Fields and Stromboli reveal intermediate compositions between arc and intraplate volcanics. It is suggested that the mantle sources beneath these volcanoes consist of a mixture of intraplate- and slab-derived components. Intraplate material was probably provided by inflow of asthenosheric mantle into the wedge above the subducting Ionian Sea plate, either from the Apulian plate and/or from the Tyrrhenian Sea region. Fluids or melts released from the sinking slab and associated sediments generated metasomatic modification of the intraplate material, whose melting gave rise to the Stromboli, Vesuvius and Phlegraean Fields magmas.

4.8.4.9 Kamafugitic Rocks Near L'aquila Abruzzo of the Umbria–Latium-Ultra-Alkaline-District (Ulud) Near Roman Province, Italy

Hovis et al. (2002) have described rare kamafugitic rocks near L'Aquila, Abruzzo from. the Umbria Latium District (Ulud) near Roman Province. They tried to establish possible genetic links between these rocks and those from the Oricola-Camerata Nuova volcanic field, occurring 20 km to the west of this field. The new kamafugitic rocks were found in a cave known as Grotta del Cervo, where these rocks are associated with epiclastic and pyroclastic materials. In the latter case, lapilli ash tuff, welded lapilli, ultramafic xenoliths, cognate lithics and pelletal lapilli have been identified. The mineralogy of the welded lapilli comprises, in order of decreasing abundance, the following minerals: diopside, leucite, hauyne, Mg-mica, andraditic garnet, apatite, magnetite, kalsilite and olivine. The rock is carbonate-free, and based on bulk-rock chemistry, it is classified, as kamafugites. The chemistry, closely approaches the composition of ULUD kamafugites (Sahama 1974). Separate lapilli and ash tuff (characterized by the same silicate mineralogy as that of the welded lapilli, plus modal carbonate exceeding 10 wt%), is classified as a carbonatitic kamafugite. Bulk-rock and trace-element compositions confirm that the Grotta del Cervo rocks closely approach the ULUD analogues.

The Grotta del Cervo occurrence partially fills the geographical and compositional gap between ULUD rocks and the rocks from the Vulture Complex, also a carbonatite and melilitite-bearing locality situated 200 km south of Grotta del Cervo, and adds considerably to the bulk chemistry of kamafugitic and related rocks lying along the Italian Apennines.

4.8.4.10 Ischia

The Ischia Island (Fig. 4.25) is located west-southwest of Naples. Together with Phlegrean Fields and Somma-Vesuvius complex, it constitutes the Neopolitan complex (Vezzoli 1988; Di Girolamo et al. 1995). The volcanic activities of Ischia has been subdivided into four main phases:

(1) The first phase (older than 150,000 years B.P.) was dominated by pyroclastic products and associated lava flows.
(2) The second phase was responsible for the formation of a lava dome formed 150,000–75,000 years B.P. Pyroclastic rocks are also intercalated with the lava dome.
(3) This phase (55,000–20,000 years B.P.) was associated with eruption of trachytic ignimbrites. The Green tuff of Mount Epomeo also erupted during this phase. The volcanic activities continued from different centres and lava flows in particular, were erupted towards the end of this phase.
(4) The last phase of volcanic activities started about 10,000 years B.P. The Arso lava flow (1302 A.D.) represents the last eruption of this island. The lavas of

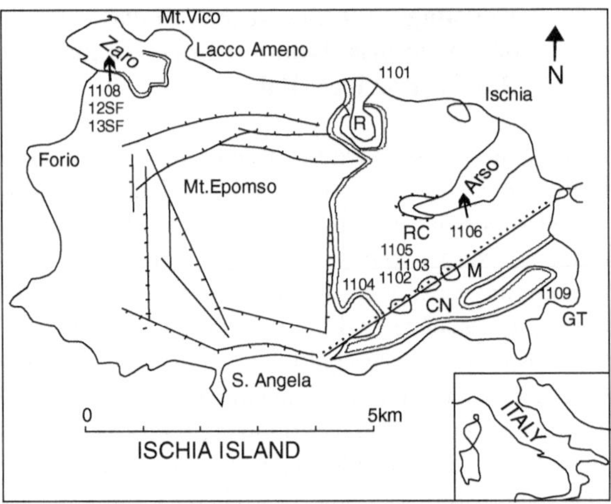

Fig. 4.25 Sketch map of Ischia complex. The *dotted line* is the limit of the volcanics younger than 10,000 year B.P. *Symbol*: *GT* Grotta di Terra dyke; *V* Valeliero; *CN* Cava Nocelle; *M* Molara; *RC* Rio Corbore; *R* Monte Rotaro (after Girolamo et al. 1995)

Arso erupted around 1302 A.D. The rocks are porphyritic with phenocrysts of sanidine, green clinopyroxene, forsteritic olivine, chromian-rich spinel inclusion, plagioclase, biotite and magnetite in a groundmass of alkali feldspar displaying a trachytic texture.

Latites are observed at Cava Nocelle. The rocks have similar texture as those erupted from Arso, but with relatively less amount of olivine and K-feldsar. Latites of Vateliero are more mafic but contain relatively more plagioclase in the groundmass.

The mafic inclusions in the rocks of Ischia represent the most primitive products found so far in the Island, as they contain relatively lowest $^{87}Sr/^{86}Sr$ ratios (0.70504) amongst the rocks of Roman province.

4.8.4.11 Mount Vulture Volcanic Complex

The volcanic complex of Mount Vulture (Lucania, Italy; Fig. 4.26) is the eastern most complex of the Roman magmatic province. This complex is situated close to the buried front, marking the overthrust of southern Apennine nappes on the Apulian foreland (De Lorenzo 1980, Melluso et al. 1996). The Mt. Vulture complex is different from other volcanic complexes of the Roman Province, which are located west of the Apennine chain, far from the Apulian foreland. Mt. Vulture consists of several volcano-clastic units and subordinate lava flows that together built up a strato volcano. The volcanic activity started in the Middle Pleistocene (La Volpe et al. 1984) and was active between 0.8 and 0.42 Ma (De Fino et al. 1986). The rocks of Mt. Vulture are predominantly basanites to trachyphonolites. It is also the only complex, where sodalite group of feldspathoids is more important than leucite.

The volcanic rocks, which were erupted first, are haüyne-bearing trachy-phonolites and phono-tephritic products. They occur on the lower and outer flanks of the

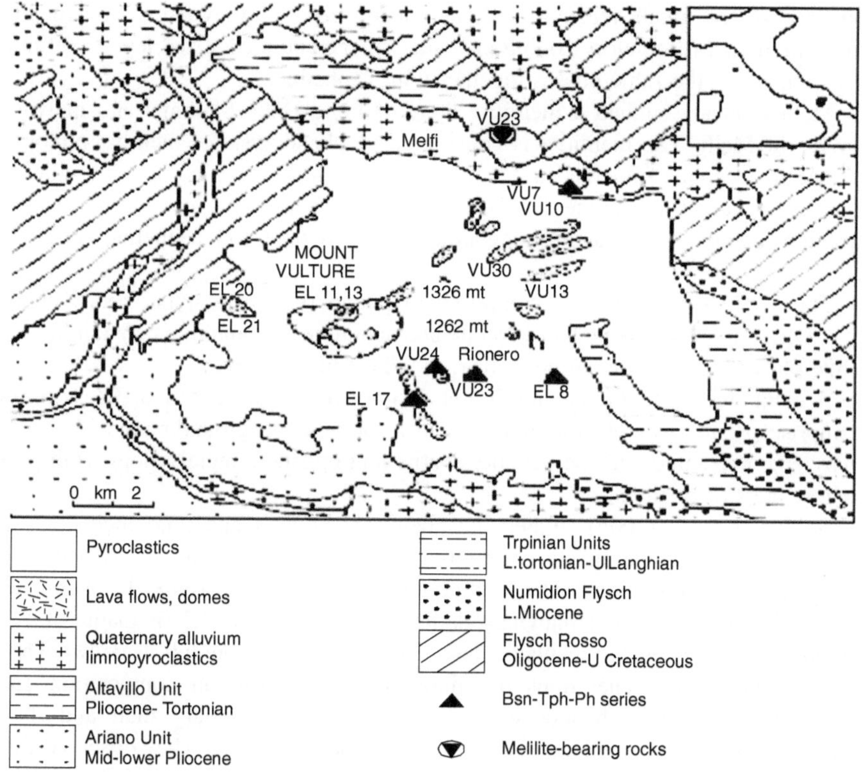

Fig. 4.26 Sketch of Mt. Vulture complex (after Melluso et al. 1996)

volcano. The bulk of the edifice is made up of haüyne- and leucite-bearing undersaturated rocks ranging in composition from Mg-rich basanites through tephrites and foidites to phonolites.

Melilite-bearing volcanic rocks were also erupted during the activity of the strato-volcano. The Scimia melilite dyke and Melfi (Fig. 4.26) haüynophyre lava flows occur outside the volcanic cone. Melilite laths are also found in the S. Caterina melafoidite lava flow (De Fino et al. 1982). The lavas of the Mt. Vulture complex are characterized by the presence of phenocrysts of clinopyroxene, haüyne, leucite and opaque minerals. Alteration of leucite to analcite is common.

Basanites contain pale green to green clinopyroxene with smaller amounts of olivine (containing inclusions of Cr- and Al-rich spinels) and haüyne. Clinopyroxenes contain apatite inclusions set in a glassy groundmass. According to Melluso et al. (1996), tephrites and phonolites are characterized by phenocrysts of zoned pyroxenes (often green in colour having sectors or inverse zoning), plagioclase, magnetite, leucite (frequently analcitized) and haüyne. Sometimes, amphibole, biotite and olivine also occur. Anorthoclase is predominantly observed in the groundmass as a microcrystalline phase together with apatite. Phonolites present phenocrysts of alkali feldspar, haüyne, deep green clinopyroxene, magnetite, melanite garnet (oscillatory zoning present), sphene and analcitized leucite set in a groundmass of the same phases together with nepheline.

Melluso et al., studied a melilitite dyke of Prete Della Scimmia and haüynophyres of Melfi. The melilitite is constituted of phenocrysts of melilite, clinopyroxene and opaque minerals in a holocrystalline groundmass comprising apatite, perovskite, garnet, nepheline, leucite and haüyne. These rocks also present ocellar structure composed of calcite, magnetite, clinopyroxene and nepheline (Melluso et al. 1996).

The Melfi haüynophyre is composed of phenocrysts of haüyne, zoned clinopyroxene, subordinate amount of leucite and apatite in a groundmass containing the same phases with additional nepheline, melilite and magnetite.

A melafoidite (melilite ankaratrite) is also noted in south Caterina. These rocks comprise phenocrysts of clinopyroxene in a fine-grained groundmass with clinopyroxene, nepheline, haüyne, leucite, apatite and opaque minerals along with pale yellow laths of melilite.

Solovova et al. (2005) studied the chemistry of melt and fluid inclusions in minerals of an olivine-leucite phonotitic nephelinite bomb from the Monticchio Lake Formation in Vulture. The rock contains 50 % clinopyroxene, 12 % leucite, 10 % alkali feldspars, 8 % hauyne/sodalite, 7.5 % nepheline, 4.5 % apatite, 3.2 % olivine, 2 % opaques, 2.6 % plagioclase, and <1 % amphibole (all in vol%). They distinguished three generations of clinopyroxene differing in composition and morphology. All the phenocrysts bear primary and secondary melt and fluid inclusions, which recorded successive stages of melt evolution. They observed that the most primitive melts contained the most magnesian olivine and the earliest clinopyroxene phenocrysts. The melts are near primary mantle liquids and are rich in Ca, Mg and incompatible and volatile elements. Thermometric experiments with the melt inclusions suggested that melt crystallization began at a temperature of about 1,200 °C. Because of the partial leakage of all primary fluid inclusions, the pressure of crystallization is constrained only to minimum of 0.35 GPa. Combined silicate-carbonate melt inclusions were found in apatite phenocrysts. According to them they are indicative of carbonate-silicate liquid immiscibility, which occurred during magma evolution. Large hydrous secondary melt inclusions were found in olivine and clinopyroxene. The inclusions in the phenocrysts recorded an open-system magma evolution during its rise towards the surface including crystallization, degassing, oxidation, and liquid immiscibility processes.

4.8.4.12 Latera Caldera

Turbeville (1993) studied petrology and genesis of the rocks of Latera Caldera. According to him the Latera Plateau is constituted of 8 km^3 of tuffs, which are exposed outside the caldera, and another 7 km^3 of intracaldera tuffs, established by drilling. ^{40}Ar and ^{36}Ar content of sanidine crystals indicate that the explosive eruption in the Latera Caldera took place in four episodes between 232 and 155 ka ago. This was preceded by another explosive event resulting in the outpouring of phonolite magma on the southern part of the plateau. Culmination in the initial period of caldera collapse began around 232 ka ago, when trachytic Plinian (A-tuffs) pumice eruption and ignimbrite emplacement took place. This was followed by eruption of thick phonotrachytic and phonolitic tuff sequence (ignimbrite-X and B tuffs). Turbeville further described that after roughly 25,000 year, there was emplacement of several coarse-grained phonolitic ignimbrite designated by the Italian volcanologists as Y and C tuffs. This was followed by phreatomagmatic eruptions that produced trachytic ignimbrites and quite a few ash flow units, the age of which is placed at 191 ka (D-tuffs). This was followed by eruption of wide spread phonolitic tuffs, tephritic phonolitic spatter and scoraceous ignimbrites (e-series ignimbrite and E-tuffs) from the northern caldera rim (Nappi et al. 1991). This event was followed by 40,000 years of mafic surge (all in vol%). Their study was also related to the development of scoria cone around the caldera rim (Poggio Pinzo Tuffs). There was also replacement of zoned tuffs around a vent northwest of the Latera Caldera. A major explosive eruption from the Latera volcano took place around 156 ka and this may be considered as the last volcanic activity. After drilling more than 350 m, a zoned syenitic pluton of 5 km width, was discovered. The seismic reflection profiles across Latera Caldera suggest that the pluton may contain melt fractions.

Turbeville (1993) described more than six different types of pumice characterized by different colours: white, pink, light gray, dark gray, green, and brown to black. According to him, the more felsic fragments contain phenocrysts of sanidine > leucite > plagioclase > clinopyroxene > biotite > magnetite with or without apatite and sphene. In some clasts he found that leucite is more abundant than sanidine. Although mineralogically different pumice types are similar, they vary in vesicularity, crystallinity and composition. Glommero-phenocrysts of plagioclase, clinopyroxene and biotite are present almost always in the trachytic and phonolitic pumice fragments. Clinopyroxenes are of two varieties: colourless diopside and dark green salite. The trachytic A and D tuffs are moderately vesicular, pink to light gray in colour and they are mainly pumice clasts. They are usually poorly porphyritic and lack leucite. Plagiocalse and sanidines crystals are unzoned containing <0.13 wt% BaO. Some diopside crystals have brown glass inclusions. The X and B tuffs are phonolitic and display large difference in colour and texture. The phenocryst content of the pumice clast is <5 to >50 vol%, but on an average it is 15–20 %. Leucite is usually absent from pumice clast in ignimbrite X, but in a few flows leucite-bearing clasts are very common at the top of a flow unit. Apatite intergrown with plagioclase and diopside is also common. Ignimbrite Y and the C-tuff contain black phonolitic pumice with fibrous translucent brown matrix. The clasts are often porphyritic and are very rich in leucite (leucite ≫ sanidine in most clasts). Aggregates of plagioclase, diopside and magnesite are embayed and show reaction relationship.

In the stratigraphically lower e-series flow units, phonolitic pumice clast lack leucite, whereas the gray green and brown clasts in the overlying deposit are rich in leucite. The flow deposits of E-Tuffs contain five different types of vesicular clasts:

(1) phonolitic pumice,
(2) brown fibrous pumice,
(3) vesicular green pumice and
(4) and (5) two types of tephri-phonolitic scorias.

The scoria includes dense to moderate vesicular leucite spatter and aphyric glass with fine vesicules. The tephriphonolitc F-Tuffs contain porphyritic scoria fragments in which leucite is often replaced by analcite. Turbville described several metres of white phonolitic pumices (PF-4) grading upward to black tephriphonolitic spatter and a welded vitriophyric Pitigliano tuffs. The PF-4 pumice contains Ba-rich sanidine phenocrysts (Ba content up to 8 wt%). Plagioclase crystals are anorthite-rich (An_{77-87}). Chemical analyses of some rocks from Latera Caldera are given in Table 4.16.

Conticelli et al. (2002) discussed about the genesis of potassic volcanism along the Tyrrhenian border of the Italian peninsula, which has been the site of intense magmatism from Pliocene to recent times. Although calc-alkaline, potassic and ultrapotassic volcanism overlaps in space and time, a decrease of alkaline character towards south is observed. Alkaline ultrapotassic and potassic volcanic rocks are characterized by variable enrichment in K and incompatible elements, coupled with consistently high LILE/HFSE values, similar to those of calc-alkaline volcanic rocks from the nearby Aeolian arc. On the basis of mineralogy and major and trace element chemistry, two different arrays can be recognized among primitive rocks; a silica- saturated trend, which resulted in the formation of leucite-free mafic rocks, and a silica undersaturated trend, characterized by leucite-bearing rocks. According to Conticellli et al. (2002), initial Sr^{87}/Sr^{86} and Nd^{143}/Nd^{144} values of Italian ultrapotassic and potassic mafic rocks range from 0.70506 to 0.71672 and from 0.51173 to 0.51273, respectively. The Pb^{206}/Pb^{204} values range between 18.50 and 19.15, Pb^{207}/Pb^{204} values range between 15.63 and 15.70, and Pb^{208}/Pb^{204} values range between 38.35 and 39.20. The general epsilon (Sr) versus epsilon (Nd) array, along with crustal lead isotopic values, clearly indicates that a continental crustal component has played an important role in the genesis of these magmas. The main question is from where this continental crustal component has been acquired by the magmas. Volcanological and petrologic data indicate continental crustal contamination to be a leading process along with fractional crystallisation and magma mixing. Considering, however, only the samples thought to represent primary magmas, which have been in equilibrium with their mantle source, a clearer picture emerges. A large variation of epsilon (Sr) versus epsilon (Nd) is still observed, with epsilon (Sr) from −2 to +180 and epsilon (Nd) from +2 to −12. A bifurcation of this array is observed in the samples that plot in the lower right quadrant, with mafic leucite-bearing Roman Province rocks buffered at epsilon (Sr) = , + 100, whereas the mafic leucite-free potassic and ultrapotassic rocks point towards strongly radiogenic Sr compositions. Conticelli et al. argued that mafic leucite-bearing Roman Province rocks have epsilon (Sr) and epsilon (Nd) values similar to those of Miocene carbonate sediments, whereas mafic leucite-free potassic and ultrapotassic rocks owe their genesis to mixing with silicate upper crust end-members with the mantle. Lead isotopes plot well inside the field of island arcs, overlapping the values of pelagic sediments.

They suggest that prior to partial melting there have been crust mantle mixing due to subduction related process. The chemical and isotopic differences observed between the northern and southern sectors of the magmatic region were possibly due to the presence of a carbonate-rich component in the south, and an ancient silicate metasedimentary slab in the north.

4.8.5 Volcanic Province of Spain

Ultrapotassic lamproitic rocks occur in the southeastern part of Spain in association with calc-alkalic, high K-alkalic and shoshonitic rocks (Fuster et al. 1967; Lopez Ruiz; Badiola 1980; Fig. 4.27). The volcanic history of the area ranged between middle Miocene and ended at the beginning of Pliocene (17–6 Ma ago). A careful look at the chemistry of the rocks shows a progressive increase in the K_2O content of the rocks towards north. The ages of these rocks also decrease progressively in the same direction.

Potassic lavas in southern Spain are composed of jumillite, fortunites and verites. Although most well-known localities include Murcia and Almaria (Fig. 4.27), they are found in many other scattered localities (Borley 1967). The type locality of verite is the village Vera, southwest of the town 'Murcia'. Fortunites occur in a locality between Fortuna and Orihuela. The later occurs towards east of Murcia (Borley 1967). Jumillites are post-Miocene lavas occurring 40 km north–northwest of Fortuna.They appear in a locality between Hellin and Jumilla, where they outcrop as small hills. Borley described jumillites as surface flows and not as intrusives. They are characterized by the presence of green diopside (containing small amounts of Cr_2O_3), forsteritic olivines ($Fo_{85.5 \pm 3}$), Ti-rich phlogopite, kataphoritic amphibole, sanidine and rare leucite. Pyroxenes are of two generations, those of later type occur as minute laths. The pleochroic scheme of mica is as follows: α = pink, β = yellow and γ = red brown.

Verites are melanocratic rocks, and often occur as breccia. Mineralogically they are similar to jumillite, but have fewer phenocrysts of forsteritc olivine ($Fo_{90.5-91.5}$), diopside, phlogopite and leucite, but amphibole is absent.

According to Lopez Ruiz and Badiola (1980), the essential difference in mineralogy amongst jumillites, cancalites, fortunites and verites lie in the relative proportion of constituent minerals, e.g. olivine, phlogopite, clinopyroxene, sanidine and K-richterite. Leucite is present only in jumillites and verites. Accessory minerals include apatite, calcite and rutile. K-analcite occurs in jumillites, verites and fortunites. Analyses of lamproitic rocks from Spain are summarized in Table 4.17.

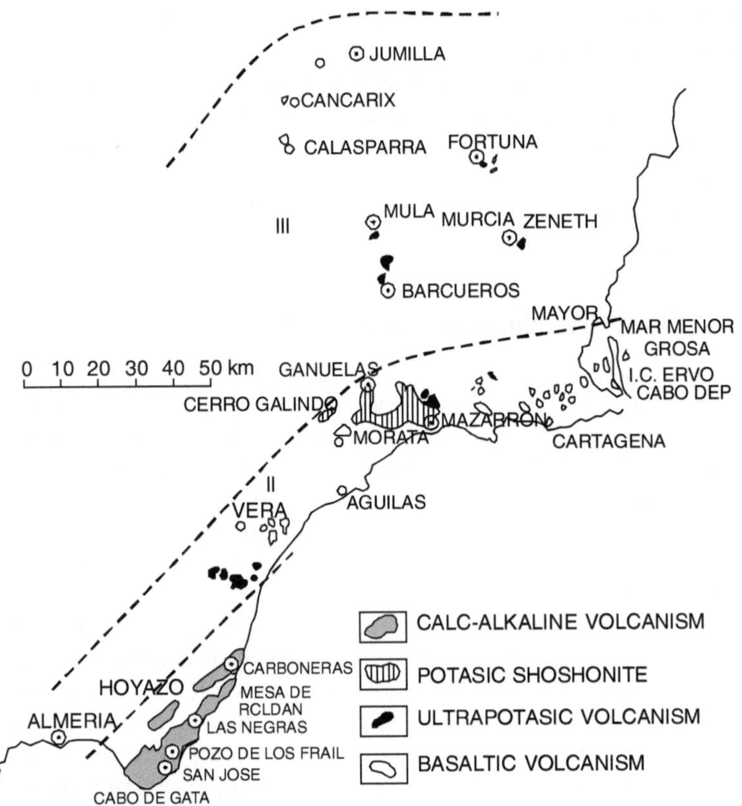

Fig. 4.27 Sketch Map of southern part of Spain showing distribution of different volcanic fields of K-rich lamproites (after Lopez Ruiz and Badiola 1980)

4.8.6 Late Cenozoic Leucite Lamproites from the East European Alpine Belt (Macedonia and Yugoslavia)

Altherr et al. (2004) described late Cenozooic leucite lamproites from the East European Alpine belt of Macedonia and Yugoslavia. They described eight localities of ultra potassic rocks within the Vardar suture zone and one locality in the Southern Carpathian fold-and-thrust belt. Most of these volcanics are characterized by high Mg# (66.6–78.6), high abundances of Ni (117–373 ppm) and Cr (144–445 ppm) as well as high primary K_2O contents (5.63–7.01 %) and K_2O/Na_2O values (1.93–4.91). Rocks with more differentiated compositions are rare. A lamproite affinity of these rocks is apparent from their relatively low contents of Al_2O_3 (9.9–14.3 wt%) and CaO (6.2–8.3 wt%) in combination with high abundances of Rb (85–967 ppm), Ba (1,027–4,189 ppm), Th (18.9–76.5 ppm), Pb (19–54 ppm), Sr (774–1,712 ppm) and F (0.16–0.52 wt%), and the general lack of plagioclase. Although eruption of the magmas took place in post-collisional

Table 4.17 Analyses of Ultrapotassic rocks from Spain

	SiO_2	TiO_2	Al_2O_3	Fe_2O_3	FeO	MnO	MgO
1	47.72	1.43	7.72	3.01	3.40	0.1	16.27
2	45.53–51.52	1.28–1.66	5.68–11.13	1.45–4.03	2.64–4.31	0.07–0.29	14.29–18.04
3	47.02	1.31	7.55	3.32	2.93	0.11	16.43
4	55.05	1.76	9.31	2.14	3.07	0.08	12.18
5	52.80–55.90	1.43–2.30	8.15–10.93	0.83–5.79	0.65–4.16	0.01–0.15	5.11–15.54
6	56.05	1.35	11.43	2.23	3.31	0.08	9.27
7	53.36–57.51	0.95–1.86	9.69–13.02	0.72–3.23	1.31–4.82	0.01–0.12	5.20–11.81

	CaO	Na_2O	K_2O	P_2O_5	H_2O	CO_2	Total
1	7.11	1.71	4.99	1.68	3.48	0.44	99.06
2	4.14–9.06	1.05–3.41	2.61–7.39	1.16–2.17	1.07–5.23	0.04–1.20	
3	7.37	1.02	5.10	1.90	4.21	–	99.23
4	3.75	1.40	8.67	0.99	1.43	0.07	99.90
5	1.61–5.61	0.80–3.08	7.56–9.92	0.36–1.25	0.44–3.93	0.13–0.59	
6	4.21	2.10	6.14	0.81	2.91	0.06	99.95
7	1.41–7.01	1.35–4.07	3.52–9.15	0.30–1.57	1.15–7.01	0.07–0.70	

1–2 Jumillite from Spain (Lopez Ruiz and Badiola 1980)
3 Jumillite from Spain (Carmichael 1967). It also includes 0.13 wt% ZrO_2, 0.09 wt% Cr_2O_3, 0.29 wt% Sr O, 0.39 wt% BaO and 0.06 wt% SO_3^-
4–5 K-rich rocks from southeastern Spain (Lopez Ruiz and Badiola 1980)
6–7 Fortunites from southeastern Spain (Lopez Ruiz and Rodriguez Badiola 1980)

extensional settings, significant depletions of Nb and Ta relative to Th and La, low TiO_2 contents (0.92–2.17 %), low ratios of Rb/Cs, K/Rb and Ce/Pb as well as high ratios of Ba/La and Ba/Th suggest close genetic relationships to subduction-related processes. Whereas Sr and Nd isotope ratios show relatively large variations (Sr^{87}/Sr^{86} = 0.7078–0.7105, Nd^{143}/Nd^{144} = 0.51242–0.51215), Pb isotope ratios display a very restricted range with Pb^{206}/Pb^{204} = 18.68–18.88 and variable. The observed petrographic, geochemical and isotopic characteristics are best explained by a genetic model involving preferential melting of phlogopite-rich veins in an originally depleted lithospheric mantle source, whereby the metasomatic enrichment of the mantle source is tentatively related to the addition of components from subducted sediments during consumption of Tethyan oceanic lithosphere.

4.9 Lamproitic Rocks from Greenland

4.9.1 Batbjerg Complex

The complex (Fig. 4.28) is located 15 km beyond the head of Kangerdlugssuaq Fjord (Gittins et al. 1982). The complex is 7 km long with an elevation of 1 km and is exposed along the eastern side of the Kangerdlugssuaq glacier. It intrudes Precambrian gneiss, marble and quartzite. These metasediments are thermally and metasometically altered, and includes Na-rich melilite, pectolite, wollastonite and ulvospinel. The main complex is composed of pyroxenite and jacupirangite with minor amounts of ijolite and urtite; which are intruded by syenite, quartz syenite and granite. According to Gittins et al. pyroxenes are present as a predominant homogeneous mineral, but sometimes zoned forming acmitic rims. The other phases present in the pyroxenites are phlogopite (sometimes biotite), olivine and amphibole. The salic minerals include leucite, nepheline, K-feldspar and pseudoleucite with intergrowth of nepheline and K-feldspar. In many pyroxenites, patchy intergrowth of nepheline and kalsilite are also observed. When such intergrowths are observed, leucite is usually found to be absent. Discrete grains of pyroxene are rare and fresh leucite crystals are found as an interstitial mineral. The pyroxenite body is intruded by coarse-grained phlogopite peridotite dykes. The phlogopites show reverse pleochroism, and olivines are sometimes up to 1 m across.

4.9.2 Holtsteinberg Lamproit from Greenland

Scott (1980) made detailed investigation of kimberlite and lamproite dykes of Holtsteinberg, located in the central part of west Greenland (Fig. 4.29). The majority of the dykes occur in the coastal area between Holdsteinberg and Itivleq but a few area found north of Sukkertoppen, Iskappe (Fig. 4.29).

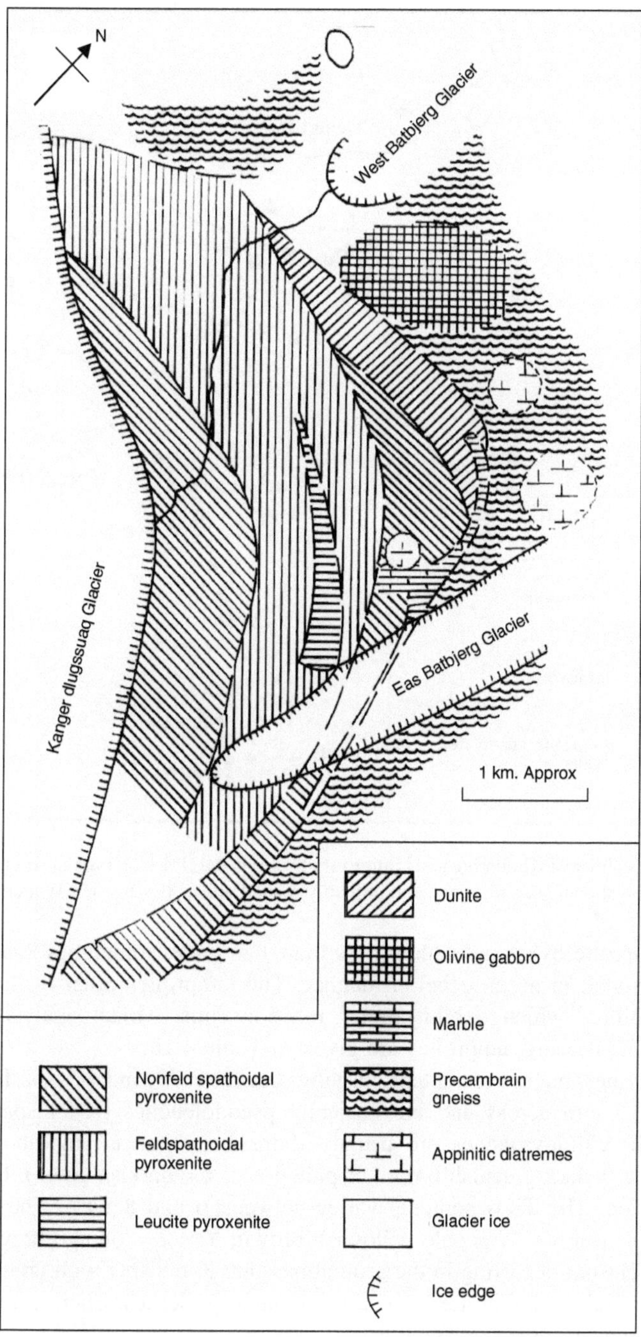

Fig. 4.28 The Batbjerg complex (after Brooks et al. 1981)

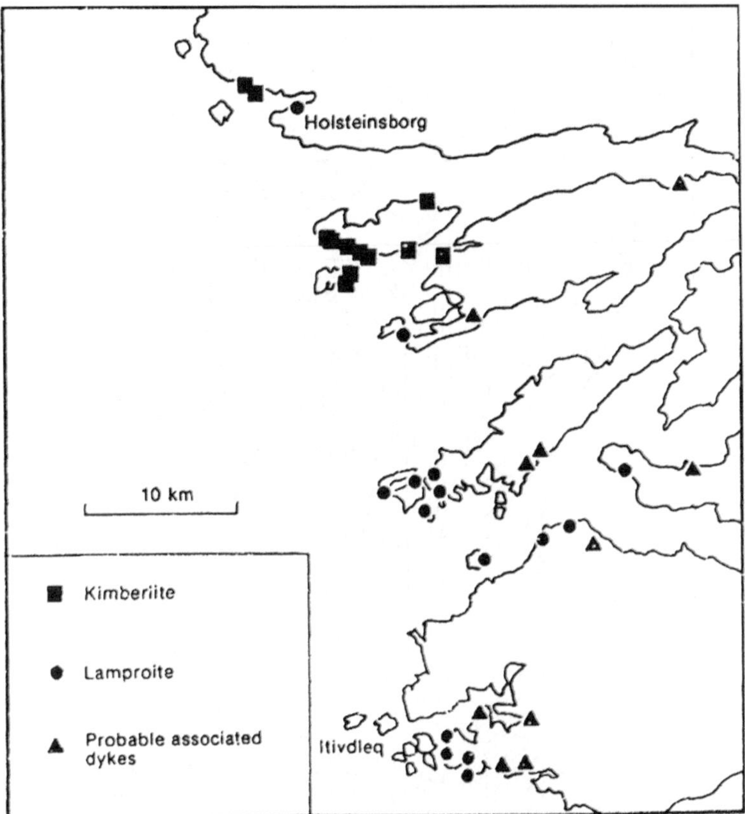

Fig. 4.29 Localities of Kimberlite and Lamproites Dykes in the Holesteinborg District, Centra West Greenland also shown are additional, and probably related, dykes (Escher and Watterson 1973)

The lamproite dykes are usually less than 1 m wide, sometimes zoned and are usually intruded in an en echelon manner. The lamproites often include country rocks xenoliths, which exhibit minor reaction rims. Modal analysis of some pseudoleucite-bearing lamproites are given in Table 4.18.

The olivines have Fo content varying between 75 and 93 %. Leucite are sometimes 5 mm across and are basically pseudoleucites (K-feldspar-nephiline intergrowth) clinopyroxenes are usually diopsidic, whereas amphiboles are K-richterites (x = dark greenish brown or pale green, z = darkish green). Pholophites are TiO_2 rich. The TiO_2 contents varies between 6 and 8 wt%. The pleochroic scheme is as follows: X = pale yellowish brown, Y = Z = orange brown.

Alkali feldspar occurring in the groundmass are K-feldspar with up to 3.84 wt% FeO. Holesteinberg lamproites sometimes contain priderites. Chemical analyses of the lamproites are given in Table 4.19.

The Rb-Sr isotopic age determination of phlogophite concentrates suggest an age of 1227 ± 12 Ma. The age of dyes is 587 ± 24 Ma this suggest that lamproites are older than the kimberlites.

Table 4.18 Modal composition of Holsteinberg Lamporites

Sample number	5647	5659	5622	5692
Olivine	2	1	11	9
Leucite	17	16	16	16
Phlogopite	31	55	48	32
Clinopyroxene	4		8	10
Richterite	15	12	6	16
K-Feldspar	13	8		13
Opaque	3	3		1
Serpentite	1		4	
Carbonate	13	10	6	3
Apatite	1	3	1	1

Table 4.19 Chemical analysis of Holesteinberg Lamproites

Sample number	5622	5632	5643	5692
SiO_2	39.29	42.75	46.48	43.79
TiO_2	2.17	3.18	3.38	2.67
Al_2O_3	4.90	8.47	7.66	7.04
Fe_2O_3	1.35	1.74	3.14	1.35
FeO	6.25	5.01	4.09	5.73
MnO	0.10	0.09	0.07	0.09
CaO	4.73	7.38	5.54	5.21
Na_2O	1.00	0.90	1.37	1.30
K_2O	6.16	7.99	8.14	7.06
H_2O^+	2.82	2.65	2.92	3.19
P_2O_5	0.94	0.73	1.43	1.16
CO_2	5.14	4.96	3.05	2.94
Total	97.40	98.01	98.01	97.12

4.9.3 Kap Dalton (69°24'N, 24°10'W)

Woolley (1987) described the occurrence of malignites, monchiquites and leucite nephelinite in association with Tertiary basalts near Kap Dalton (also see Wager 1935). The malignite comprising nearly 33 vol% of nephiline, aegirine augite, orthoclase, abundant, analcime, tinguaite and altered nosean. Closely associated with these rocks are leucite, nephilinites (nephiline + clinopyroxene + leucite ± perovskite). Some leucite nephilinites contain about 75 vol% altered leucite.

4.10 K-Rich Feldspathoidal Rocks from Colima, Mexico

Colima (19°34′; 103°36′W, Fig. 4.30) is a composite volcano located towards the western part of the Mexican volcanic belt and to the south of the Colima graben. Although the eruptive lava flows of Colima are andesitic, there are seven cinder cones towards north of the main volcanic centre. The cones are constituted of bedded ash, scoria and bombs. There are also lava flows of basanites, which are essentially nepheline-normative (Luhr and Carmichael 1981). There are also leucite basanites containing modal leucite and analcite; these rocks have been termed minettes and monchiquites. All the rock types are constituted of olivine, augite and chromite. The basanites are characterized by the presence of labradorite, Titano-magnetite and glass. The leucite basanites are constituted of plagioclase, sodic sanidine, leucite, titanomagnetite, apatite and Ti–F-rich phlogopite. In the minettes phlogopite is present both as phenocryst and groundmass minerals. Other groundmass phases include apatite, labradorite, anorthoclase, albite–rich sanidine, analcite and titanomagnetite. Allan and Carmichael (1984) described a phlogopite- and kalsilite-bearing ankaratrite from a locality 25 km north of Colima. This rock is characterized by phenocrystal phlogopite in a groundmass of diopsidic pyroxene, nepheline, kalsilite, titanomagnetite, phlogopite, analcite, zeolites and glass. The estimated ages of the cinder cones are between 1,500 years and >20,000 years.

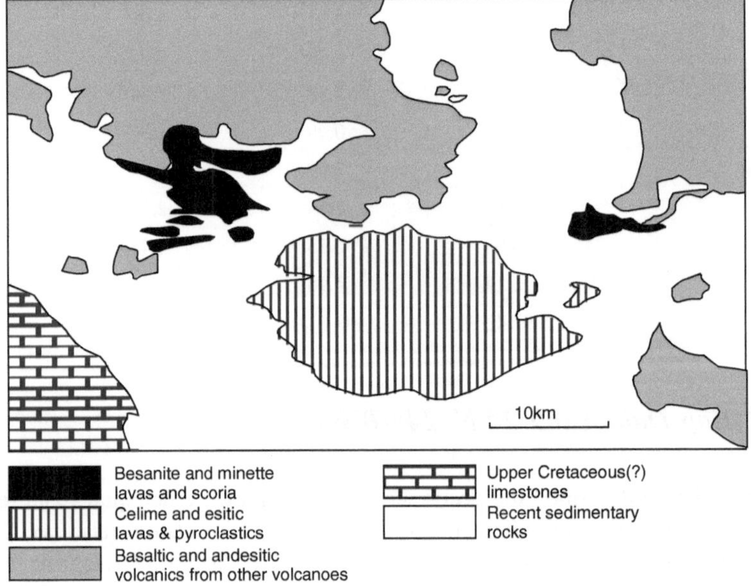

Fig. 4.30 Colima volcanic complex (after Luhr and Carmichael 1981)

4.11 K-Rich Rocks from Paraguay

The Serra Geral formation of eastern Paraguay has been the site of alkaline magmatism since Mesozoic times (Comin-Chiaramonti et al. 1997). These igneous rocks include:

(1) Late Permian-Early Triassic sodic intrusions and lavas,
(2) Early Cretaceous potassic igneous rocks and very rarely-occurring sodic lavas,
(3) Late Cretaceous-Oligocene sodic lavas.

Two distinct magmatic events are dominant in the Asuncion-Sapucai graben (ASU) of eastern Paraguay: (a) widespread potassic magmatism and eruption of tholeiites of Serra Geral formation (Early Cretaceous) and (b) Asuncion sodic magmatism (Late Cretaceous-Oligocene). The potassic rocks form a compositional continuum from moderately to strongly potassic rock series. Comin-Chiaramonti et al. suggested the presence of two potassic suites i.e. (i) basanite to phonolite and (ii) alkali basalt to trachyte and their intrusive analogues. The sodic rocks include ankaratrites, nephelinites and phonolites. Two similar but distinct parental magmas have been inferred for the potassic suites, both characterized by strongly fractionated REE and negative Ta–Nb–Ti anomalies. Slightly positive Ta and Nb anomalies distinguish the sodic rocks from their potassic counterparts. The rocks of Asuncion-Sapucai graben have been divided by them into three groups:

(1) Potassic type (K_2O/Na_2O varying between 1 and 2,
(2) Highly potassic type ($K_2O/Na_2O > 2$) and
(3) Transitional type ($K_2O/Na_2O < 1$) (Table 4.13).

There are varieties of rock types such as theralites, essexites, gabbros and ijolites. These rocks are holocrystalline with diopsidic pyroxene ($Wo_{44-51}Fs_{8-17}$-Hd_{31-41}), olivine (Fo_{75-82} to Fo_{44-66}), mica (Ti-phlogopite to Ti-biotite), Ti-magnetite, alkali feldspar, nepheline ($Ne_{64-80}Ks_{20-36}$) ± leucite ± amphibole. Leucite crystals pseudomorphed by analcite and plagioclase are common in potassic and transitional group of rocks.

Basanites, tephrites and phonotephrites typically show porphyritic textures with phenocrysts of clinopyroxene ($Wo_{40-50}Fs_{10-19}Hd_{31-41}$), olivine ($Fo_{60-85}$), and leucite pseudomorphed by sanidine + nepheline, in glassy groundmass containing microlites of clinopyroxene ± olivine, Ti-magnetite ± ilmenite, Ti-phlogopite-biotite, alkali feldspar (Or_{15-88}), nepheline$_{ss}$ ($Ne_{44-59}Ks_{17-26}$) and analcite. Phenocrystal plagioclase (up to An_{74}) is present in potassic and transitional rock type. Accessory phases are amphibole (pargasite-kaersutite), apatite with or without zircon.

Phonolites are characterized by phenocrysts of leucite pseudomorphs, alkali feldspar (Or_{47-75}), clinopyroxene ($Wo_{48-50}Fs_{11-34}$), Fe-pargasite, nepheline ± biotite ± sphene ± melanite (Ti- andradite up to 68 wt%) ± magnetite or hematite. Glassy groundmass contains microlites of alkali feldspar, nepheline, clinopyroxene ± Ti- andradite ± magnetite or hematite. Analyses of rocks from Paraguay are given in Table 4.13.

4.12 K-Rich Feldspathoid-Bearing Rocks from the Former U.S.S.R

Ultrapotassic feldspathoid-bearing rocks are found in widely scattered localities throughout the former U.S.S.R. (Bazarova and Kazaryan 1977). Of these localities the rocks occurring at Elpinsky, Talyshskii and Tezhsar (all from Caucasus), Andriyanovka (Kamchatka province), Tokko and Anomaly (Aldan), Artem (Primorye province) and Pastistennyl (Omolon province) are genuinely leucite-bearing.

4.12.1 Tezhsar (40°41′N, 44°39′E)

The alkaline complex of Tezhsar (37–39 Ma, Abovyan et al. 1981) covers an area of 80 km^2 and occurs at the Prisevanskaya zone of the Pambaksky range (Kogarko et al. 1995). The complex is ring-shaped comprising both intrusive and extrusive rocks. The effusive series occurs in a circular area (15 km^2), and is 600 m thick. There are three pyroclastic formations (from bottom to top): (a) trachyandesites, (b) trachytes, and (c) leucite phonolites or pseudoleucite porphyries. Leucite tephrites, shoshonites, sanidine- and leucite-bearing trachytes, phonolites, leucitophyres and italites (containing essentially leucite) also occur in this complex. The intrusive rocks of concentric zones are as follows: (1) alkaline syenites, (2) pseudoleucite syenites, (3) nepheline syenites, (4) nordmarkites, (5) alkali syenites and (6) quartz syenites. Nepheline-rich pseudoleucite syenites are confined to the marginal part of the core of the complex. These syenites comprise pseudoleucites (30–40 vol%, 1–6 cm in diameter) in a matrix of nepheline (25–35 wt%) and orthoclase ($Or_{90}Ab_{10}$). The central part of the intrusive is composed of alkaline to peralkaline syenite grading to peralkaline monzonite, nordmarkite and pulaskites.

4.12.2 Elpinskii (39°27′N, 46°09′E)

The Elpinskii complex was emplaced 12.50–14.53 Ma ago. They occur in the tectonically depressed area of Zavozhen (Kogarko et al. 1995) near Akhavnadzorskaya, south of Armenia (Bazarova and Kazaryan 1986). The igneous activity started with eruption of analcite-bearing alkalic basalt (25–30 m). Analcite-bearing tephrites and explosive breccias occur in the upper part of the volcanic section underlain by trachyandesites, trachytes and further below by leucite tephrites and dykes of haüyne-bearing mafic rocks.

4.12.3 Pkhrutskii (38°51′N, 48°10′E)

The Pkhrutskii massif (37–40 Ma, Bagdasaryan 1966) is extended over an area of 10 km^2. The entire body represents a single intrusive phase and is constituted of nepheline-kalsilite monzonite, essexite and medium to fine-grained syenite. In addition to nepheline and kalsilite, orthoclase, andesine, augite, aegirine-augite and biotite are other constituent minerals. While sodalite is an accessory phase, albitization associated with a metasomatic process is also encountered.

4.12.4 Talyshskii (38°45′N, 48°22′E)

The alkaline volcanic complex of Talyshskii was emplaced in a folded belt, located in the southeastern part of Azerbai'an during Eocene (Azizbekov et al. 1979). The complex is characterized by the presence of a series of leucite- and sanidine-bearing tuffs, leucite tephrites and basanites having a total thickness of 500 m. These volcanic series were erupted during early, middle and late Eocene in the southern part of the Talyshskii zone and are extended up to Iran. According to Azizbekov et al. (1979), leucite-bearing tuffs, comprising sanidine, plagioclase, leucite, augite and hornblende set in a glassy groundmass, occur as volcanic domes at Shandan-Kalasy and Green Hill. The leucite tephrites and basanites are products of volcanoes at Geldar, Govery, Kiz-Kalashym Pashgra and Rasgov. These tephrites are characterized by the presence of phenocrystal leucite, plagioclase, olivine and augite set in a groundmass of the same phases but with additional magnetite. Leucite tephrite also occurs as sills in Dimansky. The large intrusive complex at Kalakhanskaya occurs as a cone with a height of 1732 m, comprising trachyandesite, gabbro, teschenite, picrobasalt and leucite-bearing rocks.

4.12.5 Ishimskii Complex (51°17′N, 66°33′E)

The Ishimskii Massif (Mineyeva 1972) is located at the periphery of Cuba-Sadir Mountain in central Kazakhstan. The complex is characterized by both intrusive and extrusive rocks comprising leucite-bearing tepherites, augite porphyries, pseudo-leucite porphyries. There are also two volcanic phases of pseudoleucite-bearing trachytes and tinguaites.

4.12.6 Daubabinskoe (42°28′N, 70°07′E)

The complex is located in the southern part of Kazakhstan. According to Kogarko et al. (1995) there are two volcanic sequences, one comprising leucite and analcite-bearing trachybasalts and tuffs and the other with lava flows, breccias and tuffs of leucite tephrites and minor amount of picritic lava flows. Leucite tephrites and leucitites occur in the central part of the section (450 m). The volcanic section at the top is composed of agglomerates of leucite and biotite-bearing phonolites and tuffs of latite and trachyandesites. Volcanic necks and dykes, comprising trachybasalts, leucite tephrites and trachytes, shonkinite porphyry and biotite syenites are observed within the extrusive rocks and in the vicinity of the complex. According to Eremeev (1984), the extrusive rocks were erupted 322 Ma ago.

4.12.7 Kaindy (42°21′N, 70°35′E)

The Kaindy complex located at Talassky Alatau is a pipe-like stock comprising psuedoleucite-bearing pyroxenite. There are also associated intrusive rocks of monzonites, syenites (with or without pseudoleucite) and nepheline syenites (Nurlibayev 1973).

4.12.8 Irisu (42°20′N, 70°27′E)

Stocks with ring dykes (Fig. 4.31) intrude limestone-bearing sedimentary formation at Irisu (Kogarko et al. 1995). The complex extends over an area of (9 × 3 km^2). The rocks from youngest to oldest age in progression include, the following rock types: (1) olivine leucitite and leucitite, (2) nepheline porphyry, shonkinite porphyry and

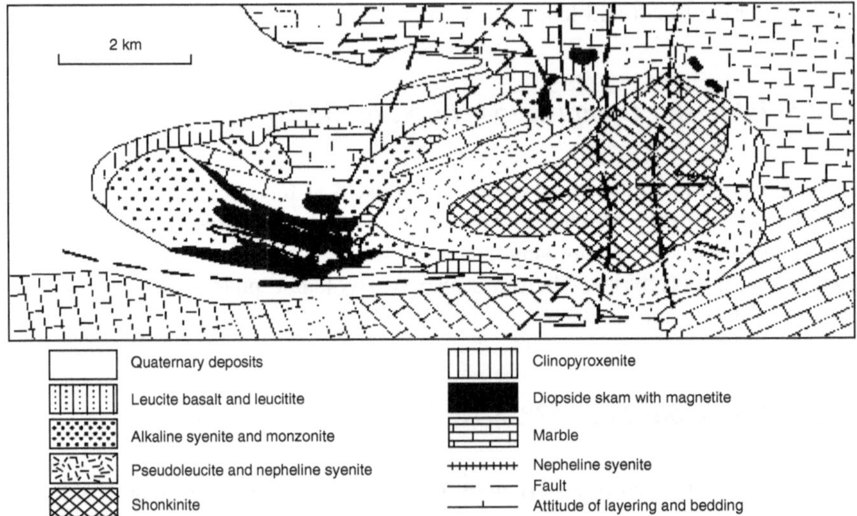

Fig. 4.31 Sketch map of Irisu complex, Kazakhstan (after Nurlibayev 1973)

phonolites and (3) a subvolcanic layer comprising olivine, clinopyroxenite, leucite and orthoclase-bearing pyroxenite (first phase), alkali monzonite, shonkinite and syenite (second phase), pyroxene-bearing syenite and nepheline syenite (third phase) and finally syenite porphyry (fourth phase of volcanic activity).

4.12.9 Kolbashinskii (42°20′N, 73°44′E)

Leucite-bearing mafic rocks including leucite tephrite, alkali trachytes, tuffs, breccia and peralkaline trachyte (270–309 Ma) occur in a volcanic field in an east–west-trending zone (40–50 km long, 5 km wide) located in the Kirgiz shield of the northern Tyan-Shan region, west of lake Issyk-Kul (Bagdasarov et al. 1974).

4.12.10 Synnyr (56°55′N, 111°20′E)

The Synnyr pluton (Baikal province) is an oval-shaped body intruding into a sedimentary sequence of Cambrian limestone, conglomerate and sandstone and Palaeozoic granitoids. From core to the rim following succession of rocks occur in following order: alkaline syenite, pseudoleucite syenite, microcline pseudoleucite syenite and nepheline syenite. At the central part the oval-shaped stock (100 km^2), is an igneous complex comprising trachyte and pyroxene-syenite (pulaskite). One of the varieties of pseudoleucite-bearing rocks is synnyrite, consisting of microcline and intergrowth of orthoclase, nepheline and kalsilite. In the intergrowth, the ratio of K-feldspar to kalsilite (± nepheline) varies between 67 and 33 vol%. The K_2O content of these rocks may be as high as 18–20 wt%. The vol% of pseudoleucite may be as high as 50–60 %. The nepheline syenites are miaskitic and 311 Ma year old (Zhidkov 1990). During the final stage, dykes of tinguaite, shonkinite, camptonite and monchiquite were emplaced.

4.12.11 Yaksha (56°55′N, 111°48′E)

A plutonic dome comprising silica-undersaturated alkalic rocks occur at Yaksha. The larger part of the complex is made up of syenite, comprising K-feldspar and micro-perthite (60–80 %), pyroxene (5–15 %) and biotite (3–12 %), with an admixture of nepheline and albite. The accessory phases include apatite and sphene. Nepheline-kalsilite intergrowth and kalsilite-bearing syenites are closely mixed with each other. Nepheline-kalsilite syenite or kalsilite-syenite (synnyrites) is rare. According to Kogarko et al. (1995), the pseudoleucitic rocks with preferred orientation of dark-coloured minerals are predominant. The complex was emplaced 313 ± 11 Ma year ago (Zhidkov 1990).

4.12.12 Molbo (59°05′N, 118°49′E)

A phlogopite-bearing leucite lamproite (thickness: 1.3–1.5 m) occurs at Molbo (Aldan Alkaline Province). It cuts lower Cambrian marls, which overlie the Precambrian basement. The dyke has a zoned structure, with the central portion being porphyritic and the periphery is aphyric. The core is composed of phenocrysts of olivine (often altered), leucite and lath-shaped phlogopite. The fine-grained portion comprises phlogopite, pseudoleucite and microlites of olivine, clinopyroxene, magnetite and apatite set in a groundmass of phlogopite, chlorite and alkali amphibole. The lamproite was emplaced 122–92 Ma years ago (Makhotkin et al. 1989). Lamproitic rocks comprising olivine, phlogopite and diopside also occur at Khani (57°54′N, 120°22′E), Lyzhnaya (58°42′N, 125°46′E) and Ryabinovaya (58°40′N, 125°51′E). The lamproitic pipe of Ryabinaovaya is associated with phlogopite-bearing shonkinite and phlogopite minette. The breccia pipe at Kaila (600–700 m) is one of the largest breccia pipes. It intrudes Precambrian crystalline rocks of the Aldan shield basement complex and dolomites of lower Cambrian age. The lower part of the pipe is constituted of crystal tuffs, breccias and xenolith-bearing tuff breccias, which are made up of small fragments of intrusive or autolithic fine-grained lamproite along with large megacrysts of olivine and feldspar aggregates. The fragmental lamproites are fine-grained porphyritic rocks with phenocrysts of olivine (10–12 %), chrome-diopside (7–15 %) and phlogopite (5–12 %) in a fine-grained groundmass of diopside, fine laths of phlogopite and K- feldspar. Pseudoleucite is present in the fine-grained breccia fragments.

4.12.13 Tommot (58°23′N, 125°13′E)

Both extrusive and intrusive potassic rocks (167 ± 5–125 ± 3 Ma old; Orlova 1990) were emplaced in Archaean granite and granites gneiss. The extrusive rocks are found in a depression up to 950 m deep in the western part of the massif. The volcanic rocks comprise three segments: (1) the lowest one is feldspar-bearing leucitite, overlapped by (2) peralkaline trachytic lavas and breccias; and finally (3) the peralkaline trachytic ignimbrites, constituting the uppermost part. Leucitite and leucite tephrite dykes cut across ignimbrites towards the eastern part of the complex. Parallel to the margin of the complex there is a larvikitic body, comprising strongly oriented tablets of alkali feldspars (50–70 vol%) and clinopyroxene (10–25 vol%) with such accessory phases as amphibole, biotite, sphene etc.

4.12.14 Yakokut (58°27′N, 125°29′E)

The Yakokut complex (Kanukov et al. 1991) intrudes lower Cambrian dolerites and Jurassic formation. The complex is characterized by dykes of alkaline and moderately alkaline co-magmatic rocks (Fig. 4.32). The volcanic rocks occur as relics within a deeply eroded faulted caldera. The lowermost section (200 m thick) is composed of feldspar-bearing epileucitite and leucitite. The middle unit (200–250 m thick) is constituted of lavas, braccias and tuffs of leucitites, biotite leucitites, and melaphonolites. The upper most section is constituted of lavas and lava breccias of leucite phonolites and trachyphonolites. Lamproites of similar compositions have been found in the contact zone of the complex. Phologopite-bearing leucitite is characterized by the presence of phlogopite (20 vol%) and diopside (10 vol%) in a groundmass of orthoclase and plagioclase. Biotite-bearing leucite melaphonolites have phenocyrsts of pseudoleucite and diopside. Leucite phonolites are characterized by the presence of orthoclase, mica and pseudoleucite.

Microshonkinites constitute the earliest intrusive series. They were emplaced before the eruption of volcanic units. Nepheline and pseudoleucite-bearing syenites and peralkaline syenites are also observed in the central part of the massif. Within the central complex, a large number of north-south-trending dykes (1–3 m long and several small stocks of about 100 m diameter) also occur. They comprise olivine leucitite, leucitite, leucite and aegirine-bearing phonolites, olivine-leucite-phlogopite phonolites, syenite porphyries and lamproites.

4.12.15 Rododendron (58°22′N, 125°36′E)

A circular stock (200–300 m diameter) intruding Jurassic sandstones, is found in Rododendron. The central part of the complex comprises shonkinites and syenites at the centre of the complex. It is surrounded by shonkinite porphyries and syenite porphyrites. Volcanic rocks also occur at the margin of the intrusive rocks, which are constituted of leucite-bearing trachytes, comprising phenocrysts of biotite, orthoclase, sodic plagiocalse (An_{20-30}) and apatite in a fine-grained matrix. The shonkinite porphyries are melanocratic with augite phenocrysts (30–40 %) and a small amount of biotite (up to 10 vol%) with large amount of pseudoleucite, olivine and alkaline amphibole. The groundmass is constituted of K-feldspar, biotite, augite, plagioclase (An_{40-50}) and small grains of leucite, apatite and magnetite. The central part of the shonkinite is of two types: one portion containing nepheline, the other part has pseudoleucite. The age of the stock was determined to be 166 ± 5 Ma (Eremeyev 1987).

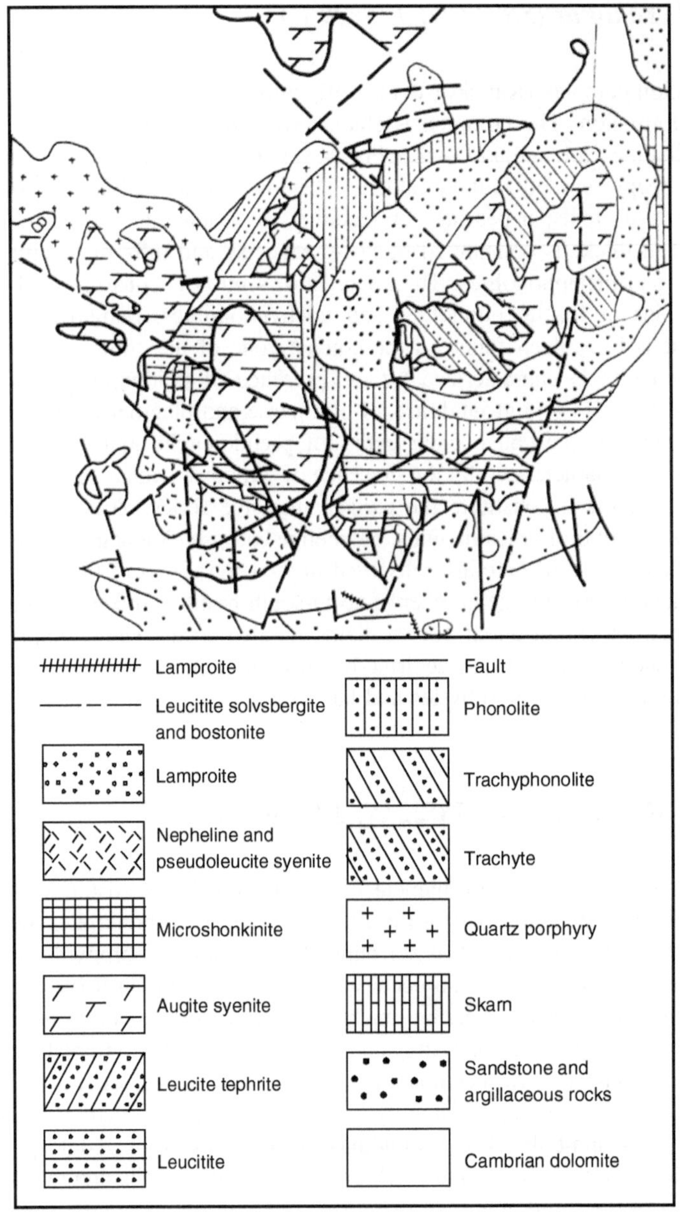

Fig. 4.32 Sketch map of Yakokut complex (after Maksunov 1973)

4.12.16 Lomam (57°07′N, 128°05′E)

Several stocks cutting across Cambrian and lower Jurassic country rocks are observed at Lomam. The stocks are constituted of mica peridotite, missourite, shonkinite and fergusite. The last two rock types are constituted of olivine, diopside, biotite or phlogopitic mica, leucite, pseudoleucite, K-feldspar and K-richterite in different proportions. Kalsilite and rarely nepheline are found within leucite. Apatite, sphene, chromian spinel and magnetite are present as accessory minerals.

The mica peridotites are constituted of olivine, clinopyroxene and phlogopite in equal amounts. Inclusions of leucite are found within pyroxenes. In missourites, the content of mafic minerals gradually decreases. Apart from leucite inclusions in pyroxene up to 30 vol%, this mineral is observed in areas between mafic minerals. Shonkinites often grade to melanocratic pseudoleucite syenites. The shonkinites are of variable compositions, consisting of pseudoleucite, feldspar-kalsilite intergrowth and mica in the following proportion of 6:2:1. They also contain unaltered leucite crystals with well-defined crystalline outline.

4.12.17 Tokko (55°36′N, 130°00′E)

Mafic lava flows (180 m thick) covering an area of 200 km^2 occur at Tokko on both sides of the Stanovoy ridge. The lower part of the mafic sequence comprises alkali basalts followed upward by peralkaline and weakly alkaline mafic rocks including hawaiite, limburgite, nepheline and leucite-bearing mafic rocks (Semenova et al. 1987). These rocks are associated with melanephelinite and phonolitic nephelinite. The alkaline volcanic rocks contain mantle xenoliths of spinel lherzolite, pyroxenite, wehrlite, websterite and xenocrysts of sanidine, olivine, orthopyroxene, spinel and clinopyroxene. The peralkaline mafic rocks contain Ti-phlogopite, kaersutite and Al-augite. Their $^{87}Sr/^{86}Sr$ ratio ranges between 0.7035 and 0.7045.

4.12.18 Dezhnevski Complex (66°05′N, 169°47′W)

The Dezhnevski complex (Perchuk 1964) covering an area of 125 km^2 occurs in the extreme northeast of Russia in the Chukotka province near the Berring Strait. The igneous body intrudes Carboniferous limestone and is of Palaeocene age. Syenites with or without nepheline occur in the periphery of a granitic body. Pseudoleucite-bearing shonkinites are reported within nepheline syenite.

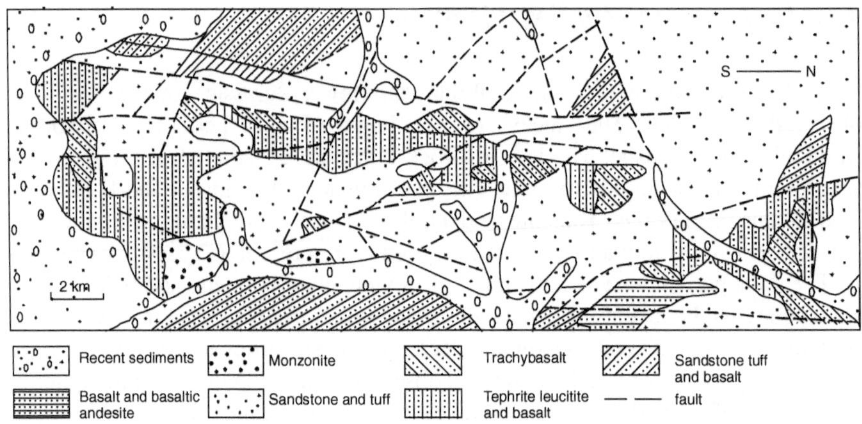

Fig. 4.33 Sketch map of Andriyanovka, Kamchatka (after Flerov and Koloskov 1976)

4.12.19 Andriyanovka (54°45′N, 158°30′E)

Trachybasalts, tephrites and trachyandesites with occasional occurrence of leucitite and absarokite are reported at Andriyanovka in the Kamchatka peninsula (Fig. 4.33) Tuffaceous rocks as well as monzonite stocks are also reported in this locality. The leucitites are characterized by the presence of phenocrystal augite (20–25 vol%), magnetite and rarely olivine in a groundmass comprising aegirine, leucite and magnetite. K–Ar age of leucite crystals yield an age of 20 ± 3 Ma (Flerov and Koloskov 1976).

4.12.20 Pyatistennyl (67°52′N, 161°36′E)

Dykes and lava flows of alkaline mafic rocks (Palaeocene) occur at Pyatistennyl of Omolon province in the northeastern part of Russia (Bazarova et al. 1981). The lavas are porphyritic with phenocrysts of clinopyroxenes, olivine, biotite (often altered) and magnetite in a glassy groundmass. Both leucite and analcite occur as inclusions within clinopyroxene.

4.12.21 Artem (43°46′N, 132°28′E)

Lavas and sills of trachydolerite, shoshonite, absarokite and leucite- and pseudo-leucite-bearing shoshonitic absorkites are wide spread at Artem of Primorye province. These rocks of Neogene age also occur as dykes, stocks and laccoliths.

One of the laccoliths is constituted of fergusite with phenocrysts of biotite, sanidine, anorthoclase, pseudoleucite and small amount of clinopyroxene (diopside, aegrine and augite). Calcite and apatite are also present in the fine-grained matrix.

4.13 Potassium-Rich Rocks from Oceanic Islands

4.13.1 Volcanic Activity in the Aeolian Arc Region

The potassic volcanism of southern Italy (including the Roman Province) is well-known and eruption of potassic lavas have been interpreted in terms of subduction-related processes (Ellam 1986). The Roman Province lavas are characterized by highly potassic volcanism, but the lavas of Aeolian islands located in the Tyrrhenian sea is known for various types of rocks from calc-alkalic series to more potassic shoshonitic lavas erupted within a period of 1 million years (Barberi et al. 1974; Dupuy et al. 1981). The archipelago is constituted of seven islands: Lipari, Salina, Stromboli, Vulcano, Panarea, Filicudi, and Alicudi (Fig. 4.20a). The archipelago is located near the northern continental slope of Sicily. According to Ellam et al. (1989) there is extensive variation in the K_2O content of the lavas of southern Italy. The Aeolian Arc lavas do not belong to the high K-series variety and may be described as low K_2O series rocks, and are also silica-saturated. They noted however, that two volcanoes of Aeolian island namely Stromboli and Vulcano erupted potassic-lavas. Rocks of Aeolian island form a continuous series from strongly silica-undersaturated ultrapotassic lavas to hypersthene-normative calc-alkaic rock series. It must be emphasized here that the majority of the Aeolian magmatism is calc-alkalic to high K- calc-alkalic in affinity.

Ellam et al. (1989) noted that the trace element pattern of K-rich mafic lavas of Roman and Aeolian lavas present negative anomalies for Nb and Ta. The rocks of Stromboli however, present relatively smaller anomaly. All the lavas show enrichment in LILE over REE. The trace element geochemistry of the Aeolian lavas is consistent with the view that they owe their origin to subduction-related volcanism.

The $^{143}Nd/^{144}Nd$ versus $^{87}Sr/^{86}Sr$ plot of variety of rocks from Aeolian Arc, Etna and Roman Province, shows continuous variation from high Nd isotopic ratios (Etna and Salina to Roccamonfina and Vulsini lavas). In fact they display a linear trend. Analyses of some of the rocks are given in Table 4.20.

According to Barberi et al. the volcanism in this area is characterized by a marked evolution in a restricted period of 1 million years, from typical cal-alkalic to shoshonitic rocks. The first volcanic stage was characterized by calc-alkalic series, ranging from high alumina basalt to dacite. At the beginning of the second stage, there was also eruption of potassic andesites at Lipari, whereas in Salina the last stage of calc-alkalic volcanism took place. The most recent volcanism is characterized by further increase of the potassium content, indicating their shoshonitic association. They comprise shoshonitic basalts, grading to latites and trachytes. Barberi et al. considered that rhyolites from Vulcanelo represent the final stage of this evolution. In Vulcanelo, leucite tephrite and related rock type were produced by fractionation of a shoshonitic magma under a f (O_2) condition of 10^{-2} bar. They thought that the evolution of Aeolian volcanism might have been related to rapid deepening of the Benioff zone. The occurrence of shoshonitic rocks and the continental nature of the crust on both sides of the plate boundaries may suggest that the Aeolian Arc is in a senile stage of evolution. From their Sr isotopic data, they considered a mantle source for Aeolian Arc volcanism.

4.13.2 K-Rich Rocks of the Tristan da Cunha Islands

The Island of Tristan da Cunha (Baker et al. 1964; Le Roxe 1990) is located at 37°S, 12°W in the South Atlantic (Fig. 4.34a). Tristen Da Cunha is the largest of a group of three islands. It is nearly circular in shape having a diameter of about 12 km and rises from a depth of −3,700 m to a height of 2,060 m. The outpouring of lava came from a central vent (Fig. 4.34b), but there are numerous parallel channels, through which some eruptions took place. The island has a near-perfect conical form and is bounded by a sea cliff, which rises nearly 600 m from the sea level. Of the two islands, one is called Nightingale and the other is named as Inaccessible. The rock types of Tristan da Cunha vary from ankaramatic basanite through phonotephrite, tephriphonolite to phonolite (Le Roxe et al. 1990). Earlier, Baker et al. (1964) thought that these rocks belong to alkali basalt-trachyte series, but Le Roxe et al. (1990) plotted composition of the rocks in a total alkali-SiO_2 diagram (le Bas et al. 1986) and found the rocks to belong to basanite-phonotephrite-phonolite series. The mineralogy of the rocks also support the nomenclature of the rock type (Fig. 4.34c). Basanite is the dominant rock type ($\sim$80 %) of the island. Basanites are characterized by the presence of 20–30 vol% euhedral to sudhedral clinopyroxene (Wo 47–50 En_{42-35} Fs_{11-15}), 5–20 vol% of olivine (Fo_{79-68}) and titano-magnetite phenocrysts. Small plagioclase (An_{90-34}), phenocrysts occur in a fine-grained matrix. The plagioclase grans show complex zoning. The basanites are characterized by the presence of normative nephilines, hence the name basanite. Some basanite samples contain additional Ti-rich kaersutites as a phenocrystal phase.

The phonotephrites range from moderately plagioclase-phyric to aphyric in texture comprising phenocrystal plagioclase (An_{88-17}), clinopyroxene (Wo_{49-47} En_{41-37} Fs_{11-15}), titano-magnetite (Usp_{56-76}) and kaersutitic amphibole. They often contain olivine (Fo_{77-42}), and alkali feldspar. In some samples leucite occurs as a late crystalline phase in the groundmass.

Tephriphonolites range in texture from microprophyritic to porphyritic with micro-phenocrysts of kaersutitic amphibole, titano-magnetite (Usp_{38-52}), plagioclase (An_{62-46}), clinopyroxene (Wo_{49-51}, En_{36-30}, Fs_{50-19}), sphene and apatite. Sometimes, clinopyroxene are aegirine-augite with hauine and sodalite.

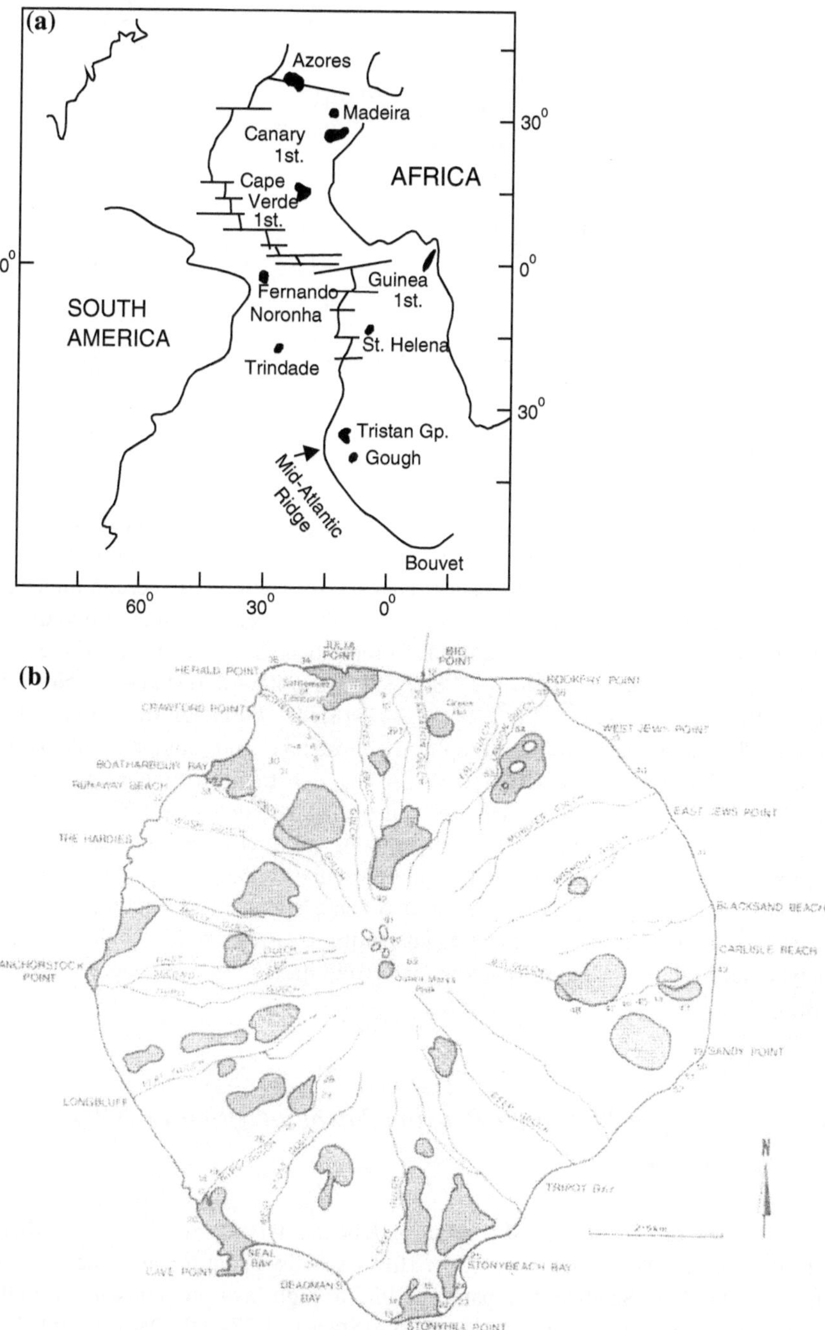

Fig. 4.34 **a** Location of Tristan da Cunha Island (after Borley 1974). **b** Map of Tristan da Cunha modified after baker et al. (1964), the *shaded* refers to Scoria deposits. **c** Plot of chemical composition of Tristan Da Cunha rocks in a total alkali versus SiO_2 diagram (after Brenna et al. 2014)

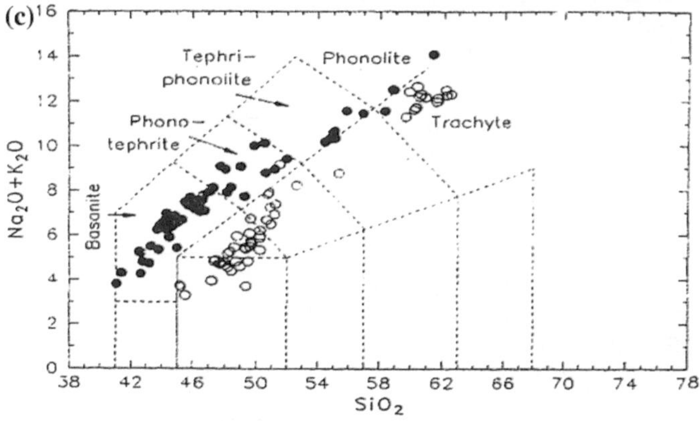

Fig. 4.34 continued

Major and trace element variations in the lavas present well-defined trends with increasing differentiation, which are consistent with fractional crystallization. The mg-numbers range from 40 to 62. There is also co-variation with certain minor and trace elements (e.g. Sr, Ba and P). This suggests the presence of at least two distinct fractionation trends. The Sr, Nd and Pb isotopic analyses of a subset of the lavas show the following ratios, $^{87}Sr/^{86}Sr$ (0.70495–0.70517), $^{143}Nd/^{144}Nd$ (0.51259–0.51247), and $^{206}Pb/^{204}Pb$ (18.47–18.74). Phonotephrites range from moderately plagioclase-phyric to aphyric in texture and comprise plagioclase (An_{88-17}), clinopyroxene (Wo_{49-47} En_{41-37} Fs_{11-15}), titano-magnetite and kaersutite. Apatite is an abundant accessory mineral. Alkali feldspar (Or_{12-19} Ab_{72-64} An_{16-17}) and leucite occurs as late crystallizing product as a groundmass phase. Tephri-phonolites are characterized by the presence of kaersutite, titano-magnetite (Usp_{38-51}), plagioclase (An_{62-46}), clinopyroxene, sphene and apatite. Phonolites comprise alkali feldspar, plagioclase, kaersutite, clinopyroxene, apatite, ilmenite, haüyne and sodalite. Melting relations of Tristan da Cunha rocks were studied by Tilley et al. (1965).

4.13.3 Trachyte-Phonolite-Bearing Lavas of Ulleung Island, South Korea

Brenna et al. (2014) studied the lavas of Ulleung Island, South Korea. Ulleung Island refers to the top of a 3,000 m (from sea floor) intraplate alkali volcanic edifice in the East Sea/Sea of Japan. The 950 m high lava pile consists of a alkali basaltic lavas and agglomeratic succession (Stage1, 1.37–0.97 Ma), intruded and overlain by a sequence of trachytic lavas and domes, which erupted in two episodes (Stage 2, 0.83–0.77 Ma; Stage 3, 0.73–0.24 Ma). Geological setting and digital elevation model for the Ulleung Island (900 m above the sea level is shown in

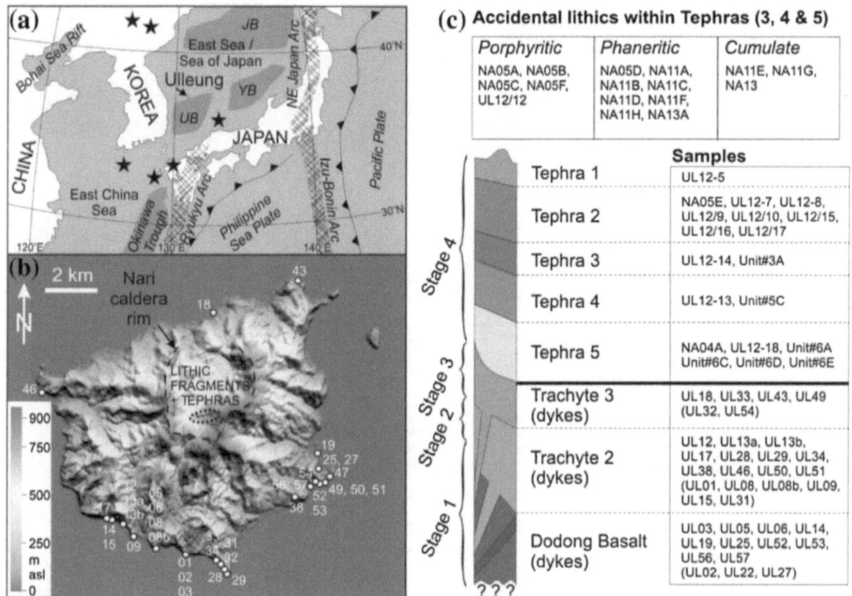

Fig. 4.35 **a**, **b**, **c** Geological setting and, **b** digital elevation model (DEM) for Ulleung Island (950 m above sea level). *Stars* in (**a**) represent sites of Quarternary intraplate volcanic activity and *cross-hatching* indicates volcanic arcs. *UB*, *YB* and *JB* are the Ulleung, Yamato and Japan Basins. The *JB* is the only part of the East Sea/Sea of Japan to have experienced new oceanic crust formation, the rest being variously stretched continental bloks (Tamaki et al. 1992). The *numbers* in (**b**) indicates locations of UL series whole-rock samples. The tephra samples and associated accidental lithics were collected within the *dotted* region inside the Nari caiidera. **c** Schematic stratigraphic column (not to scale) and nomenclature with sample groupings; simplified after Kim et al. (1999), Brenna et al. (2014)

Fig. 4.35 a and b, whereas stratigraphic succession of various tephras and trachytic dykes and basalts are Fig. 4.35c. The youngest eruptions, post 20 ka BP, were explosive, generating thick tephra sequences of phonolitic composition (Stage 4), which also entrained phaneritic, porphyritic and accidental cumulate-bearing rock types. Major element chemistry of the evolved products show a continuous spectrum of trachyte to phonolite compositions, but these have discordant trace element trends and distinct isotopic characteristics, excluding a direct genetic relationship between the two end-members. Despite this, the Stage 3 trachytes and some porphyritic accidental lithics have chemical characteristics transitional between Stage 2 trachytes and Stage 4 phonolites. Within the phonolitic Stage 4 tephras three subgroups can be distinguished. The oldest, Tephra 5, is considerably enriched in incompatible elements and chondrite-normalized rare earth element (REE) patterns, which display negative Eu anomalies. The later tephras, have compositions intermediate between the early units and the trachyte samples, and their REE pattern do not have significant Eu anomalies. The last erupted, Tephra 1, from a small intra-caldera structure, has a distinct tephriphonolite composition. Trace element and

Table 4.20 Analyses of K-Rich rocks from Oceanic Islands

	SiO_2	TiO_2	Al_2O_3	Fe_2O_3	FeO	MnO	MgO	CaO	Na_2O	K_2O	P_2O_5	H_2O^-	LOI	Total
1	54.80	1.59	19.13	–	5.37	0.21	1.43	5.48	5.49	4.89	0.37	0.11	0.62	99.49
2	56.93	1.33	19.50	–	4.64	0.15	1.31	3.84	6.08	5.39	0.31	0.20	1.03	100.70
3	61.38	0.51	19.78	–	2.30	0.18	0.30	1.31	7.28	6.81	0.06	0.19	0.74	100.80
4	58.34	1.13	19.90	–	3.69	0.13	0.76	2.91	5.70	5.88	0.20	0.48	1.18	100.30
5	57.91	0.65	18.22	1.90	3.20	0.27	1.01	3.58	4.98	6.69	0.48	0.38	–	100.20
6	51.79	0.66	14.96	9.03	–	0.17	4.78	8.60	3.51	4.53	0.44	–	0.36	98.83
7	57.32	0.53	15.14	7.40	–	0.14	3.49	6.38	3.80	5.04	0.34	–	0.05	99.63
8	50.86	0.77	16.15	10.06	–	0.17	3.69	8.06	2.71	4.22	0.51	–	1.41	98.61
9	50.44	0.90	16.21	8.62	–	0.17	5.80	9.92	2.29	3.45	0.69	–	1.27	99.76

1–4 Tristan da Cunha (after Le Roxe et al. 1990)
5 The bulk chemical composition of Leucite-bearing rock from Utsuryoto Island (after Tsuboi 1920). It also includes 0.07 wt% ZrO_2 and 0.82 wt% H_2O^+
6–9 Leucite-bearing lavas from Vulcano Island, Aeolian Arc (Ellam et al. 1989)

isotopic chemistry as well as textural characteristics suggest a genetic relationship between the phaneritic lithics and their host phonolitic pumices.The Stage 4 tephras are not related to earlier phases of basaltic to trachytic magmatism (Stages 1–3). They have distinct isotopic compositions and cannot be reliably modelled by fractional crystallization processes.The differences between the explosive phonolitic (Stage 4) and effusive trachytic (Stage 2–3) eruptions are mainly due to different pre-eruptive pressures and temperatures, causing closed- versus open-system degassing. Based on thermodynamic and thermobarometric modelling, the phonolites were derived from deeper (subcrustal) magma storage and rose quickly, with volatiles trapped until eruption. By contrast, the trachytes were stored at shallower crustal levels for longer periods, allowing open-system volatile exsolution and degassing before eruption.

The volcanic history of the island, has been described by Tsuboi (1920), who described the island as Utusryo or Ullungdo. According to volcanologists, the eruption of lave started first with the eruption of alkali basalt followed by a long period of erosion, then there was out-pouring of trachytic lavas. This was followed by collapse of the apical part of the trachytic body. In the final stage there was eruption of leucite-bearing lavas.

The alkali basaltic lavas are characterized by the presence of phenocrystal olivine, titanaugite and plagioclase in a groundmass of plagioclase, titanaugite, magnetite and glass. The second stage of evolution of the volcano was accompanied by eruption of 47 distinct flows of latite, trachyte, hornblende-bearing trachyte, aegirine-augite-bearing phonolite, aegirine-augite trachyte, amphibole-bearing trachyte and biotite trachyte. The last stage of evolution was associated with eruption of leucite-bearing tephritic phonolites through a triangular depression near the top of the volcano. The lavas were poured out at the beginning from the bottom part of the caldera and afterward from its northwest corner forming a small cone at a place called Arpong; where there was accumulation of lapilli and pumice around the vent through which they were ejected. The last lava flow was discharged from this vent to flow northward. This vent is 800 m long, 300 m wide and 20 m thick. Under thin-section the lavas from Arpong shows the presence of phenocrystal pyroxene, amphibole, sanidine and labradorite and microphenocrysts of anorthoclase, olivine and apatite in glass. Pyroxene phenocrysts are of two types: titanaugite and aegirine-augite. The aegirine-augite is weakly pleochroic, whereas titanaugite has purplish tint with hour-glass structure. The rocks also show presence of tabular biotite crystals, apatite needles and rare amount of olivine in the groundmass. Minute laths of K-feldspar and plagioclase are associated with round crystals of leucite, aegirine-augite and magnetite. Chemical analyses of a K-rich rock from Ulleung Island are given in Table 4.20 (Fig. 4.35).

Chapter 5
Minor and Rare Earth Element Geochemistry of K-Rich Silica-Undersaturated Igneous Rocks

The silica–undersaturated K-rich igneous rocks are highly enriched in the lighter REE. Their La abundance for example, is often 1,000–1,400 times greater than that of chondrites (Mitchell et al. 1987). Abundance of the heavy REE is not as high, but still 6–20 times that of chondrites. The La/Yb ratio of these rocks is significantly high and range between 160 and 280, average being 209. In those varieties with high mg-number, Ni, CO and Cr contents are also high. As a consequence of high K content, the rocks are naturally enriched in Rb, Ba and Sr contents. The LILE/LREE ratios in these rocks are in general low, but their $^{87}Sr/^{86}Sr$ ratios are high. In lamproites from Spain, this ratio could be as high as 0.725–0.730. In contrast, the same ratio in basaltic rocks varies between 0.703 and 0.705. The ε_{Nd} content of K-rich rocks is always negative, compared to that of average basaltic rocks, which are characterized by positive values. In the following sections the trace element and REE characteristics of the K-rich silica–undersaturated lavas from various localities are described.

5.1 The Minor and Rare Earth Element Characteristics of Lamproites from Damodar Valley Coal Fields

Five intracratonic basins, developed in eastern India, during rifting in the Damodar valley coal fields region (from west to east), are: (1) Karanpura, (2) Bokaro, (3) Ramgarh; (4) Jharia; and (5) Raniganj (Fig. 4.3). The basal unit in each basin is the Talchir boulder bed of Carboniferous to Asselian age, succeeded by thick coal-bearing silica-rich clastic sequences (Ghosh and Mukhopadhyay 1985). The basins are characterized by synsedimentary extension faulting, and are demarcated to the south by the southern boundary fault. Precambrian gneisses are present to the north and south. Large-scale intrusions of lamproites and lamprophyres are present as east-west-trending dykes and sills confined to the basins. Contemporaneous dolerite dykes also occur within and outside the basins.

From geochemical data on ultrapotassic rocks in the Jharia basin, Rock et al. (1992) concluded that these were a comagmatic suite of olivine lamproites, lamproites, lamprophyres, and minettes. Jia et al. (2003) analysed K-rich intrusive rocks from the

A.K. Gupta, *Origin of Potassium-rich Silica-deficient Igneous Rocks*,
Springer Geology, DOI 10.1007/978-81-322-2083-1_5

Jharia, Ranganj, and Bokaro basins and two lamprophyres from the Ranganj basin. All the samples have been characterized petrographically by Gupta et al. (2002).

As there is a large variation in the isotopic composition of the geologic reservoir (Clayton 1981), to increase the database on ultrapotassic rocks, Jia et al. (2003) also included a lamproite from dyke swarms at Chelima in the Andhra Pradesh district of south India, which were emplaced in the Cuddapah basin on the eastern margin of the Archean Dharwar craton. They intrude a rock sequence of quartzite and shales intercalated with limestone, dolomite, and cherts emplaced at about 1,200 Ma ago (Bergman and Gas 1984).

Several petrographic and geochemical studies have been conducted on the Gondwana ultrapotassic suites (Rock et al. 1992) and Basu et al. (1997). To summarize, the rocks are characterized by variable modal proportions of major phases, principally micro-phenocrysts of Ti–rich phlogopite, and Mg–olivine pseudomorphed by talc and serpentine. The following minor minerals have been reported: aegirine, amphibole, apatite, both primary and secondary ankerite and Mg–calcite, chlorite, Cr–spinel, diopside, fluorite, K–feldspar, hematite, ilmenite, perovskite, priderite, pyrite, rutile, titanomagnetite, and zeolites (Rock et al. 1992). Least altered rocks feature typical lamprophyric textures, with euhedral mafic minerals and micas, globular structures, and lack of felsic phenocrysts. They are distinct from kimberlites in having euhedral rather than rounded olivine, and in the presence of amphibole and feldspar.

5.1.1 Nitrogen Content of Gondwana Potassic Rocks

The total range of δ^{15} N values for the Gondwana ultrapotassic suite (Jia et al. 2003) is from +1.6 to +8.7‰ with and average of +3.8 %. Ranges for the different basins overlap, and lamprophyres have similar N–isotope compositions to lamproites. Nitrogen content averages 107 ppm. Results for phlogopite separates from these ultrapotassic rocks yield a similar range of δ^{15}N values from 1.4 to 5.2 %. Bulk rock—phlogopite N–isotope fractionations ($\Delta^{15}N = \delta^{15}N_{rock}–\delta^{15}N_{phlogopite}$) are –0.8 to + 0.7 %. The Proterozoic lamproite from Cuddapah is also ^{15}N-enriched (Table 5.1 and Fig. 5.1). The Sr isotopic characteristics as a function of 1/Sr is shown in Fig. 5.2. The co-variation of K_2O and N content of ultrapotassic rocks from Damodar Valley region is shown in Fig. 5.3, whereas chondrite-normalized spidergrams for the same rocks (Table 5.1) are shown in Fig. 5.4.

5.1.2 Major and Trace Elements

Lamproites from the Raniganj and Jahira basins are characterized by extremely high contents of TiO_2 (4.2–10.1 wt%), K_2O (2.4–7.6 wt%), and P_2O_5 (2.3–6.2 wt%) over a range of Mg # from 80 to 75, in keeping with the compositions of lamproites

Table 5.1 $\delta^{15}N$ contents and $\varepsilon^{15}N$ values, and N/K molar ratios of selected samples and phlogopite separates in Gondwana ultrapotassic suite (Jia et al. 2003)

Basin	Sample no.	Localities	Rock type	N (ppm)	$\delta^{15}N$ (%)	K_2O	N/K ($\times 10^{-3}$) molar ratio	N[a] (ppm)	$\delta^{15}N$[a] (%)
Damodar valley, East India									
Bokaro	CP60/93	Jarangdih	Lamproite	163	4.4	2.89	18.9	252	5.2
	D-4	Bokaro	Lamproite	88	3.9	5.32	5.6		
Jharia	SD-1	Sudamdih	Lamproite	209	4.3	7.09	9.9	244	3.8
	SD-1	Sudamdih	Lamproite	63	6.2				
	SD-3	Sudamdih	Lamproite	232	1.5	7.14	10.9		
	P-1	Pathardih	Lamproite	277	1.6	7.31	12.7	405	1.4
	P-2	Pathardih	Lamproite	294	2.2	7.56	13.1		
Raniganj	M2/1	Mohanpur	Lamproite	119	4.1	3.42	11.7	260	3.4
	M2/2	Mohanpur	Lamproite	394	5.5	4.69	28.2		
	KH-1	Khudia	Lamprophyre	82	1.6	4.91	5.6	450	3.7
	KH-2	Khudia	Lamprophyre	371	3.3	4.96	25.1		
Andhra Pradesh, South India									
Cuddapah	R8339A	Chelima	Lamproite	21	8.7	2.41	2.9		
Primitive mantle[b]				1–2		0.0301	112–22.3		
Continental crust[c]				60		2.14	9.4		

[a] Values were measured on phlogopite separates
[b] N and K_2O contents of mantle are from Marty (1995) and Sun and McDonough (1989), respectively
[c] N and K_2O contents of continental crust are from Wedepohl (1995)

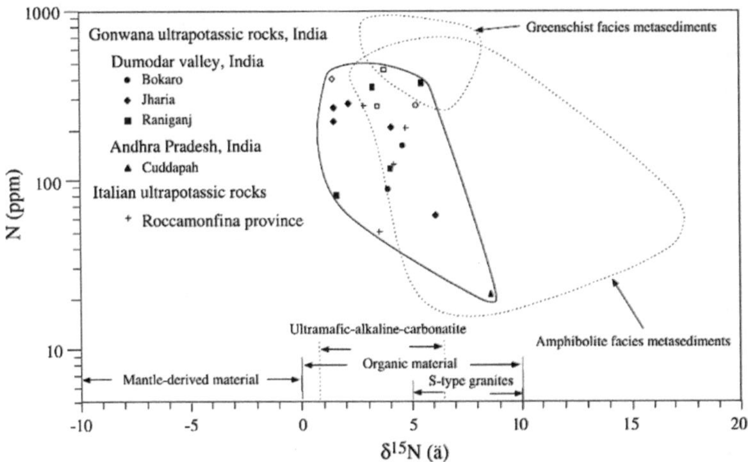

Fig. 5.1 N contents and δ^{15}N value of lamproites and lamprophyres (closed symbols) and phlogopite separates (open symbols) from the Gondwana basin and of ultrapotassic rocks from Italy (cross symbols; data from Gupta et al. (2002). Field sources: mantle-derived material from Javoy et al. (1984), Boyd et al. (1987), Cattigny et al. (1997), Marty and Humbert (1997); metasedimentary rocks from Jenner et al. (1990); S-type granites from Boyd et al. (1993); organic material from Peters et al. (1978); and ultramafic-alkaline-carbonatite of the Kola Peninsula from Dauphas and Marty (1999, after Jia et al. 2003)

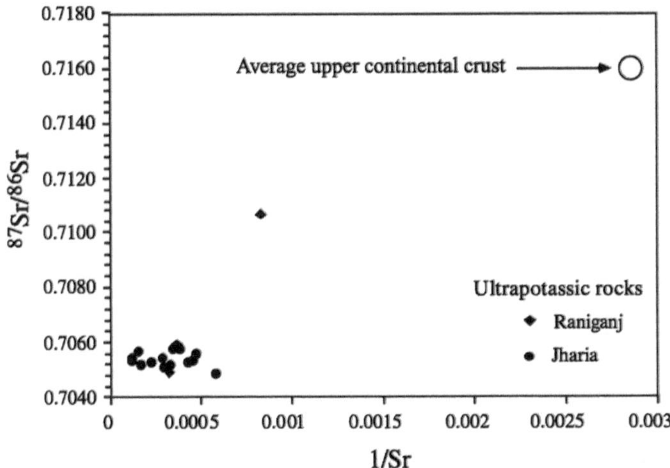

Fig. 5.2 Plot of ^{87}Sr/^{86}Sr versus 1/Sr for Raniganj and Jharia ultrapotassic rocks in eastern India. *Sources* Raniganj data (Middlemost et al. 1988), Jharia data (Rock et al. 1992), and average upper continental crust (Taylor and Mclennan 1985; Faure 1977)

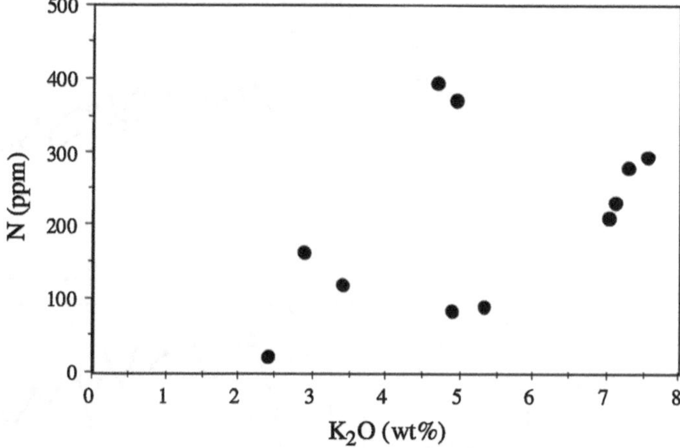

Fig. 5.3 Covariation of K_2O and N content of whole ultrapotassic rocks from the Damodar valley. Data in Table 5.1

worldwide (Bergman 1987). Ultramafic lamprophyre from the Raniganj basin feature greater SiO_2 but lower TiO_2, P_2O_5, and K_2O. Lamproites and lamprophyres from the Damodar valley, have the combination of extreme REE abundances (1,019–5,400 ppm), with strongly fractionated REE, where $La/Sm_{cn} = 3.5–5.6$ and $GD/Yb_{cn} = 7.3–18$, consistent with residual garnet in the source (Table 5.2, one outlier KH-2). There is an inflection in the REE patterns at Nd–Sm, as observed in other studies of lamproites from the Damodar valley (Rock et al. 1992). On primitive mantle-normalized diagrams, the Damodar Ultrapotassic suite shows: (1) normalized depletions of Cs and Rb relative to Ba and Th; (2) negative anomalies at Nb–Ta, Pb, P, Zr–Hf, and Ti; and (3) Nb/Ta and Zr/Hf ratios that are in general variably greater than the respective primitive mantle values of 17 and 36, as is seen in mantle alkaline and carbonic magmas. The Andhra Pradesh lamproite is distinctive in greater abundances of Cs and Rb, with no or diminished negative anomalies at Nb–Ta, Zr–Hf, and Ti (Table 5.2 and Fig. 5.4)

5.2 The REE and Minor Element Geochemistry of Birunga and Toro-ankole Rocks

The REE and trace element geochemistry of K-rich volcanic rocks from Birunga and Toro Ankole was studied in detail by Higazy (1954), Bell and Powell (1969), De Mulder (1985), Rogers et al. (1985) and Rogers (1992). It is observed that these rocks are enriched in such mantle incompatible elements as Ba (100–7,000 ppm), Rb (150–3,000 ppm), Sr (4,500–9,500 ppm) and Zr (800–1,200 ppm). The more ultramafic variety is rich in such compatible elements as Ni (100–230 ppm), Co (65–85 ppm) and Cr (290–650 ppm). In some of the Mg–poor melilite leucitite and

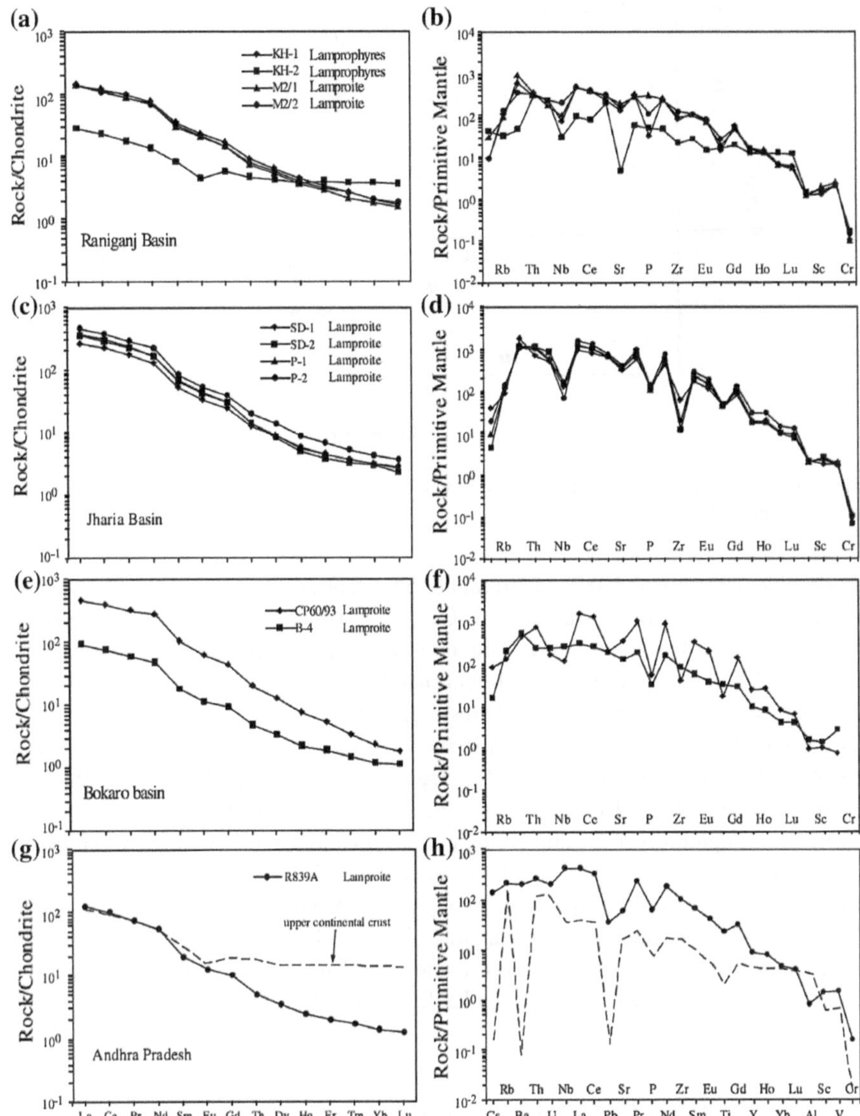

Fig. 5.4 Rare Earth element and primitive mantle-normalized multi-element diagrams for the ultrapotassic intrusive suite from the Damodar valley (**a**–**f**) and Andhra Pradesh (**g**–**h**.) Normalizing factors are from Sun and Mc Donough (1989; after Jia et al. 2003)

melilite-nepheline leucitite, the content of Ni (30–40 ppm), Co (40–70 ppm) and Cr (1–70 ppm) is significantly lower.

The $^{87}Sr/^{86}Sr$ ratio of mafurite (0.7050), leucitite (0.7058), kivite (0.7067), katungite (0.7047), K-rich basanite (0.7043) and absarokite (0.7076) is high. The ratio is comparable to K-rich rocks of Leucite Hills and Navajo Hopi volcanic provinces

Table 5.2 Major and trace elements in Gondwana lamprophyres and lamproites from Damodar Valley Coal Fields and Andhra Pradesh

	Raniganj basin				Jharia basin				Bokaro		Andhra Pradesh
	Lamprophyres		Lamproites		Lamproites				Lamproites		Lamproites
	KH-1	KH-2	M2/1	M2/2	SD-1	SD-2	P-1	P-2	CP60/93	B-4	R839A
SiO_2	47.62	43.21	35.29	35.7	32.37	33.21	33.41	34.21	30.30	41.10	40.08
TiO_2	3.20	3.68	4.25	5.94	9.32	10.12	9.21	9.62	3.60	7.20	5.21
Al_2O_3	5.80	6.41	5.54	5.99	9.84	9.21	9.56	10.12	4.32	7.34	3.99
Fe_2O_3	8.51	9.11	11.2	11.22	8.68	8.75	9.1	8.21	8.58	12.44	9.76
Mno	0.14	0.2	0.286	0.137	0.12	0.11	0.32	0.42	0.15	0.16	0.11
Mgo	8.01	12.12	8.31	10.05	8.09	8.31	8.11	8.42	19.89	10.65	15.93
CaO	8.40	6.98	7.18	5.85	8.26	8.31	8.29	7.18	10.25	2.89	7.51
Na_2O	0.11	0.14	0.11	0.08	0.13	0.25	0.16	0.31	0.15	0.13	0.06
K_2O	4.91	4.96	3.42	4.69	7.09	7.14	7.31	7.56	2.89	5.32	2.41
P_2O_5	0.71	1.11	6.32	2.42	2.93	2.45	2.32	2.81	1.20	0.75	1.41
LoI	12.54	12.12	14.77	15.61	11.95	11.35	12.31	11.85	2.86	2.18	12.05
Total	99.95	100.04	98.48	98.99	98.78	99.21	100.10	100.71	84.19	90.16	98.52
Mg#	78.86	84.05	74.62	78.02	78.69	79.00	77.93	80.25	90.15	77.23	86.61
Al(ppm)	30,697	33,925	29,320	3,1702	5,2078	4,8744	50,596	53,560	22,864	38,847	21,117
Ti	19,181	22,058	25,475	35,604	55,864	60,659	55,205	57,662	21,578	43,157	31,229
P	3,098	4,844	27,580	10,561	12,787	10,692	10,124	12,263	5,237	327.	6,153
Cr	310	483	308	424	342	205	342	274			479
Co			49	49	83	81					
Ni	214	373	425	298	336	350					
Rb	67.5	20.4	57.0	85.2	54.3	81.8	83.2	91.5	85.5	133	137
Sr	2,818	99	4,047	3,170	6,311	7,011	7,344	8,394	7,631	2,816	1,275

(continued)

Table 5.2 (continued)

	Raniganj basin				Jharia basin				Bokaro		Andhra Pradesh
	Lamprophyres		Lamproites		Lamproites				Lamproites		Lamproites
	KH-1	KH-2	M2/1	M2/2	SD-1	SD-2	P-1	P-2	CP60/93	B-4	R839A
Cs	0.30	1.32	0.99	0.31	10.26	0.15	0.30	0.60	2.73	0.50	4.51
Ba	4,151	337	6,610	2,585	12,149	7,598	8,126	7,443	3,305	3,838	1,439
Sc	24.6	27.6	34.0	23.5	31.0	44.4	41.2	40.4	17.7	23.5	25.4
V	194	176	201	179	151	150	161	137	62	221	123
Ta	2.80	1.33	3.28	5.57	4.22	4.43	5.17	1.92	5.09	7.80	13.5
Nb	54.7	22.2	72.3	147	87.8	107	118	46.8	82.6	186	308
Zr	865	252.3	1,062	1,420	679	133	144	215	449	946	1,186
Hf	24.1	7.30	18.8	29.3	18.3	4.39	5.73	5.63	12.5	25.2	28.4
Th	25.4	24.63	30.5	28.8	59.6	94.8	94.4	86.4	64.4	20.9	22.6
U	4.73	4.76	3.59	5	10.3	14.4	11.9	10.8	3.72	5.14	4.20
Y	59.6	59.6	69.0	73	77.1	77.3	86.8	134	114	44.6	41
La	340	66	337	334	627	827	864	1,059	1,097	219	297
Ce722	722	142	687	671	1,337	1,765	1,812	2,259	2,440	461	597
Nd	342	64	330	318	564	728	746	1,008	1,241	217	249
Sm	51.6	12.1	45.5	47.7	78.6	102	97.1	126	156	7.0	29.91
Eu	13.2	2.5	11.7	12.2	19.2	24.3	23.4	30.5	35.0	6.40	7.18
Gd	34.6	11.7	29.4	29.7	49.0	59.5	59.1	76.8	87.5	18.2	19.74
Tb	3.34	1.7	2.73	2.94	4.52	5.05	5.15	7.13	7.40	1.80	1.86
Dy	16.0	10.6	13.2	14.4	20.9	20.6	21.7	33.4	32.2	8.35	8.77
Ho	2.45	2.14	2.05	2.21	3.17	2.83	3.13	4.86	4.39	1.29	1.36
Er	5.50	6.45	4.64	5.07	7.25	6.19	7.03	11.1	8.56	3.04	3.21

(continued)

Table 5.2 (continued)

	Raniganj basin				Jharia basin				Bokaro		Andhra Pradesh
	Lamprophyres		Lamproites		Lamproites				Lamproites		Lamproites
	KH-1	KH-2	M2/1	M2/2	SD-1	SD-2	P-1	P-2	CP60/93	B-4	R839A
Tm	0.64	0.94	0.54	0.64	09	0.81	0.91	1.30	0.84	0.37	0.42
Yb	3.44	6.21	3.09	3.35	5.06	4.75	5.19	6.90	3.97	2.03	2.34
Lu	0.41	0.90	0.40	0.46	0.69	0.57	0.67	0.90	0.45	0.29	0.31
Li	11.6	53.3	21.0	39.2	32.0	10.9	11.7	22.1	22.4	9.55	69.0
Be	10.0	2.7	8.49	12.7	7.87	7.85	11.2	8.15	5.68	7.04	10.3
Tl	0.14	0.23	0.41	0.21	0.70	0.46	0.48	0.43	0.25	0.35	0.08
Pb	41.6	38.57	55.1	57.9	115	119	129	135	38.3	35.9	6.51
Bi	0.08	0.27	0.10	0.08	0.20	0.20	0.28	0.24	0.05	0.07	0.07
LA/Yb cn	68	7	75	69	86	120	115	106	191	75	88
La/Sm cn	4.1	3.5	4.6	4.4	5.0	5.1	5.6	5.3	4.4	5.1	6.2
Gd/Ub cn	8.3	1.6	7.9	7.3	8.0	10.3	9.4	9.2	18.2	7.4	4.0
Ce/Ce*	1.01	1.03	1.01	0.66	1.03	1.04	1.03	1.03	1.04	1.02	1.03
Eu/Eu*	0.947	0646	0.968	0982	0.938	0946	0.938	0944	0.914	0.877	0.897
Al_2O_3/TiO_2	1.81	1.74	1.30	1.01	1.06	0.91	1.04	1.05	1.20	1.02	0.77
Ti/Zr	22	87	24	25	82	457	385	268	48	46	26
Zr/Y	14.53	4.23	15.38	19.45	8.81	1.72	1.65	1.61	3.94	21.24	29.23
P/Nd	9.05	75.64	83.51	33.19	22.67	14.67	13.57	12.17	4.22	15.10	24.73
Zr/Hf	36	35	57	48	37	30	25	38	36	38	42
Nb/Ta	19.54	16.70	22.05	26.34	20.79	24.06	22.77	24.35	16.22	23.83	22.86

Note The major element analyses of Cp 60/93 and B-4 are from Basu et al. (1997)

of the U.S.A. and the Phlegrean Fields (Italy), but lower than that of such rocks from West Kimberley, Jumilla (Spain) and Alban Hills (Italy).

The lavas from the eastern Birunga regions are strongly enriched in the light rare earth elements (LREE), for example, La content varies from 65 to 132 ppm (Rogers et al. 1985). These lavas are similarly enriched in other incompatible elements and the convex-upward mantle-normalized abundance curves are essentially similar to those of other intra-plate mafic alkaline rocks. The La/Yb ratios of the eastern Birunga rocks range from 30 to >55. Within the low-silica lavas from Muhavura, the REE fractionation varies from La/Yb = 28 at ~ 47 wt% SiO_2 to La/Yb = 46 in samples with ~43 wt% SiO_2. Other indices of incompatible element enrichment, such as Nb/Y and Zr/Nb, also vary systematically with silica. This behaviour contrasts with the opposite trend shown by the Sabinyo latites, within which La/Yb ratio increases from 48 to 62 wt%). The mantle normalized abundance patterns for various elements observed in cases of lavas of Muhavura, Gahinga and Sabimyo are shown in Fig. 5.5a–d. All the lavas are shoshonitic with $1 < K_2O/Na_2O < 2$, and are strongly enriched in incompatible elements. The $^{87}Sr/^{86}Sr$ ratio varies from 0.70586 in the K–basanites to 0.70990 in the latites. The $^{143}Nd/^{144}Nd$ ratio ranges from

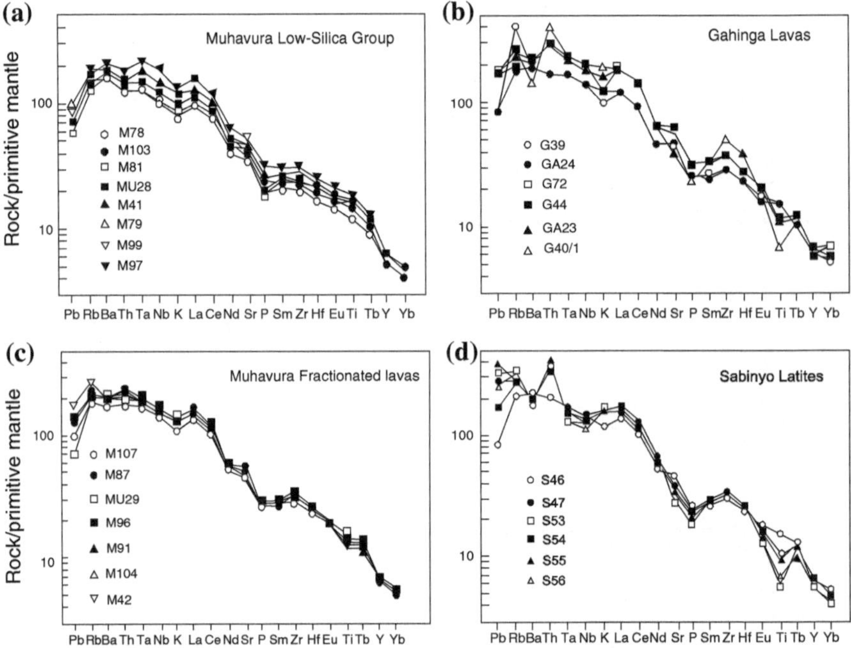

Fig. 5.5 Mantle-normalized (Sun and McDonough 1989) abundance patterns for selected lavas from the eastern Birunga province showing smooth and convex-upward profiles of the K-basanite series lavas from Muhavura and Gahinga. Occasional anomalies at Ba and P fractionation of K-feldspar and titano-magnetite, respectively, in more evolved lavas. The Sabinyo latites are distinct from the K-basanites showing pronounced enrichment of Th, Pb and Rb and slight enrichement of K and La over Ta and Nb. Negative Ti anomalies are more pronounced than in the K-basanite (after Rogers et al. 1998)

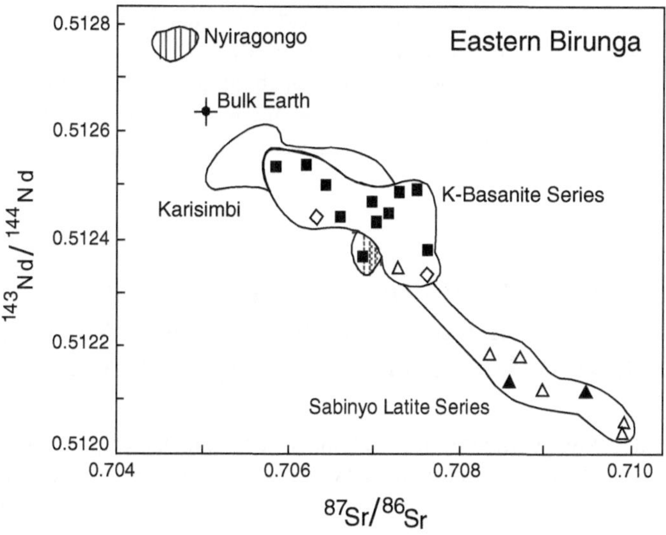

Fig. 5.6 Conventional Sr–Nd isotope plot for the eastern Birunga lavas compared with the field defined by K-basanites from Karisimbi and Nyiragongo in the western parts of the Birunga province (Vollmer and Norry 1983a). All of the data plot in the enriched quadrant of the diagram, implying source regions with higher Rb/Sr and lower Sm/Nd ratios than bulk Earth. The sabinyo latites plot at more extreme values of the isotope ratios than the K-basantie series, in agreement with two analyses of similar rocks from Vollmer and Norry (1983b) (after Rogers et al. 1985)

0.5124 to 0.51206. The Pb isotope ratios range in the following manner: $^{206}Pb/^{204}Pb$: 19.30–19.51; $^{207}Pb/^{204}Pb$: 15.69–15.93 and $^{208}Pb/^{204}Pb$ 40.28–41.5). The low-silica K–basanites reflect interaction between a mafic K–basanitic melt ($^{143}Nd/^{144}Nd$ ~ 0.51204, $^{87}Sr/^{26}Sr$ ~ 0.707) and a nephelinite ($^{143}Nd/^{144}Nd$ ~ 0.51267 and $^{87}Sr/^{86}Sr$ ~ 0.7045). The isotope ratios of Sr and Nd of selected eastern Birunga lavas are illustrated in a conventional isotope diagram (Fig. 5.6).

Rogers (1992) compared the mantle-normalized abundance of incompatible elements of leucite-basanites from Karisambi (Birunga volcanic Province (De Mulders et al. 1986) and Vulsini (Italy; Rogers et al. 1985). Despite similar abundance, Vulsini leucite basanites are lower in Ta, and Nb and to a lesser extent Zr and Nb contents than similar rocks from Karsimbi. Rogers (1992) thought that K-rich basanites, representing average analyses of Karsimbi rocks is similar to magmas from in-plate environment.

The Rb/Sr ratio in Karsimbi lavas is ~1 and significantly higher than non-potassic basanites and basalts, which have this ratio close to 0.03. The Th /U ratio in leucite basanites from Karsimbi is very high having an average of 5.5 compared with the values of 4 or less in most crustal rocks and mantle-derived magmas.

The K-rich volcanic rocks from Toro-Ankole and Birunga have initial $^{87}Sr/^{86}Sr$ ratio varying between 0.7036 and 0.7076. The K–feldspar-bearing absarokite

(0.7076) and kivite (0.7067) have higher $^{87}Sr/^{86}Sr$ ratios than feldspathoid-bearing katungite (0.7047), leucitite and mafurite (0.7050).

Ultrapotassic lavas from the western branch of the East African Rift display similar Pb isotope characteristics, although with a wider range of $^{207}Pb/^{204}Pb$ ratios, overlapping with the MORB-OIB field, suggesting that they may have experienced some crustal contamination.

5.3 The Rare Earth Element and Trace Element Geochemistry of Lamproites from Western Australia, Leucite Hills (U.S.A.) and Gaussberg (Antarctica)

Nelson et al. (1986) compared trace element geochemistry of potassic rocks from Western Australia (McCulloch et al. 1983; Neslon et al. 1986; Jaques et al. 1984), Presley Peak and Gaussberg (Antarctica, Sheraton and England 1980; Sheraton and Cundari 1980) and Leucite Hills (Kuehner et al. 1981; Volmer et al. 1984). The data compiled by Nelson et al. are summarized in Fig. 5.7, where the trace element geochemical pattern involving such elements as Pb, Rb, Ba, Th, U, K, Nb, La, Ce,

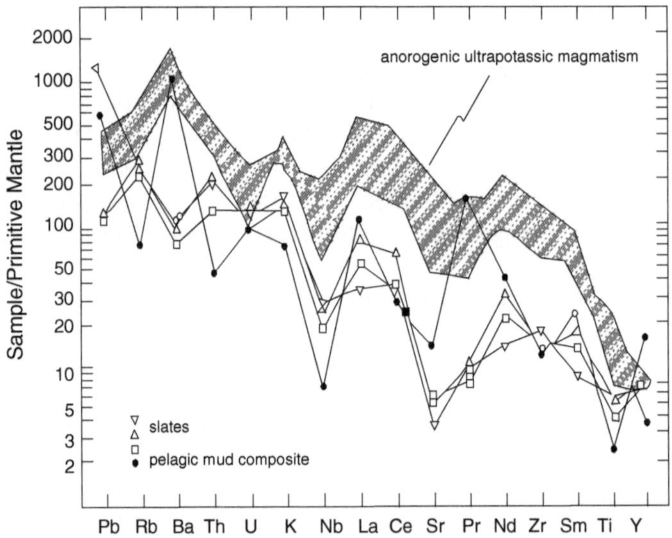

Fig. 5.7 Trace element patterns normalized to estimated primitive mantle (abundances of average potassic magma) hatched (from Western Australia) (McCulloch et al. 1983; Nelson et al. 1986; Jaques et al. 1984; Nixon et al. 1984), Priestly peak and Gaussberg, Antarctica (Sheraton and England 1980; Sheraton and Cundari 1980) and Leucite Hills (orendites and wyomingites from Kuehner et al. 1981; Volmer et al. 1984) compared with some examples of modern sediments (Thompson et al. 1984; after Nelson et al. 1986)

Sr, P, Nd, Zr, Sn, Ti and Y is shown (hatched area) in a chondrite-normalized spider diagram, which demonstrates extremely high concentration of Sr, Pb, and REE.

There is also extreme degree of LREE enrichment. In a plot of ε_{Nd} versus $^{87}Sr/^{86}Sr$ ratio, it is observed that the rocks of Leucite hills and Presley Peak has Sr isotopic ratio > 0.705, and lamproites of Western Australia show extreme enrichment with respect to the same ratio.

The Gaussberg lamproites have initial values ranging between 0.7105 and 0.7120. All these rocks are characterized by ε_{Nd} values lower than −8. The MORB basalts have positive ε_{Nd} values between 5 and 13 and lower Sr isotropic ratios. Initial ε_{Nd} values for Manning Massif tristanite, Mt. Bayliss alkali mela-syenite and Bunger Hills trachybasalt dykes are −9.3, −12.3 and <−16.3, respectively.

The Pb–Pb isotope diagram showing composition of lamproites from Western Australia, Gaussberg (Nelson et al. 1986), Leucite Hills and Antarctica dykes are shown in Fig. 5.8a. The Pb isotopic signature of the Western Australian and Gaussberg lamproites has been interpreted by Nelson et al. (1986) in terms of involvement of an extremely ancient source component, which had high U/Pb ratio, early in its history, followed by a mixing with more recent rock materials, which lowered the U/Pb ratio. They estimate a minimum of 2.1 Ga for this differentiation event. The high $^{207}Pb/^{204}Pb$ and low $^{206}Pb/^{204}Pb$ ratios of these rocks contrast with

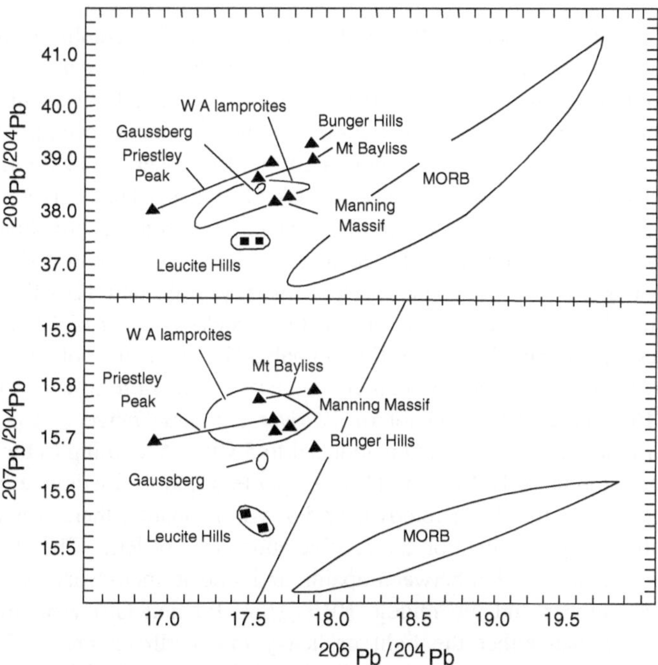

Fig. 5.8 **a** and **b** $^{208}Pb/^{204}Pb$ versus $^{206}Pb/^{204}Pb$ and $^{207}Pb/^{204}Pb$ versus $^{206}Pb/^{204}Pb$ isotope data for K-rich rocks from various localities. The data on Mid-Oceanic Ridge basalts are also plotted for comparison (after Nelson et al. 1986)

those of oceanic-island basalts, which define a linear array of positive slope, extending from the unradiogenic Pb field of MORB to the high $^{207}Pb/^{204}Pb$ and $^{206}Pb/^{204}Pb$ ratios of ocean island basalt of St. Helena and Tubuaii.

The position of the MORB-OIB array to the right of the geochron (Fig. 5.8b) has been attributed to the progressive loss of Pb from the mantle to the core. However, if the lamproites of Western Australia and Gaussberg were generated from low U/Pb source components within the subcontinental lithosphere, the existence of such a reservoir could also compensate for a general increase of U/Pb ratio in the OIB mantle source rock (Nelson et al. 1986). Lamproites from Smoky Butte have the most extreme nonradiogenic Pb isotopic compositions recorded for ultrapotassic rocks. Their isotopic characteristics are similar to those of old granulite facies rocks from the lower crust. However, on the basis of their trace element characteristics, it is extremely unlikely that crustal contamination could have modified the isotopic characteristics of these magmas. Thus, their isotopic composition suggests their mantle source.

5.4 Minor Element Geo-chemistry of Potassium–Rich Silica-Deficient Volcanic Rocks from Italy

Hawkesworth and Volmer (1979) divided the Roman petrographic province into three separate zones: 1) The northern province of Tuscany, where the rocks are predominantly calc-alkalic, 2) the second zone, which includes Vulsini, Vico and Sabatini, and is characterized by hybrid magma of mantle origin, and 3) the southern zone comprising primitive potassic suite. Rogers et al. (1985) investigated the trace element content of the potassic rock of Vulsini. Their chondrite-normalised REE patterns are summarized in Fig. 5.9a–c. These figures show that in leucite basanite, Mg–leucitite and tephritic leucitite, the REE concentration shows a considerable variation with La content, which is very high, and 120–420 times the value of chondrites. The La_N/Yb_N ratio of these rocks is equal to 22.5. The rocks also have consistent small negative Eu anomaly. The abundance of Th, U, Pb, and Nb vary by a factor of 3, as CaO content decreases from 14 to 3 wt%. The content of light REE increases by a factor of 12 over the same range of CaO and this enrichment took place before CaO is reduced to 5 wt%. According to the fractional crystallization model of Holm et al. (1982), apatite began to fractionate first from a highly potassic magma. With removal of 5–8 vol % apatite, there was paucity of P_2O_5 and this had great effect on the relative abundance of REE. They thought that the partitioning of the REE between apatite and silicate melts vary systematically with atomic number and the middle REEs (Sm, Eu and Gd) were more easily accommodated than either the light or heavy rare earth elements. Thus, when perovskite starts fractionating at nearly 8 wt% CaO, the middle REEs are extracted from the melt at greater rate than other elements resulting in the increase of La/Sm ratio. This further results in the cross-cutting REE profiles in $^{87}Sr/^{86}Sr$ isotopic ratio

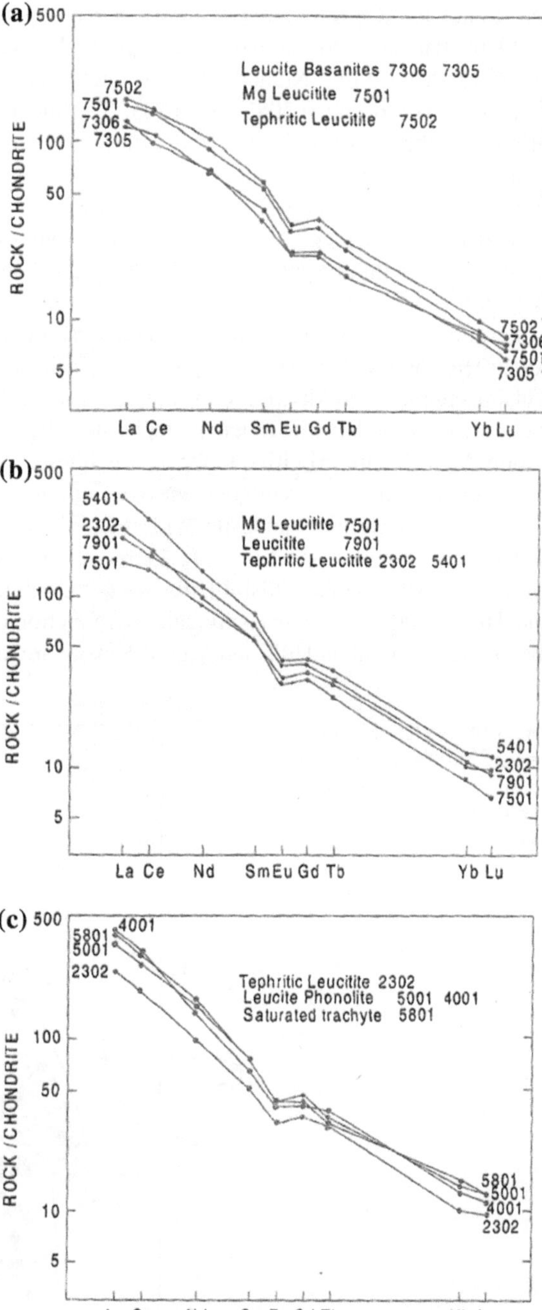

Fig. 5.9 Chondrite-normalized REE diagrams for the analysed high K-series lava of Italy (after Rogers et al. 1985)

was found by Roger et al. (1985) to increase progressively with fractionation. Available data showed marked increase in the low K-series than the high K-series. In general the $^{143}Nd/^{144}Nd$ and $^{87}Sr/^{86}Sr$ ratios have following ranges, respectively: 0.51209–0.51216 and 0.70879–0.71105. The incompatible trace element characteristics of the rocks of Vulsini and adjoining complexes are shown in Fig. 5.9, which indicates that lavas of Vulsini, Roccamonfina and Ernici have coherent variation in radiogenic isotopes. The K-rich rocks of in-plate regions have a smoother trace element distribution pattern, but those from Italian lavas show strong spikes with depletion in Ti, Ta, Nb, and P and they are also high in Rb, K and Th content.

Ferrara et al. (1985, 1986) made Sr and O isotopic studies of potassic rocks from Alban Hills and Vulsini volcanic province. Their data are plotted in Fig. 5.10. The Vulsini samples were divided into four subgroups: A*, A, B, and C. The Group A rocks are tephritic leucitite to leucite tepherite ($SiO_2 = 53$ wt%, $K_2O > 6.0$ wt%), the Group A* rocks are primitive tephritic leucitites.

Group B and C are saturated or oversaturated with respect to SiO_2 (olivine shoshonite, shoshonitic andesite and trachytic ignimbrite). All A samples are undersaturated with respect to SiO_2. Ferrara et al. (1986) plotted the $^{87}Sr/^{86}Sr$ ratio of rocks of Alban Hills, Vulsini Hills, Mount Cimini, Mount Amiata, San Venanzo and Tuscan (Fig. 5.11) metasediments as a function of 1/Sr (Fig. 5.11). They noted that the lavas of Alban Hills and Mount Vulsini are primitive and may represent the

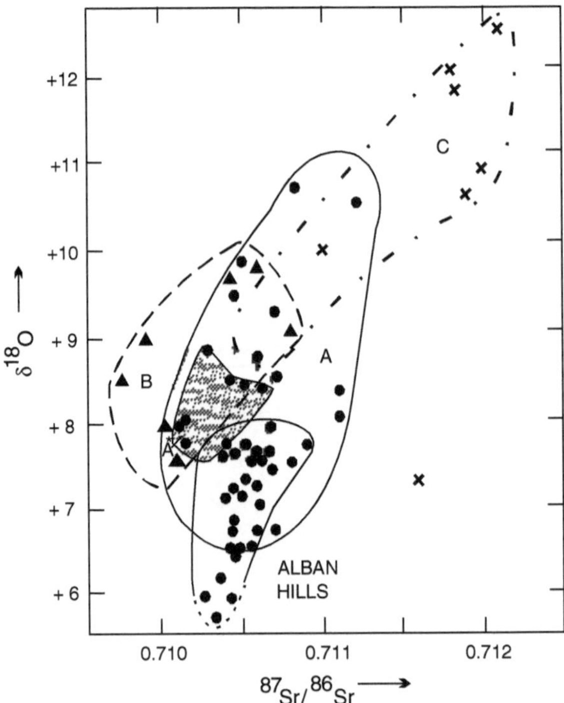

Fig. 5.10 A plot of $\delta^{18}O$ versus $^{87}Sr/^{86}Sr$ ratios for the Vulsini samples (analysed by Ferrara et al. for Holm and Munksgaard 1982)

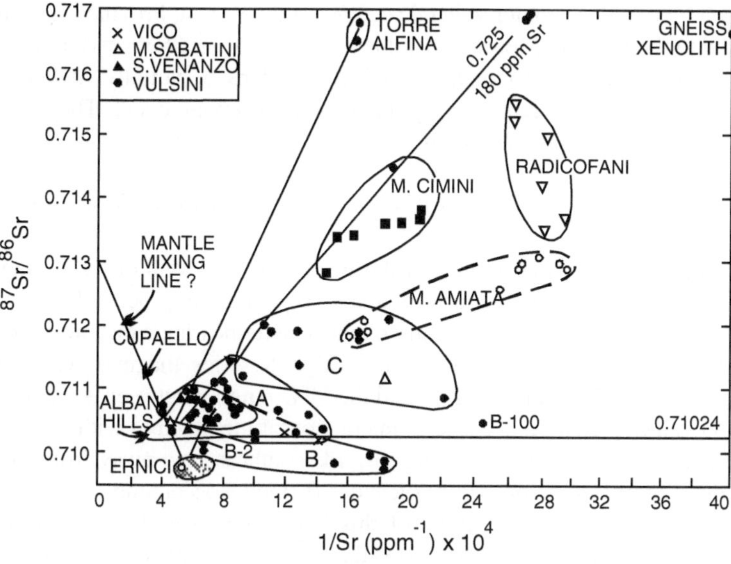

Fig. 5.11 A plot of $^{87}Sr/^{86}Sr$ ratios versus [1/Sr (ppm)] X 10^4 of various K-rich Tuscan lavas of Mt Cimini, Radicofani and Mt. Amiata (Hawkesworth and Volmer 1979; Volmer 1976; Poli et al. 1984) including data from San Venanzo, Vico, Mt Sabatini, Alban Hills (Stippled pattern) and Mt. Ernici (diagonal line pattern) (from various sources, see Ferrara et al. 1986). The lines are related to mixing. A and B refer to the highest and lowest K_2O content, respectively. C refers to rocks of transitional characteristics; B-2 and B-100 represent tephritic leucitite and a drill core sample beneath Latera Caldera

primary high K magma of central Italy. They further thought that because of high initial Sr contents of the Mount Vulsini lavas, the $^{87}Sr/^{86}Sr$ probably were not affected by wall rock exchange reactions or by low temperature processes that raised the $\delta^{18}O$ values. They thought that the $^{87}Sr/^{86}Sr$ ratios of K-rich magma of Mount Vulsini, Cupaello and San Venanzo are quite high, though they are mantle-derived. They thought that these magmas were derived from a zone in the upper mantle which was previously metasomatically enriched in ^{87}Sr. According to them, it is possible that the high K-series parent magma at Mount Vulsini had a uniform $^{87}Sr/^{86}Sr$ ratios (0.7102–0.7103), which are identical to those of the parant magma of Alban Hills volcanic centre, located 120 km to the south. Even though the Sr isotopic data at Vico and Mount Sabatini are sparsely determined, they thought that same conclusion could be derived for the rocks of these intervening centres also. Some of the low K-series rocks however, were erupted at Mt. Vulsini and were derived from a mantle source with a lower $^{87}Sr/^{86}Sr$ ratio than the rocks of high K-series rocks (0.7097). They observed that from Mt. Vulsini to Naples, the rocks belonging to low-K series end members of all the localities have lower $^{87}Sr/^{86}Sr$ ratio compared to high K-series end members. They further observed that the lavas that erupted at Mt. Vulsini have $\delta^{18}O$ values (+ 6.5–13.8) and $^{87}Sr/^{86}Sr$ ratios (0.7097–0.7168) that are in general higher and have a much broader range than any other lava type of these major volcanic centres of the Roman Province. This may be attributed to the fact that Mount Vulsini is located at the northern end of the Roman Province, where this petrographic province overlaps a partially melted terrain, which was extensively heated during the Tuscan magmatic event. They envisaged both mixing of magmas and assimilation of Tuscan metasedimentary basement rocks. It took place in conjunction with fractional crystallization of high K and low K-series parent magmas, which resulted in making the isotopic effect at Mt. Vulsini more complex.

According to Ferrera et al. (1986) during the last few million years, the upper mantle beneath central Italy underwent a large scale mixing process that introduced radiogenic Sr into the source regions of the leucite-bearing volcanic rocks. This resulted in the production of voluminous high K-series magmas with $^{87}Sr/^{86}Sr$ varying between 0.7101 and 0.7103, and the $\delta^{18}O$ ranging between + 5.5 and + 7.5. The $\delta^{18}O$ values increased because of interaction of the magmas with the overlying continental crust. According to them, towards southeast and away from the linear belt there is another zone of major volcanoes (Ernici, Roccamonfina, Vesuvius and Phlegrean Fields), where the effect of this interactions was less insignificant. In these volcanoes towards southeast, the $^{87}Sr/^{86}Sr$ ratios are much lower (0.7060–0.7095), but the $\delta^{18}O$ values do not change significantly. If this uniformity in $^{87}Sr/^{86}Sr$ ratio of this high K-series end member rocks (0.71025) is proved to be true by future investigators then it would mean that a large volume of the upper mantle beneath Italy was associated with continued magmatic and metasomatic processes. These processes took place during the last few million years. The other alternative explanation could be related to varying Rb/Sr ratios in the upper mantle reservoir, which might have led to much greater heterogeneity in the $^{87}Sr/^{86}Sr$ ratio.

The evolved phonolitic and trachytic tuffs were studied by Turbeville (1993), and found that they have high chondrite-normalized Th, Zr, Rb and REE abundances relative to Nb, Ta, Ba and Sr (Fig. 5.12). According to him the large

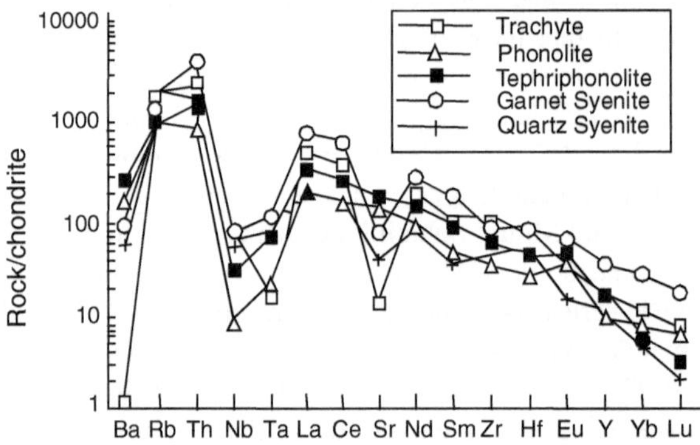

Fig. 5.12 Spidergram plots of chondirte-normalized trace element concentrations in selected samples of Latera pumice, syenites and lava (normalized values are from other sources) (after Turbeville 1993)

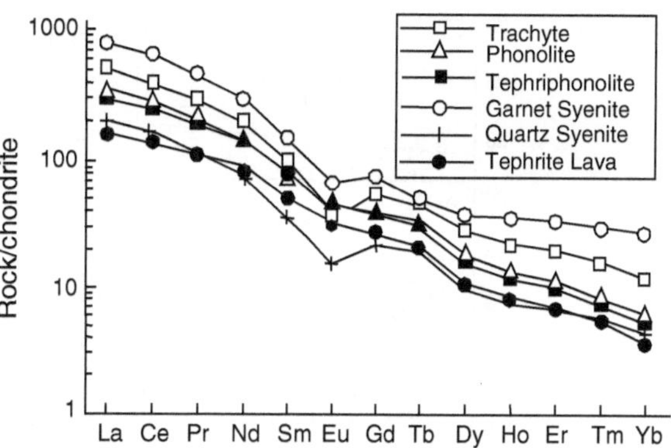

Fig. 5.13 Spidergram plots of chondrite-normalized REE concentration in selected samples of Latera pumice, syenite and lava (after Turbeville 1993)

differences in Rb/Th, Nb/Ta, Zr/Hf and La/Yb ratios suggest efficient fractionation of incompatible trace element within the evolved liquids. He observed that Na/Ta ratio increases with differentiation from tephriphonolite to phonolite and trachyte. This is related to progressive fractionation of sphene that is common in evolved pumices and syenites. In his chondrite-normalized REE plots, negative Eu anomalies are very much prevalent in the felsic tuffs (Fig. 5.13), which indicate that they are derivative liquids after plagiocase fractionation.

The tephritic phonolite tuffs and tephritic lavas do not have Eu anomaly, which according to Rogers et al. (1985) are inherent source characteristic of unevolved Vulsini magmas. Turbeville found that the syenites have widest range in trace element content amongst Latera rocks and their REE content spans the entire concentration range of these elements in the rocks of Vulsini complex. He observed that syenites have especially high Th, Zr, and Rb contents with more variable REE patterns and there is stronger decoupling among the LIL elements than is displayed by the composition of the pumice. He thought that the elemental abundances of the syenites reflect high modal concentration of sphene, apatite, zircon and amphibole. Turi and Taylor (1976) measured the $^{18}O/^{16}O$ ratios of leucite-bearing lavas, pyroclastics, and other related rocks from the Roman Province of Italy. They observed a general increase northward in the $\delta^{18}O$ values in the following sequence: Ischia (5.8–7.0), Somma-Vesuvius and Phlegrean Fields (7.3–8.3), Mount Sabatini (7.3–9.7), Vico Volcano (7.4–10.2) and Monti Vulsini (8.7–11.7). The Sr isotopic ratios of the rocks of the Roman Province also show a similar northward increase. A marked increase in the $\delta^{18}O$ value was noted by them just north of Rome, where the rocks of the Roman Province just begin to overlap the Tuscan calc-alkaline rocks. This observation led them to believe that this increase in the $\delta^{18}O$ contents and $^{87}Sr/^{86}Sr$ ratios might be due to direct mixing of the parent magma with the Tuscan rocks, which are high in the $\delta^{18}O$ content, or to the fact that the high $\delta^{18}O$ country rocks underwent heating during 2 million years of the Tuscan igneous activity. They estimated that the strongly undersaturated magma, which was probably derived from the upper mantle had a $\delta^{18}O$ value of + 6, and the Sr isotopic ratio of 0.704–0.705 and SiO_2 content of less than 44 wt%.

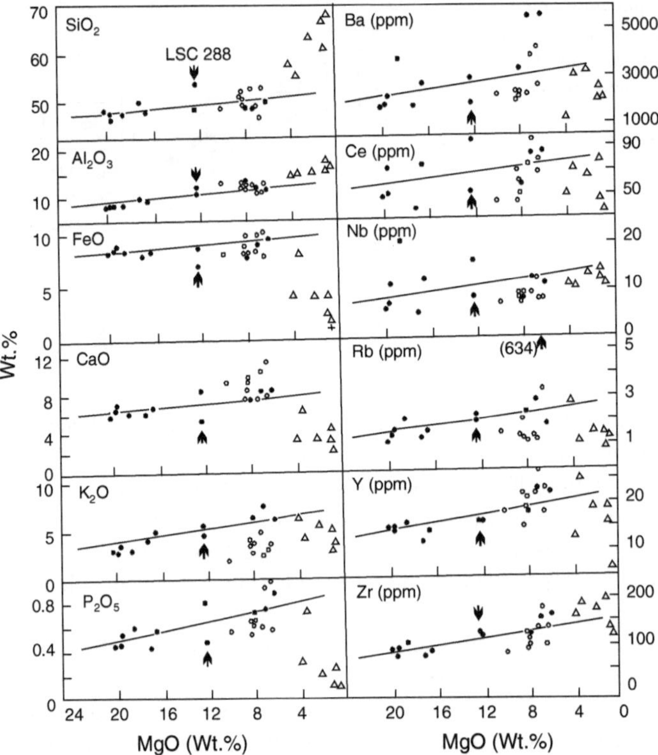

Fig. 5.14 MgO variation diagram for selected major and trace elements. The *solid line* is an olivine-clinopyroxene control, calculated by combining them in 4:1 ratio. The core composition of olivine (Fo_{93}) and diopside ($Ca_{45}Mg_{46}Fe_{9}$) in high K-minettes are after Macdonald et al. (1992)

Interpretation of Turi and Taylor (1976) was however, based on determination of oxygen isotope content of more evolved rocks. The studies of Rogers et al. (1985) and Holm and Munksgaard (1982) provide data relative to more primitive K-rich lavas (magnesian leucitites and leucite basanites), which might have been in equilibrium with a K-rich mantle source. The data of Roger et al. suggest that primary Vulsini high K-series lavas were characterized by following oxygen and Sr isotopic ratios, respectively: ~8 and ≥0.7088 and the values of $^{143}Nd/^{144}Nd$ ratios range between 0.51209 and 0.51228.5.5. The trace element and REE abundance of K-rich silica deficient Rocks of Bearpaw Mountains are shown in Fig. 5.14.

It was shown by Macdonald et al. (1992) that some of the highly magnesium-rich minettes have high Ni and Cr contents (700 and 1,200 ppm) and their high potassic character is reflected by their high Ba (up to 5,491 ppm), Rb (283 ppm), Cs (54 ppm) and Sr (2,291 ppm) content. They found that the Bearpaw rocks are more enriched in Rb contents compared to K in highly potassic minettes. The K/Rb ratio in these rocks range from 122 to 291 (mean 221). The Zr values are up to 163 ppm and Nb 20 ppm. Thus, LILE/HFSE ratio is high, which is typical of K-rich rocks

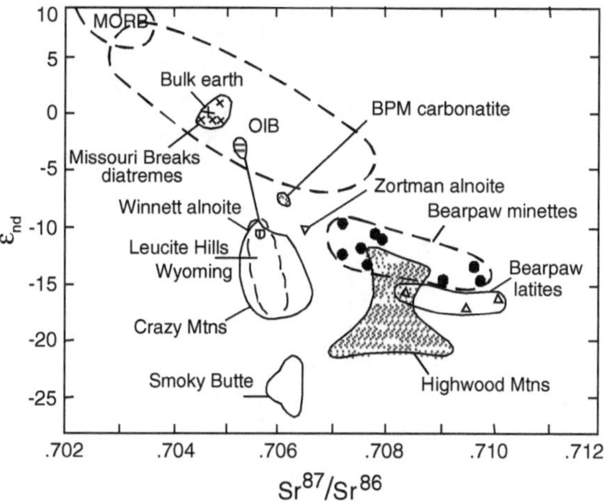

Fig. 5.15 A plot of ε_{Nd} versus $^{87}Sr/^{86}Sr$ data from rocks of Wyoming province and comparison with mid-oceanic ridges basalts (MORB), ocean island basalt (OIB) and bulk earth adopted from different sources. *Solid dots* High-K minettes; *Open circles* Crazy Mountains and Bearpaw Mountain carbonatites (Dudas et al. 1987); Leucite Hills (Volmer et al. 1984), Highwood Mountains (O'Brain et al. 1991) and Smoky Butte (Fraser et al. 1985) (after McDonand et al. 1992)

from both continental extensional zones and subduction-related settings. Such high ratios distinguish the minettes from the lamproites of Montana province. Usually, lamproites have higher concentration of such elements as Nb, Ti, and Zr and lower LILE/HFSE ratio. In a plot of MgO versus SiO_2, Macdonald et al. noted that in case of K-rich mafic lavas of the Bearpaw Mountains, there is decrease in the values of Cr and Ni and increase in Ba, Ce, Nb, Rb, Y, Zr contents with respect to silica. The MgO content has a negative correlation with respect to SiO_2. Likewise Al_2O_3, FeO_t, CaO, K_2O and P_2O_5 systematically increase with the decrease of MgO (Fig. 5.14). The Sr and Nd isotopic data for rocks from Missouri Breaks, Winnett alnoite, Leucite Hills, Crazy Mountains, Bearpaw Mountains, Highwood Mountains and Smoky Butte are summarized in Figs. 5.15 and 5.16. The minettes and lattites show considerable variation in $^{87}Sr/^{86}Sr$ ratio (0.70716–0.70978) and (0.70841–0.71008), respectively acdonald et al. observed that ε_{Nd} ranges between −9.27 and −14.53 in the minettes, and between −15.94 and −16.73 in the latites. Differential elemental ratios of the K-rich rocks (La/Nb, Sm/Nd) with respect to ε_{Nd} are shown in Fig. 5.16. They observed that the $^{87}Sr/^{86}Sr$ ratios have similar span as those from other alkaline centres in Montana, such as K-rich rocks of Leucite Hills, Crazy Mountains and Highwood Mountains but ε_{Nd} tend to be more radiogenic.

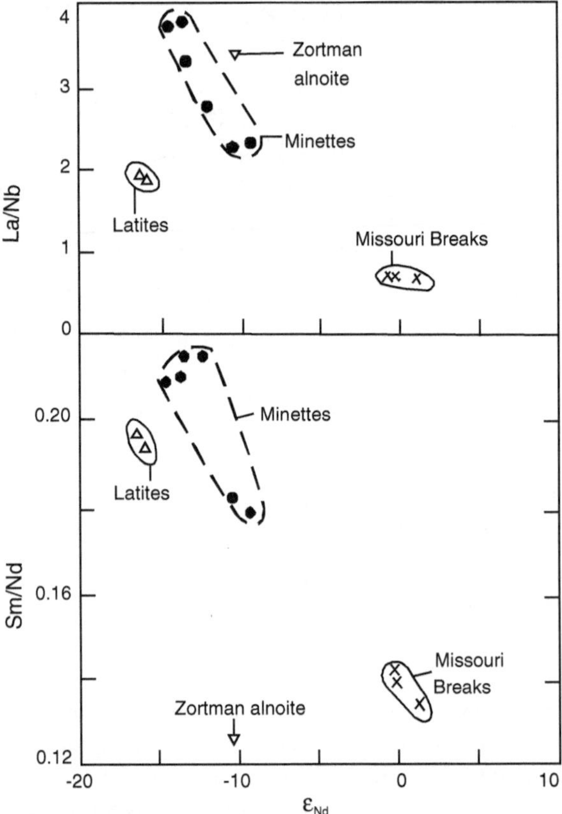

Fig. 5.16 La/Nb and Sm/ Nd ratios in Bearpaw rocks plotted against ε_{Nd}. Close circles: high K-minettes (after Macdonald et al. 1992)

5.5 The REE and Trace Element Geochemistry of K-Rich Volcanic Rocks of Smoky Butte

The chondrite-normalized REE distribution pattern for the Smoky Butte rocks were determined by Mitchell et al. (1987, Fig. 5.17). All the samples show similar distribution patterns. The rocks are strongly enriched in the LREE with La abundance being 1,000–1,400 times more than that of chondrite. The concentration of heavy REE is low 6–20 times that of chondrite values and hence La/Yb ratios are high, and range between 162 and 280, average being 209 and there is no Eu anomaly. The Smoky Butte rocks have higher La/Yb ratios compared to lamproitic rocks from Western Australia (La/Yb = 83 − 246) and wyomingites from Leucite Hills, Wyoming (121–127) but similar La/Yb ratio as that of madupites from Leucite Hills and Hillspond (Kansas) (La/Yb = 165 − 377). Spanish rocks have relatively low La/Yb ratio and lower total REE contents. Smoky Butte lamproites are richer in Sm content (27–38 ppm) for a given La/Yb ratio and compared to other lamproites. The $^{87}Sr/^{86}Sr$ ratios are low (0.70587–0.70633) and δ_{Nd} values vary between −21.6 and −25.9.

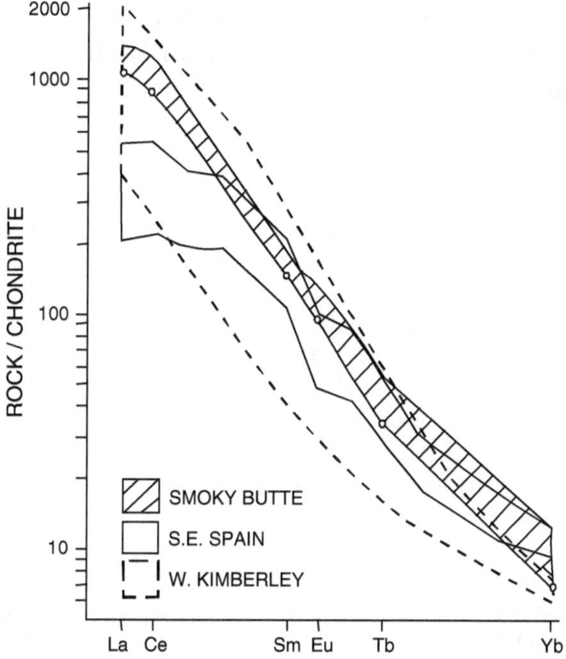

Fig. 5.17 The chondrite-normalized REE patterns of K-rich rocks from Smokey Butte compared with those of rocks from S.E. Spain and W. Kimberley, Australia (after Mitchell et al. 1987)

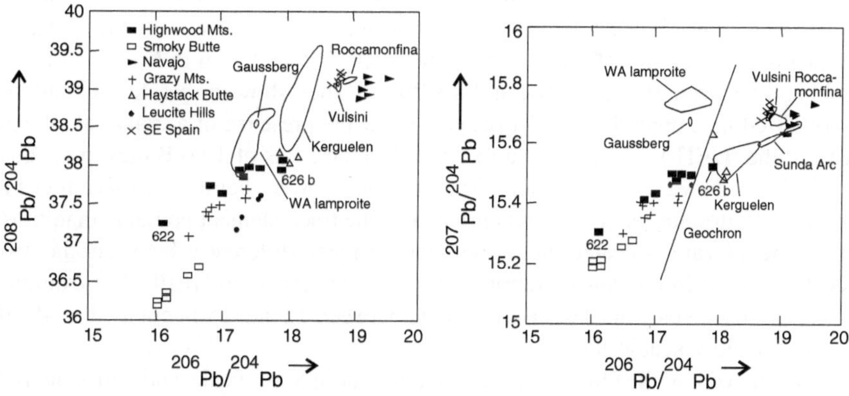

Fig. 5.18 Pb isotopic compositions for alkalic magmatic rocks from Highwood Mountains and other localities (after O'Brien et al. 1991)

5.6 Minor and REE Geochemistry of K-Rich Silica–Deficient Volcanic Rocks from Highwood Mountains

O'Brien et al. (1991) noted an overall increase in the REE abundance from primitive to more evolved K-rich rock types. This is in agreement with the subtraction of mafic phases having low D-values for REE. According to them due to extensive apatite fractionation from the magma, overall abundance is low and the patterns are concave upward, reflecting higher apatite D-values for the middle REE.

On the basis of geochemical studies of K-rich magmatic rocks from Highwood Mountains, Smoky Butte, Haystack Butte and Crazy Mountains, O`Brien et. al. suggested that in the generation of K-rich ultramafic Highwood Mountain rocks, phlogopite in the minettes probably played an important role in the generation of minettes. According to O'Brien et al. the high enrichment of Ba (2,000–500 ppm) and Sr (850–1,200 ppm) are in agreement with the idea that mica and clinopyroxene were present in the mantle source region. Presence of relatively high La/Yb ratio might indicate that garnet probably was an important phase in the residuum. This view is based on available crystal/liquid partition coefficient of these elements. They found that the rocks from Highwood and Bearpaw Mountains have high $^{87}Sr/^{86}Sr$ ratios related to the Nd–Sm trends (Fig. 4.12, Chap. 4) of the rocks of Missouri Breaks, Leucite Hills and Smoky Butte. Figure 5.18 related to different Pb isotopic ratios of that region refer to heterogeneity of the mantle (O'Brien et al. 1991). Figure 5.20 refers to the minor element and REE concentration of K-rich volcanic Rocks Associated with Rio Grande Rift.

The concentration of incompatible trace elements at Navajo-Hopi volcanic field located towards the west of the ridge and in the two Buttes are very similar (Gibson et al. 1993; Fig. 5.19). Samples of these localities are enriched in incompatible elements; for example, $(Ba/Y)_N$ ranges between 17.5 and 27.6 in Navajo volcanic field and between 10.3 and 20.4 at Two Buttes. They showed that their chondrite-normalized incompatible element patterns show the presence of a trough at Nb and Ta and the $(La/Th)_N$ ratio is equal to 1.5 and in the cases of Two Buttes, it is equal to 1.6 in the Navajo volcanic field. They further showed that the intrusive rocks at the Two Buttes are relatively less enriched in the trace element contents than those from the Navajo volcanic field, but geochemical difference between the two localities occur in the concentration of LREE with respect to HREE. For example, the $(La/Yb)_N$ ratio of the Two Buttes range between 12 and 25 in contrast to 45–86 in the Navajo volcanic field.

They noted that the Oligo–Miocene magmas at Spanish Peaks and Dulce are rich in LILE and depleted Ti. The dykes here have K_2O/Na_2O ratio > 1, and are enriched in LREE with respect to HREE [$(La/Yb)_N = 7.46$] and depleted in HFSE such as Nb and Ta relative to LREE [$(La/Ta)_N = 1.5$]. The dyke from Spanish Peak has higher K_2O content (7.13 wt%) compared to other dykes in the rift shoulders, but the La/Ta ratio (1.0) is similar to that of Ocean Island basalt. Gibson et al. (1992) found that the $^{87}Sr/^{86}Sr$ ratios of the rocks of the Two Buttes range between

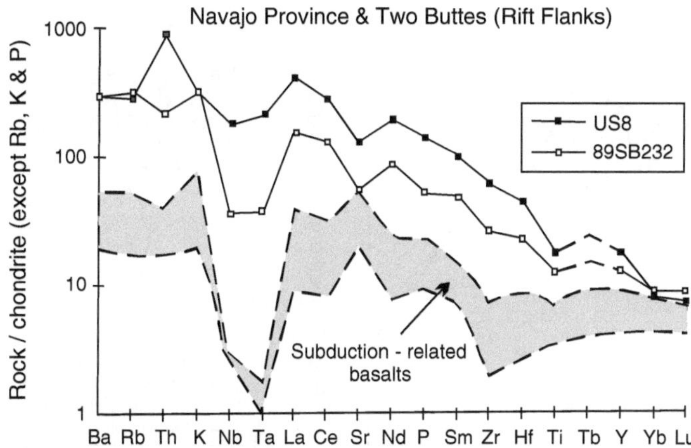

Fig. 5.19 Chondrite-normalized incompatible-element abundance patterns for representative K-rich magmas from Rio Grande Rift flanks. Gibbson et al. plotted representative samples from Navajo Volcanic (US-8) field and samples from Two Buttes (89SB232) in this diagram (after Gibbson et al. 1993)

0.70533 and 0.706901 in the Navajo volcanic field. The $^{143}Nd/^{144}Nd$ isotopic ratios for the K-rich mafic rocks at the rift flanks are lower than basalts at the rift axis, but the ratio is higher in Navajo volcanic field (0.512558–0.512713) and at Two Buttes (0.51231–0.51239).

Some of the lavas in the Espanola Basin have high $^{87}Sr/^{86}Sr$ ratio (Gibson et al. 1992), and that may be related to contamination of the magma with ^{87}Sr–rich crust. The Sr–isotopic and $^{143}Nd/^{144}Nd$ ratios in the samples from the rift flanks and shoulder, could be interpreted in the same way. The high $^{86}Sr/^{87}Sr$ and low $^{143}Nd/^{144}Nd$ ratios of the Two Buttes may also indicate crustal contamination. The K-rich mafic magmas, which were emplaced on the rift flanks and shoulders are characterized by high Sr, Nd and Pb. According to Gibson et al. very large quantities of crust need to be contaminated to obtain these values.

5.7 Minor Element Contents of Potassic Volcanic Rocks from N.E. China

The trace element geochemistry of potassic rocks from Wudalianchi, Erkeshan and Keluo volcanic fields of north–east China have been studied by Zhang et al. (1995). According to them these rocks are characterized by high LILE and LREE contents. They have relatively low U–Th, HFSE and HREE contents with normalized La/Ya ratio of 50.0–6.8. For most incompatible elements, the concentration ranges between olivine leucitites on the one hand and leucite basanites and trachybasalts on the other, with the exception of Zr, Hf and Ga, which are lower in olivine leucitites.

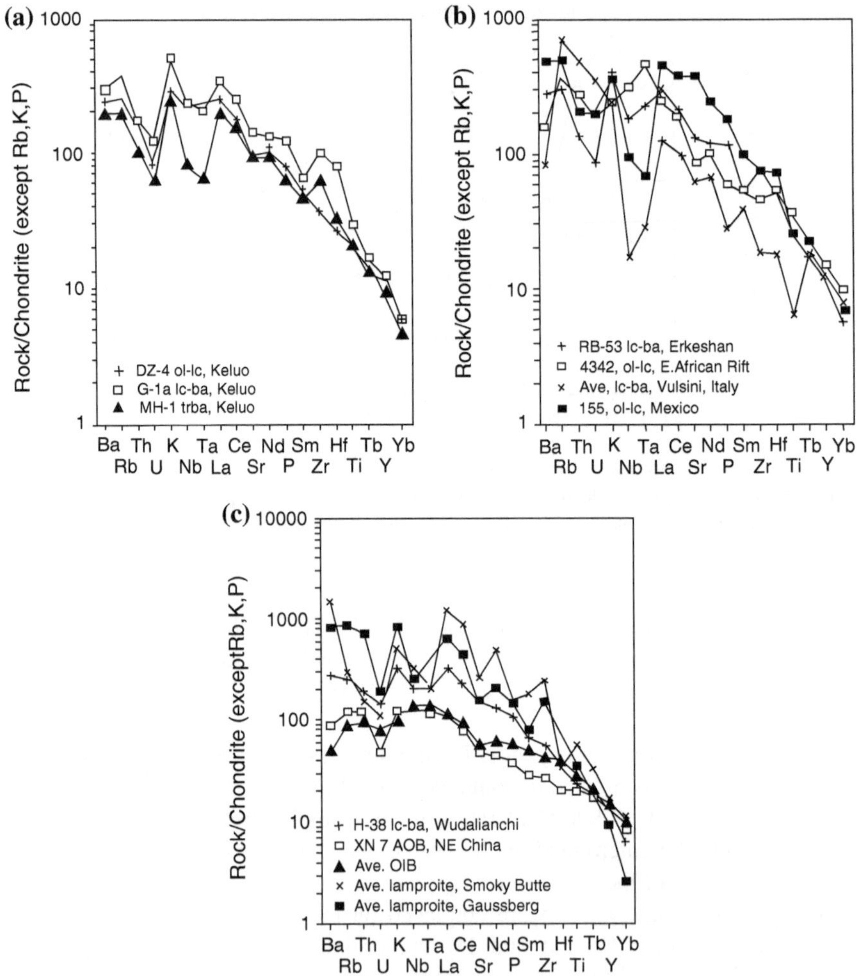

Fig. 5.20 Chondrite-normalized incompatible element plots for K-rich volcanic rocks from North East China i.e. WEK volcanic rocks and other mantle-derived lavas (*source* Zhang et al. 1995)

All potassic rocks from northeast China have similar incompatible element patterns (Fig. 5.20a–c), the right side of which exhibit subparallel smooth curves from La to Yb with slight Zr–Hf becoming more prominent with decreasing MgO (e.g. from DZ-4 to G-la and MH-1). The sample MH-1 is distinct from the other samples by its marked depletion of Nb–Ta in combination with the lowest K_2O (3.5 wt%) and the highest SiO_2 (55 wt%) contents among the north-east China samples.

The slight to moderate depletion in Nb–Ta displayed by north-east China potassic rocks is quite different from East African potassic olivine leucitites in that the rocks from the latter area present Nb–Ta enrichment similar to ocean island basalts (OIB). The Chinese K-rich rocks distinguish themselves from those of the

Italian and Mexican potassic rocks, which are as strongly depleted in Nb–Ta as calc-alkaline rocks (Fig. 5.20b). On the other hand, the Chinese potassic rocks have a similar configuration in spidergram patterns to lamproites from Gaussberg and Smoky Butte. Both display moderate depletion in Nb–Ta and U–Th and enrichment in Ba, K, LREE and Zr–Hf (Fig. 5.20c), although Chinese lavas are in general less enriched in strongly incompatible elements than the lamproites.

Remarkable differences also exist between potassic rocks from Wudalienchih, Erkeshan and Keluo and typical late-Cenozoic Chinese alkali basalts. The former have higher K_2O, LILE and LREE than the latter, which in contrast to potassic rocks, also display OIB-like incompatible element patterns (Fig. 5.20c).

The $^{87}Sr/^{86}Sr$ ratios of potassic rocks from northeastern China range from 0.70500 to 0.70565. Olivine leucitites, some leucite basanites and trachybasalts, which contain mantle xenoliths generally have lower ratios (0.70500–0.70527) than xenolith-free leucite basanites and trachybasalts (0.70533–0.70549). Six samples from this region have $^{143}Nd/^{144}Nd$ ratios of 0.51250–0.51232 (ε_{Nd} of −2.8 to −6.3) and display a negative correlation with $^{87}Sr/^{86}Sr$ (Fig. 5.21).

The Pb isotopic compositions of potassic rocks from China have the following ranges: 16.607–17.058 ($^{206}Pb/^{204}Pb$), 15.352–15.447 ($^{207}Pb/^{204}Pb$) and 36.511–37.204 ($^{208}Pb/^{204}Pb$). They plot on the left side of the ‘Geochron and above the Northern Hemisphere Reference Line (NHRL; Hart 1984, Fig. 5.22). Both $^{207}Pb/^{204}Pb$ and $^{208}Pb/^{204}Pb$ ratios of olivine leucitites and xenolith-bearing leucite basanites and trachybasalts usually have higher $^{206}Pb/^{204}Pb$ ratios (17.058–16.789) than those which are xenolith free (16.900–16.607).

On a $^{87}Sr/^{86}Sr$ versus ε_{Nd} diagram (Fig. 5.21) the Chinese potassic rocks plot on the less enriched end of a trend defined by potassic rocks with moderately high

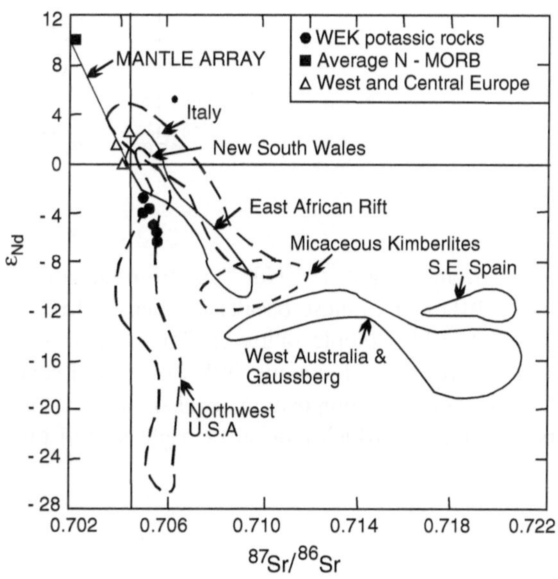

Fig. 5.21 The ε_{Nd} versus $^{87}Sr/^{86}Sr$ for K-rich rocks from northeast China and other potassic rocks. (N–MORB data from White et al. 1987; after Zhang et al. 1995)

Fig. 5.22 **a** The $^{208}Pb/^{204}Pb$ vs $^{206}Pb/^{204}Pb$ ratios and **b** $^{207}Pb/^{204}Pb$ vs $^{206}Pb/^{204}Pb$ for WEK and other potassic rocks. (N–Morb, after White et al. 1987; NHRL after Hart, 1984; after Zhang et al. 1995)

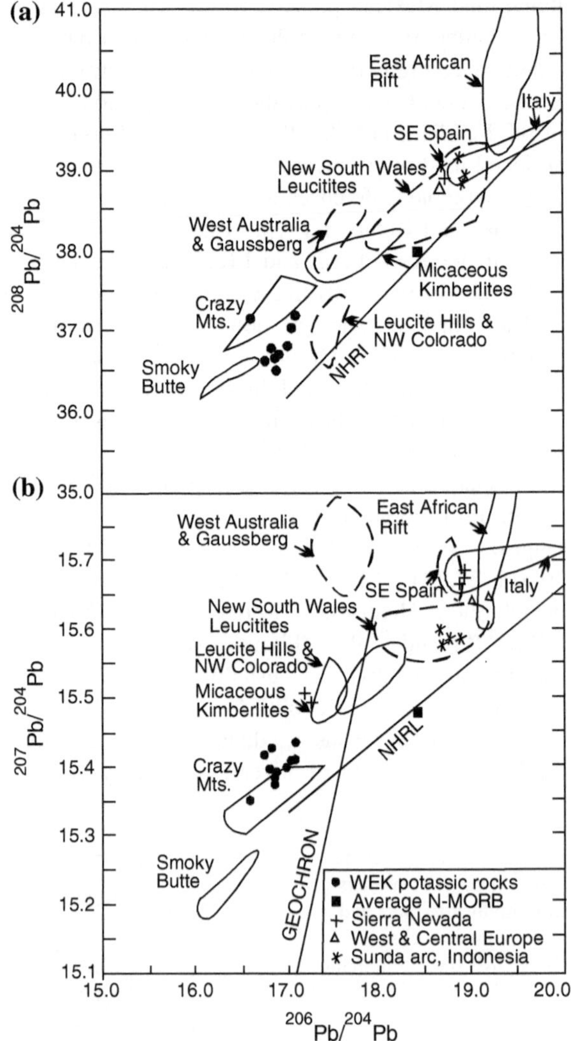

$^{87}Sr/^{86}Sr$ and very low $^{143}Nd/^{144}Nd$ from the north western USA. They differ from most OIBs and Chinese basalts by their higher $^{87}Sr/^{86}Sr$ and lower $^{143}Nd/^{144}Nd$ ratios. The Pb isotopic ratios of Chinese potassic rocks overlap those of K-rich rocks from Crazy Mountains, NW Colorado, Leucite Hills and Smoky Butte. This unradiogenic Pb composition is also distinct form OIBs and Chinese alkaline basalts, most of which plot on the right side of the 'Geochron'.

5.8 Trace Element Geochemistry for Ringgit-Beser Complex (Indonesia)

The arc lavas from Ringgit-Beser complex have been classified into calc-alkaline, high-K calc-alkaline, shoshonitic and leucititic rock types by Edwards et al. (1994). The classification of these four rock types is shown in a K_2O versus SiO_2 diagram (Fig. 5.23). The incompatible trace element data for the mafic and evolved lavas belonging to excess K (EK), high K (HK) and high K-calc-alkaline (HK calc) series are summarized in a chondrite-normalized trace and minor element plots (Fig.5.24a–c; Edwards et al. 1994). These data are similar to typical island arc element patterns in having LILE and LREE enrichments related to HFSE and HREE.

The EK series lavas show little change with MgO variation in the incompatible trace element content. The Rb content in these lavas is higher relative to Ba and Th. The chondrite-normalized lavas have similar or slightly lower LILE contents than those of the in-plate lavas of Cameroon ocean island basalts. The K-series lavas of the complex have low Sr/Nd ratios similar to those of EK series. Edwards et al. observed that all lava series have negative Ti and Zr anomalies in addition to Nb anomalies i.e. typical of arc lavas (Fig. 5.24a–c). They further observed that the negative HFSE anamolies are present in the high magnesian lavas and are not related to magnetite and/or hornblende fractionation. They observed that the transition from calc-alkalic to high K-lava composition is related to LILE, LREE and HFSE contents, and a reduction of negative Nb anomaly and increase in the LREE/HREE ratios.

The $^{87}Sr/^{86}Sr$ ratios of the lava series range from 0.70419 to 0.70549 and the $^{143}Nd/^{144}Nd$ ratios from 0.512674 to 0.512821 (Fig. 5.25). The highly potassic

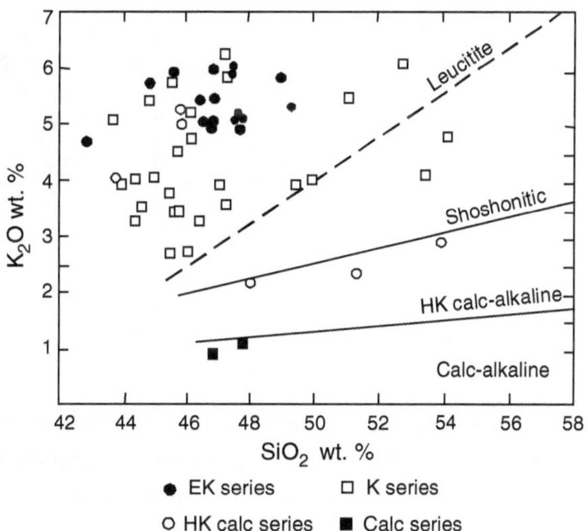

Fig. 5.23 A plot of K_2O versus SiO_2 for the calc, HK calc, K. and EK series rocks of Ringgit-Beser (after Edwards et al. 1994)

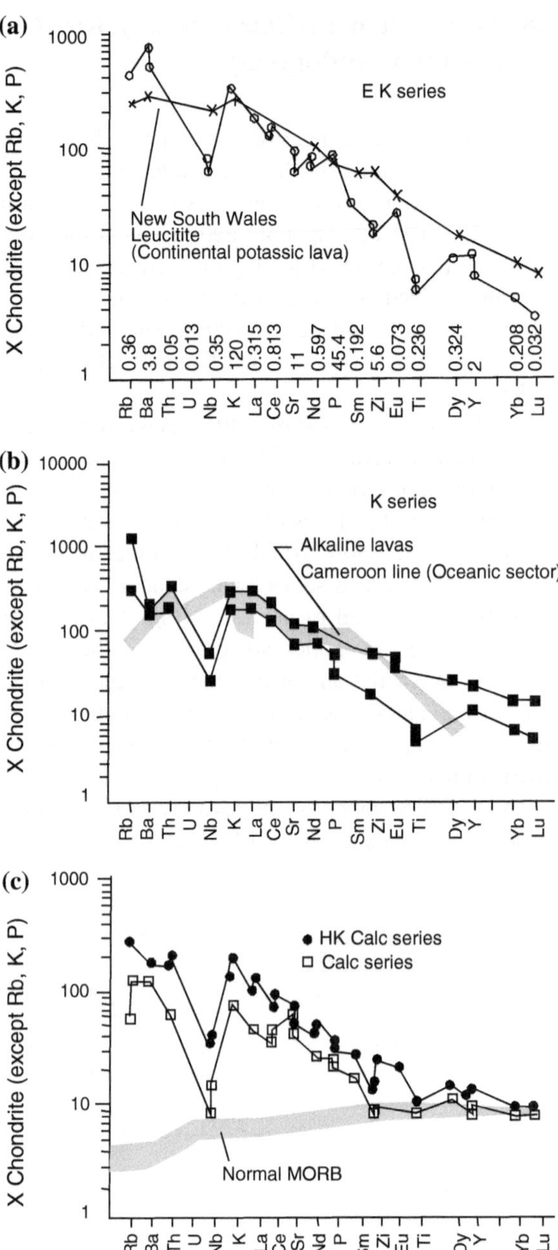

Fig. 5.24 Chondrite-normalized trace element diagrams for the most basic and most evolved samples from the EK (**a**), K (**b**) HK calc and calc-alkalic series (**c**) from Ringgit-Beser (after Edwards et al. 1993)

lavas have the highest $^{87}Sr/^{86}Sr$ ratios (0.70445–0.70549) and lowest $^{143}Nd/^{144}Nd$ ratios (0.512674–0.512806). According to Edwards et al. the EK and high K-calc series have high Sr and Nd isotope ratios which are within the range of K-series

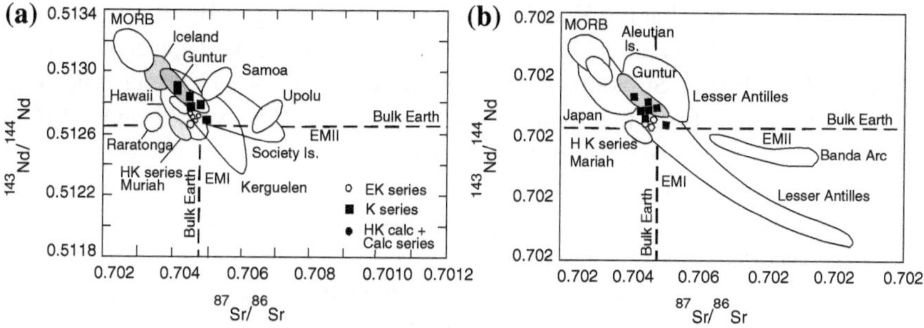

Fig. 5.25 $^{143}Nd/^{144}Nd$ versus $^{87}Sr/^{86}Sr$ isotope diagram showing the Ringgit-Beser data relative to Guntur and HK series lavas from Muriah (after Edwards et al. 1991)

lavas. The Sr and Nd isotopic data however, differ from the K-rich lavas of Muriah, which have lower $^{143}Nd/^{144}Nd$ ratios at similar $^{87}Sr/^{86}Sr$ ratios.

A plot of oxygen isotope data ($\delta^{18}O\%$) with respect to $^{87}Sr/^{86}Sr$ ratios for the lavas of Ringgit-Beser is given in Fig. 5.26. It may be noted that the K-rich lavas of the area (+ 6.8 ± 0.2 %) lie within the mantle range for MORB, seamounts, and hot-spot basalts. The $\delta^{18}O$ versus $^{87}Sr/^{86}Sr$ data suggest that only limited amounts of crustal contamination is possible.

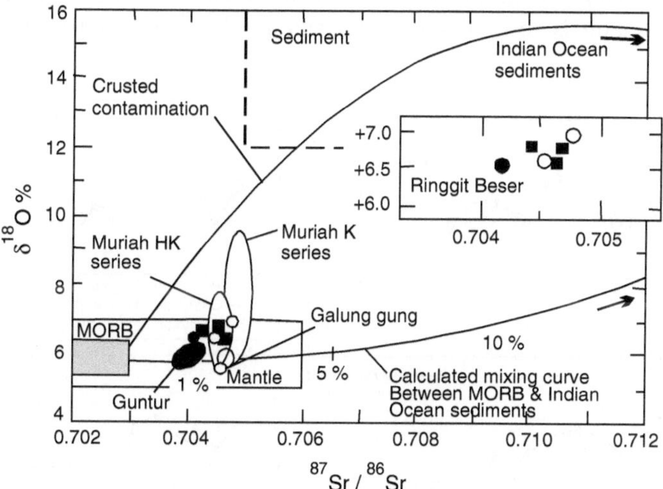

Fig. 5.26 The $\delta^{18}O$ versus $^{87}Sr/^{86}Sr$ diagram showing the whole rock data from Ringgit-Beser relative to Guntur, MORB and mantle hot-spot basalts and seamounts (after Edwards et al. 1994). The $^{206}Pb/^{204}Pb$ ratios of the lavas from Ringgit-Beser Complex range from 18.493 to 18.915, $^{207}Pb/^{204}Pb$ form 15.617 to 15.672 and $^{208}Pb/^{204}Pb$ from 38.975 to 39.191. They also found that the potassic lavas of Ringgit-Beser have less radiogenic Pb than calc-alkalic lavas. The EK series lavas have higher $^{208}Pb/^{204}Pb$ ratio than those of the K-series lavas of Ringgit-Beser Complex, dominated by leucite-bearing rocks

Figure 5.26 shows that the K and EK series lavas of Ringgit Beser plot in a close cluster on the ^{18}O versus $^{87}Sr/^{86}Sr$ diagrams, and slightly displaced from the MORB field to higher Sr–isotopic ratios. Edwards et al. found that the calc series samples plot away from the K-rich rocks to a lower $\delta^{18}O$ values and $^{87}Sr/^{86}Sr$ ratios. The Sr and O oxygen isotopic values of K-rich rocks of Ringgit-Beser Compex however contrast with those of Muriah located in a back-arc extensional setting. The REE and Trace Element geochemistry of K–rich silica undersaturated Rocks of Alto Paraniba Igneous Province.

The geochemistry of Potassic rocks of Alto Paranaiba Igneous Province has been studied by Gibson et al. (1992). They found that Cr and Ni content of these rocks range between 25 and 2000 and 70–1,530 ppm, respectively. As expected these two elements have positive correlation with MgO. In the kamafugitic rocks, Sc content (25–145 ppm) shows high abundance. The other incompatible elements such as Ba (1,500–6,000) and Th (12–40 ppm) show high concentration.

Gibson et al. plotted the composition of the Alto Paranaiba Igneous Province in a chondrite-normalized multi-element-diagram (Fig. 5.27). They observed that these rocks have relatively smooth patterns with a peak at La, and less H_2O–bearing samples with $(Ba/La)_N$ ratios being < 1. The patterns are similar in case of both extrusive and intrusives rocks with most samples having large troughs at K and Sr. The lavas of Mata da Corda formation commonly have troughs at P, whereas some of the intrusives at Limeria (I) and Indaia (I) have trough at Ti. These rocks are depleted in Ta relative to La (Fig. 5.27). The mafic potassic rocks of the Alto Paranaiba igneous province have high enrichment with respect to REE. The La and Lu concentration is respectively, 1,300 and 11 times the values of chondrites,

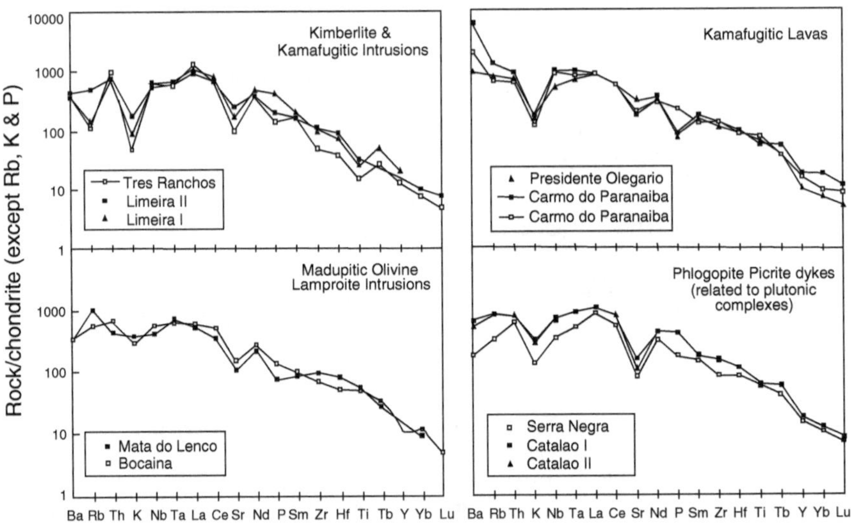

Fig. 5.27 Chondrite-normalized trace element diagram for K-rich rocks of Alto Paranaiba (after Gibson et al. 1992)

respectively. In some of the most MgO–rich rocks of Alto Paranaiba igneous province (e.g. Tres Ranches intrusions), the highest concentration of LREE (La = 402 ppm) is observed. The highest enrichment of HREE (Lu = 0.49 ppm) is obtained in rocks of Sao Gotardo, which has the lowest MgO content. This high HREE indicates the effect of fractional crystallization in the evolution of the rocks. Gibson et al. observed that all the mafic potassic rocks are more enriched in LREE compared to MREE with La/Sm ratio lying in the range of 6–12. They observed that the rocks of Tress Ranchos intrusion exhibit the most extreme enrichment in LREE related to MREE and has the highest La/Yb ratios (230). The lamproitic rocks however, have the lowest La/ Yb ratios (70–100) of all the mafic potassic rocks.

The negative Sr anomaly in this province is very similar to that of the rocks in many lamproitic localities, but presence of a trough at K is quite distinct from the patterns displayed by many lamproitic, kimberlitic and kamafugitic provinces (Fig. 5.27). The rocks of this province show absence of any significant depletion in such HFS elements as Nb and Ta in the samples of Alto Paranaiba igneous province, which contrasts with the rocks of Roman province and Leucite Hills. The Nb content (96–364 ppm) of the K-rich rocks of this province is similar to those of the West Kimberley lamproites and East African kamafugitic rocks. The Zr content of the rocks of Alto Paranaiba Igneous Province (322–1,451) is lower than that of the potassic rocks of Leucite Hills and West Kimberley. The rocks of this province do not show any distinct peaks of Zr and Hf on the normalized multi-element diagram. Gibson et al. observed that the abundance of REE of the Alto Paranaiba igneous province is similar to that in most enriched lamproites of West Kimberley and the Italian kamafugitic rocks. Compared to the rocks of this province, kamafugitic rocks of East Africa have much lower REE abundance.

5.9 Synthesis of Trace Element and Isotopic Data by Nelson (1992)

The potassium-rich silica-undersaturated mafic and ultramafic igneous rocks can be subdivided into a subduction-related "orogenic" sub-group and an "anorogenic" sub-group that are confined to stable continental settings (Nelson 1992). According to him, representatives of both sub-groups possess trace element and isotopic signature consistent with the contamination of their magma sources by incompatible element-rich and isotopically evolved "metasomatic" components. He argued that these metasomatic components are principally derived from subducted lithosphere, including subducted sediments. Nelson concluded that examples of orogenic potassic magmatism (e.g. Italian potassic rocks, Spanish lamproites, Sunda arc leucitites) have trace element and Sr, Nd and Pb isotopic characteristics, consistent with the contamination of their mantle sources by a component derived from marine sediments. He thought potassic magmas belonging to anorogenic subgroups have generally comparable trace element and Sr and Nb isotopic characteristics to those of orogenic potassic magmas, but many of them have unusual Pb isotopic compositions with respect to unradiogenic $^{206}Pb/^{204}Pb$ ratio. He pointed out that modern marine sediments characteristically have low U/Pb ratios and the unradiogenic $^{206}Pb/^{204}Pb$ ratio of anorogenic potassic magmas may have evolved during long-term storage of subduction sediments (or components derived from them) within the subcontinental lithosphere. According to him, these unusual Pb isotopic compositions require substantial time periods (>1 Ga) to have elapsed between the fractionation events lowering the U/Pb ratio through erosion and sedimentation at the earth's surface and subsequent potassic magmatism and it is therefore, not surprising that most examples of anorogenic potassic magmatism are not associated with recent subduction process. This idea seems to be correct. He concludes that the eruption of potassic magmas is commonly related to rifting and hotspot activity, and these processes do not necessarily play an important role in the genesis of the unusual sources from which potassic magmas are derived.

Chapter 6
Chemical and Physical Constraints for Crystallization of Feldspathoids and Melilite in Potassium-Rich Rocks

Leucite appears as a mineral in K–rich lavas, if the SiO_2 content of the rock varies between 44 and 56 wt% (Washington 1906). When the SiO_2 content is close to 56 wt%, it may appear but then the SiO_2/K_2O ratio should be below 10; in that case the K_2O content of the rock is close to 6 wt%. In silica-deficient rocks ($SiO_2 < 45$ wt%), leucite may appear even though the K_2O content is slightly higher than 1 wt%. The K_2O/Na_2O ratio is always greater than one in leucite-bearing rocks, but the prohibitive tendency of high soda may be eliminated if the SiO_2/K_2O ratio is low.

6.1 P–T Conditions Related to Leucite Stability

The P–T stability of leucite, kalsilite (as a break down product of leucite) and melilite, which are typically found in kamafujitic rocks, is discussed in the following sections. The $f(O_2)$ condition of formation of these rocks as determined by different investigators is also described.

6.1.1 Stability of Leucite

The breakdown of leucite to form K-feldspar and kalsilite was studied first by Scarf et al. (1966) and later by Fasshauer et al. (1998, Fig. 6.1). Breakdown of leucite containing analcite takes place at a lower pressure than that of its pure synthetic equivalent. The study of Scarf et al. and Fasshauer et al. (1998) explains the absence of leucite under plutonic conditions.

Schairer (1950) studied the system nepheline–kalsilite–SiO_2, and established that sanidine melts incongruently to leucite and a silica-rich liquid under atmospheric pressure and there is a large field of leucite in the system nepheline–kalsilite–SiO_2 (to be discussed later, Fig. 7.1). Later study by Tuttle and Bowen (1958, Fig. 6.2) showed that the field of leucite, produced as an incongruent melting phase of sanidine,

A.K. Gupta, *Origin of Potassium-rich Silica-deficient Igneous Rocks*,
Springer Geology, DOI 10.1007/978-81-322-2083-1_6

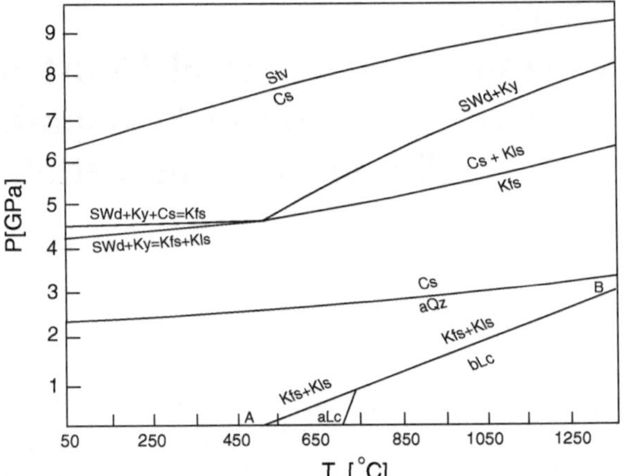

Fig. 6.1 Breakdown of leucite to K-feldspar and kalsilite under pressure (modified after Fasshauer et al. 1998)

decreases with progressive increase of pressure. Figure 6.2 shows that compared to the field of leucite, produced as an incongruent melting phase at 1 atmosphere, that observed for the K-feldspar–albite join studied at 0.1 and 0.2 GPa [P(H_2O) = P (Total)] progressively decreases. In case of the 0.2 GPa isobaric section of the join K-feldspar–albite, it is vanishingly small. Morse (1969) observed that under 0.5 GPa water pressure, the field of leucite is completely eliminated. Incongruent melting of sanidine (Fig. 6.3) was studied up to 4.0 GPa under dry condition by Lindsley (1966), who found that the field of leucite + liquid, persists up to 1.9 ± 0.1 GPa at 1,440 °C; above this pressure sanidine melts congruently. It can be seen from the negative slope that the melting of leucite produces a liquid of higher density.

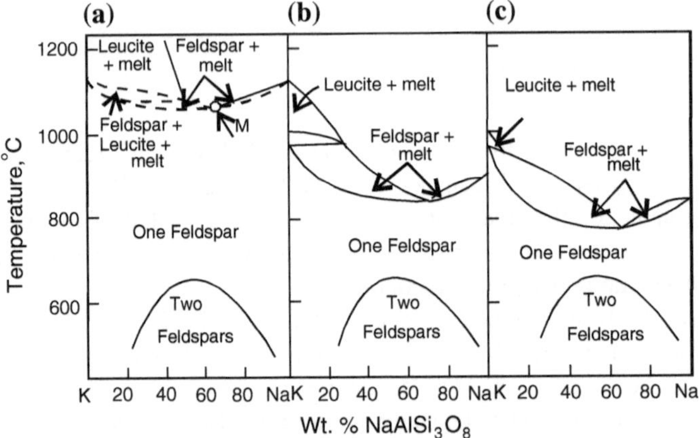

Fig. 6.2 Decrease in the field of leucite formed as an incongruent melting product of sanidine (modified after Tuttle and Bowen 1958). **a** Dry, **b** PH_2O = 0.1 GPa, c PH_2O = 0.2 GPa

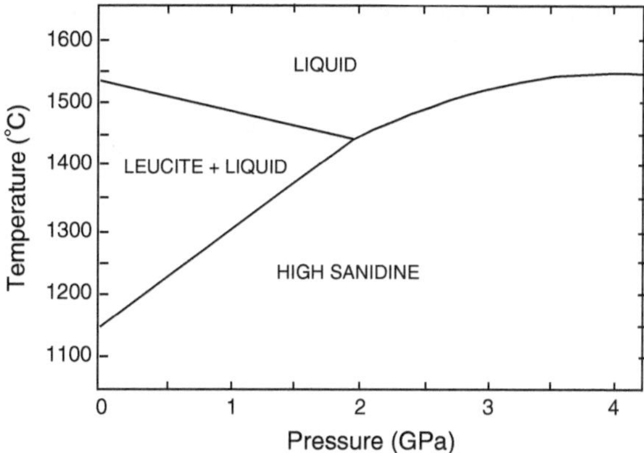

Fig. 6.3 High pressure study on melting of sanidine (modified after Lindsley 1966)

6.1.2 Melilite Stability

Melilite is often present in association with leucite in ultrapotassic rocks Lower Austria, Rift valley of East Africa, Roman Province (Italy) and Alto Paranaiba province of Brazil. The P–T stability of melilite is therefore, important to know the condition of formation of ultrapotassic rocks containing melilite.

Yoder (1973) studied the stability of pure akermanite in presence of excess water pressure, and established that between 680 °C and 1 bar, and 750 °C and 0.63 GPa, akermanite breaks down to wollastonite, monticellite and vapour (Fig. 6.4).

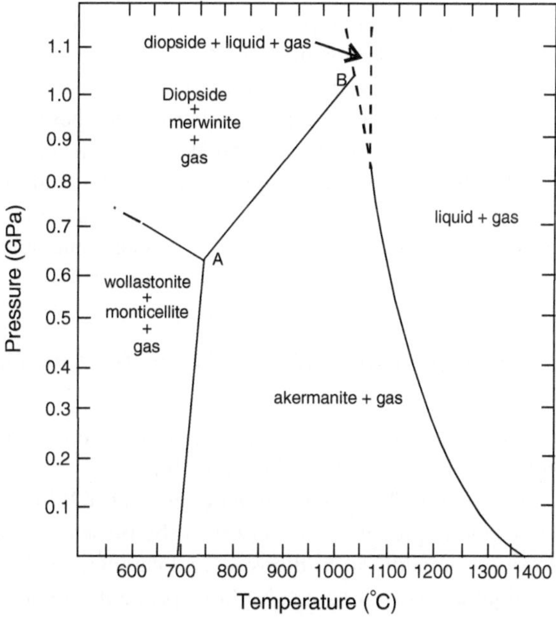

Fig. 6.4 Akermanite–H_2O system under variable P–T conditions (modified after Yoder 1973)

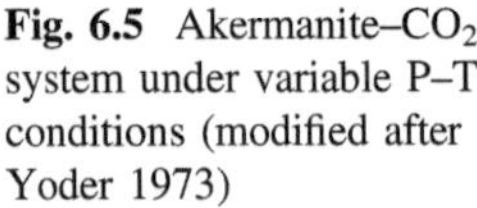

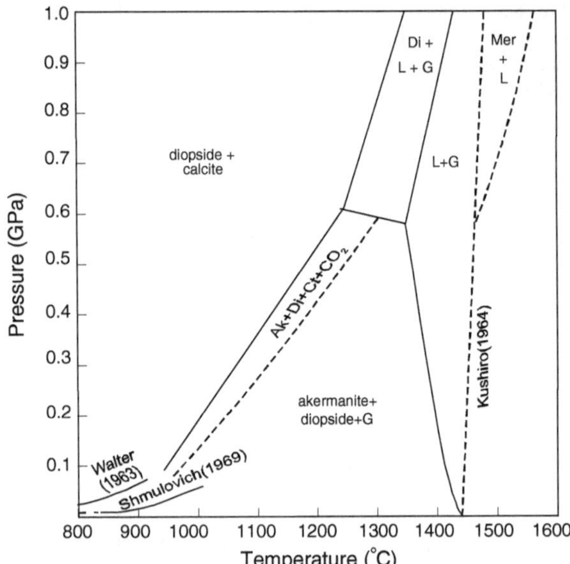

Fig. 6.5 Akermanite–CO_2 system under variable P–T conditions (modified after Yoder 1973)

Between 750 °C and 0.63 GPa, and 1,050 °C and 1.3 GPa akermanite breaks down to diopside + merwinite + vapour. In presence of forsterite, the akermanite stability is further restricted. Breakdown of akermanite under CO_2 pressure (Fig. 6.5) has also been studied by Yoder (1973).

Willemse and Bensch (1964) provided field evidence of the low-temperature and low pressure breakdown of akermanitic melilite from the Bushveld Complex. In this area they found graphic intergrowth of monticellite and wollastonite, indicating breakdown of akermanite. Field evidence of the intergrowth of kalsilite and K-feldspar is also reported from the same area, along with an unidentified phase, probably leucite, which may indicate that these two phases represent the low-temperature and low-pressure breakdown of leucite.

The above discussion suggests that leucite, along with forsterite and akermanite, must be restricted to conditions of high temperatures and low pressure. Leucite-bearing rocks are thus, products of volcanic and subvolcanic activities.

6.1.3 Appearance of Melilite in the Join Diopside–Nepheline

In natural leucite-bearing rocks, melilite, whenever present, is found to be a solid solution of akermanite and soda-melilite.

Bowen (1922) and later Schairer et al. (1962, Fig. 6.6) studied the join under one atmospheric pressure, and observed the presence of a large primary phase field of forsterite in addition to diopside and carnegieite. Melilite appears as an important subliquidus as well as a subsolidus phase in this join.

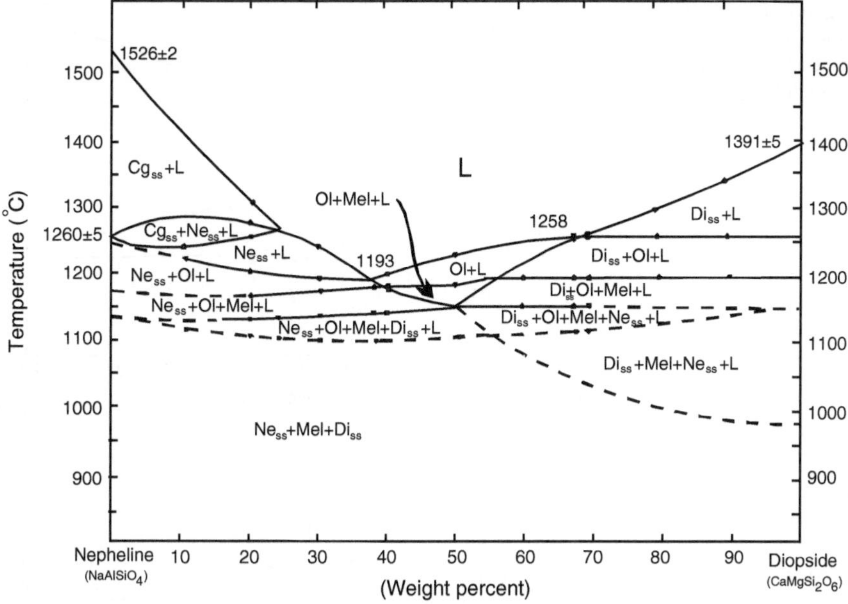

Fig. 6.6 Phase relation in the system diopside–nepheline under one atmospheric pressure (after Schairer et al. 1962)

The presence of melilite and olivine in K-rich nephelinites can be explained by the following reaction (Yoder and Kushiro 1972):

$$\underset{\text{diopside}}{3CaMgSi_2O_6} + \underset{\text{nepheline}}{2NaAlSiO_4}(\underset{\text{akermanite}}{Ca_2MgSi_2O_7} + \underset{\text{sodamelilite}}{CaNaAlSi_2O_7}) + \underset{\text{forsterite}}{Mg_2SiO_4} + \underset{\text{albite}}{NaAlSi_3O_8}$$

Albite did not appear in the system diopside–nepheline (Bowen 1928; Schairer et al. 1962). It is possible that it remains as a dissolved species in the melt at high temperatures, and at low temperature it is included in nepheline as solid solution.

Singh et al. (2000) studied the join diopside–nepheline at variable pressures and temperatures, and established that the reaction of diopside and nepheline to produce olivine and melilite is eliminated at 1.9 ± 0.3 GPa, and above this pressure melilite and olivine did not appear. Phase relations of the diopside–nepheline join studied under 2.8 GPa is shown in Fig. 6.7, which shows complete disappearance of melilite, and presence of only diopside$_{ss}$ and nepheline$_{ss}$ in the subsolidus region.

Singh and Gupta (1992) and Sing et al. (2000) studied the system diopside–nepheline at 1.0 GPa (Fig. 6.8) and variable temperatures in presence of excess water. They noted that under hydrous condition subsolidus assemblage is

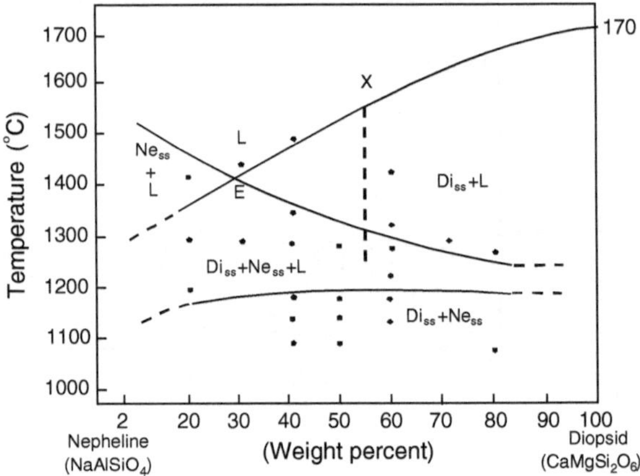

Fig. 6.7 Phase relation in the diopside–nepheline system under 2.8 GPa (after Singh et al. 2000)

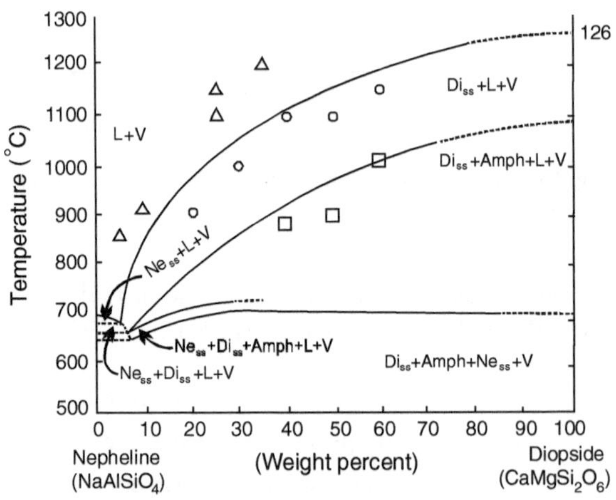

Fig. 6.8 Phase relation in the diopside–nepheline system in presence of H_2O at 1.0 GPa (after Singh et al. 2000)

nepheline$_{ss}$ + diopside$_{ss}$ + amphibole. Melilite and olivine, which appear as a reaction product of diopside and nepheline at atmospheric pressure are absent.

Singh et al. (2000) studied a composition $Di_{40}Ne_{60}$ in presence of excess water pressure (Fig. 6.9). They fused this starting material to glass and crystallized it at 800 °C. The crystalline material contained diopside$_{ss}$, melilite and nepheline. Their results are summarized in Fig. 6.9. The line A–B–C denotes the solidus for the studied composition. Reference to the study of the join under atmospheric pressure (Schairer et al. 1962) suggests that there is a drastic decrease in the solidus from 1,100 ° at 1 bar to 750 °C at 0.17 GPa, a drop of 350 °C. Above 0.3 GPa there is no change in the temperature of the solidus and it is essentially constant at 700 °C. Above the line D–E, melilite and forsterite$_{ss}$ completely disappear. Below the curve B–C, forsterite reacts with the liquid and is eliminated. The curve F–G marks the disappearance of amphibole at higher temperatures. The curve I–H–J marks the

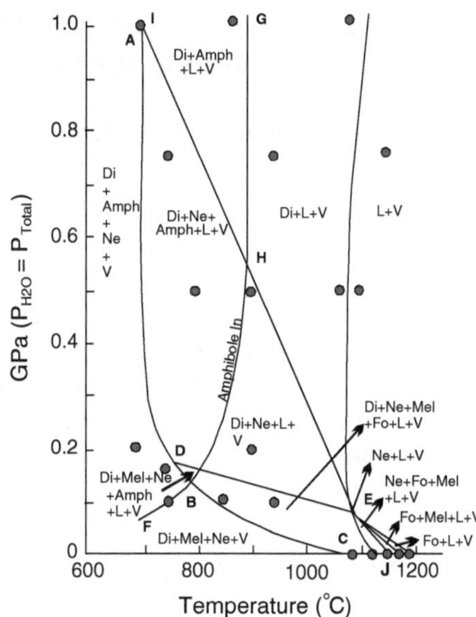

Fig. 6.9 Study of the composition $Di_{40}Ne_{60}$ under variable P–T condition in presence of excess water (after Singh et al. 2000)

appearance of nepheline at a lower temperature. At high pressure it may be observed that the equilibrium assemblage is diopside$_{ss}$ + amphibole + nepheline$_{ss}$ below the solidus. At low pressures between 0.1 and 0.15 GPa the assemblages obtained progressively at lower temperatures are as follows:

(1) forsterite$_{ss}$ + L,
(2) forsterite$_{ss}$ + melilite + L,
(3) nepheline$_{ss}$ + forsterite$_{ss}$ + melilite + L,
(4) diopside$_{ss}$ + melilite + forsterite$_{ss}$ + nepheline$_{ss}$ + L, and
(5) diopside$_{ss}$ + melilite + nepheline$_{ss}$.

Below 0.57 GPa as the temperature is systematically lowered the assemblages are:

1. diopside$_{ss}$ + L,
2. diopside$_{ss}$ + nepheline$_{ss}$ + L,
3. diopside$_{ss}$ + nepheline$_{ss}$ + amphibole + L, and
4. diopside$_{ss}$ + amphibole + nepheline$_{ss}$.

It is interesting to note that whereas forsterite appears as a primary phase below 0.1 GPa, diopside becomes the liquidus phase above this pressure. But there is a small field of Diopside + nepheline + L above the D–E line at high temperatures at a limited range of pressure. Melilite and forsterite are totally eliminated at 0.15 ± 0.05 GPa. The assemblage, characteristic of nephelinites (diopside$_{ss}$ + nepheline$_{ss}$ + L), is restricted to a pressure below 0.57 but above 0.15 GPa and at a temperature range of 880 and 805 °C.

6.2 Partial Pressure of Oxygen Related to Genesis of K-Rich Volcanic Rocks

6.2.1 Oxygen Fugacity Related to Stability of Annite

The stability of annite was studied by Eugster and Wones (1962). Their study is summarized in Fig. 6.10, which shows the coexistence of iron-rich olivine, leucite, magnetite and vapour in equilibrium with iron-rich mica. In natural K-rich rocks, mica coexisting with leucite, olivine, magnetite and vapour is phlogopite-rich containing annite molecule. Phlogopite containing 20–30 mol% annite will have higher P–T stability and hence, the assemblage, olivine + leucite + magnetite + vapour in equilibrium with mica will be stable at oxygen fugacity possibly between MW and QFM buffer.

6.2.2 The Fe^{3+}/Fe^{2+} Ratio for Determination of Oxygen Fugacity in Potassic Rocks

Turbevile (1993) determined the oxygen fugacity of Latera Caldera rocks (Fig. 6.11) containing phenocrysts of sanidine ± leucite with plagioclase,

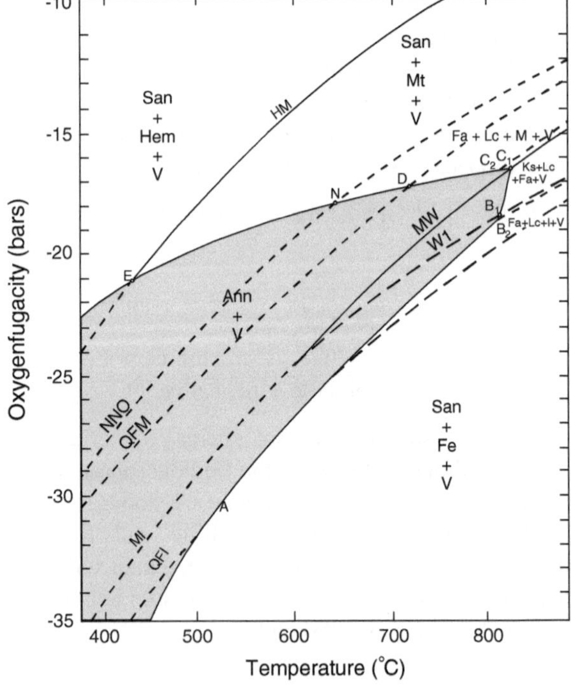

Fig. 6.10 Phase relation of annite (+vapour) bulk composition as a function of oxygen fugacity (after Eugster and Wones 1962)

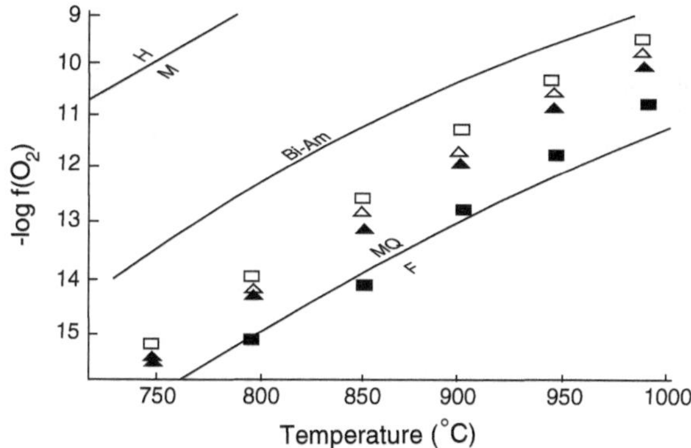

Fig. 6.11 Plots of $-\log f(O_2)$ versus temperature for selected Latera pumice and scoria samples calculated from whole-rock data; also shown are buffer curves for magnetite–hematite, fayalite–magnetite–quartz and stability of biotite–amphibole. *Open square* trachyte, *open triangle* evolved phonolite, *solid triangle* unevolved phonolite; *solid square* tephriphonolite (after Turbeville 1993)

clinopyroxene, biotite, magnetite, apatite and sphene in the groundmass. On the basis of Fe^{3+}/Fe^{2+} ratio of the whole rock, he calculated the oxygen fugacity of the rocks. His plot of the rocks from Latera Caldera in a $-\log f(O_2)$ versus temperature diagram shows that oxygen fugacity of these Latera Pumice plots above FMQ but below biotite–amphibolite curve.

6.2.3 *Oxide Phases as an Indicator for $f(O_2)$ Condition of Formation of Potassic Rocks*

The oxygen fugacity condition of formation of an ilmenite-bearing minette, was calculated by Eggler (1987). According to him the $f(O_2)$ data cluster between QFM and MW buffers at the estimated pressures of origin (2.5 GPa and above).

The analyses of lamproites suggests that they are relatively more H_2O-rich but CO_2-poor. Crystallization of phlogopite from lamproitic magma suggests that the genesis of the rock is related to hydrous melting. The presence of diamond in some olivine lamproites, led Foley (1985) and Foley et al. (1986b) to believe that oxygen fugacity condition of the source rock for some lamprophyres at least is best represented by CW (carbon–water) buffer (Fig. 6.12). The buffer curve is the locus of points on the carbon-saturation surface, where $X(H_2O)$ is maximum. This buffer curve lies roughly mid-way between IW and WM buffers at temperatures and pressures likely to represent diamond stability curve. They think that it is possible that melt generation is induced by dehydration at the beginning and introduction of a water-rich volatile phase, which may lead to a condition more oxidizing than CW

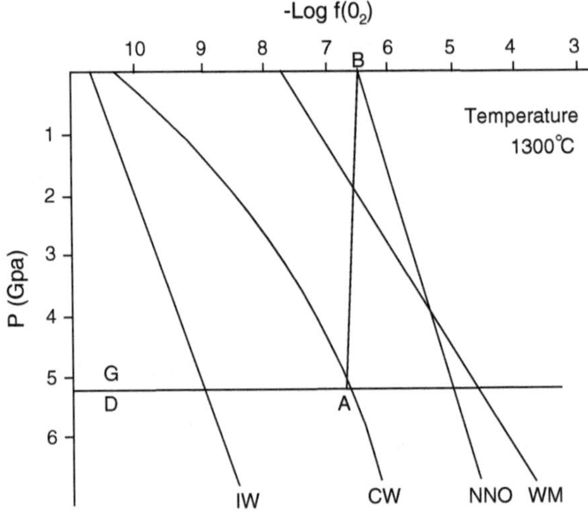

Fig. 6.12 A schematic P-f (O_2) diagram illustrating increase in oxygen fugacity of a lamproitic liquid as it ascends from the mantle (after Foley et al. 1986b)

(carbon–water) buffer, if not constrained by carbon saturation. They also think that survival of diamond in an olivine lamproite would be related to sluggish diamond breakdown reaction. Foley (1985) studied chrome spinels found in olivine phenocrysts in lamproites and experimentally calibrated the ferric number [100 X $Fe^{3+}/(Fe^{3+} + Fe^{2+})$] of spinel as oxygen fugacity sensors. Foley concluded that the Gaussberg lamproites crystallized under $f(O_2)$-T condition, which is close to $10^{-6.5}$ and 1,300 °C, which corresponds to NNO buffer condition.

If the Gaussberg lamproitic melt originated at a depth > 52 km under C–H buffer condition, then the primary melt probably originated under a much more reducing condition. If the melt travelled along the line AB (at a constant $f(O_2)$ of $10^{-6.5}$ to reach B, then according to their estimate 0.09 wt% H_2O must dissociate to maintain a constant $f(O_2)$ path from C–H to FMQ buffer condition near the surface at 1,300 °C.

The reaction, $H_2O \Leftrightarrow H_2 + 1/2O_2$, coupled with H_2 loss by diffusion was suggested by Sato (1972) for oxidation mechanism in a magma. Foley et al. (1986b) suggested that the intrinsic $f(O_2)$ of the system may be maintained between the path A–B during the ascent by diffusion of H_2 loss via following reaction,

$$FeO + 1/2\ H_2O = 1/2\ H_2 \uparrow + FeO_{1.5}$$

They thought that the Gaussberg composition has a median value for primary lamproitic magma (FeO in Gaussberg lamproite is 6 wt% compared to a range of 4–8 % for primary lamproites, Barton and Hamilton 1978).

6.3 Determination of Oxygen Fugacity in Potassic Rocks Based on the Presence of Picroilmenite

Picroilmenite has been reported to occur in phlogopite-bearing minette from Buell Park, Arizona. Esperanca and Holloway (1987) observed that picroilmenite in these minettes are stable only at QMF buffer and temperatures below 1,100 °C. Under this $f(O_2)$ condition at 1 GPa, it coexists with olivine, clinopyroxene and phlogopite, and contain lower amounts of TiO_2 and FeO than at 1.5 GPa, where it coexist with clinopyroxene and phlogopite.

Roden and Smith (1979) studied ilmenites and oxidized spinels from ring dykes at Buell Park. They employed the method of Buddington and Lindsley (1964) and obtained a temperature of 815 °C and $f(O_2)$ condition of $10^{-10.4}$. These values also correspond to QM buffer condition, and subsolidus temperature of re-equilibration condition. Titanomagnites also occur in Agathapa Park and the Thumbs minnetes (Jones and Smith 1983). Luhr and Carmichael (1981) reported the occurence of chromian spinel from K-rich basanites, and minettes of Colima graben, where it occurs as an early crystallization phase. Titano-magnetite appears as a groundmass phase. Estimated $f(O_2)$ values for Colima graben potassic suite range between NNO and HM buffer conditions (Sack et al. 1980). Esperanca and Holloway thought that partial re-equilibration in volatile-rich magmas at variable $f(O_2)$ conditions during ascent could have been responsible for the scatter in the chemical composition of micropheno crystal Fe–Ti oxides in the K-rich lavas of different localities.

6.4 Oxidation Path of a Leucitite Magma with Respect to CO_2 Solubility

The solubility of CO_2 in a Ca-rich leucitite was determined by Thibault and Holloway (1994). In this connection, they calculated the oxidation path (Fig. 6.13) of an ascending leucitite magma originated at a depth of 75–80 km (2.5 GPa). They estimated a temperature of 1,300 °C for the magma at the source region. They considered an isothermal ascent at this temperature. If no carbon is dissolved and there is no oxidation through degassing, then the magma will ascend through path (a). If the magma generated again at 75–80 km and dissolves 5 wt% CO_2 (i.e. the liquid is not saturated with respect to CO_2), then it will follow path (b). If the magma is saturated with respect to CO_2 after its generation at same depth, it would then take an ascending path denoted by the curve (c). The path would then show maximum deviation with respect to ΔNNO[log $f(O_2)$ CCO − log $f(O_2)$ NNO]. The heavy lines denoted by NNO and CCO represent respectively the $f(O_2)$ values of the NNO oxygen buffer and a graphite-saturated CO fluid.

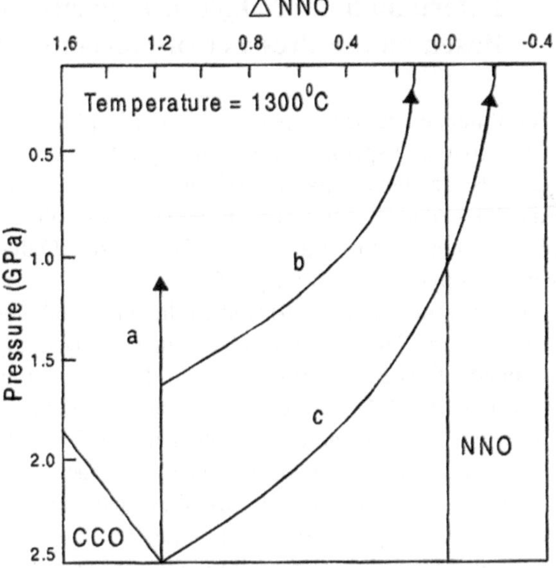

Fig. 6.13 Calculation oxidation paths for leucitite magma (after Thubault and Holloway 1994)

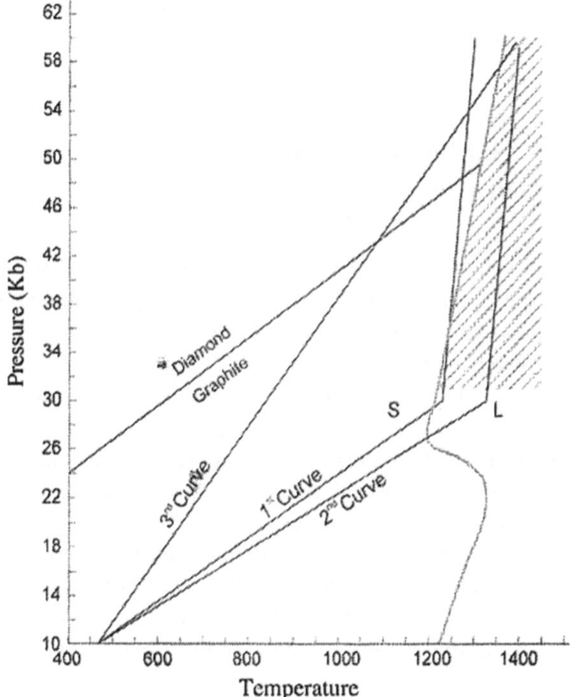

Fig. 6.14 The kink *curved line* refers to the peridotite *solidus curved line* of Wyllie (1979). *Curves 1, 2, and 3* are cooling curves in a P–T space as conducted in the experiments by Bearua et al. (2013). *Shaded area* was considered by Wyllie (1979) to be the regime, where kimberlitic or carbonotitic magma is originated

6.5 The Ascent Rate of Diamond and Phlogopite-Bearing Olivine Lamproite or a Kimberlitic Magma

Barua, Gupta, Mandal and Singh (2013) estimated the ascent rate of a phlogopite and dolomite-bearing ultramaic magma (the assemblage contained olivine, enstetite and diopside in appropriate proportion mimicking a model kimberlitic magma) containing diamond. Their experiments were conducted using a Walker-type multi-anvil apparatus. The experiments were made with a piece of diamond, placed within such a synthetic assemblage in each set of an experiment, keeping an initial pressure and temperature of 6 GPa and 1,350 °C. The assemblage was then cooled at different rates (along 1st, 2nd and 3rd curve, see Fig. 6.14). The topology of the solidus of peridotite is after Wyllie (1979). It was observed that the volume fraction of diamond to graphite conversion strongly depended on the ascent rates. Using electrical resistivity and X-ray diffraction studies, Barua et al. measured the degree of graphitization as a function of the ascent rate (u). For $u < 3$ m/s, diamond underwent almost complete graphitization (conversion > 90 %), whereas it remained nearly intact (conversion of 10 %) when u is >10 m/s. Their theoretical calculations of the settling velocity of mantle xenoliths again confirm that diamond cannot exist when u is as small as 3 m/s. They also performed numerical experiments with finite element (visco-elastic) models to analyse the dynamics of tensile failure at the tip of magma pools, leading to dilatational vertical fractures for magma transport. Considering the tensile strength of mantle in the order of 0.5 kb, their models show this failure process as a function of the critical shape (A_r ratio of vertical and horizontal dimensions) and density contrast ($\Delta\rho$) of magma pools. The critical $\Delta\rho$ was estimated to be nearly 200 kg/m^3 when A_r is considered to be very large (>4) (Fig. 6.14).

Chapter 7
Ternary Systems with Feldspathoids

In the fractionation scheme of basaltic magmas, CaO, MgO and total iron oxide ($FeO + Fe_2O_3$) contents decrease progressively, and those of Al_2O_3, Na_2O and K_2O increase with respect to silica (Harker 1909), as a result of which the residual liquid becomes enriched with respect to $NaAlSiO_4$ (nepheline), $KAlSiO_4$ (kalsilite), and SiO_2 components. Bowen (1928), therefore, referred to the system $NaAlSiO_4$–$KAISIO_4$–SiO_2 as petrogny's residua system, in which the $NaAISi_3O_8$ (albite)–$KAlSi_3O_8$ (K-feldspar) join acts as a thermal barrier at low pressures. The alkali feldspar join thus divides the system into a silica-deficient and a silica-saturated portion. The albite-K-feldspar-SiO_2 has relevance to the rocks of granite family, and the silica-undersaturated portion has bearing on the genesis of feldspathoid-bearing ijolites, nepheline syenites ultrapotassic rocks and pseudoleucite.

7.1 The System Nepheline–Kalsilite–SiO_2 Under Variable P–T Conditions at or Below 5 Kb in Presence of Excess Water

Schairer and Bowen (1935) and Schairer (1950, Fig. 7.1) studied this system under atmospheric pressure. They established extensive solid solution relationship between albite and orthoclase under one atmosphere, and the join does not have any eutectic, but has a temperature minimum; hence in the ternary join albite-orthoclase-SiO_2, there is a cotectic line F–H and instead of a eutectic, there is also a thermal minimum (point G, Fig. 7.1). This minimum is also popularly known as the "granite minimum". With the increase of pressure in presence of H_2O (up to 0.3 GPa), the granite minimum shifts toward albite (Tuttle and Bowen 1958). Slightly above 0.35 GPa, the albite-orthoclase join has a eutectic and the study of Morse (1969) showed the presence of a eutectic in this join at $Or_{28}Ab_{72}$ under 0.5 GPa. Since sanidinemelts incongruently at high temperature to formleucite and a silica-saturated liquid, there is a field of leucite, even in the silica-saturated side of the albite-K-feldspar join. Since this volume is related to the genesis of silica-

A.K. Gupta, *Origin of Potassium-rich Silica-deficient Igneous Rocks*,
Springer Geology, DOI 10.1007/978-81-322-2083-1_7

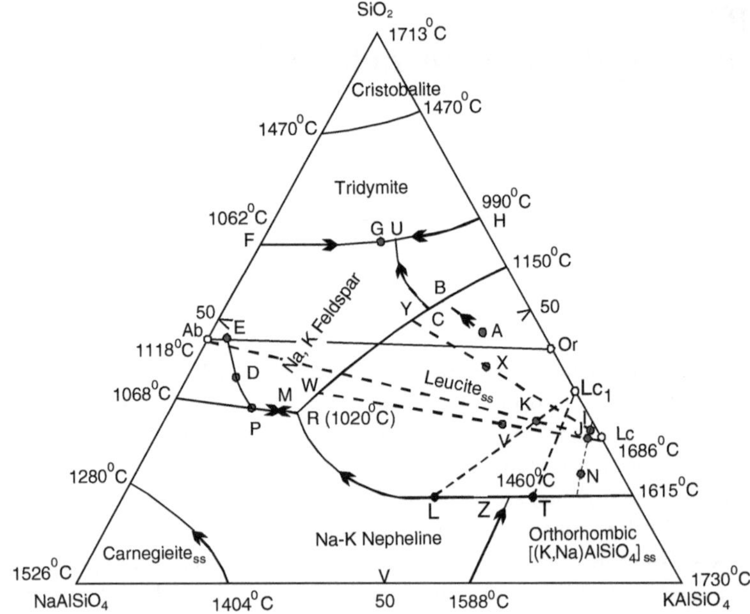

Fig. 7.1 Phase relation in the system nepheline–kalsilite–SiO_2 under atmospheric pressure (after Schairer 1950)

undersaturated potassium-rich rocks, further discussion on the silica-saturated portion of the system will not be made.

In the silica-undersaturated portion of the system, nepheline–kalsilite–SiO_2, there is also a temperature minimum M, and a reaction point R. If the composition of an initial liquid 'X' lies within the triangle, Lc–Ab–Or (Fig. 7.1), after crystallization of leucite, the liquid composition should cross over the thermal barrier Ab–Or, till it reaches the univariant curve B–Y–R at Y (the point Y is determined by joining the initial bulk composition leucite and X). Feldspar will co-precipitate with leucite at point Y and then the composition of liquid should move along Y–W to R. At R leucite will co-precipitate with Na–K feldspar. Crystallization should continue till the liquid is exhausted at the point R. Thus, even though the liquid composition lying within the triangle, Lc–Or–Ab, crosses over temporarily to the silica-saturated side, they ultimately return to the silica-deficient portion. In case the liquid composition lies above the thermal barrier Ab–Or, but within the leucite field (say point A), the liquid should yield leucite first and its composition should move along A–B. The line A–B once again has been derived by joining the composition of leucite with point A, and extending Lc–A toward the B–Y–W–R curve. At B, leucite and K–Na feldspar co-precipitate from the liquid. Then, with lowering of temperature the liquid composition moves along the path B–C with gradual elimination of leucite, which reacts with the liquid. At C, all leucite is exhausted and the liquid composition moves along the curve C–U (the composition of the path can be determined by electron microprobe analyses). With further lowering of temperature the liquid moves along the curve U–G to G. Tridymite co-precipitates with feldspar along the curve U–G. Final composition of the liquid is represented by the minimum point G.

When the liquid composition lies in the silica-undersaturated portion of the system within the triangle Ab–Or–nepheline (say point D), with lowering of temperature, the liquid composition should move along the curve E–D–P–M to M (the minimum point). The point E represents the composition of feldspar in equilibrium with the liquid D. The point E (the composition of feldspar) has to be determined by electron microprobe analysis. The curved path D–P is obtained by determining the compositions of liquids by microprobe analyses.

When the bulk composition of the liquid lies within this triangle Lc–Or–Ne (say V), leucite should precipitate first and the composition of the liquid should move along Lc–V–W. At W co-precipitation of alkali feldspar and leucite should take place. When the liquid composition reaches the invariant point R, $nepheline_{ss}$ should also appear. Leucite reacts out at R and then the liquid composition should move along R–M to M ($Ne_{52}Ks_{15}Qz_{33}$), where $nepheline_{ss}$ will co-precipitate with alkali feldspar. If the liquid composition lies within the triangle nepheline-Lc-kalsilite (at point N), with lowering of temperature, leucite should precipitate first. The liquid composition then moves along the line Lc–N–T–Z. Along the line T–Z leucite and orthorhombic kalsilite should co-precipitate from the liquid. At Z (1460 °C) crystallization should stop and the liquid should be exhausted. In the above discussion it has been assumed that leucite contains little or no solid solution at one atmospheric pressure.

Hamilton and MacKenzie (1965) studied the system nepheline–kalsilite–SiO_2 in presence of excess water under 0.1 GPa (Fig. 7.2). Their study shows that at 0.1 GPa, the field of leucite, produced due to incongruent melting of sanidine, decreases. Hamilton and MacKenzie noted that although the temperature of the minimum melting point is lowered from 1,050 (established under atmospheric pressure) to 750 °C under 0.1 GPa P (H_2O), the composition of the minimum ($Ne_{50}Ks_{19}Qz_{31}$) is very similar to that found by Schairer (1950).

Luhr and Giannetti (1987) recalculated the composition of Brown Leucitic Tuffs in terms of nepheline–kalsilite–SiO_2 system (Fig. 7.3). The asterisk symbols represent compositions, which are phonolitic with low CaO content (<5.6 wt%). The compositions with CaO content (>5.6 wt%) are indicated by open circle (>5.6 wt%, Fig. 7.3). Luhr and Giannetti (1987) also plotted the composition of primitive lava (marked by stars) in the same diagram. The 0.1 GPa cotectic line is shown by solid line but dashed line indicates one atmosphere cotectic line. When the anorthite content is high the liquid composition should not lie in the ternary join and the course of crystallization of liquid cannot be depicted in this join. Assuming that before eruption, the melt equilibrated at shallow depth, it should precipitate leucite first from the bulk composition denoted by the unfilled triangles. With progressive decrease in temperature, alkali feldspar should coprecipitate with leucite, as the liquid reaches the cotectic line. As nepheline did not appear, it may be interpreted

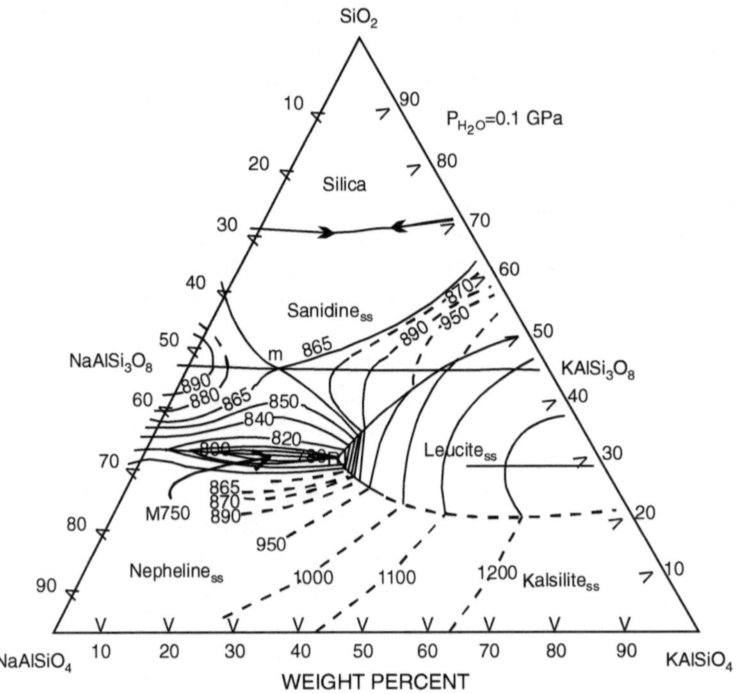

Fig. 7.2 Phase relation in the system nepheline–kalsilite–SiO_2 under 0.1 GPa [$P(H_2O) = P(Total)$] (after Hamilton and MacKenzie 1965)

that the liquid probably got frozen, as crystallization was hindered because of eruption.

Cundari and Mattias (1974) plotted the composition of lavas of Somma-Vesuvius and Phlegrean Field containing very low amount of MgO and other components in terms of nepheline, kalsilite and SiO_2 (Fig. 7.3), they found that the most evolved rocks have their compositions clustering around M.

The Vesuvius rocks of more salic type were plotted by Cundari and Le Maitre (1970, Fig. 7.3) and found that they plot -near the 0.1 GPa [$P(H_2O) = P(Total)$] minimum representing phonolitic composition (Qz = 27–34, Ne = 35–47, Ks = 25–27).

The silica–undersaturated part of the system, nepheline–kalsilite–SiO_2 was studied under 0.2 GPa [$P(H_2O) = P(Total)$] by Taylor and MacKenzie (1975, Fig. 7.4) and showed that the minimum melting point occurs at $Ne_{51}Ks_{20}Qz_{29}$ and 710 ± 7 °C. Comparison of Figs. 7.1, 7.2, 7.3 and 7.4 show a significant reduction of leucite field under 0.2 GPa $P(H_2O)$.

Zhang and Mackenzie (1984) studied the system $NaAlSiO_4$–$KAlSiO_4$–SiO_2–H_2O under 0.5 GPa. Their results are summarized in Fig. 7.5, which shows that in contrast to Figs. 7.1, 7.2 and 7.3 and 7.4 leucite field has been significantly reduced, because this phase produced as an incongruent melting product of sanidine, is

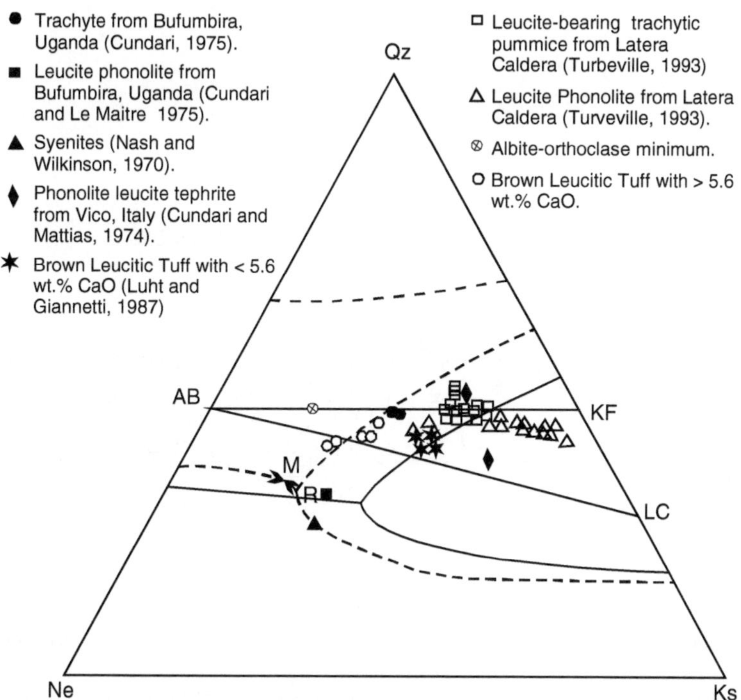

Fig. 7.3 Plot of bulk composition of Brown leucitic Tuffs (after Luhr and Giannetti 1987) in the nepheline–kalsilite–SiO_2 ternary diagram, studied under 0.1 GPa [$P(H_2O) = P(Total)$]

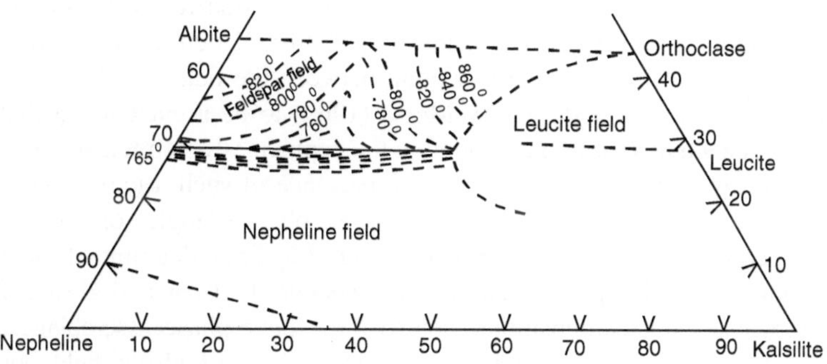

Fig. 7.4 Phase relation in the system nepheline–kalsilite–SiO_2 under 0.2 GPa [$P(H_2O) = P(Total)$] (after Taylor and MacKenzie, 1975)

progressively eliminated. At 0.5 GPa, the field of leucite does not encompass sanidine composition, and sanidine melts congruently. The Fig. 7.5 further shows that the "granite minimum" becomes a eutectic. Likewise, "the nepheline syenite minimum" has also become a eutectic. According to them there is a boundary between the fields of nepheline and kalsilite in the ternary system. This in turn

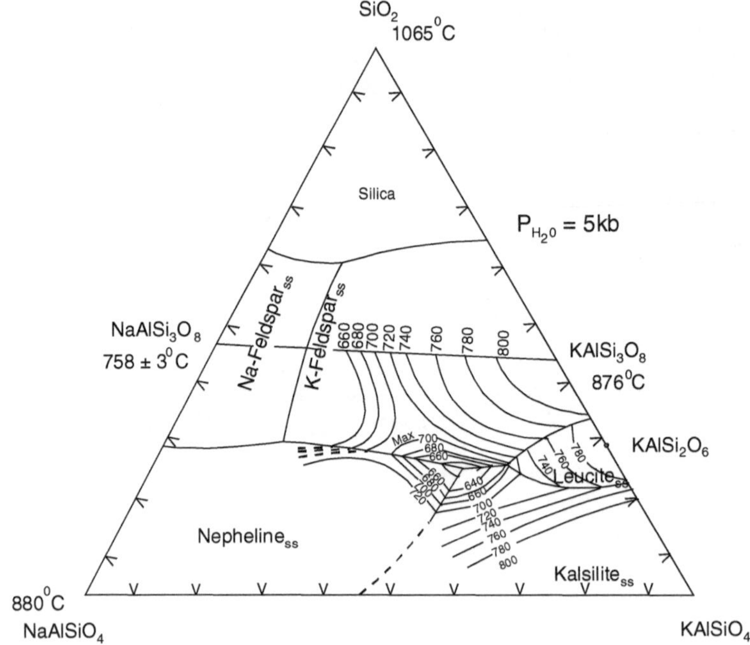

Fig. 7.5 The system nepheline–kalsilite–SiO_2 under 0.5 GPa [P(H_2O) = P(Total)] (after Zhang and MacKenzie 1984)

intersects thenepheline-feldspar field boundary at $Ne_{25}Ks_{51}Q_{24}$ and 630 °C. Before they determined the liquidus of this system Zhang and MacKenzie (1984) establish that the invariant point between the fields of nepheline, K-rich feldspar and kalsilite to be a eutectic. Earlier, the invariant point between the fields of K-rich feldspar, Na-rich feldspar was also shown by Morse (1969) to be a eutectic. In that case, there should be a temperature maximum on the nepheline-K-rich feldspar boundary as shown in Fig. 7.5. In order to prove the presence of such a temperature maximum, they determined the positions of two three-phase triangles on either side of the presumed eutectic, which are shown in Fig. 7.6. They determined these three phase triangles by microprobe analyses of nepheline, feldspar and glass from run products, whose bulk compositions yielded these three-phase assemblages. They argued that the temperature maximum on the nepheline-feldspar field boundary should lie between the trailing edges on the three phase triangles, and they established the maximum at $Ne_{37}Ks_{37}Q_{26}$. In the diagram (Fig. 7.6) they showed a number of experimentally-determined conjugation lines especially in the K-rich feldspar and kalsilite field. These conjugation lines demonstrate the directions toward which the liquid should move with fractional crystallization.

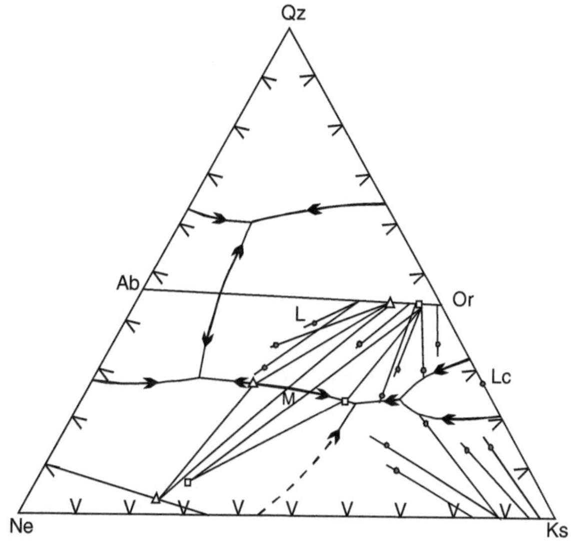

Fig. 7.6 The phase triangle showing that there is a maximum on the field boundary between nepheline and feldspar. A few conjugation lines in the K-feldspar and kalsilite field in the system $P(H_2O) = 0.5$ GPa are also projected on to the anhydrous base (after Zhang and Mackenzie 1984)

7.2 Phase Relations in the System Nepheline–Kalsilite–SiO_2 at 2 Gpa [$P(H_2O) = P(\text{Total})$]

Gupta et al. (2010) studied the system nepheline–kalsilite–SiO_2 at P (H_2O) = 2GPa and various temperatures (Fig. 7.7). It may be noted that with the increase of pressure to 2GPa, the field of leucite$_{ss}$ is completely eliminated, and there is an additional field of jadeite$_{ss}$, as suggested by Huang and Wyllie (1975).

The study of Gupta et al. (2010) demonstrates that there are two eutectics:

(1) E_1 occurs at $Kls_{18}Ne_{52}$ Qtz_{30} and 620 ± 10 °C, where jadeite$_{ss}$, sanidine$_{ss}$, nepheline$_{ss}$ liquid and vapour are in equilibrium, and (2) E_2 occurs at $Kls_{49}Ne_{25}Qtz_{26}$ and 615 ± 10 °C, where nepheline$_{ss}$, kalsilite$_{ss}$ and sanidine$_{ss}$ coexist with liquid and vapour.

They established three-phase triangles (Fig. 7.8) to mark the courses of the compositions of the liquids in equilibrium with crystalline phases. For example, a liquid of composition 1 (lying on the cotectic) is in equilibrium with nepheline of composition 3; likewise, a liquid of composition 10a is in equilibrium with feldspar of composition 10b (Fig. 7.8). A temperature maximum (M) occurs at $Kls_{35}Ne_{38}Qtz_{27}$ and 635 °C (Fig. 7.8). From the maximum, the cotectic moves toward E_1or E_2 as established by the three-phase triangles (points 4–5–6 and points 7–8–9). Similarly, a melt (11) coexists with feldspar (13) and feldspathoid (12). Furthermore, feldspathoid (12) does not fall on the nepheline–kalsilite join because kalsilite incorporates both nepheline and excess silica in solid solution. The location of the three-phase triangles was established by electron–microprobe data. The maximum is related to the intersection of the nepheline–K-feldspar join with the nepheline + feldspar cotectic phase boundary.

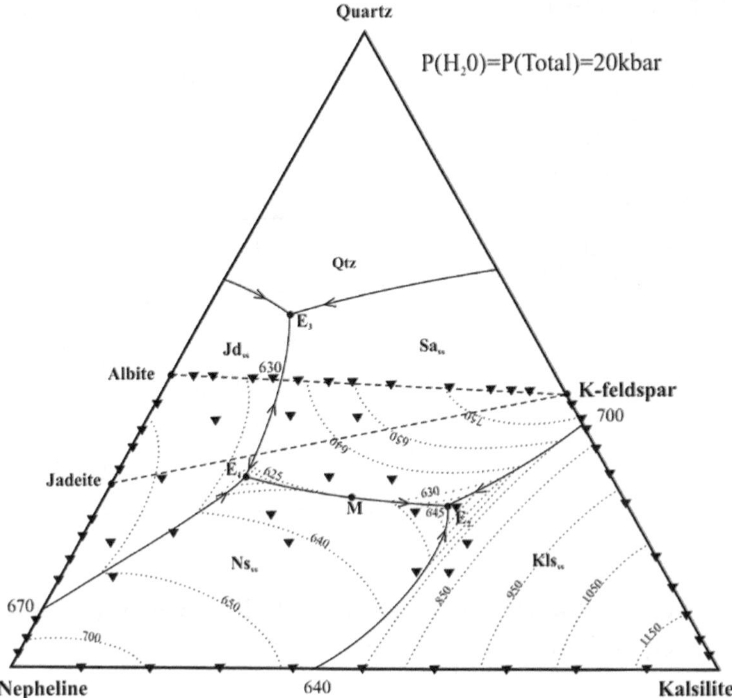

Fig. 7.7 Phase relations in the pseudoternary system nepheline–kalsilite–SiO_2 determined at 2 GPa [$P(H_2O)$ = P(Total)] and various temperatures (present study). Plot of the ternary eutectic E3 is after Huang and Wyllie (1975). The symbol "▼" shows the compositions of the starting material made to study the four binary joins Kls-Sa, Kls-Ne, Ne-Ab, Ab-Sa and the silica-undersaturated portion of the system Ne-Kls-SiO_2 (Gupta et al. 2010)

Because of the reaction between nepheline and albite to produce jadeite at 2 GPa in the presence of excess H_2O, K-feldspar—jadeite becomes a thermal barrier.

Albite also breaks down to jadeite + quartz at 2 GPa [P (H_2O) = P(Total)] in the temperature range investigated. Thus, the albite—sanidine thermal barrier is no longer valid at 2 GPa in the presence of excess H_2O. Optical and EPMA study of the compositions $Ne_{40}Kls_{20}Qtz_{40}$, $Ne_{25}Kls_{35}Qtz_{40}$ and $Ne_{50}Kls_{10}Qtz_{40}$ show that the subsolidus assemblage at $P(H_2O)$ = 2 GPa is alkali feldspar + $jadeite_{ss}$ + quartz, but in the same three compositions yield $nepheline_{ss}$, $leucite_{ss}$, alkali feldspar, liquid and vapour below the solidus at low pressure, as albite-sanidine is a valid thermal barrier under these P–T conditions.

The bulk composition of nepheline syenites from the Morro Redondo alkaline complex, Brazil, has been plotted in the nepheline–kalsilite–silica system (curve b–u in Fig. 7.9). Different rock-types from the same petrographic province plot along a curved path, as shown by the arrow. As the compositions of feldspars are solid solutions, the trend of crystallization should follow a curved path, as was obtained by Brotzu et al. (1989). If the tangents to the curve at various points are

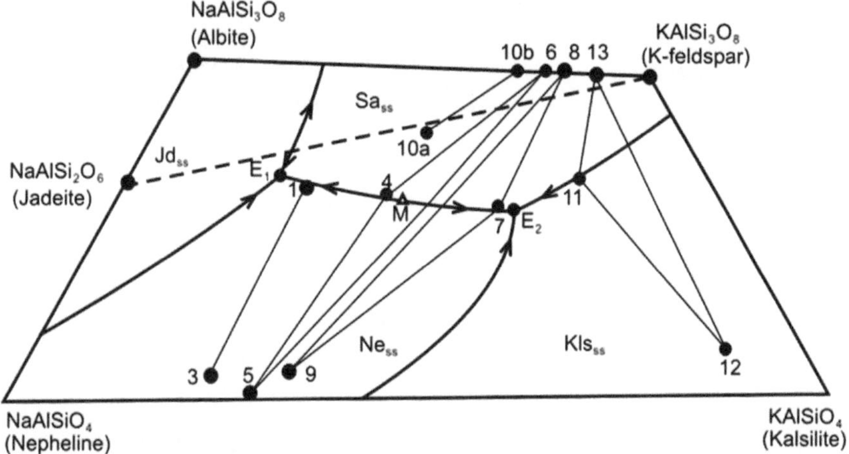

Fig. 7.8 Three-phase triangles showing the composition of liquid in equilibrium with feldspar$_{ss}$ and feldpathoids. For example, in case of triangles 4,5 and 6, the composition of liquid 4 is in equilibrium with the composition of feldspar 6 and triangles show the movement of liquid either toward the eutectic E_1 or the eutectic E_2. M denotes the thermal maximum produced by the intersection of the line nepheline–K–feldspar with the nepheline$_{ss}$—sanidine$_{ss}$ cotectic Line (Gupta et al. 2010)

extended toward the albite—K-feldspar join, these should give the compositions of the primary alkali feldspar. Two such tangents aa and bb show the composition of feldspar along the line a–b. Brotzu et al. (1989) observed that the feldspar compositions really lie within this range. With reference to their system (Fig. 7.8), the coetectic should intersect the curve ME_1 at 'u', after which a liquid should move to the eutectic E_1, as there is no jadeite-bearing syenite at that locality, Brotzu et al. concluded that the rock types of Morro Redondo, Brazil crystallized at low-pressure conditions, and the final liquid should have composition close to the minimum point, It may be noted that compositions of the eutectic E_1 and the minimum point at 0.1 GPa are close to each other. Even if the crystallization took place at higher pressure, and jadeite did crystallize, it is possible that at lower pressure jadeitic pyroxene reverted back to nepheline and alkali feldspar.

Points V, W and X (Fig. 7.9) represent the bulk composition of nepheline syenite from the khibiny massif, Kola peninsula, Russia, plotted onto the nepheline—kalsilite—silica plane. A liquid of these bulk compositions would follow i-V, j-W and k-X curves (the nature of the curve has to be determined from the crystallization trend of liquids in that petrographic province). In the case of curve i-V, sanidine should crystallize first, followed either by nepheline or kalsilite depending on the curvature of the path of crystallization. The final assemblage in the case of both i-V and j-W should comprise sanidine$_{ss}$ + nepheline$_{ss}$ + kalsilite$_{ss}$ (nepheline–kalsilite syenite); that is what is found in these rocks. Likewise, from liquid of composition X, first kalsilite$_{ss}$, then sanidine$_{ss}$ and finally nepheline$_{ss}$ will precipitate. The crystallization trend X-k (dotted curve in Fig. 7.9) is hypothetical, and

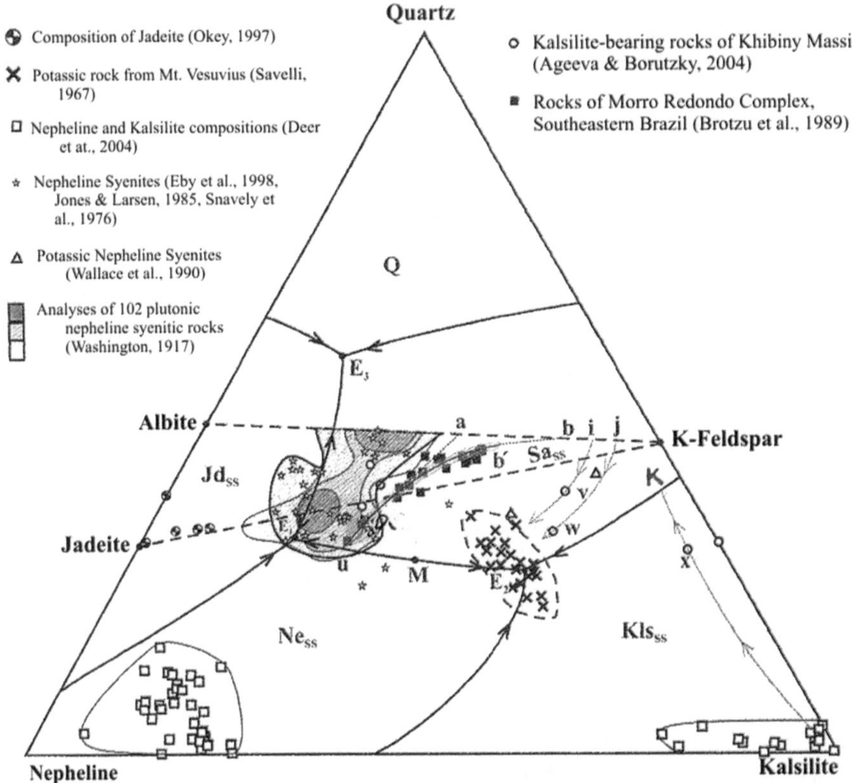

Fig. 7.9 Plots of the rock compositions in the albite—K-feldspar–nepheline–kalsilite quadrilateral. The compositional plots near the eutectic E_1 are taken from Washington (1917), Eby et al. (1998), Jones and Larsen (1985) and Snavely et al. (1976). The compositional plot near eutectic E_2 is after Savelli (1967), Wallace and Martin (1990) and Ageeva et al. (2004). The compositional plots near the jadeite end are from Okey (1997). Compositions of rock samples from Brasil are taken from Brotzu et al. (1989). The position of the ternary eutectic E3 is after Huang And Wyllie (1975)

convex toward nepheline. In the khibiny massif of Kola Peninsula, the final assemblage is nepheline–kalsilite syenite. According to Ageeva and Borutzky (2004), these rocks show an intergrowth of kalsilite with nepheline or orthoclase, and in which, the cubic habit of primary leucite is absent. The absence of Leucite and the presence of an intergrowth of kalsilite and K-feldspar suggest that these rocks crystallized at least at a depth of 60–65 km.

Hamilton and Mackenzie (1965) selected 102 compositions of plutonic nepheline syenitic rocks from Washington (1917). These compositions were recalculated to 100 % in terms of nepheline, kalsilite and quartz (total modal per cent of which exceeds 80 %) by them. Analytical data for nepheline syenites published by various other authors were considered by Gupta et al. The analytical data of following investigators were included: Eby et al. (1998), Jones and Larsen (1985), Snavely et al. (1976). These, together with the data of Washington (1917), have been plotted in Fig. 7.9, superimposed on the 2 GPa isobaric polythermal projection. One may see that eutectic E_1 of the system nepheline–kalsilite–SiO_2 lies close to the maximum concentration of the shaded area of the compositional plot. It should be noted that the eutectic E_1 related to the study of Gupta et al. (2010) is very close to the minimum established by Hamilton and Mackenzie (1965).

Savelli's (1967) compositions of potassic lavas from Mt. Vesuvius are plotted in terms of nepheline–kalsilite–SiO_2 system (Fig. 7.9); they lie near the eutectic E_2. This observation suggests that although most of the feldspathoid-bearing mineral phases crystallized near the surface, the lava might have originated at a depth of 60–65 km (2GPa).

Gittins et al. (1980) studied the potassic rocks of Batbjerg intrusion of east Greenland in which nepheline and leucite are the important constituents. In these rocks, they noted the occurrence of vermicular intergrowths of nepheline and K-feldspar, and patchy to micrographic intergrowth of kalsilite and K-feldspar. The assemblage near the eutectic E_2 comprises $nepheline_{ss}$, $kalsilite_{ss}$ and $K\text{-}feldspar_{ss}$ (Fig. 7.9). With the lowering of the temperature, initial composition of the liquid, lies in the nepheline or feldspar field, should move toward the cotectic M-E_2 with co-precipitation of $nepheline_{ss}$ and alkali feldspar (the assemblage corresponds to nepheline syenite). When the liquid composition reaches point E_2, then simultaneous precipitation of $nepheline_{ss}$, $feldspar_{ss}$ and $kalsilite_{ss}$ will take place. It may be noted that the 2 GPa (equivalent to 60 to 65 km depth) eutectic $E_{2,}$ plots in the $leucite_{ss}$ field of the low pressure isobaric diagrams of the Ne–Kls–SiO_2 (Figs. 7.1, 7.2, 7.3, 7.4, 7.5, 7.6). Thus if a melt containing the assemblage $nepheline_{ss}$ $kalsilite_{ss}$ liquid and vapour (E_2, Fig. 7.8) ascend rapidly toward the surface, $leucite_{ss}$ starts to precipitate in the intergranular spaces surrounding $nepheline_{ss}$, kalsilite and sanidine phenocrysts, as $leucite_{ss}$ is a stable phase at low pressure. This may be associated with concomitant resorption of the high-pressure phases. Relics of the intergrowths of kalsilite–K-feldspar and kalsilite-nepheline with leucite in the interstitial spaces should still survive, as in the case of Batbjerg complex.

Kalsilite has been reported to occur in fine-grained symplectites of a K-rich phase including possibly sanidine, replacing leucite as a product of high–pressure breakdown of the latter phase (Sandiford and Santosh 1991). This intergrowth of sanidine and kalsilite replacing leucite is consistent with a high-pressure origin (Scarfe et al. 1966). In the Synnyr alkaline pluton, northern Baikal (Zalutskii and Chulkov 1971) and in lamproite dykes of Napoleon Bay, Baffin Island (Hogarth 1997), a sanidine-kalsilite intergrowth (presumably in pseudoleucite) also has been reported.

In the nepheline–kalsilite–SiO_2 system, the albite-sanidine join no longer acts as a thermal barrier at 2 Gpa in the presence of excess H_2O; instead, the jadeite-sanidine join becomes the stable thermal divide. The compositions within the triangle albite—sanidine—jadeite yield an assemblage comprising quartz + $sanidine_{ss}$ + $jadeite_{ss}$ + liquid + vapour. At low pressure, the same compositions would result in the crystallization of feldspathoid- and feldspar-bearing alkaline lavas. In many petrographic provinces, such as Fort Portal and Birunga of Uganda

(Holmes 1950; Rogers et al. 1998), the Colli Albani complexes (Aurisicchio et al. 1988) and in Hawaii (Yoder and Tilley 1962), feldspathoid-bearing rocks are associated with quartz-bearing trachyandesite or latite series of rocks. These SiO_2–saturated varieties and feldspathoid-bearing rocks are apparently part of different volcanic cycles issued from the some magma chamber. The genesis of these two contrasting varieties of lava is possibly controlled by the shifting of the thermal barrier as a function of pressure and temperature (also see Yoder and Tilley 1962).

Jadeite-bearing amygdules have been recorded from a mafic extrusive rock occurring northwest of Kamchatka by Ponomareva and Dobretsov (1965). The occurrence of jadeite suggests a deep-seated origin for the magma. Jadeite occurs in association with nepheline in a rock from Tibet (Bauer 1896; Lacroix 1930). The rock was also studied by Tilley (1956, also see Deer et al. 1997) , who reported that jadeite crystals were traversed by nepheline. The latter also occurs together with albite and jadeite in a troctolite from equatorial Pacific.

A jadeite-, K-feldspar- and quartz bearing assemblage similar to the subsolidus assemblage jadeite—albite—K-feldspar has been reported from meta-morphosed felsic magmatic rocks (e.g. metagranite) by Compagnoni and Maffeo (1973) from Mt. Mucrone in the western Italian Alps. Such assemblages have also been reported from the Alpine region of Acceglio, in Italy, by Lefèvre and Michard (1965). Okey (1997) described jadeite—and K-feldspar-bearing rocks from the Bektaslar region, located west of Hormancik town, 60 km south of Busla, in Turkey (also see Deer et al. 1997). The rocks underwent metamorphism, the P–T conditions of which are similar to those of blue-schist facies. The modal percentage of jadeite, K-feldspar and quartz with or without albite is plotted in Fig. 7.9. The occurrence of jadeite along with K-feldspar and quartz indicates the high-pressure crystallization of the rocks, the bulk composition of which lies in the jadeite—sanidine—silica subsystem. The results of Gupta et al. (2010) at 2 GPa [P (H_2O) = P (Total)] suggest that the observed *subsolidus* assemblage justifies their presence along with glaucophane in the blueschist facies. Osborne (1930) and Gummer and Burr (1946) have reported the occurrence of nepheline-albite together with jadeite in the gneissic rocks in the southern Grenville province of the Canadian shield (see also Appleyard 1969). In the Franciscan blueschist terrane, California, Ernst et al. (1970) reported the occurrence of albite and quartz along with sodium-rich pyroxene containing 70–90 mol. % jadeite. With reference to the diagram of Gupta et al. (2010, Fig. 7.8), the assemblage jadeite + alkali feldspar + quartz occurs as a *subsolidus assemblage.*

McBirney et al. (1967) observed the occurrence of jadeite along with albite and quartz from the Motagua fault zone of Guatemala. Such jadeite may be related to the breakdown of albite at higher pressures. Chihara (1971) described the occurrence of jadeite from Mt. Osha at Oya, in Hyogo Prefecture, and Wakasa, in Tottori prefecture, Japan. At Sambagawa in the Kamuiketan belt of Hokkaido, Seki (1960) observed the occurrence of jadeite along with albite. Such an occurrence has also been reported from Kotaki, Japan, by Iwao (1953) in a narrow (20 cm) zone around an albite-rich core. The occurrence of jadeite and albite in these localities is considered by those authors to be of metamorphic origin.

7.3 Genesis of Pseudoleucite with Reference to Nepheline–Kalsilite–Silica System

The name pseudoleucite refers to an intergrowth of nepheline and K-feldspar within a trapezohedron crystal in the shape of leucite. Bowen and Ellestad (1937) explained the genesis of pseudoleucite, with reference to the system, nepheline-kalsilite-SiO_2 (Fig. 7.10). They plotted the composition of a typical leucite with 1.5 wt% Na_2O (point X, Fig. 7.10). They thought that after its formation, if a leucite of composition X, reacts with the liquid of composition R (Fig. 7.10), it should yield a solid product of composition similar to the pseudoleucite as its composition lies close to the line XR near point Y, the ratio of the two reactants, liquid and leucite where should be equal to the ratio of XY/YR. Nepheline and K-feldspar produced by the reaction should have a composition given by points U and V, respectively. If from a liquid of composition X some leucite is subtracted as cooling continues the composition of the liquid may not reach point R, but may reach either curve WR or else curve MR. If the liquid reaches WR, leucite should be transformed by reaction to K-feldspar only, and if it reaches the curve MR it would react with the liquid to form nepheline.

Larsen and Buie (1938) studied pseudoleucite crystals from Highwood Mountains), Montana. They thought that the hypothesis of Bowen and Ellestad should involve a slow reaction process; but in the Highwood Mountains, some of the dykes, where pseudoleucites are found, the groundmass is very fine- grained, and indicated rapid crystallization. They further thought that in such a reaction process between leucite and liquid, the trapezohedral form with such a sharp outline will be completely destroyed.

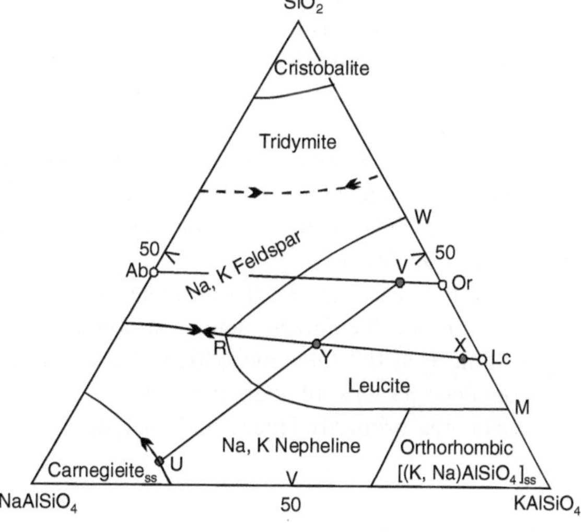

Fig. 7.10 Plot of the compositions of pseudoleucite in the system nepheline–kalsilite–SiO_2 (after Bowen and Ellestad 1937)

Fudali (1963) studied a selected part of the system nepheline–kalsilite–SiO_2 at 0.1 GPa P(H_2O) and elevated temperatures. His study was confined to $KAlSi_2O_6$-$NaAlSi_2O_6$ join of the system in presence of excess water at 0.1 GPa. The phase equilibrium relation studied by him is summarized in Fig. 2.1 (Chap. 2), which demonstrates that the join cuts through the primary phase volumes of $leucite_{ss}$ and $nepheline_{ss}$. In the leucite-rich section, he found that between 800 and 820 °C there is a field of $leucite_{ss}$ + $nepheline_{ss}$ + L; and below 800 °C the liquid phase is eliminated (Fig. 2.1, Chap. 2). In the nepheline-rich side of the join, there is a primary phase field of $nepheline_{ss}$, which is followed at low temperature by the appearance of the assemblage, $nepheline_{ss}$ + K–Na feldspar + liquid, and below the curve a-b, the liquid phase is eliminated. Reference to Fig. 2.1, it may be found that from a liquid of composition X, $leucite_{ss}$ of composition A should crystallize at 1100 °C, as the temperature is lowered to 815 °C, the liquid phase is eliminated, and only $leucite_{ss}$ should appear. However, below 700 °C it should yield nepheline, K-feldspar and a leucite phase should also coexist in equilibrium. Fudali (1963) considered that a pseudoleucite is formed by breakdown of a sodium-rich $leucite_{ss}$. He criticized the view of Bowen and Ellestad and considered that if pseudoleucite formation is related to solid-liquid reaction, then the replacement should be from the rims inward; however, the reverse case was actually observed by him. Based on experimental observations, he even thought that presence of viscous liquid coexisting with leucite may inhibit the formation of pseudoleucite (Fudali 1963, p. 1110).

Watkinson (1973) studied pseudoleucites from Lakner Lake and Prairie Lake in Ontario, where they are associated with nepheline and feldspar. He supported the hypothesis of Fudali (1963), and concluded that these pseudoleucites were produced by the breakdown of a sodium-rich leucite at low P (H_2O).

A mixture of composition $Ne_{25}Ks_{45}Qz_{30}$ was studied at 840 °C and 0.2 GPa in presence of excess water by Taylor and Mackenzie (1975), which was cooled under different conditions:

i. At first, they cooled the pressure vessel with sample capsule in presence of compressed air. This led to the formation of leucite with glass and vapour. The leucites were found to have exsolved phases, which they concluded to be analcitic in composition. Such leucites have 11 wt% soda leucite molecule in solid solution.
ii. In the second instance, cooling of the vessel was done slowly with fall of temperature and pressure. This yielded zoned leucite crystals in equilibrium with glass and vapour. According to them the rim had composition similar to that of pseudoleucite.
iii. In the last case, the vessel was cooled in two stages first in air or keeping the vessel inside the furnace and then cooling it from 840 to 630 °C and then holding it at 0.2 GPa and 630 °C for 2–4 h. In this case the rim of leucite contained 33 wt% of sodium-rich leucite, but the core had exsolution texture containing primarily leucite and analcite.

7.4 Survival of Leucite; Alteration to Analcite

Leucite is usually found in the rocks of Tertiary or younger age. Absence of)leucite in the older rocks may be attributed to its alteration to analcite. From their experimental and theoretical studies Gupta and Fyfe (1975) demonstrated that given sufficient time, leucite will not survive diagenetic processes, as the leucite-bearing rocks will convert to their analcite-bearing analogues before any major metamorphism occurs.

A natural leucite from Roccamonfina, Italy (SiO_2: 54.60 %, Al_2O_3: 22.8 %, Fe_2O_3: 2.72 %, MgO: 0.03 %, CaO: 0.16 %, Na_2O: 1.18 %, K_2O: 18.46 %; total 99.95 %, Anal. B. Kronberg) was used by them for such a study. The leucite was ground to 100 mesh and was allowed to react with salt solutions. In one set of experiments 20 mg of sodium chloride with 20 % water was placed in sealed gold capsules and the reaction was studied at 0.1 GPa total pressure. In this case the salt solution was saturated.

In a second set of experiments, 0.1 g of leucite was mixed with 2 cm^3 of synthetic sea water (a salt mix of major species prepared for biological studies). These solutions contain sodium and potassium in normal sea water concentration. The results of the experiments are summarized in Fig. 7.11. The amount of conversion was estimated from X-ray diffraction patterns calibrated with known mixtures of analcite and leucite.

From the experimental results it is clear that the reaction is very fast and can be measured easily down to 150 °C. As the reaction is easily studied over nearly 100 °C range, it follows that the activation energy is small (Fyfe, 1973). If we assume that for the early part of the reaction the rate equation is of zero order, that is

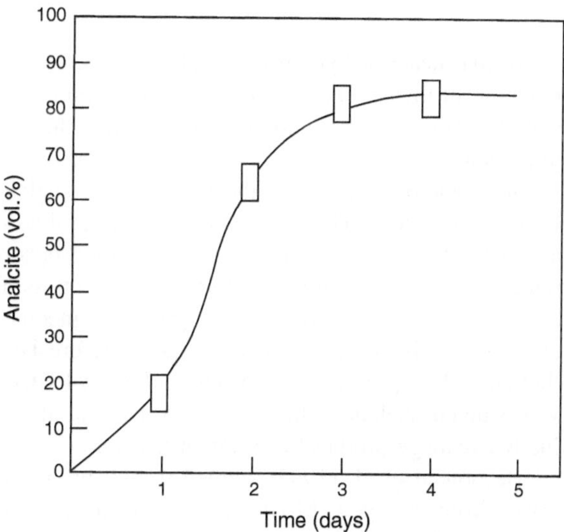

Fig. 7.11 Alteration of leucite to analcite (after Gupta and Fyfe 1975)

$$\frac{dc}{dt} \approx K.$$

It may be noted that the process is almost linear up to 80 % conversion, then the activation energy is in the order of 8 kcal mol^{-1} and certainly less than 10 kcal. This is very low activation energy and the reaction will be rather temperature-insensitive. If we consider that the time constants of diagenetic (burial) process is in the order of 10^5–10^7 years then even at 25 °C given appropriate fluids, the reaction would go to completion.

The process studied was:

$$\mathrm{KAlSi_2O_6(solid) + Na^+aq + H_2O \Leftrightarrow NaAlSi_2O_6.H_2O(solid) + K^+aq}$$

$$\Delta G^o_{298} = -1093\ \mathrm{mol^{-1}}(\text{Robie and Waldbaum, 1968}).$$

While there is considerable uncertainty in ΔG^o, this figure suggests that analcite is stable even when the concentration of potassium exceeds the concentration of sodium ($K^+_{aq}/Na^+_{aq} = 6.4$). This implies that even normal river water ($K^+/Na^+ = 0.3$) could cause the reaction.

Although no nucleation is involved in the reaction (but a 10 % volume expansion occurs) a number of steps could be rate-determining which might include:

1. diffusion rates of Na^+ or K^+ in the crystal,
2. dehydration rates of ions before entry, and
3. diffusion rate of water into the crystal.

The overall H (cf. E) of processes like:

$$K^+_{solid} + Na^+_{aq} \rightarrow Na^+_{solid} + K^+_{aq} \text{are small (Larimer 1964).}$$

Normal heats of hydration of solids are similar to the activation energy determination in their work. The fact that more dilute seawater appears to react faster than the saturated salt solution, it may indicate that the water activity in diffusion is important.

Other authors also noted the facile leucite-analcite reaction (Deer et al. 1963; Bragg et al. 1965). The chances for leucite survival over a long geological period are therefore, small and in fact most reported occurrence of volcanic leucite are rather in recent rocks. It is therefore very much possible that this process may occur at low temperatures when other primary igneous materials might show little alteration. This possible reaction must therefore be considered before analcite is thought to be a primary igneous phase (Pearce 1970). Nakamura and Yoder (1974), who studied analcite "phenocrysts" in basalts, also concluded that they are most likely exchange products of original leucite.

Macdonald et al. (1992) studied turbid, brown rounded pseudoleucite phenocrysts from Bearpaw Mountains, Montana. They found that the pseudoleucite consisted of a fine-grained inter-growths of nepheline and sanidine. According to them, fresh leucites occur as inclusions within olivine and clinopyroxene grains. This supports the idea that leucite appeared from these rocks as primary crystals, but were converted to pseudoleucite due to metasomatic processes. As soda-rich fluid could not diffuse through clinopyroxenes and olivines, those leucite grains occurring as inclusion within these crystals survived.

Luhr and Giannetti (1987) studied leucite crystals from the Brown Leucitic Tuffs of Roccamonfina, and observed that leucites are typically less than 2 mm across, but reach 0.25 mm across in some pumices. Primarily leucite is rarely preserved in the cores of larger crystals but secondary analcite crystals can retain primary radial inclusion pattern. According to them analcitization of leucite is a common geological phenomenon in these tuffs.

Scott Smith et al. (1987) studied pseudoleucite from Kapamba lamproites of Luwanga (Chap. 4), and observed that all the grains analysed, are composed of sanidine and brown turbid grain with chemistry similar to that of analcite (SiO_2: 52. 04 %, Al_2O_3:23.58 %, Na_2O:11.89 wt%). A few pseudoleucites however, consisted of sanidine and nepheline (SiO_2:46.75 %, Al_2O_3: 25.77, Na_2O: 25.53 wt%). These minerals may occur in one grain showing similarity toward pseudoleucite, and all these are pseudomorphous after leucite.

Mitchell et al. (1987) observed the presence of pseudoleucite from Smoky Butte, Montana. Here, they observed the presence of analcite, which they think, are not primary but are pseudomorphous after primary leucite phenocrysts. They supported the experimental study of Gupta and Fyfe (1975), and considered that they were formed by metasomatism of Na-rich fluid by diagenetic process.

The phonolitic and tephritic lavas of Azerbaizan volcanics from northwestern Iran were studied by Comin-Chiramonti et al. (1997). Analcite occurs in these extrusive rocks as a devitrification product or as a secondary mineral. They considered their formation due to complete transformation of possibly leucite by metasomatic process. The fact that originally they were leucite is borne by the fact that a trace of leucite could be found in the rim of the analcites, which were predominately potassic with 15 % K_2O, and having essentially analcitized core. Wilkinson (1977) considered that analcite phenocrysts of vitrophyric analcitite (similar to Razy glassy dike of Azerbaijan) are possibly ion-exchanged leucites and conversion of leucite to analcite must have been complete during hydration. According to Comin-Chiramonti et al. the analcites occurring in the groundmass of tephri-phonolitic rocks of Sehjafarlu, Azerbaijan were originally leucites, crystallized at low pressure of H_2O. The bulk composition of these analcites is similar to that of leucite.

Pseudoleucites from the Crowsnest formation of Canada and Colima volcanic complex in Mexico were studied by Karlsson and Clayton (1991). They obtained stable isotope (H, N, O), electron microprobe and ion microprobe data for analcites from these two localities. They obtained isotope ratios of framework ($\delta^{18}O_f$) and the channel water ($\delta^{18}O_{cw}$, δD) for two Crowsnest samples and one Colima sample. Both O and H isotopic ratios of channel water in all three samples, fall in the meteoric water line, and definitely are not magmatic. The $\delta^{18}O_f$ values for Crowsnest (13.6 and 14.2 %) and Colima (8.7 %) indicate that these analcite samples have undergone exchange reaction with external fluids at subsolidus temperatures or formed from a pre-existing mineral such as leucite. According to Karlsson and Clayton (1991) the whole rock $\delta^{18}O$ value (7.6 %) is significantly higher than those of basanites and leucite basanites, and the high N content (6 ppm) and low $\delta^{15}N$ value imply interaction with water either during magma genesis, transport or post-extrusion. They observed that the glass matrix in the Colima minette is also unusually H_2O-rich (4 wt%) suggesting post-eruption, glass-fluid interaction. On the basis of above observation, Karlsson and Clayton (1991) concluded that the analcites from Crowsnest and Colima are of secondary origin.

An experimental study of the replacement of leucite by analcime was made by Putnis et al. (2007). Leucite and analcime have open framework aluminosilicate structures, where ion exchange by cation substitution has been previously used to explain the replacement of one phase by another. Using O^{18}-enriched NaCl solutions in a container, they made hydrothermal reactions and analysed the run products using a scanning electron microscope, infrared and Raman spectrometer. They also employed a time-of-flight secondary ion mass spectrometer, involving cation exchange by volume diffusion. They showed that the replacement of leucite by analcime takes place by solid-state reaction involving cation exchange by volume diffusion. Textural features such as nano-pores and cluster, as well as the detection of high amounts of O^{18} in the framework are exchanged and a new analcime structure forms at the moving interface through the parent leucite crystal. The characteristic high porosity (on a nano-scale) in the product leucite phase, results in diffusion and some of the parent phase is lost to the solution to give a volume deficit reaction. However, pseudo-morphic replacement of an open framework alumino-silicate structure takes place by a coupled dissolution and reprecipitation mechanism.

Chapter 8
Incompatible Mineral Pairs in K-Rich Rocks

The mineral pairs orthopyroxene–leucite, leucite–albite, and melilite–plagioclase are not observed together in natural K-rich silica-deficient igneous rocks formed under volcanic to subvolcanic conditions. The probable cause of incompatible relationship for these pairs are described in this chapter.

8.1 Incompatibility Between Leucite and Orthopyroxene

Wendlandt and Eggler (1980b) studied the reaction, sanidine + forsterite ≈ leucite + enstatite. They observed that the univariant equilibrium curve is stable at the following pressure–temperature conditions: (0.5 GPa, 1,150 °C), (0.25 GPa, 1,040 °C) and (0.1 GPa, 1,040 °C). They also established that the reaction leucite + enstatite ≈ forsterite + liquid may set the upper limit of coexistence of leucite and enstatite. This univariant reaction occurs at following pressures and temperatures: (0.25 GPa, 1,175 °C) and (0.5 GPa, 1,150 °C). Thus, under volatile-free condition, orthopyroxene and leucite may coexist in an extremely narrow field, but investigation of Wendlandt and Eggler shows that below 1,000 °C leucite and enstatite reacts to form sanidine and forsterite; and this reaction may inhibit the presence of leucite and enstatite in volcanic rocks crystallised at temperature below 1,000 °C, when the lava cools down slowly to room temperature.

In their study of the system forsterite–kalsilite–SiO_2–H_2O–CO_2, Wendlandt and Eggler (1980b) observed that in presence of volatile, following reactions would take place:

1. phlogopite + enstatite + sanidine + vapour → forsterite + liquid (0.2 GPa and 950 °C; 0.5 GPa and 975 °C),
2. phlogopite + sanidine + vapour → forsterite + (leucite + liquid (0.5 GPa and 1,000 °C and 0.2 GPa and 975 °C).

Thus, in presence of H_2O and CO_2 the above two reactions are stable and leucite and orthopyroxene should not coexist under equilibrium condition. On the basis of the data for the plane forsterite–kalsilite–quartz provided by Wendlandt and Eggler

A.K. Gupta, *Origin of Potassium-rich Silica-deficient Igneous Rocks*,
Springer Geology, DOI 10.1007/978-81-322-2083-1_8

(1980, p. 397, Fig. 3), Yoder (1989) discussed various assemblages in the system Fo-Ks-Qz at 950 °C, which are displayed in Fig. 8.1. As the incongruent melting of sanidine persists to 2.0 GPa (Lindsley 1966), the join forsterite–sanidine does not become a thermal barrier until the pressure exceeds 2.0 GPa. Yoder further stated that as the incongruent melting of enstatite terminates at about 0.14 GPa, the principal thermal barrier in the system is forsterite–leucite up to 1.9 GPa. It is essential to establish the thermal barrier in order to ascertain the limits of composition within which K-rich magmas may be generated. Figure 8.2 demonstrates that leucite–sanidine–enstatite-bearing assemblage is stable below <0.3 GPa.

In this connection mention may be made about the study of Yoder (1989) on the system forsterite–kalsilite–SiO_2–H_2O (Fig. 8.1). It may be observed that the stable assemblages in the silica-undersaturated part of the system are phlogopite + leucite + liquid, phlogopite + liquid, phlogopite + enstatite + liquid, leucite + liquid, leucite + sanidine + liquid or sanidine + liquid (Fig. 8.1). Figure 8.2a–d demonstrates stable assemblages under dry conditions. This figure suggests a narrow range of stability for leucite and enstatite (Fig 8.2b).

The above discussion shows that in volatile-free system leucite and enstatite may co-exist at low pressure within a limited *P–T* field, but at lower temperatures <1,000 °C, it is eliminated by a reaction to produce sanidine and olivine. Under vapour-present condition leucite reacts with mafic components to produce phlogopite + enstatite + liquid or phlogopite + leucite + liquid (depending on bulk composition) and they do not coexist in equilibrium.

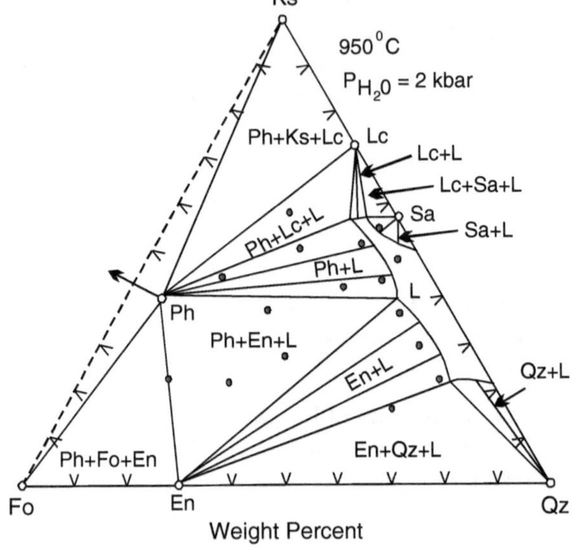

Fig. 8.1 Phase relation in the system forsterite–kalsilite–quartz at 950 °C under 0.2 GPa in presence of excess water (after Yoder 1989)

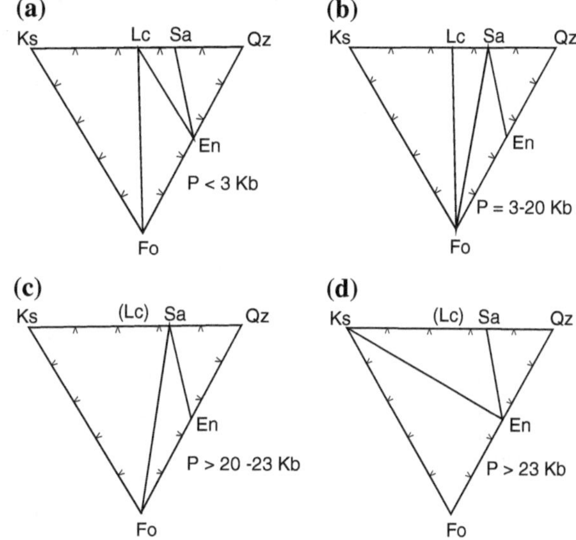

Fig. 8.2 Joins in the Ks–Fo–Qz system as a function of pressure for T = 1,100 °C. Deduced by Yoder (1989) from data of Schairer (1954), Luth (1967) and Wendlandt and Eggler (1980)

8.2 Incompatible Relation Between Leucite and Sodic-Plagioclase

Feldspar coexisting with leucite is either calcium-rich plagioclase or a potassium-rich feldspar (Shand 1943). Albite and leucite are not found to coexist in equilibrium in ultrapotassic rocks (Cross et al. 1902).

On 'thermochemical grounds', however, Miyashiro (1960) postulated that the apparent incompatibility between leucite and albite might not exist at high temperatures. MacKenzie and Rahman (1968) also described veins of leucite and sodium-rich feldspar in a basanite from the Massif Central, France, but they did not confirm this as a stable assemblage.

Gupta and Edgar (1975) studied the phase relations in the system leucite–albite and leucite–albite–anorthite at atmospheric pressure. The following discussion should highlight the causes of their incompatibility.

8.2.1 *Phase Relations in the Join Leucite–Albite under Atmospheric Pressure*

Phase relations in the leucite–albite join (Gupta and Edgar 1975) is summarised in Fig. 8.3, which indicates that the join is a pseudo-binary, and is part of the ternary system nepheline–kalsilite–SiO_2. Point A ($Lc_{41}Ab_{59}$, 1,068 ± 5 °C) represents a pseudoeutectic point, where leucite and feldspar coexist with liquid. The temperature of the pseudoeutectic point is in good agreement with that of Schairer (1950)

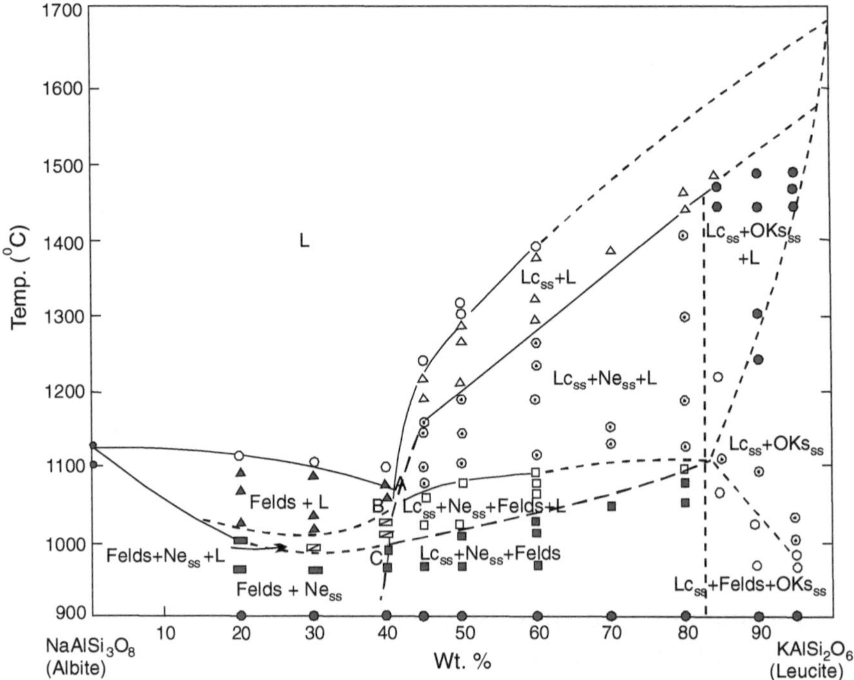

Fig. 8.3 Phase diagram of the Join leucite–albite at 1 atm (after Gupta and Edgar 1975)

for the same chemical composition in the $NaAlSiO_4$–$KAlSiO_4$–SiO_2 system, where leucite, nepheline, and feldspar coexist with liquid. Point C (990 ± 5 °C) is the intersection of solidus in the system and the boundary curve of the field, where leucite and feldspar are naturally stable phases. Reference to Fig. 8.3 shows that nepheline$_{ss}$ and alkali feldspar coexist in equilibrium in the compositional range of Ab_{100} and $Ab_{59}Lc_{41}$, and leucite does not coexist with albite; but leucite$_{ss}$, feldspar and nepheline$_{ss}$ are in equilibrium between $Ab_{59}Lc_{41}$ and $Ab_{17}Lc_{83}$. Calibration of alkali feldspar composition as a function of cell volume was determined using X-ray crystallographic technique (Orville 1967); it is observed that feldspars coexisting with leucite$_{ss}$ are monoclinic, and maximum amount of albite in the feldspars, crystallizing from a mixture $Ab_{59}Lc_{41}$, is 55 mol%. With progressive increase in leucite content in the bulk compositions, albite content of these feldspars further decreases. Thus, in the compositional range, where leucite and feldspar are in equilibrium the feldspar composition varies between $Or_{55}Ab_{45}$.

As shown in Fig. 8.3, the extent of solid solution of albite in leucite is very small, but increases with increasing temperature up to $Lc_{83}Ab_{17}$ and 1,100 °C, where leucite$_{ss}$ coexist with orthorhombic kalsilite$_{ss}$. A slight increase in the interplanar spacings (211, 004, 400) of leucite, crystallised from bulk composition $Lc_{95}Ab_5$ and $Lc_{90}Ab_{10}$ compared to those of its pure synthetic equivalent, and a decrease in those of (132 and 202) in kalsilite from the same compositions relative to pure kalsilite (Smith and Tuttle 1957); indicate that both these minerals are solid solutions. However, the exact compositions of these two phases were not determined. Appearance of kalsilite$_{ss}$ and nepheline$_{ss}$ in the leucite–albite join can be explained, if it is considered that leucite contains small amounts of excess silica and $NaAlSiO_4$ in solid solution, as is usually observed in nature. If the crystallization of a liquid of bulk composition J is considered (Fig. 7.1, Chap. 7), with drop in temperature, leucite$_{ss}$ of composition Lc_1 should crystallise and the liquid should move along the line J–T towards the leucite$_{ss}$–orthorhombic kalsilite$_{ss}$ boundary and at T, orthorhombic kalsilite$_{ss}$ should crystallise. If the bulk composition lies at K, with temperature decrease leucite$_{ss}$ of composition Lc_1 should again appear as the primary phase and the composition of the liquid should move along the line Lc_1–K and at L, nepheline$_{ss}$ should appear. With further drop in temperature, liquid should move to R and following phases should appear: Lc_{ss} + Ne_{ss} + Na–K feldspar.

Compositions of nepheline$_{ss}$, crystallised at the same temperature from the bulk compositions $Ab_{60}Lc_{40}$ and $Ab_{50}Lc_{50}$ (where leucite is an additional phase) was found to be $Ne_{73}Ks_{25}Qz_3$. Nepheline$_{ss}$ coexisting with feldspar at 960 °C in the mixture $Ab_{70}Lc_{30}$ has following composition: $Ne_{68}Ks_{30}Qz_2$. When the composition is plotted along with coexisting feldspar composition ($Ab_{59}Or_4$) in the system $NaAlSiO_4$–$KAlSiO_4$–SiO_2 the tie line passes very close to the bulk composition $Ab_{70}Lc_{30}$ indicating the absence of any other phase.

8.2.2 The Leucite–Albite–Anorthite Join

Phase relations in this pseudo-ternary join of the system $KAlSiO_4$–$NaAlSiO_4$–$CaAl_2Si_2O_8$–SiO_2 are shown in Fig. 8.4. Only the liquidus relations within the system have been studied. However, it appears from the bulk compositions lying close to the $KAlSi_2O_6$–$CaAl_2Si_2O_8$ join leucite$_{ss}$ and plagioclase should be the only two crystalline phases. Figure 8.4, however, suggests that with the increase in the $NaAlSi_3O_8$ content in the bulk composition, nepheline$_{ss}$ and alkali feldspar should be additional phases at low temperatures and when the starting material is considerably rich in the albite content, leucite should cease to appear.

Due to the pseudoternary nature of this join compositions of liquids do not lie in the leucite–albite–anorthite join. Estimation of the nature of these liquids can be obtained from the determination of the composition of feldspars, crystallised at or near the boundary line A–B (Fig. 8.4). Microprobe analyses of feldspar were made on four bulk compositions, crystallised at temperatures close to liquidus, ranging from 1,320 to 1,180 °C (Table 8.1). For each feldspar the contents of CaO, Al_2O_3, and SiO_2 were determined by microprobe, assuming that Na_2O + K_2O form the remainder. In Fig. 8.5 tie lines have been drawn between the bulk compositions and the corresponding feldspar compositions projected on to the leucite–albite–anorthite join. These tie lines show progressive enrichment in the albite–orthoclase contents of the feldspar with decreasing temperature, but indicate that the plagioclase,

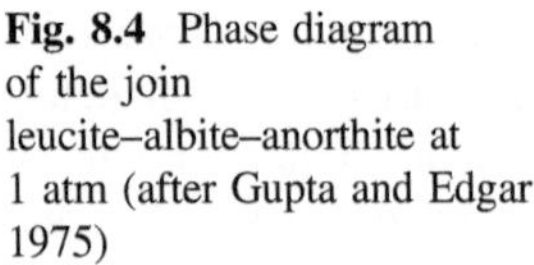

Fig. 8.4 Phase diagram of the join leucite–albite–anorthite at 1 atm (after Gupta and Edgar 1975)

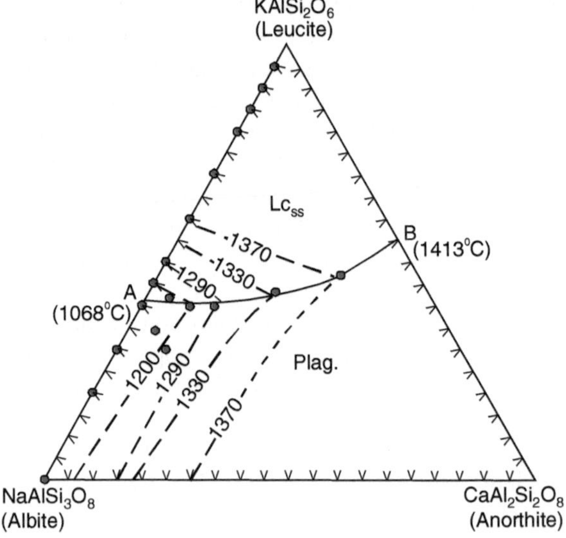

Table 8.1 Composition of ternary feldspars crystallizing in the system leucite–albite–anorthite

Composition of mixtures			Temp (°C)	SiO_2 (wt%)	$Na_2O + K_2O$	Al_2O_3	CaO (wt %)	An (wt %)	Ab + Or (wt%)
Lc	Ab	An							
43	31	26	1,320	45.92	4.54	33.32	16.22	80.40	9.60
40	50	10	1,180	51.87	5.72	29.64	12.77	63.15	36.85
53	42	5	1,180	50.35	7.05	30.29	12.31	60.93	39.07
30	60	10	1,200	57.09	5.44	28.23	9.24	45.63	54.37

coexisting with a liquid low in anorthite content ($Lc_{53}Ab_{42}An_5$), is rich in anorthite (An_{61}, Ab–Or_{39}). This implies that the residual liquid after crystallization of such a feldspar, must be enriched in $NaAlSi_3O_8$, although small amounts of soda may be incorporated in leucite. These results demonstrate that the addition of anorthite does not change the leucite–albite incompatibility.

8.2.3 Petrological Implications

Study of phase relations in the join leucite–albite and leucite–albite–anorthite at atmospheric pressure supports the conclusion of field studies that leucite and albite are incompatible. In the absence of the anorthite molecule, the most sodium-rich feldspar that can coexist with leucite contains, 46 wt% orthoclase molecule. In the presence of the anorthite molecule, leucite coexists with a ternary feldspar,

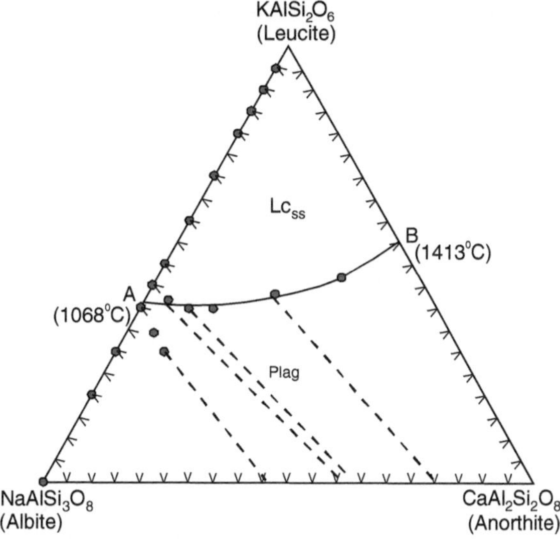

Fig. 8.5 Tie lines showing bulk compositions and corresponding feldspar compositions in equilibrium with leucite in the system leucite–albite–anorthite (after Gupta and Edgar 1975)

containing high proportion of anorthite molecule (approximately $An_{50}Ab–Or_{50}$). The incompatibility of leucite and albitic feldspar in the absence of anorthite molecule is probably due to the following reaction:

$$2NaAlSi_3O_8 + KAlSi_2O_6 \rightarrow KAlSi_3O_8 \cdot (1 - X)$$

leucite (Alkali feldspar)

$$(X)\ NaAlSi_3O_8 + 2\ NaAlSiO_4 \cdot X\ NaAlSi_3O_8.$$

$(nepheline_{ss})$

The above reaction explains the formation of K–Na feldspar and $nepheline_{ss}$ by the reaction between leucite and albite. In the leucite–albite–anorthite system, liquids closely representing natural magmas from which leucite and feldspar-bearing assemblage crystallise, the products of crystallization at low temperatures near the surface are anorthite-rich ternary feldspars, leucite, and a residual liquid, enriched in Na_2O and SiO_2. Some of the Na_2O may be incorporated into leucite. MacKenzie and Rahman (1968) noted that leucite rims in leucite-sodic feldspar veins in the Massif Central basanite became enriched in Na_2O with falling temperature. The pressure of a residual liquid might also explain the albitic nature of the feldspar in these veins; the Na-rich feldspar having formed from an original K-rich feldspar by a process of alkali ion exchange. Such a mechanism should explain the maximum orthoclase content of 20-mol% (MacKenzie and Rahman 1968), relative to the albite–orthoclase content of 50 mol% from the direct primary crystallization of low temperature liquids in the simplified leucite–albite system.

The effect of $P(H_2O)$ on the incompatibility of leucite and albitic feldspar may be estimated by comparing the results of this study with that of Fudali (1963) for

Table 8.2 Composition of feldspars in the join leucite–albite at 1 atm and 900 °C

Bulk composition		Cell volume (a^3)	Composition as wt% determined from cell volumes, following the method of Orville (1967)	
(Ab)	(Lc)		(Ab)	(Or)
100	0	666.51	100	–
80	20	677.19	84.5	15.2
70	30	694.49	59	41
50	50	700.19	50.5	49.5
40	60	704.26	44	56
10	90	709.05	35	65

the 850 °C isothermal section of the $NaAlSiO_4$–$KAlSiO_4$–SiO_2 at $P(H_2O)$ = 0.0256 GPa, where the stable assemblage for a bulk composition of approximately $Ab_{50}Lc_{50}$ is leucite$_{ss}$, nepheline$_{ss}$ and feldspar of composition about $Or_{60}Ab_{40}$. Compositions of feldspar crystallizing from different bulk compositions in the leucite–albite join are summarised in Table 8.2, which suggests that maximum albite content in alkali feldspar in equilibrium with leucite is 50.5 wt%. Thus, crystallization under moderate [$P(H_2O) = P$(Total)] conditions appears to decrease the chances of compatibility of leucite and albite. Under higher $P(H_2O)$ the stability of leucite-feldspar assemblage will be restricted to a very potassium-rich compositions of feldspar.

The compositions of feldspar coexisting with leucite (Fig. 8.5) at temperatures above 1,000 °C do not support the suggestions of Miyashiro (1960) and MacKenzie and Rahman (1968), who considered that leucite and Na-feldspar may be compatible at high temperatures.

8.3 Incompatibility Between Melilite–Plagioclase in Leucite-Bearing Lavas

In the lavas of alkali suites there is an incompatible relationship between melilite and plagioclase (Yoder and Schairer 1969; Yoder 1973). From volcanic centres of the same petrographic province, either melilite-bearing lavas are extruded or tephritic pyroclastics are ejected. Petrologic study of the melilite-bearing potassic rocks and leucite-bearing basanites from the Eifel area by the present author supports the conclusion of Yoder and Schairer (1969). From his study of natural rocks from various areas Yoder (1973) established that in alkaline lavas melilite and plagioclase do not appear to coexist. Coexistence of these two minerals is, however known in metamorphic rocks (Yoder 1973).

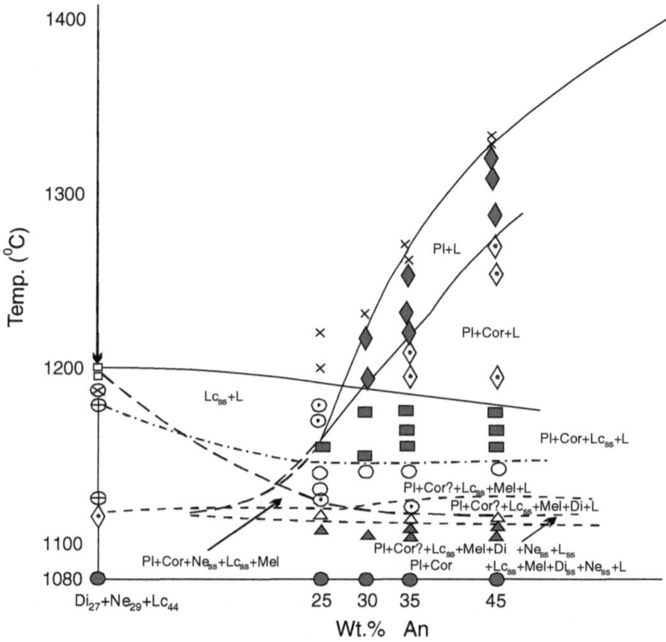

Fig. 8.6 Phase diagram of the join $(Di_{27}Ne_{29}Lc_{44})_{100\text{-}x}An_x$ at 1 atmosphere (after Gupta and Yagi 1980)

Gupta and Lidiak (1973) studied the system diopside–nepheline–leucite, and found that the final assemblage consisted of $diopside_{ss}$, $nepheline_{ss}$, $leucite_{ss}$ and melilite. A mixture $(Di_{27}Ne_{29}Lc_{44})$, containing the same assemblage, was chosen as one of the end members of a pseudobinary system to which anorthite was added to study the incompatibility between the mineral pair melilite and plagioclase.

The phase diagram of the system $(Di_{27}Ne_{29}Lc_{44})_{100\text{-}x}$–$An_x$ is given in Fig. 8.6, which shows the final assemblage in the system to be $diopside_{ss}$ + $nepheline_{ss}$ + $leucite_{ss}$ + plagioclase + corundum. Composition of melilite, as determined from the mixture $Di_{27}Ne_{29}Lc_{44}$ at 1,100 °C by the method of Hamilton and MacKenzie (1960), is $Ak_{64}Sm_{26}Geh_{10}$. The microprobe analysis of plagioclase from a mixture of bulk composition $(Di_{27}Ne_{29}Lc_{44})_{75}An_{25}$ gave the composition $An_{92}Ab_6Or_2$. In leucite-bearing rocks, plagioclase is found to be calcium-rich (Shand 1943; Savelli 1967), as found in this synthetic system. Schairer et al. (1965) plotted the compositions of natural melilites in a composition triangle, where the three end members were $CaNaAlSi_2O_7$, $Ca_2Al_2SiO_7$, and $Ca_2MgSi_2O_7$. When the composition of the melilite in the mixture of composition $(Di_{27}Ne_{29}Lc_{44})_{75}An_{25}$ is plotted in this triangle, it falls in the field of natural melilite. Yoder and Schairer (1969) studied the system akermanite–albite–anorthite at 1 atm and found that plagioclase and melilite are compatible with liquid over much of the field of plagioclase, with the exception of compositions on the join akermanite–albite. In a diagram of ln $a(SiO_2)$, versus temperature (°C), Carmichael et al. (1970) plotted

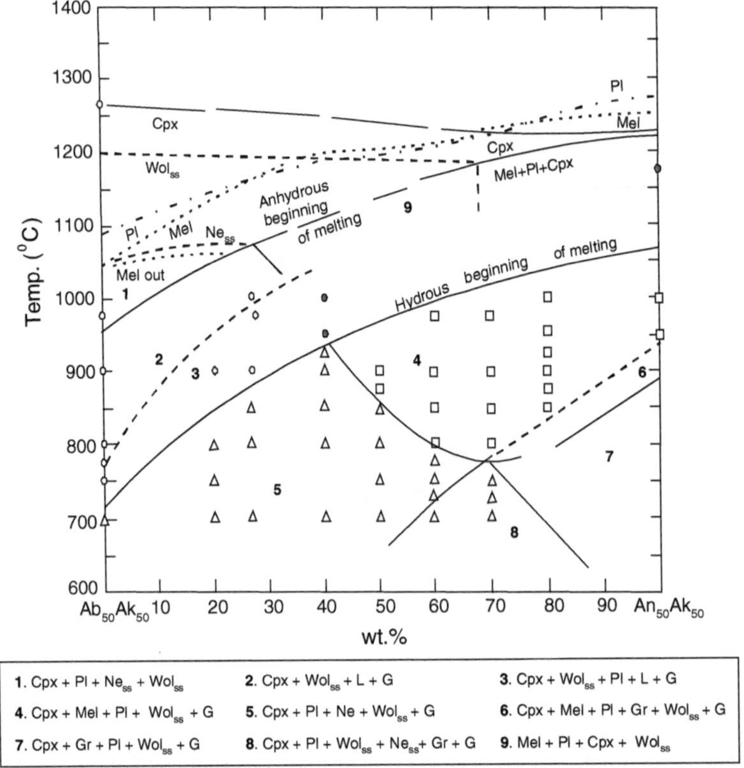

Fig. 8.7 Phase relation in the join $Ab_{50}Ak_{50}$–$An_{50}Ak_{50}$ at 1 atmosphere showing that melilite can coexist with anorthite-rich plagioclase under atmospheric pressures (after Yoder 1969b). At lower temperatures are presented the hydrothermal relations in an excess of water at $P(H_2O)$ = 0.2 GPa for the same composition. The *curves* mark the boundaries between assemblages

various silication reactions, which indicated that silication of forsterite + akermanite to produce diopside takes place much before the silication of nepheline to form albite.

The phase diagram of the join albite$_{50}$akermanite$_{50}$–anorthite$_{50}$akermanite$_{50}$, studied at 1 atm. by Yoder (1969; Fig. 8.7) shows the melilite–plagioclase coexistence over a broad temperature-composition range. He also studied a natural melilite (akermanite:sodamelilite is 2:1) and natural plagioclase (close to anorthite$_{50}$) mixed in equal proportions, and noted that above 850 °C, melilite and plagioclase coexisted with diopside$_{ss}$, wollastonite$_{ss}$, and liquid however, under 0.2 GPa $P(H_2O)$, the assemblage consisted of diop-side$_{ss}$ + plagioclase + nepheline$_{ss}$ + wollastonite$_{ss}$ at 700, 800 and 900 °C.

A mixture of composition $(Di_{27}Ne_{29}Lc_{44})_{75}$ An_{25} was studied under water pressures [$P(H_2O)$ = P(Total)] by Gupta and Yagi (1980) to see if the presence, of water in the system has any effect on the melilite–plagioclase coexistence. The study showed that below 0.25 GPa, at 700 and 800 °C the assemblage consisted of diopside$_{ss}$ + nepheline$_{ss}$ + melilite + anorthite$_{ss}$ + liquid + vapour. Above 0.25 GPa phlogopite appeared as an important phase and leucite$_{ss}$ disappeared, but melilite coexisted with anorthite$_{ss}$, phlogopite$_{ss}$, nepheline$_{ss}$, liquid and vapour. Above 0.28 GPa in presence of water, melilite disappeared and garnet appeared. The equilibrium assemblage at 700 and 750 °C and 0.5 GPa consists of large quantities of grossularite$_{ss}$, phlogopite$_{ss}$, small amounts of diopside$_{ss}$ and rare anorthite$_{ss}$.

Yoder (1969) studied the join $Ab_{50}Ak_{50}$–$An_{50}Ak_{50}$ (Fig. 8.7). It may be observed that in the anorthite-rich portion of the join at low temperature (assemblage 4 and 6), calcium-rich plagioclase and melilite can coexist in equilibrium but in the albite-rich portion this two minerals do not coexist.

Schairer and Yoder (1970) studied the system CaO–MgO–Al_2O_3–SiO_2, and found that in the volume enclosed by anorthite–akermanite–diopside–forsterite–spinel, akermanite coexists with spinel, anorthite and diopside in the absence of forsterite. Such a calcium-rich assemblage is found only in metamorphic rocks, whereas the assemblage forsterite–akermanite–diopside–spinel has representatives among igneous rocks. They indicated that olivine may be a deciding factor in the melilite–plagioclase incompatibility in the lavas. Thus, in the absence of olivine, melilite and plagioclase may coexist in some hybrid contact zone, and even in the absence of olivine and presence of water, there is a limit to the anorthite content of plagioclase that can coexist with melilite.

Chapter 9
Leucite- and Feldspar-Bearing Systems

Nephelinites have been considered to be the most perplexing group of igneous rocks by Bailey (1974). Such rocks predominantly contain nepheline and augite with other accessory minerals. They are usually associated with rocks of melteigite–ijolite–urtite series in close proximity to either leucite-bearing mafic lavas, or carbonatites. Sometimes, they are also observed as a highly fractionated group of rocks derived from lavas of alkali basalt–phonolite series; in such cases, sanidine becomes an important phase in addition to augite and nepheline.

Addition of sanidine to the diopside–nepheline join has a very important petrogenetic significance, because bulk compositions of nepheline-bearing phonolites, mela-nephelinite, trachyte and alkali feldspar-bearing urtite lie on the diopside–nepheline–sanidine join of the nepheline–kalsilite–CaO–MgO–SiO_2 system. As sanidine melts incongruently to leucite + liquid, there is a primary phase field of leucite, which is known to be unstable at high pressure (Scarfe et al. 1966; Fasshauer et al. 1998). Besides, the incongruent melting phenomenon of sanidine is eliminated above 0.3 GPa [$P(H_2O)$ = P(Total); Tuttle and Bowen 1958]. Thus, experimental studies of the join under 0.1 GPa (in presence of excess water) should establish the paragenetic relationship between olivine nephelinite, melanephelinite and nepheline-bearing phonolite (with or without leucite). At low pressure (<0.3 GPa in presence of excess water), leucite produced as an incongruent melting phase should be eliminated at low temperatures to generate leucite-free phonolites. It should be interesting to see whether melilite is present near the nepheline–diopside join at high water pressure, and amphibole co-precipitates as an additional phase in the subsolidus region from more mafic starting compositions in this join. The diopside–nepheline–sanidine join was, therefore, studied under 0.1, 1.0 and 2.0 GPa and variable temperatures by Gupta et al. (2006).

A.K. Gupta, *Origin of Potassium-rich Silica-deficient Igneous Rocks*,
Springer Geology, DOI 10.1007/978-81-322-2083-1_9

9.1 Study of the System Diopside–Nepheline–Leucite The Join Diopside–Nepheline–Sanidine Under Atmospheric Pressure

The system diopside–nepheline–sanidine studied by of Platt and Edgar (1971; Fig. 9.1) is important in establishing the paragenetic relationship between the rocks belonging to melanephelinite series and phonolites. They established that all phases in this system are solid solutions. If the presence of small amounts of possible Ca-Tschermak's molecule is ignored, the join can be treated as quinary, i.e., a part of the system $NaAlSiO_4$–$KAlSiO_4$–CaO–MgO–SiO_2. This join cuts through the primary phase volumes of forsterite$_{ss}$, diopside$_{ss}$, nepheline$_{ss}$ and leucite$_{ss}$. Forsterite$_{ss}$ has a reaction relation with the liquid. The field of leucite$_{ss}$ is related to the incongruent melting of sanidine$_{ss}$, thus the former should react with the silica-rich liquid at low temperatures to form alkali feldspar. In the $KAlSi_3O_8$-poor part of the system there is a primary phase field of forsterite$_{ss}$. The assemblages near the four phase points and below the solidus, consist of the following phases:

1. nepheline$_{ss}$–diopside$_{ss}$–leucite$_{ss}$–l (point B; Fig. 9.1),
2. forsterite$_{ss}$–nepheline$_{ss}$–diopside$_{ss}$–l (point A; Fig. 9.1),
3. nepheline$_{ss}$ + diopside$_{ss}$ + leucite$_{ss}$ + alkali feldspar (between 950 and 1,000 °C central portion) and below 900 °C leucite is eliminated.
4. nepheline$_{ss}$ + diopside$_{ss}$ + alkali feldspar (below 900 °C), and
5. nepheline$_{ss}$ + diopside$_{ss}$ + melilite (near the nepheline–diopside join containing <5 wt% sanidine).

Assemblages (1–5) correspond to nepheline leucitite, olivine nephelinite, nepheline–leucite phonolite, nepheline phonolite, and melilite–nephelinite, respectively. After elimination of leucite and appearance of K–Na feldspar, the

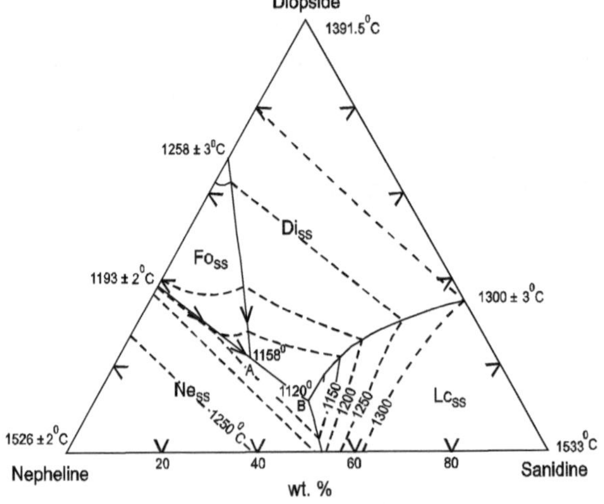

Fig. 9.1 Phase relation in the system diopside–nepheline–sanidine under atmospheric pressure (after Platt and Edgar 1971)

solidus assemblage corresponds to pyroxene-rich nepheline phonolite. The study of this system therefore, suggests that a pyroxene-rich nepheline phonolite can be produced from a nepheline leucitite melt, which is a derivative of a melilite–nephelinite magma. The last magma type itself is a product of an olivine–melilite–nephelinite magma. Their study thus supports the conclusions of King (1965), who suggested that nephelinites and phonolites from the Napak and Elgon volcanoes (eastern Uganda) were derived from mela-nephelinitic magma by early separation of forsterite, melilite, and pyroxene. It also supports the conclusion of Wright (1963), who considered that the differentiation of a mela-nephelinitic magma should explain the genesis of the Tertiary lava series in the western part of the rift system in Kenya. These lavas have variable compositions, which correspond to phonolites, nephelinites, and melanephelinites.

9.2 Experimental Study of the System Diopside–Nepheline–Sanidine at 0.1, 1 and 2 GPa and Variable Temperatures

Bulk compositions of nephelinite, mela–nephelinite, trachyte and phonolite can be represented in the join, diopside–nepheline–sanidine, which is a part of the quinary system, nepheline–Kalsilite–CaO–Mgo–SiO_2. Gupta et al. (2006) therefore studied the system diopside–nepheline–sanidine at 0.1, 1 and 2 GPa [$P(H_2O) = P(\text{Total})$] and variable temperature to determine the significance of the system in the genesis of nepheline-bearing phonolites (with or without leucite) in association with melteigite, urtite and ijotite series of rocks.

Nephelinites predominantly contain nepheline and augite. They are usually associated with mela–nephelinite, phonolite (nepheline syenite) and trachyte (syenite) in association with carbonatite (Dawson 1966). They are also observed as a highly fractionated series derived from lavas of alkali basalt–phonolite series (Yoder and Tilley 1962). Association of nephelinites with leucitites, phonolites and trachytes are also well known from the Birunga volcanic field located near the East African rift Valley (Sahama 1973; Rogers et al. 1998). The bulk compositions of the above mentioned sodium or potassium-rich silica-undersaturated basic and ultrabasic igneous rocks lie on the diopside–nepheline–sanidine join of the nepheline–kalsilite–CaO–Mgo–SiO_2 system. Forsterite appears as a reaction product of nepheline and diopside in the diopside–nepheline join at 1 atm (Schairer et al. 1962).

Sanidine melts incongruently to produce leucite$_{ss}$ plus a liquid below 0.3 GPa in the presence of excess water (Tuttle and Bowen 1958). According to them at low temperature, leucite$_{ss}$ disappears by reaction with liquid, and sanidine$_{ss}$ reappears near the solidus. Thus, various combination of such phases as leucite$_{ss}$, nepheline$_{ss}$, forsterite$_{ss}$ and diopside$_{ss}$ are expected to appear in their study of the system diopside–nepheline–sanidine at low pressure under a condition of [$P(H_2O) = P(\text{Total})$]. In the present study, phase relations in the join diopside–nepheline–sanidine at 0.1 GPa (in the presence of excess water) were examined first to establish the paragenetic relationship between olivine nephelinite, melanephelinite, nepheline leucitite, trachyte and phonolite (with or without $leucite_{ss}$). Phonolite by definition is a rock containing alkali feldspar and $nepheline_{ss}$ with accessory clinopyroxene, but in potassium-rich silica-undersaturated rocks nepheline may not appear and leucite may take its place instead. Such rocks have also been termed as phonolite by Holmes and Harwood (1937). In some phonolites both $nepheline_{ss}$ and $leucite_{ss}$ may be present. In the present study phonolite has been designated as the alkali feldspar-bearing volcanic rock containing nepheline with accessory clinopyroxene. If both $leucite_{ss}$ and $nepheline_{ss}$ are present it has been described in the present study as leucite phonolite to emphasize the presence of both nepheline and leucite. The plutonic equivalents of the volcanic rocks relevant to this study are shown within brackets in italics.

The study of this join was further extended to higher pressure (1 and 2 GPa) under the condition of $P(H_2O) = P(\text{Total})$, as the magmas related to alkali-rich, silica-undersaturated feldspathoid-bearing basic and ultrabasic magma are believed to be generated in the upper (Tilley and Yoder 1968; Clague and Frey 1982). Conversion of nephelinite to amphibolites has been observed by Tilley and Yoder (1968), White et al. (1972), Woolley and Jones (1987) and Downes (1987). It was considered by White et al. (1972) that at suitable pressure [$P(H_2O) = P(\text{Total})$] conditions rocks of these series should be converted to amphibolites by subsolidus hydration. Direct crystallization of amphibole above solidus was also observed by Singh et al. (2000) in the diopside–nepheline system in the presence of excess water. The vol.% of alkali amphibole in this join in the subsolidus region obtained by them was more than 80 %. The diopside–nepheline–sanidine join was, therefore, studied in the presence of excess water to check the possibility of generation of amphibole-bearing nephelinite, phonolite (nepheline syenite) and trachyte (syenite) at low temperature and pressure from this system.

9.2.1 Phase Relations in the System Diopside–Nepheline–Sanidine at 0.1 GPa [P(H_2O) = P(Total)]

Chattopadhyay et al. (1999) presented only the liquidus phase relations in the system diopside–nepheline–sanidine. The subliquidus and subsolidus data for the same system was not determined earlier. Experimental results for the system at 0.1 GPa in the presence of excess water and variable temperatures are summarized in Fig. 9.2, which describes liquidus phase relation only. Amphibole always appears as a phase appearing above or below the solidus. Its presence cannot therefore be deciphered from the liquidus phase relations (Fig. 9.2). The results show that the system intersects the phase volumes of $diopside_{ss}$, $nepheline_{ss}$, $leucite_{ss}$ and $forsterite_{ss}$. As $forsterite_{ss}$ and $leucite_{ss}$ appear in this system, it is not a ternary. The electron microprobe analyses also show that diopside contains Al_2O_3 as a $CaAl_2SiO_6$ molecule. The join

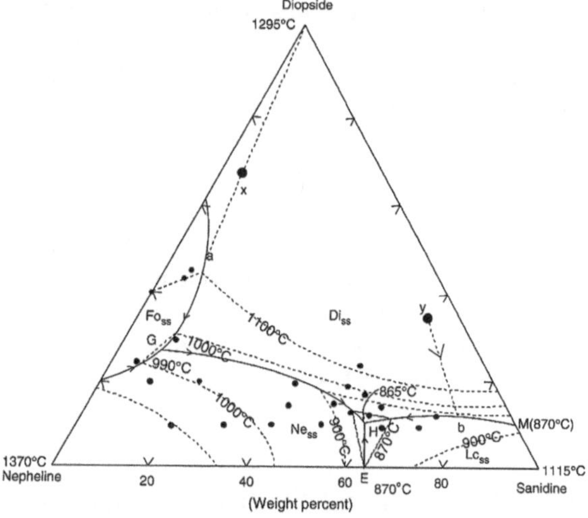

Fig. 9.2 Phase relation in the system diopside–nepheline–sanidine at variable temperatures under 0.1 GPa in presence of water (after Chattopadhyay et al. 1999)

diopside–nepheline–sanidine is, therefore, a part of the system Na_2O–K_2O–CaO–Al_2O_3–MgO–SiO_2. If the small amount of Al_2O_3 in diopside is ignored, the system may be described as a quinary join of the system $NaAlSiO_4$–$KAlSiO_4$–CaO–MgO–SiO_2. Appearance of leucite in this join is related to the incongruent melting of sanidine (Schairer and Bowen 1938; Tuttle and Bowen 1958). As pointed out earlier, in the join diopside–nepheline, the appearance of forsterite$_{ss}$ and melilite is related to a reaction relationship between diopside and nepheline. These phases were also observed in a mixture ($Di_{40}Ne_{60}$) crystallized at 950 °C and 0.1 GPa in the presence of excess water. It may be observed that in the mixture $Di_{45}Ne_{50}San_5$, crystallized at 900 °C and 0.1 GPa [$P(H_2O) = P(\text{Total})$], forsterite and diopside co-precipitate, but there is no melilite. At lower temperature under similar pressure, although melilite appears in the diopside-nepheline join, there is no melilite within the diopside–nepheline–sanidine system at temperatures between 550 and 600 °C at 0.1, 1 and 2 GPa [$P(H_2O) = P(\text{Total})$]. This is primarily because addition of sanidine causes increase in silica activity (Carmichael et al. 1970). Thermodynamic calculation of Carmichael et al. predicts that in a silica activity versus temperature diagram, silica combines with melilite to form pyroxene and wollastonite before leucite reacts with silica to form sanidine. A small amount of wollastonite supposed to appear by desilication of melilite is possibly incorporated in the pyroxene structure as solid solution, and thus it is not observed.

Reference to Fig. 9.2 suggests that the four-phase assemblage of $Di_{ss} + Ne_{ss} + Lc_{ss}$ + L (H) occurs at $Di_{11}Ne_{31}San_{58}$ and 865 + 3 °C, whereas that of $Fo_{ss} + Di_{ss} + Ne_{ss}$ + L(G) appears at $Di_{26}Ne_{66}San_8$ and 990 + 10 °C. The composition of point A, where diopside$_{ss}$ + leucite$_{ss}$ + liquid coexist in equilibrium, was estimated from Yoder and Upton (1971), whereas that of E, where nepheline$_{ss}$ + leucite$_{ss}$ + liquid are in equilibrium ($San_{64}Ne_{36}$, 870 °C), was obtained from Hamilton and MacKenzie

(1965). Melting temperatures of sanidine (1,115 °C), diopside (1,295 °C) and nepheline (1,370 °C) at 0.1 GPa in the presence of excess water were estimated from Hamilton and MacKenzie (1965, Luth and Boettcher (1986) and Boettcher and Wyllie (1969), respectively. Comparison of the two-phase diagrams studied at 1 bar under dry condition (Platt and Edgar 1971; Fig. 9.1) and 0.1 GPa (Fig. 9.2) further shows considerable shrinkage in the leucite and olivine fields at 0.1 GPa [$P(H_2O) = P$ (Total)]. The temperature of appearance of the four-phase point at 865 °C is also considerably less than that established by Platt and Edgar (1,120 + 5 °C) at 1 atm. It was observed that although in this system the subsolidus assemblage under 0.1 GPa at 830 °C comprises $Di_{ss} + Ne_{ss} + Lc_{ss} + San_{ss} + L + V$ at 815 + 5 °C leucite is absent because of its reactions with liquid and the final assemblage comprises $Di_{ss} + Ne_{ss} + San_{ss} + L + V$.

9.2.2 Experimental Study of the System at 1 and 2 GPa [P(H$_2$O) = P(Total)]

It was noted by Sigh et al. (2000) that in the join diopside–nepheline at 1 GPa in the presence of excess water, there is no primary phase field of forsterite$_{ss}$. The present result indicates that at 1 and 2 GPa isobaric conditions, there is a systematic shift in the positions of the Di_{ss}–Ne_{ss}–San_{ss}–l four-phase point. At 0.1 GPa and 865 °C, the four-phase points is located at $Di_{11}Ne_{31}San_{58}$ (Fig. 9.2, Chattopadhyay et al. 1999). At 1 GPa and 670 °C, the assemblage $Di_{ss} + Ne_{ss} + San_{ss}$, co-exists with liquid in the composition $Di_4San_{65}Ne_{31}$ (Fig. 9.3). The four-phase point at 2 GPa and 640 °C is located at $Di_3San_{73}Ne_{24}$ (Fig. 9.4). It may be observed that four-phase point ($Di_{ss} + Ne_{ss} + San_{ss} + L$) shifts towards sanidine with increasing pressure.

It is interesting to observe that if comparison of Fig. 9.1 (Platt and Edgar 1971) is made with respect to Fig. 9.2 the four-phase (Point B) under atmospheric pressure is shifted more towards the sanidine-rich side (point H at 0.1 GPa in the presence of excess water). The invariant points determined at 1 GPa (Point T, Fig. 9.3) and 2 GPa (Point P, Fig. 9.4) are shifted even farther towards the sanidine-rich side and away from the diopside end due to extensive solubility of nepheline and sanidine in the alkali silicate aqueous fluid.

Yoder (1958) studied the system nepheline–water, whereas Yoder and Upton (1971) investigated the system diopside–sanidine. These studies also show relatively high degree of dissolution of sanidine compared to nepheline with the increase of pressure. Thus shift in the composition of the four-phase points ($Di_{ss} + Ne_{ss} + San_{ss} + L$) towards sanidine (Figs. 9.2, 9.3 and 9.4) in the join diopside–nepheline–sanidine with the increase of pressure is expected. The liquid composition of the four-phase points in case of 1 and 2 GPa [$P(H_2O) = P$(Total)] isobaric diagram should correspond to an extremely pyroxene-poor phonolite.

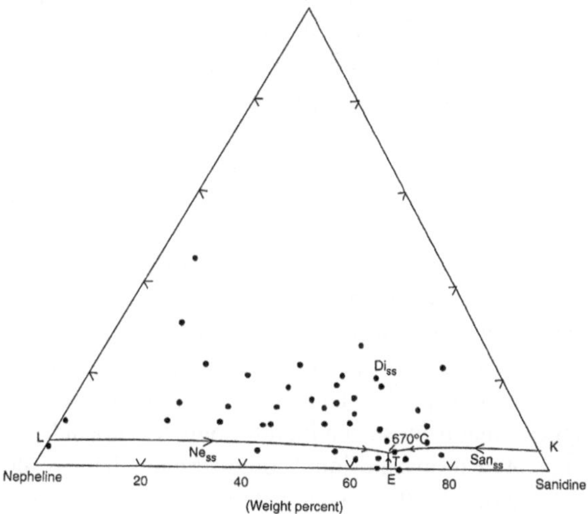

Fig. 9.3 Liquidus phase relations in the join diopside–nepheline–sanidine under 1 GPa [$P(H_2O) = P$ (Total)] and variable temperatures (Gupta et al. 2006)

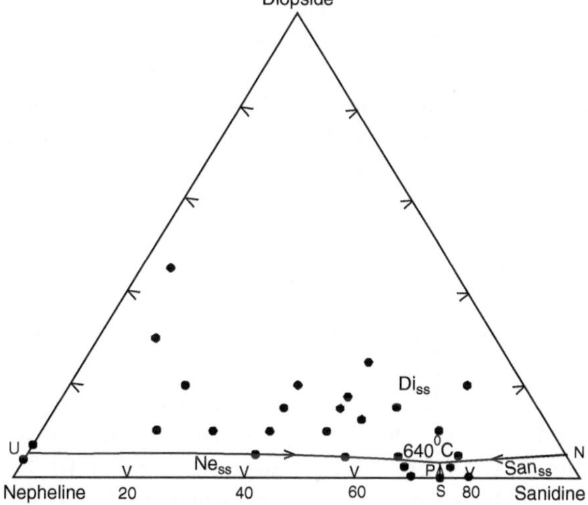

Fig. 9.4 Liquidus phase relations in join dipside–nepheline–sanidine under 2 GPa [$P(H_2O) = P$ (Total)] and variable temperatures (Gupta et al. 2006)

9.2.3 Petrological Significance

If a small amount of possible alumina (3 wt%) in diopside is ignored, then the join diopside–nepheline–sanidine may be represented as a join of the system nepheline–kalsilite–CaO–MgO–SiO_2 system. Because of the incongruent melting phenomenon of sanidine, there is a primary phase of leucite$_{ss}$. Forsterite$_{ss}$ also appears as a reaction product between diopside and nepheline at low pressure in air and 0.1 GPa (Figs. 9.1 and 9.2), but disappears below 985 ± 5 °C (point G in Fig. 9.2). Complete elimination of leucite$_{ss}$ occurs at 815 ± 10 °C, after the disappearance of leucite and forsterite, the join behaves essentially as a ternary system. The present experimental study of the join diopside–nepheline–sanidine at 0.1 GPa (Fig. 9.2) suggest that a nepheline-bearing leucitite (point H) can be derived from a Na-rich leucitite (curve M–H), a nepheline-bearing italite (point E) or an olivine nephelinite (point G).

Let us consider point 'X' ($Di_{70}Ne_{25}San_5$; Fig. 9.2). As content of normative pyroxene and nepheline is high, the bulk composition should correspond to mela-nephelinite (ijolite or melteigite; also see Fig. 9.2 for mineralogical composition). Mela-nephelinite is a melanocratic nephelinite with higher concentration of pyroxene compared to nepheline and having dark colour. Reference to the diagram of Fig. 9.2 suggests that from such a liquid (point X), diopside$_{ss}$ should precipitate first. As the temperature drops, the liquid composition will move along diopside-X to 'a'. At point 'a' forsterite$_{ss}$ will co-precipitate with diopside$_{ss}$ and the composition of the liquid should leave the diopside–nepheline–sanidine plane and move towards diopside$_{ss}$ + nepheline$_{ss}$ + forsterite$_{ss}$ liquid univariant line (olivine nephelinite). Forsterite start to react with liquid and disappears around 985 + 5 °C. When the univariant line moves to point G (Fig. 9.2) and forsterite$_{ss}$ completely reacts out, the liquid comes back to the diopside–nepheline–sanidine plane. After the disappearance of forsterite$_{ss}$ the liquid should move along the curve G–H with co-precipitation of nepheline$_{ss}$ and diopside$_{ss}$ resulting in the generation of a nephelinite. At H (865 °C), diopside$_{ss}$, nepheline$_{ss}$ and leucite$_{ss}$ (nepheline leucitite) are in equilibrium with liquid (Fig. 9.2). Leucite$_{ss}$ start to react with the liquid below 840 °C, and sanidine start to appear. The assemblage at 840 °C, therefore, corresponds to a leucite and pyroxene-bearing phonolite (Lc$_{ss}$ + Ne$_{ss}$ + San$_{ss}$ + pyroxene L). Pargasitic amphibole precipitate just before the solidus is reached and the assemblage at 820 + 10 °C corresponds to a potassic pargasite-bearing leucite–phonolite. The crystallization of a liquid of composition 'Y' should follow the course Y–b–H as mentioned in Fig. 9.2. This should correspond to following assemblages: Di$_{ss}$ + L(1) → Di$_{ss}$ + Lc$_{ss}$ + L (b, leucitite) (2) at (900 °C) → Di$_{ss}$ + Lc$_{ss}$ + Ne$_{ss}$ + L (3) at 865 °C (nepheline leucitite) → Di$_{ss}$ + Lc$_{ss}$ + Ne$_{ss}$ + San$_{ss}$ + L at 830 + 10 °C (A, leucite-bearing phonolite) (4) → Di$_{ss}$ + Ne$_{ss}$ + San$_{ss}$ + Lc$_{ss}$ + amphibole + L at 820 °C (potassic pargasite and leucite-bearing phonolite) (5) → Di$_{ss}$ + Ne$_{ss}$ + San$_{ss}$ + Amph$_{ss}$ + L at 815 + 10 °C (leucite-free potassic pargasite-bearing phonolite).

Edgar (1964) plotted the normative composition of 129 rocks containing alkali pyroxene, nepheline and alkali feldspar (Fig. 9.5). The 1 and 2 GPa isobaric four-phase points (Di$_{ss}$ + Ne$_{ss}$ + San$_{ss}$ + L) in the system diopside–nepheline–sanidine are plotted in Fig. 9.5. It is noted that the four-phase point B (1 GPa isobaric point) is not far from the area where maximum density distribution occurs, and point C (2 GPa isobaric four-phase point) is very close to maximum density distribution.

As the composition of the four-phase points (Di$_{ss}$ + Ne$_{ss}$ + San$_{ss}$ + L) for the system diopside–nepheline–sanidine at 1 and 2 GPa are $Di_4San_{65}Ne_{31}$ and $Di_3San_{73}Ne_{24}$, respectively, the first liquid at those pressures should contain 4–3 wt%

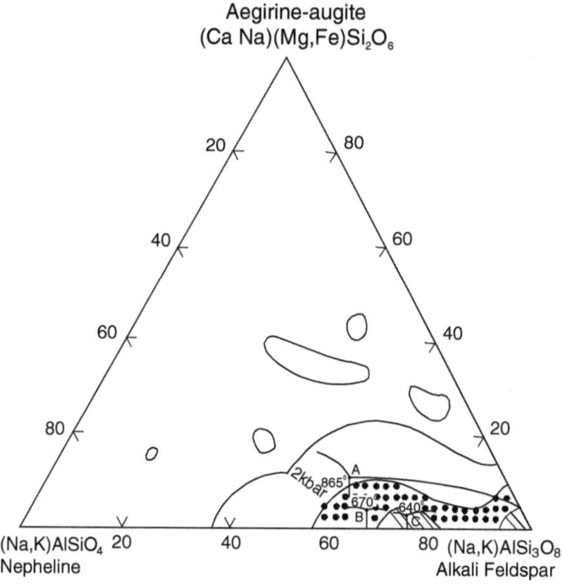

Fig. 9.5 Density distribution diagram for 129 plutonic rocks containing 80 % or more normative clinopyroxene–nepheline–alkali feldspar. The dotted area represents the locus of more than 75 % of all compositions. The shaded area contains 15 % of all the compositions, and remaining 10 % is represented by unmarked areas. *A* Di_{ss}–Ne_{ss}–San_{ss}–L (four phase point) determined under 1 atm. pressure; *B* same four phase point determined under 1 GPa; *C* same four phase point determined under 2 GPa $P(H_2O) = P(\text{Total})$ (after Edgar 1964)

clinopyroxene, 65–73 wt% sanidine and 31–24 wt% nepheline, respectively. These compositions (points B and C) plot either in the dotted or shaded area (Fig. 9.5). Such a liquid should have the mineralogical composition similar to phonolite. When a phonolitic melt is generated at 2 (60–70 km) or 1 GPa (30–35 km) they are leucite-free. When they ascend near the surface conditions, $leucite_{ss}$ appears as an incongruent mineral phase of sanidine. The crystallization sequence at 0.1 GPa, 1GPa and 2 GPa can be explained by Figs. 9.6a and 9.6b.

Mt. Etinde is a volcano situated on the southwestern flank of the large Mount Cameroon. Esch (1901; also see Nkoumbou et al. 1995) described the rock types of Mt. Etinde, a volcano that is built of leucitites, leucite–nephelinites, and nephelinite. The assemblages observed in an around Mt. Etinde volcano can be explained from the lineages established from the present investigation (Fig. 9.6a). Various crystallization trends of different rock types from Etinde are shown by curves paths marked E-I, E-II and E-III. It may be observed from Fig. 9.6a that a phonolite can be derived from either a trachyte (along the curve K–V close to the line MH) or a nephelinite (curve G–H). Some rocks of the trend E-III may hit either nephelinite (the curve GH) or a leucitite (the curve MH, Fig. 9.6a) depending upon the bulk composition.

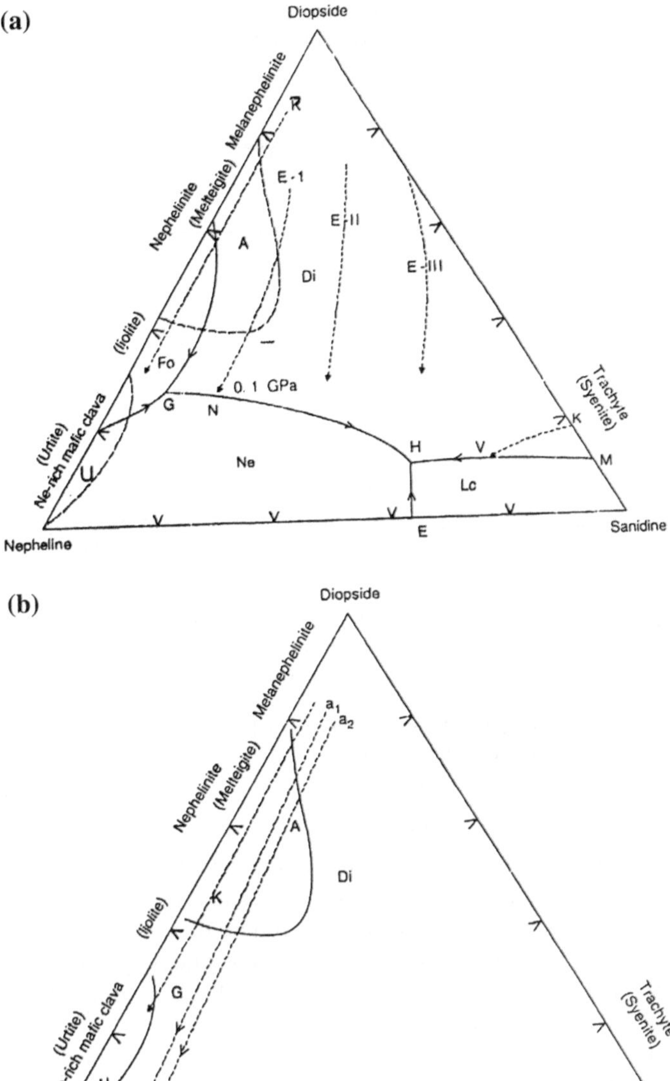

Fig. 9.6 **a** Normative composition of phonolites (Nkoumbou et al. 1995), nephelinites (Le Bas 1977, area A), urtites (Le Bas 1977; area U) and trachytes (Nkoumbou et al. 1995; among K–V) from various localities all over the world. The trends marked E-I, E-II and E-III are those of Etinde nephelinites (Nkoumbou et al. 1995). The curve K–V is the trend shown by the trachytic lava from the same locality (Nkoumbou et al. 1995). The shaded area corresponds to nephelinite; the dotted area represents pyroxene-poor nepheline-rich volcanic rocks (*urtite*). Trend K corresponds to crystallization path of phonolites from Kenya rift valley. **b** Trend of crystallization of nephelinitic liquids at 1 and 2 GPa from initial bulk compositions a1 and a2, respectively. The *shaded* and *dotted area* represent composition corresponding to nephelinites and pyroxene-poor, nepheline-rich rock (*urtite*)

If crystallization proceeds at 1 and 2 GPa, from nephelinitic liquids of composition a_1 and a_2 (Fig. 9.6b) the melt composition will reach point b_1 along the line a_1–b_1 (at 1 GPa) or b_2 along the line a_2–b_2 (at 2 GPa, Fig. 9.6b), resulting in the genesis of a pyroxene-poor nephelinite lava (point b_1 or b_2). Under plutonic conditions, if crystallization stops at these points an urtite should be generated. If the crystallization proceeds along the curve b_1–L to point L (at 1 GPa) or along b2–P to P (at 2 GPa), in either case the final assemblage at L or P should correspond to a phonolite (Fig. 9.6b, nepheline syenite). It may be noted that the four-phase point L or P contains less than 5 % diopside. Thus the first melt formed either at 1 or 2 GPa should be pyroxene-poor phonolites. If comparison of the crystallization paths for composition 'X' (Figs. 9.2 and 9.6a) is made it is found that olivine nephelinite is a low-pressure product. A plot of the phonolite bulk compositions around the four-phase points at 1 and 2 GPa suggests that many phonolitic magmas may be generated at depths of 30–60 km.

Whereas Lippard (1973) conducted field, petrographic and geochemical studies on phonolitic rocks of the Kenya rift valley, Jaggerson and Williams (1964) discussed the genesis of alkaline rocks in the north Tanganyika alkaline district. Lippard observed that bulk composition of the phonolites from the Kenya rift valley can be easily expressed in terms of the model diopside–nepheline–sanidine system, the crystallization trend (marked R), which is plotted in Fig. 9.6a. the present study is, therefore, quite relevant to understanding the paragenetic relationship amongst these rock types occurring near the Kenyan rift valley. Baker (1987) also studied the nepheline-bearing rocks from the alkali igneous province of Kenya and northern Tanzania, both of which are associated with the East African Rift Valley, where the following spatial distribution was noted: melanephelinite, nephelinites, phonolite and leucite phonolite (point H) along with trachyte (path M–H, Fig. 9.6a).

Differentiation of melanephelinitic magma (e.g. path a_1–b_1 and a_2–b_2, Fig. 9.6b) should yield nephelinites (point b_1 or b_2), then the liquid should follow the paths b_1–L or b_2–P and then at point L or P should produce nepheline syenites (Fig. 9.6b). This course of crystallization has been observed in Usaki complex, Kenya (Le Bas 1977). He also described a nephelinite volcano in Kisingiri, Western Kenya. The products of the volcano comprise a melanephelinite, nepheline-bearing melilitite and nephelinite agglomeratic tuff. Only 300 km to the north of Kisingiri is a highly eroded Napak volcano. The lavas make up 3 % of the extrusive succession, and they are melanephelinites, olivine and melilite-bearing varites and nephelinites. Figure 9.6a suggests that the trend of crystallization (marked K, Fig. 9.6b) should result in the crystallization of a melanephelinites first, followed by an olivine melilite nephelinite and then melilite nephelinite (below 1GPa). In the system diopside–nepheline–sanidine, in compositions close to the diopside–nepheline join, melilite appears in addition to forsterite, diopside and nepheline. Thus, when the K_2O content of these liquids is very negligible (<5 %), at the end of crystallization the final product is nephelinite containing minor amount of pargastitic amphibole. In case of one composition $Di_{10}Ne_{60}$ forsterite, nepheline, melilite and pyroxene-bearing assemblage appears at 950 and 850 °C, resembling the assemblage obtained at Kissigiri. It is noteworthy that a melilite, olivine and pyroxene-bearing assemblage is predominant in the diopside–nepheline join studied at 1 atm by Bowen (1928) and Schairer et al. (1962). In case of crystallization under 2 GPa a magma should not yield melilite, as this phase is eliminated (see Fig. 6.4) below 1.1 Gpa.

During the eruption of lava at Campi Flagrei (Italy), Melluso et al. (1995) noted the following paragenetic sequence: trachyte → leucite phonolite to trachytic phonolite with amphibole. Reference to Fig. 9.6a suggests that a trachytic melt (curve M–H) should yield a leucite phonolite at 865 °C. At 815 °C after leucite reacts out, at low pressure the final product is a pyroxene-poor trachytic phonolite containing pargasitic amphibole (K-feldspar + $diopside_{ss}$ + $nepheline_{ss}$ + amphibole + L. This study also supports the conclusion of King (1965), who suggested that a parental melanepheline (trend E_1, Fig. 9.6a) should yield a nephelinitic liquid (N–H), which in turn should result in the genesis of a phonolitic liquid (point H, Fig. 9.6a).

Cundari (1973) described the occurrence of leucite-bearing lava flows from New South Wales and Victoria. The dimension of the flows varies from 10 to 28,000 km^2. The variation of modal mineralogy is as follows: leucite (20–33 %), clinopyroxene (30–50 %), olivine (11–14 %), alkali feldspar + nepheline (14 %) and glass, which is very much in consonance with the mineralogy that is expected to be observed during progressive crystallization from this system at low pressure (Fig. 9.2). Cundari did not describe reaction relationship between olivine and liquid in these rocks. In the alkali igneous rocks of the Balcones volcanic province of Texas, Spencer (1969) however, observed reaction relationship between olivine and liquid in olivine nephelinites, which have paragenetic relationship with nephelinites. These nephelinites grade to phonolites (point H, Fig. 9.2).

Close association of trachyte (curve M–H, Fig. 9.2) and phonolite (point H) has been described from the Phlegrean fields located northeast of Naples (Italy) by Beccaluva et al. (1990). The lava of Vulsinian district (also of Italy) includes lava flows grading from trachyte to leucite-bearing phonolite (Holm et al. 1982). A similar trachyte–phonolite association has been observed by Baldridge et al. (1981) from the same locality. In the quaternary lavas of java ad Bali in the Sunda Arc (Indonesia), Whiteford et al. (1979) described highly potassic rocks containing various combinations of clinopyroxene, alkali feldspar, potassic pargasite, leucite and nepheline and olivine (present or absent). Trachytes (nepheline syenites), leucite phonolites (nepheline–leucite syenites), phonolite (nepheline syenite) have been described by Kogarko et al. (1995) from Yaksha, Baikal. Similar combination of the above-mentioned minerals has been described by the same authors from Artem (Primorye Province, USSR). They also reported the occurrence of an alkali complex from Irisu (Kazakistan), were rock types grade from leucite–pyroxene phonolite to phonolite, suggesting that leucite formed by incongruent melting of K-feldspar reacted out at low temperature. The present study at 0.1 GPa suggests that crystallization of rocks from Indonesia and Tezhar (Caucasus) took place at low-pressure near-surface conditions.

Wright (1963) suggested that mela-nephelinites or nephelinites are parental to phonolites, the same trend as exemplified by the trend E_1–N–H (Fig. 9.6a). Nash et al. (1969) in their study of the rock from Mount Suswa (Kenya) observed the crystallization trend from alkali trachytic liquid to phonolite (the trend M–H, Fig. 9.6a). In general, they believed that the magmas of Mount Suswa might have been generated due to partial or total fusion of silica-undersaturated syenitic plutonic parent, as xenoliths of syenite are observed in the lavas of that area.

Strong (1972) carried out a detailed petrological study of the alkaline rocks of the islands of Moheli, western Indian Ocean. Strong suggested an apparent differentiation from alkali basalt through melanephelinite and trachyte to phonolite. The present study under 0.1 GPa in presence of excess water suggest the following fractionation scheme: (1) Melanephelinite → nephelinite → phonolite, and (2) alkali basalt → trachyte → phonolite.

9.3 The System $KAlSi_3O_8$–$CaAl_2Si_2O_8$–$KAlSiO_4$ at 0.5 GPa

Liquidus phase relations in the system $KAlSi_3O_8$–$CaAl_2Si_2O_8$–$KAlSiO_4$ have been determined by Jones and MacKenzie (1991) for low anorthite contents (<20 wt% anorthite) at 0.5 GPa in presence of excess water.

The phase diagram for the system is shown in Fig. 9.7. They observed that a temperature maximum occurs on the field boundary of plagioclase and leucite. They established two ternary eutectics: (1) one occurring at $Qz_{23}An_{14}Ks_{63}$ and 735 °C where kalsilite (Ks), anorthite (An) and leucite are in equilibrium; and (2) the other at a temperature slightly higher than 785 °C, where leucite, anorthite and alkali feldspar coexist in equilibrium with liquid.

Occurrence of leucitite, tephritic leucitite, phonolitic leucite tephrite, tephritic leucite phonolite, and leucite phonolite has been reported from various localities. Appleton (1972) not only described the occurrence of these rocks in Roccamonfina, but tried to describe the genesis of these rocks in the light of the system leucite–anorthite–quartz.

Figure 9.8 (Jones and MacKenzie 1991) summarizes the phase relation in the join leucite–anorthite–quartz at 1 bar, 0.26 and 0.5 GPa. In Fig. 9.8, the solid line describes phase boundary at 1 bar, dashed line demonstrates the phase diagram derived by Appleton (1972) and the dotted line refers to the phase diagram obtained

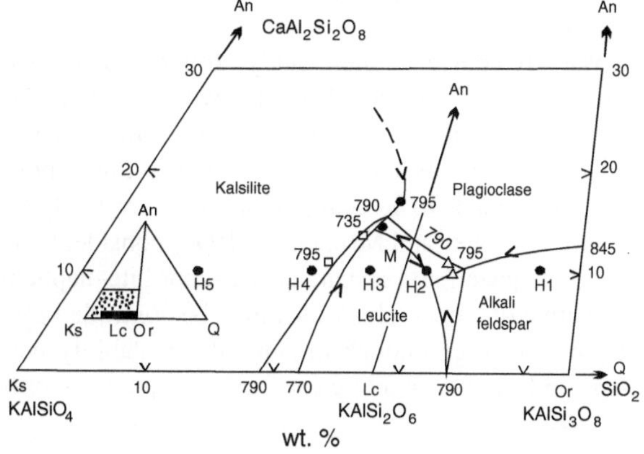

Fig. 9.7 Phase relation in the join anorthite–sanidine–kalsilite at 0.5 GPa in presence of excess water (after Jones and MacKenzie 1991). The *inset diagram* indicates the portion which has been studied

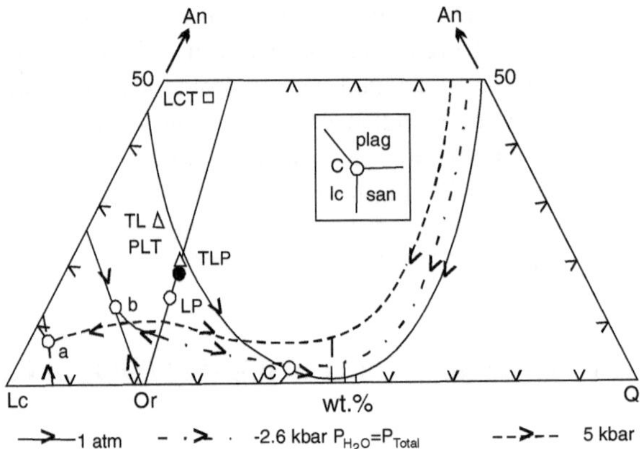

Fig. 9.8 Phase relation in the leucite–anorthite–SiO_2 at 1 bar, 0.26 and 0.5 GPa (after Jones and MacKenzie 1991). *Inset diagram* represents the equilibrium assemblage at point C. *LCT* leucite tephrite, *TL* tephritic leucitite, *PLT* phonolitic leucite tephrite, *LP* leucite phonolite, *TLP* tephritic leucite phonolite

by Jones and MacKenzie (1991). It was pointed out by them that under a condition, when $P(H_2O) < P$(total), at 0.5 GPa, the stability field of leucite expands, and liquidus temperature is higher than that under a condition, when $P(H_2O) = P$ (total) = 0.5 GPa. The fractionation trend of tephritic leucitites, leucite–trachytes and tephritic leucite phonolite has been described by Appleton (1972) in terms of anorthite–leucite–quartz diagram at 0.26 GPa in presence of excess water. Jones and MacKenzie pointed out that Appleton's deduction on phase relationship is wrong in that he showed that temperatures on the plagioclase–sanidine field boundary falling along the entire length. Experimental results have shown that within the field of feldspar in this system the ternary feldspar join acts as a thermal barrier and temperature decreases from the ternary feldspar plane on both directions. Their revised phase diagram is shown in Fig. 9.8 along with the field boundary determined at 0.5 GPa. Jones and MacKenzie extended the field boundaries schematically to silica-saturated portion of the system. In the SiO_2-rich side expansion of quartz field relative to that of feldspar with increase of pressure is expected.

With reference to their study, Jones and MacKenzie considered that a bulk composition of tephritic leucitite, which plot in the plagioclase field. It should precipitate plagioclase first followed by sanidine and then leucite or vice versa and remain silica-undersaturated. Only under lower pressure are magmas with compositions similar to Roccamonfina lavas are likely to produce silica-saturated differentiates. Jones and Mackenzie (1991) considered the same rocks compositions in terms of quaternary system quartz–anorthite–nepheline–kalsilite and showed that derivative of leucite phonolites may be controlled by fractionation of two feldspars at pressures high enough to support the stability field of leucite. This phase relationship however, is not clear from the quartz–anorthite–leucite system.

9.4 The System Forsterite–Diopside–Leucite–Anorthite

The bulk compositions of plagioclase-bearing potassium-rich silica-undersaturated rocks such as olivine leucitite (ugandite), leucite tephrites and basanite lie within the system forsterite–diopside–leucite–anorthite. These rocks occur in the Somma-Vesuvius region of Italy (Savelli 1967), the Birunga area of Uganda (Holmes and Harwood 1937; Ferguson and Cundari 1975), and the East Eifel region of West Germany (Duda and Schminke 1978). In these regions these rocks are often associated with phlogopite and/or olivine-bearing leucitites. A study of the system forsterite–diopside–leucite–anorthite may thus help to understand the paragenetic relationships between leucite-bearing tephrites) and basanites and more mafic potassic-rich lavas without plagioclase.

The system forsterie–diopside–leucite–anorthite has four bounding joins:

(1) forsterite–diopside–leucite,
(2) forsterite–anorthite–diopside,
(3) forsterite–leucite–anorthite, and
(4) diopside–leucite–anorthite.

Of these four joins, join (1) has already been discussed. The others are discussed below.

9.4.1 The Join Forsterite–Diopside–Anorthite

The join forsterite–diopside–anorthite (Osborn and Tait 1952) has two piercing points: one at $Di_{22.5}An_{57.5}Fo_{20}$ and 1,317 °C, the other at $Di_{49}An_{43.5}Fo_{7.5}$ and 1,270 ° C. At 1,317 °C forsterite$_{ss}$, spinel$_{ss}$ and anorthite$_{ss}$ coexist with liquid, whereas at 1,270 °C, forsterite$_{ss}$, diopside$_{ss}$, and anorthite$_{ss}$ are in equilibrium with liquid and the assemblage at the end of crystallization consists of diopside$_{ss}$ + forsterite$_{ss}$ + anorthite + liquid.

9.4.2 The Join Forsterite–Anorthite–Leucite

This join has not yet been studied. However, results from the bounding joins forsterite–anorthite (Anderson 1915), forsterite–leucite (Schairer 1954), and leucite–anorthite (Schairer and Bowen 1947) are already known. The phase diagram of this join can therefore be constructed as shown in Fig. 9.9. The spinel field in this case is small, judging from the study of the join forsterite–anorthite–SiO_2 (Anderson 1915), and the association of leucite and Mg-spinel is not reported in nature, although the latter occurs in xenoliths within potassic lavas of East Eifel (Duda and Schminke 1978).

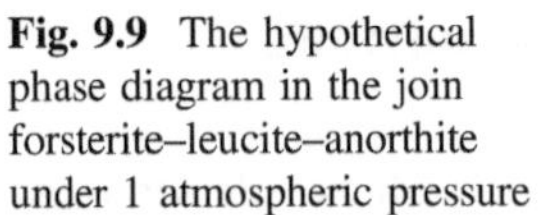

Fig. 9.9 The hypothetical phase diagram in the join forsterite–leucite–anorthite under 1 atmospheric pressure

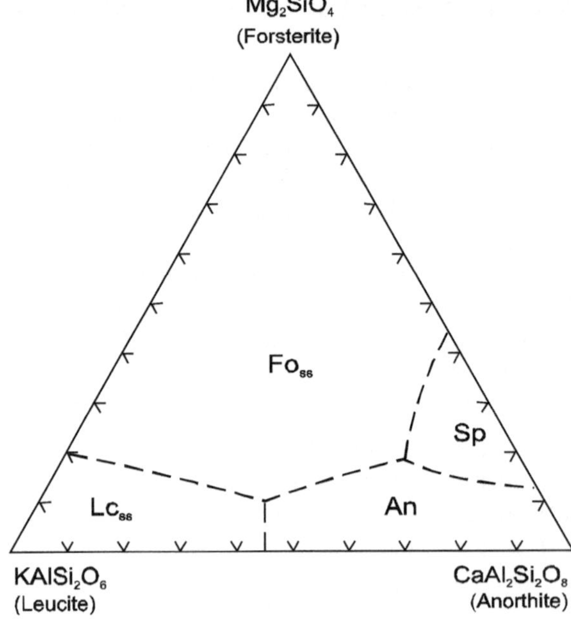

9.4.3 The Join Diopside–Leucite–Anorthite

Experimental results on this join are summarized in Fig. 9.10. At A ($Di_{43}An_{35}Lc_{22}$), diopside$_{ss}$ + leucite$_{ss}$ + anorthite$_{ss}$ + coexist with liquid. Optical and X-ray diffraction studies indicate that diopside may incorporate Ca-Tschermark's molecule, whereas leucite probably includes $K_2O{\cdot}MgO{\cdot}5SiO_2$ in solid solution. The composition of the liquid thus does not lie in this join, and point A therefore is a piercing point (1,249 ± 5 °C).

9.4.4 Paragenesis

On the basis of the phase diagrams of these four bounding joins, the course of crystallization of a liquid is the system forsterite–leucite–anorthite, can be described by a simplified flow sheet diagram as given in Fig. 9.11, in which a rock nomenclature corresponding to each assemblage is also shown. This shows that a leucite-bearing basanite can be produced from an olivine leucitite (m, Fig. 9.11), atephrite (n) or an olivine plagioclase italite (p) or a murambite (o), approaching the composition of a magma, belonging to the shoshonite absarokite series. Lavas of this type have been reported by Holmes and Harwood (1937; p. 1450) from the Kogoma area Birunga.

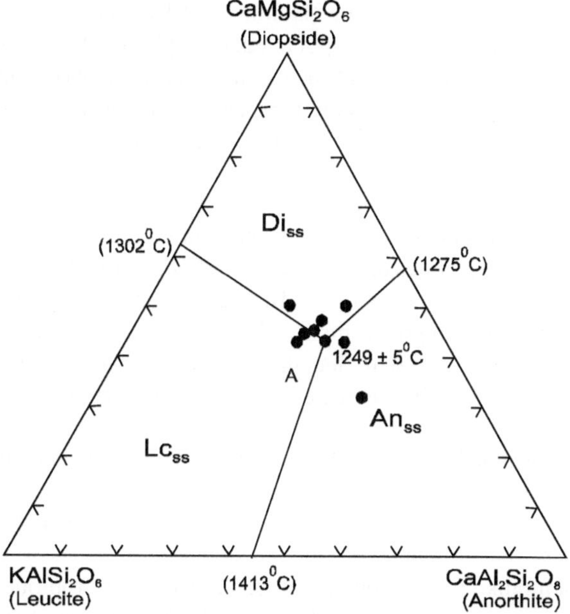

Fig. 9.10 Phase relation in the join diopside–leucite–anorthite under atmospheric pressure (after Gupta and Yagi 1980)

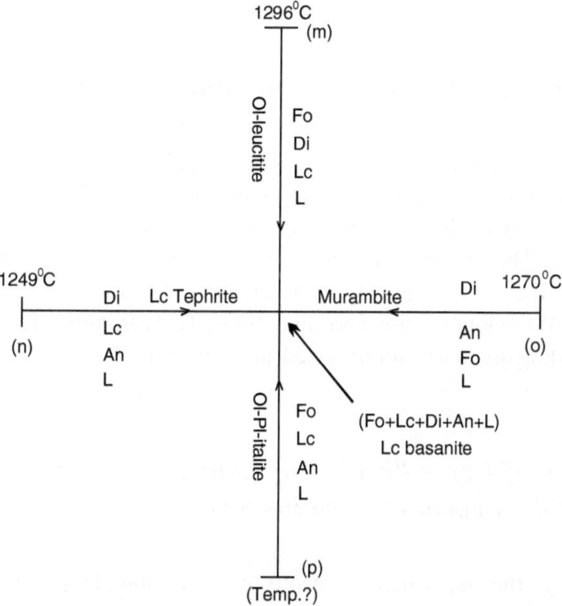

Fig. 9.11 Flow sheet diagram summarizing the course of crystallization of liquid in the system forsterite–diopside–leucite–anorthite (after Gupta and Yagi 1980)

In the Bufumbira area of Uganda (Holmes and Harwood 1937; p. 98) leucite basanite locally grades into olivine-poor leucite tephrite). In the Mabunga area, Holmes and Harwood (1937) found inclusion of leucite tephrites in leucite basanites. Reference to Fig. 9.11 shows the paragenetic relationship between these two rock types. Leucite tephrites occur at Kitale area, about 24 km to the west of Bufumbira, where leucite tephrites are found in association with leucite basanite. Figure 9.10 suggests that if the composition of liquid lies close to the diopside–leucite join inside the system diopside–leucite–anorthite, a leucitite magma will yield a leucite tephrite. Holmes and Harwood (1937) described lava flows of leucite basanite from the following areas of the Birunga (Chap. 4) petrographic province: Busumba, Mihanga, Mabungo, Kumizika, Ishozi, Nyakimanga, and Lungandable. Lavas transitional between olivine leucitite and leucite basanitie (Fig. 9.11) have been described from the Kisi area near the northwest corner of Lake Kivu in Uganda.

Many older volcanic districts to the northwest of Vesuvius are characterized by leucite tephrites, In the Vulsinian district the lava changes from leucite tephrite to leucitite and very basic basanites. In some places the later rock types are low in modal plagioclase and similar to olivine leucitite.

Duda and Schminke (1978) found that leucite basanites crop out at Kunkskopf, Veitskopf, Rothenberg, Nickenicher Hummerich, Nickenicher Sattel, Heidskopf and Satelberg in the East Eifel region of West Germany (Fig. 4.17). In Kunkskoph, the association of leucite tephrite and basanite is found. Leucite tephrite was also described from Alterberg, E. Eifel by Duda and Schminke (1978).

9.5 The System Diopside–Leucite–Anorthite–SiO_2

The paragenetic relationship between leucite tephrite, phonolite, trachyte, and potassium-rich latite can be understood from the study of the system diopside–leucite–anorthite–SiO_2. Experimental results on various joins of the system are already known. The system has four bounding joins: (1) the join diopside–leucite–SiO_2, (2) the join diopside–anorthite–SiO_2, (3) the join diopside–leucite–anorthite and (4) the join leucite–anorthite–SiO_2. Temperatures of piercing points of different bounding joins are summarized in Table 9.1.

9.5.1 Course of Crystallization of Liquid in the System Diopside–Leucite–Anorhtite–SiO_2

On the basis of the experimental results on four bounding joins the course of crystallization of liquid in the system diopside–leucite anorthite–SiO_2 can be summarized by a flow sheet diagram (Fig. 9.12).

Table 9.1 Temperatures of various piercing points within the system diopside–leucite–anorthite–SIO_2

Piercing point	Temp (°C)	System	Reference
1	1,249	Diopside–leucite–anorthite	Gupta and Yagi (1980)
2	<1,200	Diopside–leucite–sanidine.	Schairer and Bowen (1938)
3	?	Diopside–sanidine–anorthite	Not determined
4	<1,100	Diopside–sanidine–anorthite	Schairer and Bowen (1947)
5	<1,200	Diopside–leucite–SiO_2	Schairer and Bowen (1938)
6	<900	Anorthite–leucite–SiO_2	Schairer and Bowen (1947)
7	1,200	Diopside–anorthite–SiO_2	Hytonen and Schairer (1961)

The join diopside–anorthite–orthoclase (sanidine) divides the system into a silica-saturated and a silica-undersaturated portion. On the basis of the results of these ternary joins, the various phase volumes and the courses of crystallization of melts within the system can be constructed as shown in Fig. 9.12. The system has two invariant points: E_1 (silica-undersaturated) and E_2 (silica-saturated). If the initial composition of the liquid lies in the silica-deficient portion of the system, the

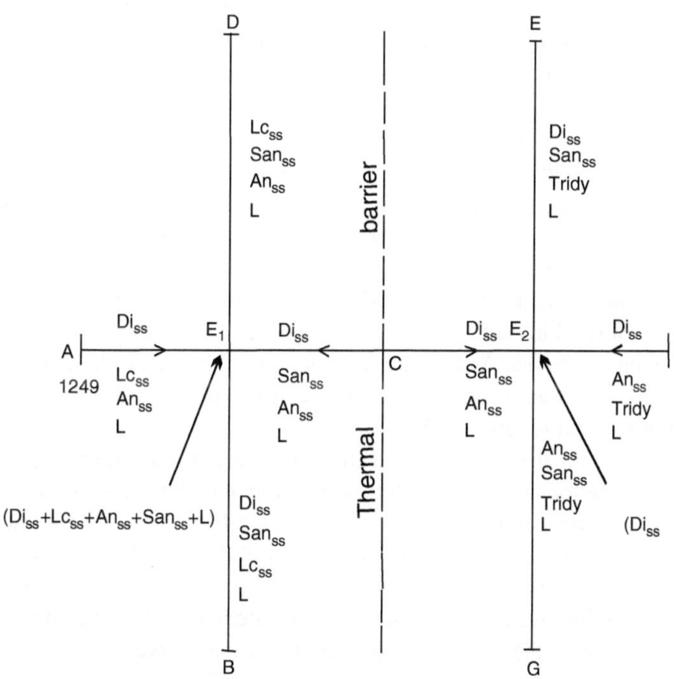

Fig. 9.12 Flow sheet diagram related to the system diopside–leucite–anorthite–SiO_2 (after Gupta and Yagi 1980)

composition of the melt should move towards E_1, whereas the liquid in the silica-undersaturated portion leads towards E_2. The coexisting assemblages at the invariant points are diopside + leucite + sanidine + anorthite + liquid (E_1), and diopside + sanidine + plagioclase + tridymite + liquid (E_2). The univariant assemblages of diopside–anorthite–leucite–liquid (A–E_1, leucite tephrite), diopside–sanidine–leucite–liquid (B–E_1, pyroxene–leucite phonolite), diopside–sanidine–anorthite–liquid (C–E_1, shoshonite), and anorthite–leucite–sanidine–liquid (D–E_1, plagioclase–leucite phonolite) lead to E_1. The univariant assemblages of diopside–sanidine–tridymite–l (E–E_2), anorthite–sanidine–tridymite–liquid (G–E_2), diopside–anorthite–tridymite–liquid (F–E_2), and diopside–sanidine–anorthite–liquid (C–E_2) move towards E_2. It should be noted here that at point C, which lies in the diopside–anorthite–sanidine join, diopside, leucite anorthite and liquid coexist. However, because of a reaction with the liquid, leucite should disappear and the resulting assemblage thus should consist of diopside + anorthite + sanidine + liquid.

The univariant assemblages in the silica-undersaturated portion of the system and the coexisting phases at E_2 correspond to the trachyandesite–shoshonite–quartz latite series. Potassium-rich rocks consisting of various combinations of leucite, K-feldspar, clinopyroxene, and plagioclase have been reported from the Somma-Vesuvius (Savelli 1967), the Vico area of Italy (Cundari and Mattias 1974) and the Bufumbira region of East Africa (Ferguson and Cundari 1975). Leucite and nepheline-bearing tephrites have been found at Rothenberg, Epfelsberg and Krufterofen in the East Eifel region by Duda and Schminke (1978).

When Cundari and Mattias (1974) plotted the mineralogical compositions of the rocks from the Vico area of Italy in a triangular diagram of alkali feldspar (A)–plagioclase (P) and feldspathoid (F), most of the rocks lie in the tephrite–leucite phonolite (TLP) field (see Fig. 7.3). This figure shows that under equilibrium conditions of crystallization the final product would be a tephritic–leucite phonolite. Cundari and Mattias (1974) observed a general gradation from tephritic leucitite (TL) through phonolitic leucitite (PLT) to alkali trachyte (LTR).

Ferguson and Cundari (1975) found that there are two series of leucite bearing-rocks in the Bufumbira area of East Africa. They underwent the following paragenetic sequences: Series A: Leucite tephrite, tephritic leucite, leucitite, phonolitic leucitite and phonolite. Series B: Leucite tephrite, phonolitic leucite tephrite, tephritic phonolite, latite, and trachyte.

In their study on the volcanism in the Eolian arc region, Barberi et al. (1974) found a close relationship between leucite tephrite and shoshonite in Vulcanello, of which the former rock type was probably produced from the latter by fractionation.

Highly vesicular lava flows of leucite and K-feldspar-bearing tephrites (E_1, Fig. 9.12), covering extensive areas have been reported by Holmes and Horwood (1937) from the Sagitwe region of Birunga. Leucite tracyte has been described from the Karsimbi area of the Birunga region by the same workers.

If the composition of the liquid lies in the silica-rich side of the join diopside–sanidine–anorthite, the final product should be quartz latite (E_2, Fig. 9.12). Holmes and Harwood (1937) described rock type of trachyandesite–latite series from the Sabyno area of Birunga volcanic province, where the prominent rock types are potassium-rich mafic and ultramafic varieties.

Chapter 10
Melilite- and Leucite-Bearing Systems

10.1 Melilite- and Leucite-Bearing Systems Without Nepheline

10.1.1 The System Forsterite–Diopside–Akermanite–Leucite

The four essential minerals of many potassium-rich mafic and ultramafic rocks are leucite, augite, olivine, and melilite. The bulk compositions of these simplified rocks lie within the phase volumes of forsterite–diopside–akermanite–leucite. Phase relation within the system is thus applicable to the genesis of potassium-rich mafic and ultramafic volcanic rocks. A detailed study of the system was made by Gupta (1972), who showed that all phases in this system are solid solutions. Forsterite may contain monticellite; and diopside and akermanite possibly incorporate Ca-Tschermak's molecule and gehlenite in solid solution, respectively. Likewise, leucite may include $K_2O{\cdot}MgO{\cdot}5SiO_2$ and forsterite probably contains monticellite. The system is bounded by four ternary joins:

(1) diopside–akermanite–leucite,
(2) forsterite–diopside–leucite,
(3) forsterite–akermanite–leucite, and
(4) forsterite–diopside–akermanite.

Phase relations in these joins are described below.

10.1.1.1 The Join Diopside–Akermanite–Leucite

Phase equilibria diagram of the join (Gupta 1972) is presented in Fig. 10.1. At point B ($Di_{39}Ak_{29}Lc_{32}$), diopside$_{ss}$ + akermanite$_{ss}$ + leucite$_{ss}$ + liquid coexist in equilibrium at 1,281 ± 3 °C. In their study of the thermal behaviour of pure synthetic akermanite, Schairer et al. (1967) found well-distributed inclusions of diopside$_{ss}$

A.K. Gupta, *Origin of Potassium-rich Silica-deficient Igneous Rocks*,
Springer Geology, DOI 10.1007/978-81-322-2083-1_10

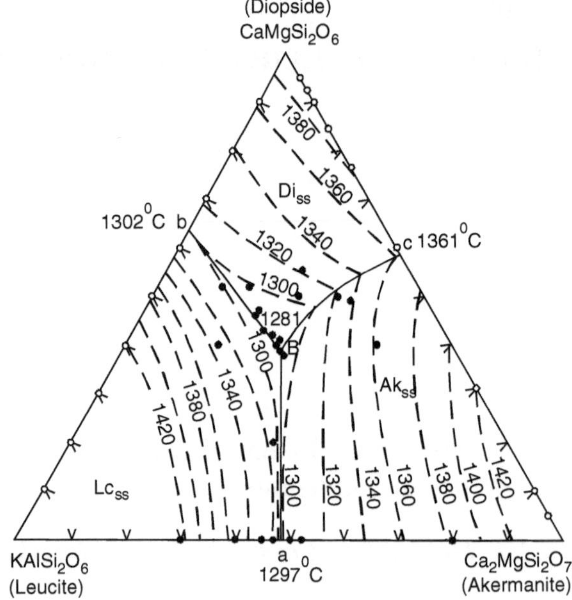

Fig. 10.1 Phase relation in the join diopside–akermanite–leucite at 1 atmosphere (after Gupta 1972)

and wollastonite$_{ss}$ within melilite and below 1,240 °C, wollastonite$_{ss}$ is absent on the liquidus of the diopside–akermanite–leucite join. As the compositions of the crystalline phases do not lie within the join, it is not ternary and point B is therefore, a piercing point and the lines dividing the primary phase fields are traces of divergent surfaces cut by the diopside–akermanite–leucite join.

10.1.1.2 The Join Forsterite–Diopside–Leucite

Figure 10.2 shows the phase diagram (Gupta 1972) for the join forsterite–diopside–leucite. At point c ($Fo_3Di_{60}Lc_{37}$), forsterite$_{ss}$ + diopside$_{ss}$ + leucite$_{ss}$ coexist in equilibrium at 1,296 ± 3 °C. Again the join is not ternary and the point c is a piercing point.

10.1.1.3 The Join Forsterite–Akermanite–Leucite

Phase diagram of the join (Gupta 1972) is presented in Fig. 10.3, which shows that it is pseudoternary and cuts the phase volume of monticellite$_{ss}$. The join has two piercing points at $Fo_{17}Ak_{78}Lc_5$ and 1,428 °C(H), where the assemblage is forsterite$_{ss}$ + akermanite$_{ss}$ + monticellite$_{ss}$ + liquid; and $Fo_9Ak_{43.5}Lc_{47.5}$ and 1,286 ± 3 °C (G), where the assemblage is forsterite$_{ss}$ + akermanite$_{ss}$ + leucite$_{ss}$. At H, monticellite$_{ss}$ reacts with the liquid to form akermanite and is completely eliminated at 1,410 ± 3 °C. Therefore, monticellite$_{ss}$ does not coexist with leucite$_{ss}$.

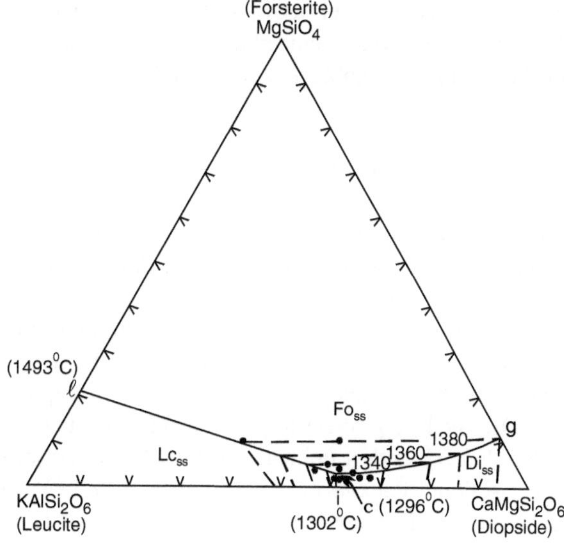

Fig. 10.2 Phase relation in the join forsterite–diopside–leucite at 1 atmosphere (after Gupta 1972)

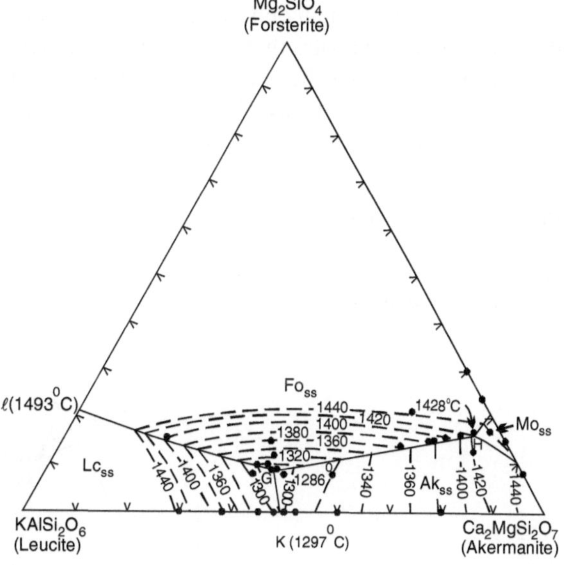

Fig. 10.3 Phase relation in the join forsterite–akermanite–leucite at 1 atmosphere (after Gupta 1972)

10.1.1.4 The Join Forsterite–Diopside–Akermanite

This join was studied by Ferguson and Merwin (1919) as a part of the system CaO–MgO–SiO_2. They found that at $Fo_{17}Ak_{77}Di_6$ and 1,430 °C (estimated from their diagram) forsterite$_{ss}$ + monticellite$_{ss}$ + akermanite$_{ss}$ + liquid are in equilibrium. Another piercing point occurs at $Fo_8Di_{50}Ak_{42}$, where forsterite$_{ss}$ + diopside$_{ss}$ + liquid coexist at 1,357 °C. In this join monticellite$_{ss}$ reacts with the liquid to form akermanite$_{ss}$ and the final crystalline assemblage is forsterite$_{ss}$ + diopside$_{ss}$ + akermanite$_{ss}$.

10.1.1.5 The Join Diopside–Akermanite–Leucite with 3 wt% Forsterite

Study of the bounding joins of the tetrahedron suggests that the invariant point, where forsterite$_{ss}$ + akermanite$_{ss}$ + diopside$_{ss}$ + leucite$_{ss}$ + liquid are in equilibrium, lies very close to the basal plane diopside–akermanite–leucite, probably containing less than 5 wt% forsterite. Accordingly, the join diopside–akermanite–leucite with 3 wt% forsterite was studied to determine the temperature of the invariant point. The equilibrium diagram for the join is represented by Fig. 10.4. At K ($Fo_3Di_{55}Ak_9Lc_{33}$), forsterite$_{ss}$ + diopside$_{ss}$ + leucite$_{ss}$ + liquid are in equilibrium at 1,295 ± 3 °C, and at point L ($Fo_3Di_{41}Ak_{24}Lc_{32}$), diopside$_{ss}$ + akermanite$_{ss}$ + leucite$_{ss}$ + liquid coexist in equilibrium at 1,280 °C, below which there is a liquid phase, which does not freeze until 1,255 °C. As the compositions of the crystalline phases lie outside the join and there is a liquid phase below 1,274 °C, the system forsterite–diopside–akermanite–leucite is not quaternary. Optical and X-ray studies, however, suggest that although they are chemically nonstoichiometric, all the phases in this join contain only small amounts of solid solution and the behaviour of the join thus approximates that of a quaternary system and points K and L behave as piercing points. From these two points liquid continues to crystallise along two different paths and at 1,274 °C a five-phase assemblage of forsterite$_{ss}$ + akermanite$_{ss}$ + leucite$_{ss}$ + liquid, is reached. Below 1,274 °C, crystallization continues as

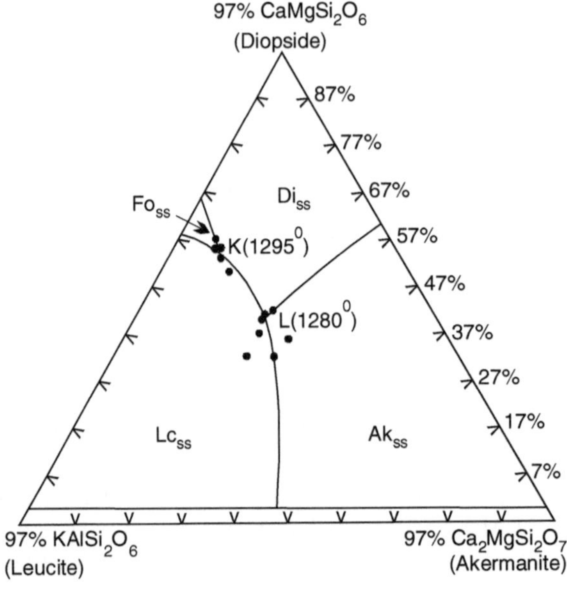

Fig. 10.4 Phase relation in the join diopside–akermanite–leucite with 3 % forsterite (after Gupta 1972)

the liquid moves towards the lowest melting point. Study of the join suggests that the lowest melting point of the tetrahedron, forsterite–diopside–akermanite–leucite, lies near the diopside–akermanite–leucite plane.

10.1.1.6 Paragenesis of Nepheline-Free Melilite and Leucite-Bearing Mafic and Ultramafic Rocks

Melilite Leucitite

The join diopside–akermanite–leucite (Fig. 10.1) shows that a melilite leucitite, corresponding to the piercing point B, can be derived from either a melilite italite (aB), leucitite (bB), or a melilitite (cB). Natural melilites contain a considerable amount of sodium in solid solution (Schairer and Yoder 1964) but are poor in K_2O. The melilitites mentioned here have compositions lying within the tetrahedron forsterite–diopside–akermanite–leucite. At Capo di Bove in Italy, Washington (1906) noted the sporadic occurrence of leucitite grading to melilite leucitite (cecilite). Results of the system diopside–akermanite–leucite indicate that melilite leucitite may be derived from a leucitite magma.

Figure 10.2 shows that olivine leucitite (piercing point C) can be derived from either potassium-rich olivine pyroxenite (gC), or olivine italite (1C), or leucitite (iC). Olivine pyroxenite ordinarily contains very small amounts of K_2O. The olivine pyroxenites referred to here have compositions lying in the tetrahedron, forsterite–diopside–akermanite–leucite. In the Bufumbira region of Uganda, Holmes (1937), described a lava flow of olivine leucitite (ugandite). At Katunga in Uganda, he noted the close association of rocks such as leucitite and olivine leucitite, and also reported complete gradation of these rocks from pyroxene-rich leucitite, which in turn grades into pyroxenite. This field evidence and the results on the system forsterite–diopside–leucite suggest that an olivine leucitite magma can be a derivative of a potassium-rich olivine pyroxenite liquid. Generation of such a liquid is possible by the partial melting of phlogopite-bearing peridotite.

Phase equilibria study of the system forsterite–akermanite–leucite (Fig. 10.3) suggests that a katungite magma can be derived from a melilite italite (kG) or an olivine italite (IG). Field evidence of a massive lava flow of katungite was reported by Holmes (1937) from Toro-Ankole, Uganda, where it occurs in association with alnoite. On the basis of this experimental study it appears possible that a katungite magma may have been derived from an alnoite, which itself may have originated from a potassium-rich monticellite alnoite.

Olivine Melilitite

Reference to Ferguson and Mervin's diagram (1919) of forsterite–diopside–akermanite system suggests that olivine melilitite may be produced from a peridotite, a melilitite, or an alnoite magma.

10.1.1.7 Course of Crystallization of Liquid in the System Forsterite–Diopside–Akermanite–Leucite

The system is pseudoquaternary, being a part of the quinary system K_2O–Al_2O_3–CaO–MgO–SiO_2. The composition of all crystalline phases lie outside the tetrahedron. However, the range of solid solution in these phases is small, and the system can thus be treated almost as quaternary. The complete course of crystallization of liquid in this system is summarised by a flow sheet diagram, shown in Fig. 10.5, in which rock nomenclatures are also given, It shows that an olivine–melilite–leucitite can be produced from a melilite leucitite, an olivine leucitite, a katungite or an olivine melilitite. Reference to Fig. 10.3 suggests that a katungite (point G) can be produced from an alnoite (point H). Schairer and Yoder (1964), in their study on the system nepheline–forsterite–larnite–silica, showed that a parental liquid for the sodium-rich undersaturated part may be represented by either an olivine melilitite or an olivine nephelinite. Olivine–melilite nephelinite is produced from liquids of these compositions. It appears that rocks called olivine melilitite can differentiate towards either an olivine–melilite nephelinite or an olivine–melilite leucitite. The Na_2O/K_2O ratio of the parental olivine melilitite melt probably controls subsequent crystallization. If the liquid is potassium-rich, leucite-bearing mafic and ultramafic rocks would be produced, and if sodium-rich, nepheline-bearing mafic and ultramafic rocks would crystallise.

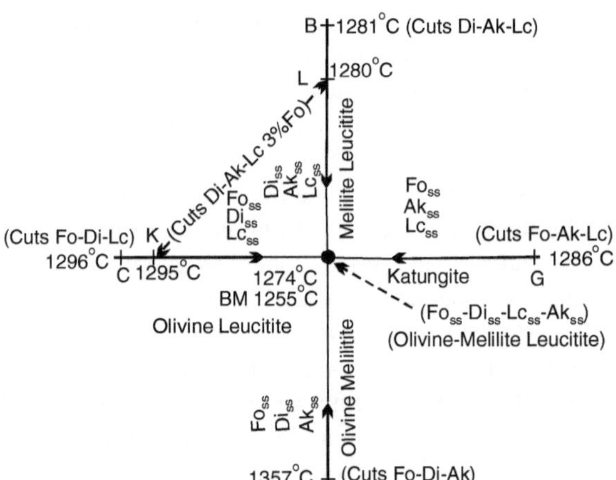

Fig. 10.5 The flow-sheet and rock-nomenclature diagram of the system forsterite–diopside–akermanite–leucite (after Gupta 1972)

10.2 Melilite- and Leucite-Bearing Mafic and Ultramafic Rocks Containing Nepheline

10.2.1 The System Diopside–Nepheline–Akermanite–Leucite

Compositions of many simplified potassium-rich nepheline-bearing rocks such as nephelinite, melilite nephelinite, leucitite, melilite italite, melilite leucitite and melilite–nepheline leucitite can be represented by the system diopside–nepheline–akermanite–leucite. In the system, forsterite appears as a phase because of a reaction relationship between diopside and nepheline (Bowen 1922; Schairer et al. 1962). Study of the system is thus important to understand the paragenesis of many other mafic and ultramafic volcanic rock types such as olivine–melilite–nepheline leucitite, nepheline-bearing katungite, and olivine leucitite. Gupta et al. (1973) studied this system and found that all phases appearing in this system are solid solutions. They established that nepheline incorporates variable amounts of kalsilite and excess silica; melilite contains soda-melilite and gehlenite; forsterite incorporates monticellite, diopside contains Ca-Tschermak`s molecule and leucite incorporates nepheline in solid solutions.

The system is bounded by four limiting joins

1. diopside–nepheline–akermanite,
2. diopside–akermanite–leucite,
3. diopside–nepheline–leucite, and
4. nepheline–akermanite–leucite.

10.2.1.1 The Join Diopside–Nepheline–Akermanite

This join (Onuma and Yagi 1967) is a pseudoternary (Fig. 10.6) with two piercing points; one is located at $Di_{55}Ak_6Ne_{39}$ and 1,212 ± 3 °C (G), where forsterite$_{ss}$, melilite, diopside$_{ss}$, and liquid are in equilibrium, and the other at $Di_{38}Ak_3Ne_{59}$ and 1,169 °C (H), where forsterite$_{ss}$ + nepheline$_{ss}$ + melilite + liquid coexist. The assemblage forsterite$_{ss}$, diopside$_{ss}$, nepheline$_{ss}$, melilite, and liquid are in equilibrium at 1,135 °C. Forsterite$_{ss}$ has a reaction relationship with liquid. The join diopside–akermanite–liquid has already been discussed (Fig. 10.1). It has a piercing point at $Di_{39}Ak_{29}Lc_{32}$ and 1,281 ± 3 °C.

10.2.1.2 The Join Diopside–Nepheline–Leucite

The join was studied by Gupta and Lidiak (1973; Fig. 10.7), who found it to be pseudoternary with two four-phase points. One at $Di_{60}Ne_8Lc_{32}$ and 1,275 ± 5 °C. (A) where forsterite$_{ss}$, diopside$_{ss}$, leucite$_{ss}$ and liquid are in equilibrium, the second at $Di_{27.5}Ne_{29.5}Lc_{43}$, and 1,194 ± 5 °C (B), where nepheline$_{ss}$, leucite$_{ss}$ and

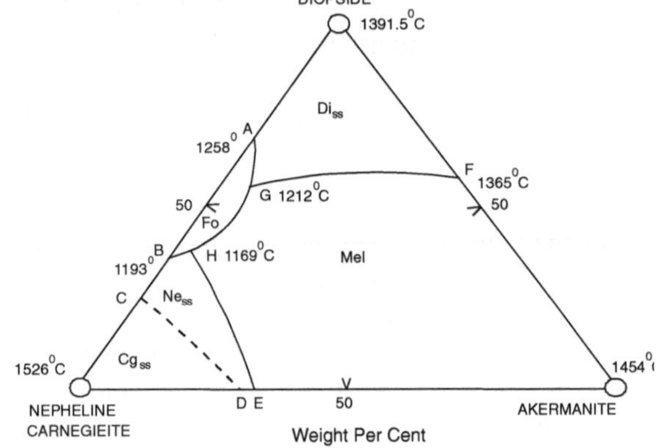

Fig. 10.6 Phase relation in the join diopside–nepheline–akermanite at 1 atmosphere (after Onuma and Yagi 1967)

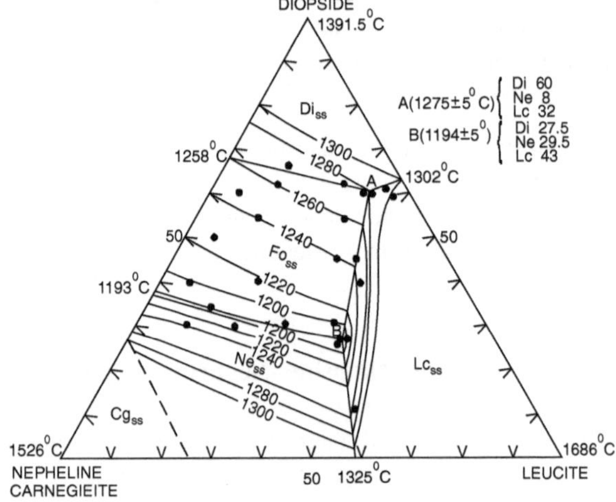

Fig. 10.7 Phase relation in the join diopside–nepheline–leucite at 1 atmosphere (after Gupta and Lidiak 1973)

forsterite$_{ss}$ coexist with liquid. In this join, the assemblage forsterite$_{ss}$ + nepheline$_{ss}$ + diopside$_{ss}$ + leucite$_{ss}$ + liquid is reached between 1,168 and 1,100 °C. Then forsterite$_{ss}$ reacts with liquid to be completely consumed. The temperature of disappearance for this phase varies: 1,135 ± 10 °C (near point A), 1,060 ± 10 °C (near point, B) and 950 ± 10 °C near the join diopside–nepheline. The final assemblage in this join consists of melilite + diopside$_{ss}$ + nepheline$_{ss}$ + leucite$_{ss}$. The course of crystallization of liquid in this join is shown by Fig. 10.8, and the rock nomenclature diagram corresponding to Fig. 10.9.

Gee and Sack (1988) performed experiments on a melilite nephelinite collected from Nyiragongo, Zaire. They added additional amount of akermanite, leucite, nepheline and olivine to this sample. The experiments were conducted under QFM

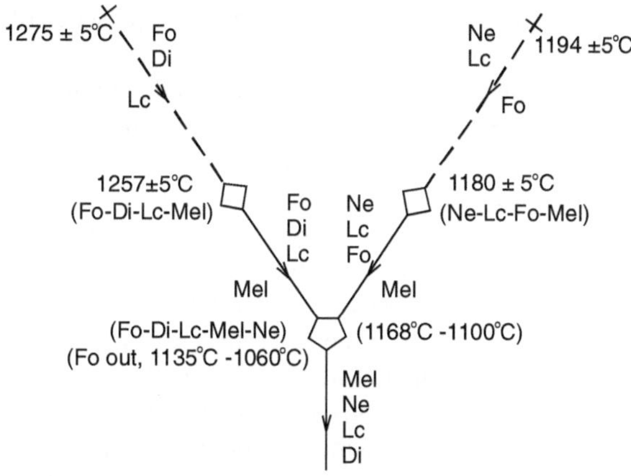

Fig. 10.8 Course of crystallization of liquid in the join diopside–nepheline–leucite (after Gupta and Lidiak 1973)

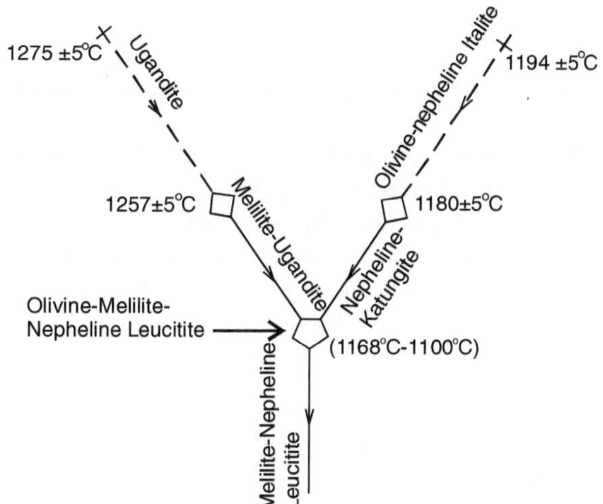

Fig. 10.9 Rock-nomenclature diagram, corresponding to Fig. 10.8 (after Gupta and Lidiak (1973)

buffer conditions in a furnace, where oxygen fugacity was maintained by passing a mixture of gases having compositions, CO_2: 95.25 % and CO: 4.75 %. They studied 19 compositions to determine phase relations and course of crystallization of an olivine–nepheline–leucitite and a katungite liquid. They concluded that both the two magma types yield an olivine–melilite–nepheline leucitite liquid at 1,090 °C under atmospheric pressure and under QFM buffer condition. They noted a reaction relationship between olivine and liquid, and found that olivine finally disappears at around 1,053 °C. The course of crystallization of liquid as established by Gee and Sack (1988) is similar to what was predicted by Gupta et al. (1973; Figs. 10.8 and 10.9).

They found cotectic crystallization of leucite + nepheline + spinel ± melilite ± high calcium pyroxene ± olivine (± whitlockite ± perovskite) in equilibrium with liquid at 1 atm (fO_2 ~ QFM) over the temperature range, 1,050–1,110 °C.

Their experimental results indicate that in presence of nepheline, leucite and spinel, olivine and melilte are stable at temperatures >1,090 °C. A peritectic relationship occurs among olivine, melilite, and liquid to produce either a olivine + high calcium pyroxene-bearing assemblage or a melilite and high-calcium pyroxene-bearing assemblage with decreasing temperature. The preferred path of liquid divergence is contingent upon bulk composition. A similar peritectic relationship has also been inferred from studies in simple systems (e.g. Onuma and Yagi 1967; Gupta et al. 1973). Relative to these simple systems, the peritectic is depressed by 30–50 °C in the natural system because of the presence of other oxides. The course of crystallization of liquid, established by Gee and Seck (1988) is very similar to that determined by Gupta et al. (1973; Fig. 10.9).

Melluso et al. (1996) plotted the Mt. Vulture melilite-bearing rocks in the diopside–nepheline–leucite pseudoternary diagram of Gupta and Lidiak (1973). They noted that the bulk compositions of these rocks plot in the central part of their diagram (10.7), and the Paragenesis of the rocks is very similar to that predicted by Gupta and Lidiak.

10.2.1.3 The Join Nepheline–Akermanite–Leucite

The join was studied by Gupta et al. (1973); and the phase diagram is shown in Fig. 10.10. At point S ($Ne_{34.5}Ak_{20.5}Lc_{45}$) and 1,170 ± 3 °C, nepheline$_{ss}$, leucite$_{ss}$ melilite, and liquid are in equilibrium. Onuma and Yagi (1967) showed that in the join diopside–nepheline–akermanite, forsterite$_{ss}$ appears as the last solid phase. This phase also appears as a subliquidus phase within the join nepheline–akermanite–leucite, but it reacts with the liquid and disappears at 1,100 ± 5 °C. As melilite contains soda-melilite, and forsterite$_{ss}$ appears as an additional phase, the join is pseudoternary.

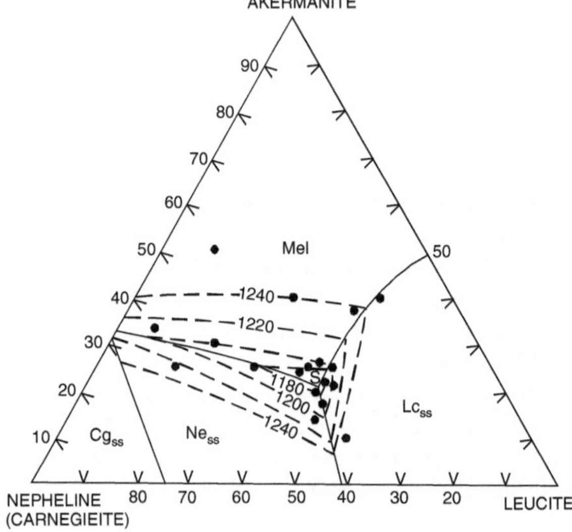

Fig. 10.10 Phase relation in the join nepheline–akermanite–leucite at 1 atmosphere (after Gupta et al. 1973)

10.2.1.4 Paragenetic Relationship in the System Diopside–Nepheline–Akermanite–Leucite under Atmospheric Pressure

Of the two piercing points in the system diopside–nepheline–akermanite (Onuma and Yagi 1967), point H (Fig. 10.6) has a relatively lower liquidus temperature. This point was therefore, chosen as one of the end members to which leucite was progressively added. The join was studied to determine the temperature of the six-phase assemblage, forsterite$_{ss}$ + diopside$_{ss}$ + akermanite$_{ss}$ + leucite$_{ss}$ + nepheline$_{ss}$ + liquid in the system diopside–nepheline–akermanite–leucite, and the phase equilibrium diagram is presented in Fig. 10.11. The join cuts the primary phase volumes of forsterite$_{ss}$ and leucite$_{ss}$. The pseudoeutectic occurs at $(Di_{38}Ak_3Ne_{59})_{61}Lc_{39}$ and 1,130 ± 5 °C, where forsterite$_{ss}$, leucite$_{ss}$ and liquid are in equilibrium. Figure 10.11 shows that at 1,105 ± 5 °C, forsterite$_{ss}$ + diopside$_{ss}$ + melilite + leucite$_{ss}$ + nepheline$_{ss}$ coexist with liquid. Forsterite starts to react with liquid and is eliminated at 1,075° ± 10 °C, and the subsolidus assemblage is Di$_{ss}$ + Mel + Ne$_{ss}$ + Lc$_{ss}$.

Experimental results on the join nepheline–akermanite–leucite, the pseudobinary join $(Di_{38}Ak_3Ne_{59})_{100\text{-}x}Lc_x$ and the other three bounding joins shows that the system diopside–nepheline–akermanite–leucite is pseudoquaternary. If however, the presence of small amounts of alumina as Ca-Tschermak's molecule in diopside and the gehlenite molecule in melilite is ignored, the system can be treated as a quaternary join of the five component system, nepheline–kalsilite–CaO–MgO–SiO_2. A flow sheet diagram of the system is shown in Fig. 10.12, where only the five-phase univariant lines are shown. At A (Fig. 10.7), forsterite$_{ss}$, diopside$_{ss}$, leucite$_{ss}$ and liquid are in equilibrium at 1,275 ± 5 °C. Gupta and Lidiak (1973) found that with further

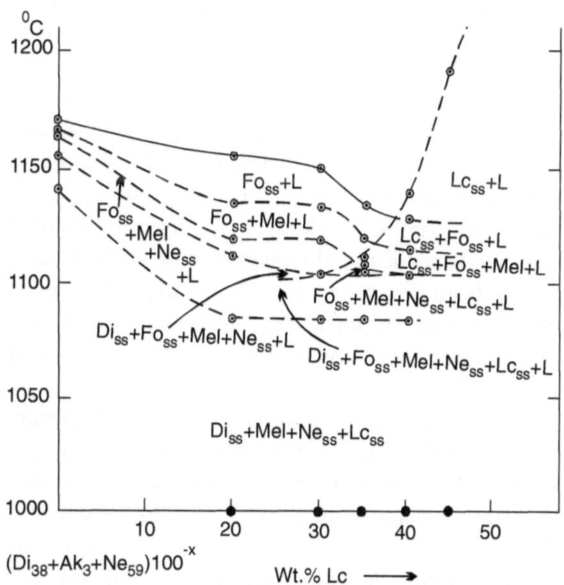

Fig. 10.11 Phase relation in the join $(Di_{38}Ak_3Ne_{59})_{100\text{-}x}$-$Lc_x$ at 1 atmosphere (after Gupta et al. 1973)

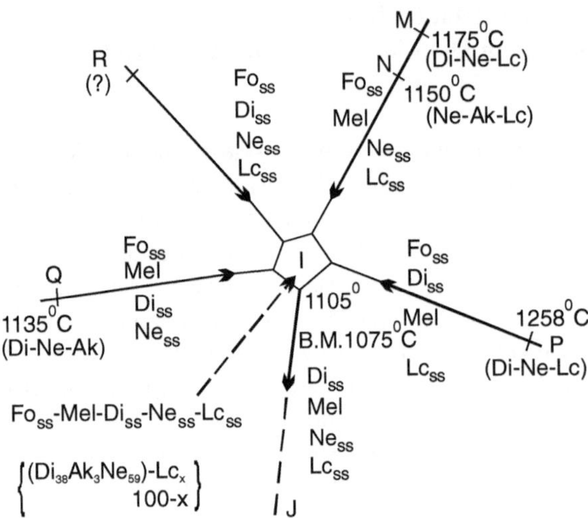

Fig. 10.12 Flowsheet diagram of the system diopside–nepheline–akermanite–leucite (after Gupta et al. 1973)

crystallization, the five phase assemblage of forsterite$_{ss}$ + diopside$_{ss}$ + leucite$_{ss}$ + melilite + liquid is reached at 1,258 ± 5 °C. The above assemblage is represented by point P of Fig. 10.12. At B (Fig. 10.7), forsterite$_{ss}$, nepheline$_{ss}$ and leucite$_{ss}$ coexist with liquid, but as the temperature drops melilite appears at 1,175 °C. The assemblage of forsterite$_{ss}$ + nepheline$_{ss}$ + leucite$_{ss}$ + melilite + liquid, is shown by point M (Fig. 10.12). In the join nepheline–akermanite–leucite (Fig. 10.10), nepheline$_{ss}$, melilite, leucite$_{ss}$ and liquid are in equilibrium at point S at 1,170 °C. As crystallization, continues the same assemblage of point M is reached at 1,150 °C (N, Fig. 10.12). Onuma and Yagi (1967) found that in the join diopside–nepheline–akermanite, the five phase assemblage of diopside$_{ss}$ + nepheline$_{ss}$ + forsterite$_{ss}$ + melilite + liquid is reached at 1,135 °C, which is represented by point Q (Fig. 10.12). The six phase invariant assemblage at I implies the possible existence of a five-phase piercing point within the system nepheline–kalsilite–CaO–MgO–SiO_2, where diopside$_{ss}$, nepheline$_{ss}$, leucite$_{ss}$, forsterite$_{ss}$ and liquid, coexist. The join, which contains this piercing point is unknown. The temperature of the six-phase assemblage at I is obtained from the pseudobinary system. After the disappearance of forsterite$_{ss}$, liquid moves from this point towards an invariant point of unknown composition. The disappearance of forsterite$_{ss}$ takes place over a temperature range of 30 °C (Fig. 10.12). A rock nomenclature diagram corresponding to Fig. 10.12 is presented in Fig. 10.13.

Paragenetic relationship based on the field occurrences of potassium-rich mafic and ultramafic lavas is not clear. However, on the basis of the phase equilibria studies on the join diopside–nepheline–akermanite–leucite, some suggestions on the genetic relationship between various rock types can be made.

Holmes (1950) described close association of such rock types as katungite (1), olivine–melilite leucitite (2), melilite leucitite (3), leucite ankaratrite (4) and

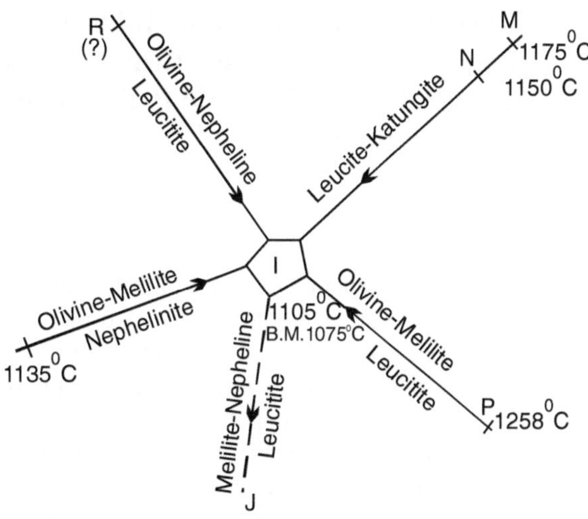

Fig. 10.13 Rock nomenclature diagram corresponding to Fig. 10.12 (after Gupta et al. 1973)

kalsilite-rich ultramafic rocks (5) from the Toro-Ankole field of East Africa. Rock types (1) and (2) may be represented by points M and P, respectively, of Figs. 10.12 and 10.13, and melilitite corresponds to point B (Gupta 1972; Fig. 10.1). The join diopside–nepheline–leucite (Fig. 10.8) of Gupta and Lidiak (1973) shows that the assemblage corresponding to nepheline–leucite katungite (Fig. 10.10) can be obtained from an olivine–nepheline italite. Holmes and Harwood (1937) found katungite in the region between Bushwaga and Goma of the Birunga area of Uganda. The assemblage of P (olivine–melilite leucitite) is obtained by crystallization of a liquid from A (Fig. 10.8). Rock types corresponding to olivine leucitite sometimes containing melilite have been reported from more than a dozen localities by Holmes and Harwood (1937; pp 81–82). Important localities include the Lutale flow, south-southwest of Lutale ridge and the Mikeno and Kisi areas (north-east corner of Lake Kivu). Olivine leucitite lavas with high concentration of olivine (ugandite) have been reported by Holmes and Harwood (1937; pp. 60–72) from various areas. Other localities include Muganza, Murehi, Musonga, and Katarara. The lava flows, corresponding to point R (olivine–nepheline–leucitite, Fig. 10.13), have been described by the same authors (p. 79) from the following areas of the same volcanic field of Uganda; Mabungo. Duzakara, Bunagana, and Hyamichunchu. They also described lava flows of melilite–nepheline leucitite (IJ, Figs. 10.12 and 10.13) from Goma (p. 83), Melilite–nepheline leucitite (Fig. 10.13) can be obtained from olivine–melilite–nepheline leucitite (I), which has been reported from the Fort Portal area of Toro-Ankole by Holmes and Harwood (1932, p. 379), which belongs to the same petrographic province of the Birunga volcanic field. Melilite–nepheline leucitite has also been reported from the Villa Senni area of Italy by Washington (1906) and Mt. Vulture (Melluso et al. 1996), the rarity of this rock type compared to other rock types represented by the flow sheet diagram (Fig. 10.12) is probably related to incomplete crystallization of the parent magma.

10.3 Experimental Study of the Joins Forsterite—Diopside—Leucite and Forsterite—Leucite—Akermanite up to 2.3 GPa [$P(H_2O) = P(\text{Total})$] and Variable Temperatures

10.3.1 Introduction

The four essential minerals of many high-K mafic and ultramafic igneous (kamafugitic) rocks are leucite, augite, olivine, and melilite. The bulk rock compositions of these rocks lie within the phase volumes of forsterite (Fo)–diopside (Di)–leucite (Lc)–akermanite (Ak), and melting phase relationships of this system is thus applicable to the understanding of the origin of high-K mafic and ultramafic igneous rocks (Yoder 1986). The phase volume of Fo–Di–Lc–Ak is a part of the larnite (La)–forsterite–kalsilite–silica tetrahedron (Fig. 10.14). Phase equilibria in this larnite-based basaltic tetrahedron have been studied for last several decades at atmospheric pressure (Schairer and Bowen 1938; Luth 1967; Gupta 1972; Yoder 1986; Veksler et al. 1998).

Phlogopite is the most abundant hydrous liquidus mineral in high-K mafic and ultramafic igneous rocks and much attention has been paid to the melting phase relationships of phlogopite-bearing mafic and ultramafic igneous rocks (ex. Barton and Hamilton 1979; Arima and Edgar 1983a, b; Foley et al. 1987; Foley 1992;

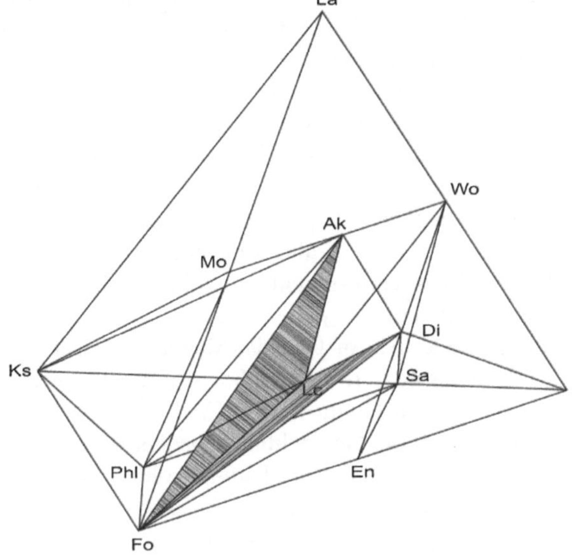

Fig. 10.14 The joins Fo–Di–Lc and Fo–Lc–Ak as a part of the La–Fo–Ks–Qz tetrahedron (after Yoder 1986). Abbreviations are given in text

Gupta et al. 2002; Conceicaoção and Green 2004). The high pressure melting experimental studies have been conducted in the join forsterite–kalsilite–quartz at different *P*–*T* conditions and in the presence of several volatile components (H_2O, CO_2, F) (Modreski and Boettcher 1972; Wendlandt and Eggler 1980a, b; Wendlandt 1984: Foley et al. 1986; Gupta and Green 1988; Melzer and Foley 1996). Recently, Conceicaoção and Green (2000) added silica to the forsterite–leucite join to elucidate the genesis of K-rich silica-saturated magmas from lherzolitic source rocks.

Nag et al. (2007) added diopside or åkermanite to the forsterite–leucite join to understand the heteromorphic relationship between the K-rich mafic and ultramafic volcanic rocks with lamprophyric or lamproitic melts originated in the upper mantle. In either of these two joins, leucite is expected to break down at high pressures and temperatures by the following reaction (Scarf et al. 1966; Fasshauer et al. 1998);

$$2\text{Lc} = \text{K-feldspar (Kf)} + \text{kalsilite (Ks)} \qquad (10.1)$$

In the cases of both the joins Fo–Di–Lc and Fo–Lc–Ak under a condition of 2.3 GPa [$P(H_2O) = P(\text{Total})$], phlogopite is an additional phase and forsterite reacts with leucite to produce the following assemblage typical of minette (Yoder 1986):

$$3\text{Fo} + 2\text{Lc} + 2\text{H}_2\text{O} = 2\text{Ph} + \text{Kf} \qquad (10.2)$$

The above discussion suggests that in the join Fo–Di–Lc at elevated pressures and temperatures, various combinations of phlogopite$_{ss}$, diopside$_{ss}$, kalsilite$_{ss}$, forsterite$_{ss}$ and K-feldspar$_{ss}$ may appear. Such assemblages are typical of lamproitic and lamprophyric rocks (Gupta and Yagi 1980; Foley et al. 1987; Rock 1991; Mitchell and Bergman 1991). In case of the join Fo–Lc–Ak following reactions are also expected under high pressures (Yoder 1973, 1986):

$$2\text{Ak} + \text{Fo} + 3\text{Lc} = 4\,\text{clinopyroxene (Cpx)} + 3\text{Ks} \qquad (10.3)$$

$$2\text{Ak} = \text{Di} + \text{merwinite (Mer)} \qquad (10.4)$$

$$\text{Ak} + 5\text{Fo} + 3\text{Lc} + 3\text{H}_2\text{O} = 3\text{Ph} + \text{Cpx} \qquad (10.5)$$

In the join Fo–Lc–Ak, reactions (10.2), (10.3), (10.4) and (10.5) lead to various assemblages found in lamprophyres and lamproitic rocks, and melting relationships in the join Fo–Lc–Ak have therefore, significant bearing on their genesis.

10.3.2 *The Join Forsterite–Diopside–Leucite at 0.1 GPa [P(H_2O) = P(Total)]*

Gupta (1972) studied this join under atmospheric pressure (Fig. 10.2), but the same join was studied by Nag et al. (2007) under 0.1 GPa [$P(H_2O) = P(\text{Total})$] at variable temperatures to learn about the effect of H_2O on phase relationships. These experimental results are summarised in Fig. 10.15. The pseudo-eutectic point for the join Fo–Lc in presence of excess water has been estimated to be $Fo_{17}Lc_{83}$ ('B' in Fig. 10.15) from Luth (1967), who studied the system Mg_2SiO_4–$KAlSiO_4$–SiO_2–H_2O at 0.1 Gpa [$P(H_2O) = P(\text{Total})$.

The join Fo–Lc in presence of excess water has been estimated to be $Fo_{17}Lc_{83}$ ('B' in Fig. 10.15) from Luth (1967), who studied the system Mg_2SiO_4–$KAlSiO_4$–SiO_2–H_2O at 0.1 Gpa [$P(H_2O) = P(\text{Total})$]. The pseudoeutectic point for the join Fo–Di was estimated to be at $Fo_{16.5}Di_{83.5}$(C, by Nag et al. 2007). From the study of the starting materials within the Fo–Di–Lc join, the pseudo-eutectic point ('D') for the Di–Lc join was determined by Nag et al. (2007) and was found to occur at $Di_{49.5}Lc_{50.5}$ at 885 ± 5 °C. The five-phase point for the join Fo–Di–Lc occurs at 'A' ($Fo_3Di_{50}Lc_{47}$) and 880 ± 5 °C, where forsterite$_{ss}$, Diopside$_{ss}$ and leucite$_{ss}$ are in equilibrium with liquid and vapour.

With reference to the study of the same join under 1 atmospheric pressure by Gupta (1972) (Fig.10.2), it is observed that in the presence of water the coexistence of $Fo_{ss} + Di_{ss} + Lc_{ss}$ + Liq occurs at 880 °C (Fig. 10.15) which contrasts the presence of the same assemblage at 1,296 ± 5 °C. (Fig. 10.2) under one atmospheric pressure. The data show a drop of about 416 °C in the temperature of coexistence for these four phases. In presence of excess water the five-phase Point ($Fo_{ss} + Di_{ss} + Lc_{ss}$ + Liq + V) moves closer towards leucite, since such mafic phases

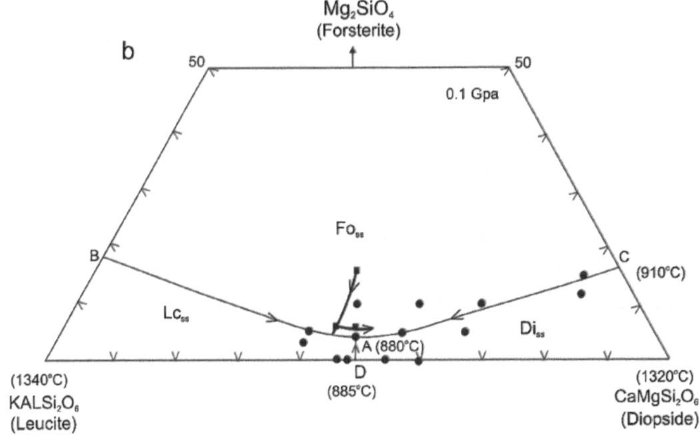

Fig. 10.15 Phase relations in the join Fo–Di–Lc studied under 0.1 GPa [$P(H_2O) = P(\text{Total})$] (after Nag et al. 2007). The bold line with squares is a projected path of the liquid as determined from the composition of glasses in equilibrium with crystalline phases

as forsterite and diopside remains relatively insoluble compared to leucite. The shift in the composition of the five-phase point towards the feldspathoid end compared to the mafic end members, has also been observed by Yoder and Upton (1971). Another interesting feature of this study is the occurrence of phlogopite_{ss} at 850 °C or below with reference to the study of Luth (1967) on the join Fo–Lc at 0.1 Gpa [*P* (H_2O) = *P*(Total)], the presence of phlogopite along with forsterite and leucite agrees well with the present study.

On the basis of composition of the glasses in equilibrium with different phases, the projected path of the liquid in this join is shown in Fig. 10.15. The projected line of descent (a bold line with solid squares) is slightly off the phase boundary. Although a liquid and a vapour phase were found to coexist with crystalline phases in the quenching run products, the vapour phase together with the liquid probably existed in the form of a fluid phase during run. Because of the appearance of phlogopite and presence of vapour (or fluid), the join may be treated as a part of a quinary system, $KAlSiO_4$–CaO–MgO–SiO_2–H_2O.

10.3.3 The Join Forsterite–Diopside–Leucite Studied under 2.3 GPa and Variable Temperatures

Seventy-two experimental runs were made by Nag et al. (2007) to study this join under a condition of [*P*(H_2O) = *P*(Total)] at variable temperatures. The results are summarised in Fig. 10.16a,

Which shows that the join penetrates the phase volumes of forsterite_{ss}, phlogopite_{ss} diopside_{ss}, K-feldspar_{ss} and kalsilite. There are three five-phase points (1) 'B' ($Fo_9Di_{49}Lc_{42}$ and 1,005 ± 10° C), where forsterite_{ss} diopside_{ss} and phlogopite_{ss} coexist with liquid and vapour, (2) 'C' ($Fo_4Di_{50}Lc_{46}$ + 990 ± 10 °C), where phlogopite_{ss}, diopside_{ss} and K-feldspar_{ss} are in equilibrium with liquid and vapour, and (3) 'D' ($Fo_3Di_{21}Lc_{76}$, 775 ± 5 °C), where phlogopite_{ss}, kalsilite_{ss} and K-feldspar_{ss} are in equilibrium with liquid and vapour. The presence of K-feldspar_{ss} and kalsilite are related to breakdown of leucite under high *P–T* conditions (Fashauer et al. 1998; reaction (10.1)). Formation of phlogopite_{ss} in this join is related to the reaction involving forsterite, leucite and vapour to produce phlogopite_{ss} and K-feldspar_{ss} (reaction (10.2)) (Yoder 1986).

In case of the Fo–Di–Lc join studied at one atmosphere, the piercing point 'I' (Fo_{SS} + Di_{SS} + Lc_{SS} + Liq), occurs at $Fo_2Di_{67}Lc_{31}$ and 1,296 °C (Gupta 1972) (Fig. 10.2). However, in their study conducted at 2.3 GPa [*P*(H_2O) = *P*(Total)] by Nag et al. 2007, Fig. 10.16a, all the three five-phase points occurring at 'B' (1,005 ° C), 'C' (990 °C) and 'D' (775 °C) have liquidus temperatures significantly lower than that of piercing pint 'C' (Fig. 10.2), because the present investigation has been conducted in the presence of excess water [*P*(H_2O) = *P*(total)]. Compared to the phase equilibrium study conducted at one atmospheric pressure (Fig. 10.2) and 0.1 GPa (Fig. 10.15), the study of the same join at 2.3 GPa shows significant

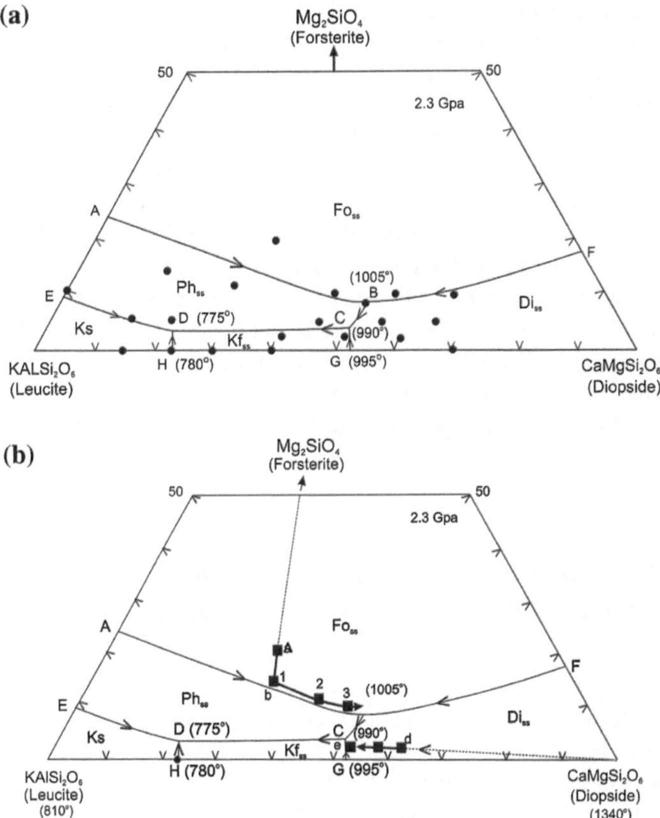

Fig. 10.16 **a** Phase relations in the join Fo–Di–Lc studied under 2.3 GPa [$P(H_2O) = P(\text{Total})$] (Nag et al. 2007). **b** Course of crystallization of liquid in the join Fo–Di–Lc studied under 2.3 GPa [$P(H_2O) = P(\text{Total})$] (after Nag et al. 2007). The *bold line with squares* is a projected path of the liquid as determined from the composition of glasses in equilibrium with crystalline phases

difference (Fig. 10.16). At 2.3 GPa phlogopite$_{ss}$, kalsilite, and K-feldspar$_{ss}$ are additional phases and leucite$_{ss}$ is completely absent. Microprobe analyses of glasses in the join Fo–Di–Lc at 2.3 GPa [$P(H_2O) = P(\text{Total})$] and variable temperatures is projected on to this join from the H_2O apex (bold line with solid squares) (Fig. 10.16b). Although projected paths of the melts are slightly off the phase boundaries, nevertheless the crystallization trends are consistent with the observed sequence of crystallization.

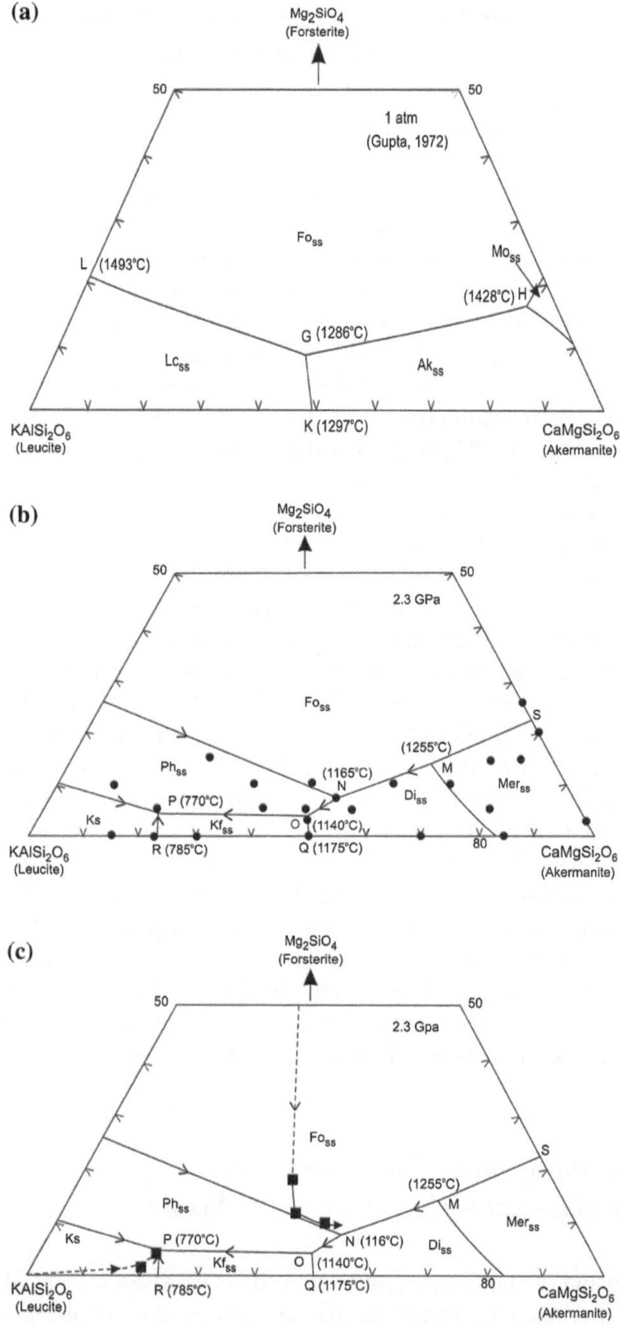

Fig. 10.17 **a** Phase relations in the join Fo–Lc–Ak at one atmospheric pressure (after Gupta 1972, in set). **b** Phase relation in the join Fo–Lc–Ak at 2.3 GPa [$P(H_2O) = P(\text{Total})$]. **c** Course of crystallization of melt in the join Fo–Lc–Ak at 2.3 GPa [$P(H_2O) = P(\text{Total})$] (Nag et al. 2007). The *bold line with squares* is a projected path of the liquid as determined from the composition of glasses in equilibrium with crystalline phases

10.3.4 The Join Forsterite–Leucite–Akermanite Studied at 2.3 GPa and Variable Temperatures

The phase relations in the join Fo–Lc–Ak were studied at 1 atmospheric pressure by Gupta (1972) (Fig. 10.17a). Nag et al. (2007), made 68 runs to investigate the join (Fig. 10.17b, c) which shows that the join intersects the primary phase volumes of forsterite$_{ss}$, phlogopite$_{ss}$, kalsilite, K-feldspar$_{ss}$, diopside$_{ss}$, and merwinite$_{ss}$. There are 4 five-phase points occurring at following compositions and temperatures: (1) 'N' ($Fo_7Lc_{42}Ak_{51}$ and 1,165 ± 10 °C), where phlogopite$_{ss}$, forsterite$_{ss}$- and diopside$_{ss}$ coexists with liquid and vapour, (2) 'O' ($Fo_3Lc_{49}Ak_{48}$ and 1,140 ± 10° C), where a liquid is in equilibrium with phlogopite$_{ss}$, K-feldspar$_{ss}$, diopside$_{ss}$ and vapour, (3) 'M' ($Fo_{13}Lc_{21}Ak_{66}$ and 1,255 ± 10 °C), where merwinite $_{ss}$ forsterite$_{ss}$ and diopside$_{ss}$ are in equilibrium with liquid and vapour, and (4) 'P' ($Fo_{5.5}Lc_{73.5}Ak_{21}$, and 775 ± 10 °C), where kalsilite, phlogopite$_{ss}$ and K-feldspar$_{ss}$ coexist with liquid and vapour (Fig. 10.17b).

When the phase relationships in the join Fo–Lc–Ak at one atmospheric pressure (Gupta 1972 (Fig. 10.17a) is compared with those at 2.3 GPa (Fig. 10.17a, b), it may be noted that there is a field of monticellite$_{ss}$ occurring near the join Fo–Ak. In contrast, Fig. 10.17b, c in set shows the presence of a field of merwinite$_{ss}$ as suggested by Yoder (1973). The piercing point 'G' (Fo$_{ss}$ + Ak$_{SS}$ +Lc$_{ss}$ + Liq) observed in this join at one at atmospheric pressure (Gupta 1972) occurs at 1,286 °C, whereas the temperatures of the five-phase points 'N' (1,165 °C), 'O' (1,140 °C), 'M' (1,255 °C), and 'P' (775 °C) are significantly lower, because the present study at 2.3 GPa was conducted under a condition of 2.3 GPa [$P(H_2O) = P(\text{Total})$]. The results are summarised in Fig. 10.17b. The comparison of the tow figures (Fig. 10.17a, b) shows that in case of the 2.3 GPa isobaric diagram there are such additional phased as phlogopite$_{ss}$, K-feldspar$_{ss}$, kalsilite and merwinite$_{ss}$, which are absent in the one atmospheric phase diagram of the same join (Figs. 10.3 and 10.17a). On the basis of microprobe analyses of liquid equilibrium with vapour phases, the trends of liquid composition (bold lines with solid squares Fig. 10.17c) are projected on to the Fo–Lc–Ak join from the H_2O apex (Fig. 10.17b). Although the trends of liquid compositions do not coincide with the exact phase boundaries, the course of crystallization is consistent with the observed sequence of crystallization.

10.3.5 The Paragenetic Sequence in the Kalsilite–CaO–MgO–SiO$_2$–H$_2$O System

If a small amount of alumina present in diopside as $CaAl_2SiO_6$ molecule is ignored, the system may then be treated as quinary having the following components, kalsilite–CaO–MgO–SiO_2–H_2O. It is further assumed that the vapour phase observed in the glasses should combine under elevated *P–T* condition as a fluid phase (f) and the remaining water is dissolved in the crystalline phases. Because of

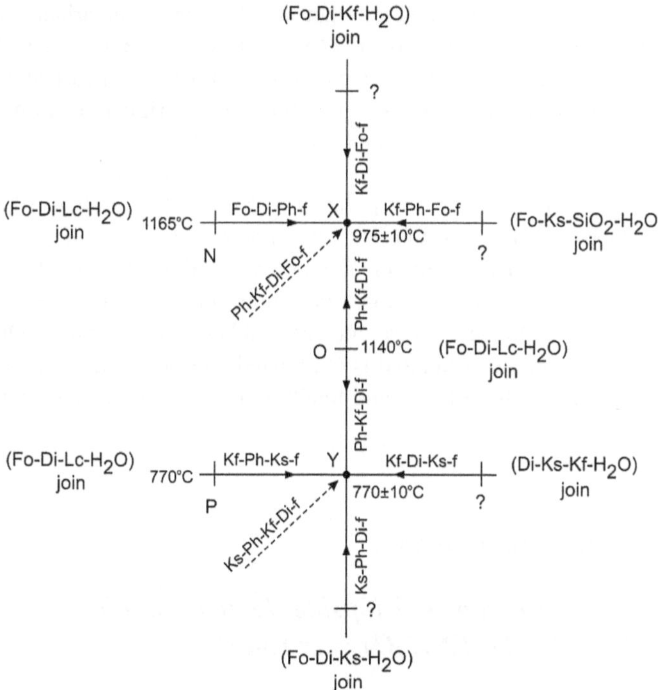

Fig. 10.18 The flow-sheet diagram showing paragenetic relationship of various assemblages in the join Fo–Di–Lc–H_2O studied under 2.3 GPa variable temperatures. In drawing the flow-sheet diagram it has been assumed that liquid + vapour observed in the products were present as fluid elevated *P*–*T* condition. It should be pointed out that the flow sheet diagram for the join forsterite–leucite–åkermanite–H_2O should be…this diagram. There is some difference in that the univariant line (Fo + Di + Ph + f) in the Fo–Lc–Ak–H_2O join is derived from the invariable (Fo + Mer + Di + f). As merwinite reacts with the liquid and is eliminated, the melt composition follows a divariant surface (Fo + Di + f) an moves along the univariant line (Fo + Ph + Di + f) and finally moves towards the point 'X' (Ph + Kf + Di + Fo + f) (after Nag et al. 2007)

this and presence of some components as solid solution in different phases, the system in effect behaves as a four-component system. On these assumptions, a flow-sheet diagram is drawn, as shown in Fig. 10.18. This helps us to draw the paragenetic sequence in the four-component system (Fig. 10.16b, c) and also near the points 'N' and 'O' (Fig. 10.17b, c). At the invariant point 'X' (975 ± 10 °C, Fig. 10.18), the number of phases is five (Fo_{ss} + Di_{ss} + Ph_{ss} + Kf_{ss} + f) as expected for a four-component system. Following the Scheinemaker's rule there should be four univariant lines leading to the five phase invariant point 'X'. The four univariant lines should be Fo_{ss}–Di_{ss}–Ph_{ss}–f, Kf_{ss}–Di_{ss}–Fo_{ss}–f, Kf_{ss}–Ph_{ss}–Fo_{ss}–f and Ph_{ss}–Kf_{ss}–Di_{ss}–f.

Likewise, the five-phase invariant assemblage near the point 'D' (Fig. 10.16) comprises Ph_{ss}, Ks_{ss}, Kf_{ss}, Di_{ss} and f, requires that there should be four univariant lines namely: (1) Kf_{ss}–Ph_{ss}–Ks_{ss}–f, (2) Ph_{ss}–Kf_{ss}–Di_{ss}–f, (3) Kf_{ss}–Di_{ss}–Ks_{ss}–f, and (4) Ks_{ss}–Ph_{ss}–Di_{ss}–f. It is noted that the five-phase invariant assemblage (Ks_{ss}–Ph_{ss}–Kf_{ss}–Di_{ss}–f) is obtained at 770 ± 10 °C. Since the four-phase univariant assemblage (Ph_{ss}–Kf_{ss}–Di_{ss}–f) common to both the invariant points 'X' and 'Y' (Fig. 10.18), there has to be a thermal barrier originating from the point "O" (1,140 ± 10 °C). This univariant assemblage Ph_{ss}–Fo_{ss}–Di_{ss}–f also originates cat point 'N' (1,165 ± 10 °C). Likewise, the univariant assemblage (Kf_{ss}–Ph_{ss}–Ks_{ss}–f) originates at the point "P" (775 ± 10 °C) (Fig. 10.17b).

The paragenetic sequence of different simplified rock types in the join Fo–Lc–ak is similar to the flow sheet diagram as shown in (Fig. 10.18). There is some difference, however, in that the univariant line (Fo_{ss}–Di_{ss}–Ph_{ss}–f) in the join Fo–Lc–Ak–H_2O, is derived from the assemblage (Fo–$_{ss}$–Mer_{ss}–Di_{ss}–f). As merwinite$_{ss}$ reacts with the liquid and is eliminated, the melt composition follows a divariant surface (Fo_{ss}–Di_{ss}–Ph_{ss}–f) and finally to the point (Fo_{ss}–Di_{ss}–Ph_{ss}–Kf_{ss}–f) (Fig. 10.18).

10.4 Petrological Significance

10.4.1 The Join Forsterite–Diopside–Leucite Studied under 0.1 GPa [P(H_2O) = P(Total)]

With reference to Fig. 10.15, it should be stated that the five-phase point A corresponds to an olivine leucitite termed as ugandite by Holmes (1942). Figure 10.15 further suggests that ugandite can be derived from an olivine italite (along B–A), a leucitite (along D–A) and a K-rich olivine–pyroxene rich magma (along C–A). With lowering of temperature, phlogopite$_{ss}$ precipitates along with olivine, leucite$_{ss}$ and pyroxene$_{ss}$ at 850 ± 10 °C. This phlogopite-bearing assemblage corresponds to a rock called missourite or kajanite [equivalent to mica-rich ugandite by Lacroix 1926]. Holmes (1942, 1950) recognised that ugandite is one of the primitive magmas in the Western branch of the African Rift. Cross, (1897) coined the term wyomingites to a rock composed of leucite, olivine, augite and phlogopite (also see Yagi and Matsumoto 1966; Carmichael 1967). The assemblage (Ph_{ss} + Fo_{ss} + Lc_{ss} + Di_{ss} + Liq) at 850 ± 10 °C is comparable to woymingite. Barton and Hamilton (1979), studied melting phase relationships of a natural)madupite composition below 0.4 GPa. At temperatures below 1,150 °C, they obtained the assemblage (Cpx + Ph + Lc + Liq). This assemblage is similar to that obtained near point 'A' and 800 ± 10 °C.

In various localities of New South Wales such as Lake Cargello, Begargo Hills, Flag staff Hills, Bygoloric, and Condobline, there are olivine leucitites containing phlogopite. Such rocks called missourite are also reported from Gaussberg, Antarctica (Sheraton and Cundari 1980) and from Mount Jugo and Colli-Albani volcanic fields of Italy (Cundari 1973). They however, described these rocks simply as leucitite ignoring the presence of olivine and phlogopite altogether.

10.4.2 *The Join Forsterite–Diopside–Leucite Studied under 2.3 GPa at Variable Temperatures*

With reference to Fig. 10.16b, the experimental study suggests that from a liquid of composition 'A' ($Fo_{20}Di_{30}Lc_{50}$) forsterite$_{ss}$ crystallises first near 'b' (melt composition '1', Fig. 10.16), phlogopite$_{ss}$ starts to precipitate in addition to forsterite$_{ss}$. The composition of liquid then moves long a multi-saturation line 'b-B' to 'B' (near melt composition '3'), where diopside$_{ss}$ co-precipitates with forsterite$_{ss}$ and phlogopite$_{ss.}$ With the lowering of temperature further, K-feldspar$_{ss}$ appears at 985 ± 10 °C. Thus, the final assemblage comprises forsterite$_{ss}$ + phlogopite$_{ss}$ + diopside$_{ss}$ + K-feldspar$_{ss}$ and liquid. The assemblage along 'b-B' corresponds to a madupite, a type locality of which is Pilot Butte at Leucite Hills. The assemblage forsterite$_{ss}$ + phlogopite$_{ss}$ +clinopyroxene corresponds to missourite or kazanite. The final crystalline product comprising phlogopite$_{ss}$, diopside$_{ss}$, forsterie$_{ss}$ and K-feldspar$_{ss}$ similar to a phenocrystal assemblage of olivine–pyroxene minette.

Makhotkin et al. (1989) described lamproites comprising olivine, phlogopite and diopside from Khani (57°54′ N, 120°46′ E) and Ryabinovaya (57°40′ N, 125° 51′ E) of Aldan alkaline province of Russia. This assemblage corresponds to that of the five-phase point 'B' (Fig. 10.16b). The lamproite of Rayabinovaya is associated with a phlogopite-bearing minette. There is also a brecciated pipe (600–700 m long) at Kaila of the same petrographic province of Aldan. The pipe is constituted of olivine, diopside and phlogopite phenocrysts with clinopyroxene, phlogopite and K-feldspar occurring in the groundmass, as is the near solidus assemblage (point 'X' in Fig. 10.18). Similarly, O'Brien et al. (1991) described olivine minettes from Highwood)Mountains, Montana, USA, which have the seriate porphyritic texture with phenocrysts of phlogopite, clinopyroxene and forsteritic olivine ($Fo_{87–92}$) in a matrix of sanidine, clinopyroxene, phlogopite and glass.

From a liquid of composition ($Fo_3Di_{60}Lc_{37}$) (Fig. 10.16b), diopside$_{ss}$ precipitates first at 1,010 °C, followed by K-feldspar as the second phase at 'e', then phlogopite appears at 990 °C (at point 'C') as the third phase. The assemblage diopside$_{ss}$ + liquid corresponds to K-rich pyroxenite, whereas the assemblage at 'e-C' (Di$_{ss}$ +Kf$_{ss}$ + Liq) correrponds to pyroxene-bearing syenite, and the assemblage at 'C' (Di$_{ss}$ + Kf$_{ss}$ + Ph$_{ss}$ + Liq) corresponds to pyroxene-bearing minette. With further lowering of temperature, the final assemblage comprises diopside$_{ss}$ K-feldspar$_{ss}$, phlogopite$_{ss}$, kalsilite$_{ss}$, and liquid. This assemblage corresponds to kalsilite–pyroxene minette. If a kalsilite–pyroxene minette magma ascends upward quickly, leucite should appear at shallower depth before K-feldspar$_{ss}$ and kalsilite$_{ss}$ completely disappear (Fig. 10.16a). Bhowmik (2000) described diopsidite veinlets in association with rocks comprising K-feldpsar, kalsilite, phlogopite, leucite, and diopsidic pyroxene from the ductile shear zones of Borra near Visakapatnam, south India. Such Kalsilite–Kfeldspar-bearing rocks have also been reported from the Punalur locality of southern India (Sandifored and Santosh 1991). This assemblage is comparable to the end product, crystallizing from a liquid composition ‘D’ in the Fo–Di–Lc–H_2O join) (Fig. 10.16b).

MacDonald et al. (1992) described minettes from Bearpaw Mountains, which are characterised by the presence of olivine and clinopyroxene phenocrysts in a groundmass of phlogopite and sanidine. This assemblage is comparable to the five-phase point ‘C’ (Ph_{ss} + Kf_{ss} + Di_{ss} + Liq + V) (Fig. 10.16b) and corresponds to pyroxene-bearing minette. As the composition closed to ‘C’ contains only 3 % forsterite, olivine is consumed in the formation of phlogopite and K-feldspar. See the reaction below:

$$3\text{Fo} + 3\ \text{Lc} = 2\text{Ph} + \text{Kf}$$

The assemblage at point ‘D’ (Fig. 10.16b) is characterised by the assemblage kalsilite$_{ss}$ K-feldspar$_{ss}$ phlogopite, and liquid. This assemblage is equivalent to kalsilite-bearing minette. Orlova et al. (1986) described phlogopite-bearing yakutite (phlogopite + K-feldspar + clinopyroxene + kalsilite) from Malomurunskii ultra-potassic complex. This assemblage is noted in a stratified lopolith. The complex has a lower layer comprising phlogopite and pyroxene at the bottom followed upward by a layer with the assemblage consisting of kalsilite, K-feldspar and clinopyroxene. Thus a Kalsilite-bearing minette magma, having an assemblage similar to the five-phase point ‘C’ cooled slowly and the phlogopite and pyroxene-rich fraction settled at the bottom leaving behind a kalsilite–clinopyroxene–K-feldspar-bearing residue (kalsilite syenite) observed in that locality.

Gupta and Green (1988) studied the system kalsilite–forsterite–silica at 2.8 GPa [$P(H_2O) = P(\text{Total})$]. They also observed the pressure of a five-phase assemblage of Ph_{ss} + Ks_{ss} + San_{ss} + L + V similar to Figs 10.16b and 10.18. Foley et at. (1986) added 4 wt% F to see the effect of F on the topology of Fo–Ks–SiO_2 phase diagram. They observed that their phase diagram broadly resembled the water saturated F-free system of Gupta and Green (1988) in having a large primary phase volumes of phlogopite, forsterite, kalsilite and sanidine. Foley et al. (1986) found that F-bearing system differs in that the fluorphlogopite has a much greater thermal stability (maximum 1,490–1,500 °C) than hydroxyphlogopite (<1,200 °C; Gupta and Green 1988).

10.4.3 The Join Forsterite–Leucite–Akermanite Studied at 2.3 GPa and Variable Temperatures

Phase relations in the leucite-rich section of the join Fo–Lc–Ak (Fig. 10.17b.) is very similar to those of the join Fo–Di–Lc studied at 2.3 GPa (Fig. 10.16b) in that there are primary phase fields of forsterite$_{ss}$, phlogopite$_{ss}$, K-feldspar$_{ss}$, kalsilite$_{ss}$ and diopside$_{ss}$ in the åkermanite-poor region of the join. Comparing Figs. 10.16b and 10.17b, it may be stated that the paragenetic relationship in the leucite-rich but åkermanite-poor (<60 wt% åkermanite) section of the join is very similar to that already described in the previous section. With reference to Fig. 10.17b, it may be observed that the assemblage at 'N' ($Fo_7Lc_{40}Ak_{53}$; 1,165 ± 10 °C) corresponds to olivine-bearing madupite. Like phase relations in the Fig. 10.16b, with the lowering of temperature at point 'N', the final assemblage near the solids is constituted of $forsterite_{ss}$, $phlogopite_{ss}$, $diopside_{ss}$, K-$feldspar_{ss}$ liquid, and vapour (olivine–pyroxene-bearing minette). Lloyd et al. (1985) investigated melting phase relationship of an olivine melilite glass under variable *P–T* conditions, and noted that the above-mentioned assemblage is similar to that obtained near the solidus close to point 'N' (Fig. 10.17b). If such a liquid coexisting with $forsterite_{ss}$, $phlogopite_{ss}$, clinopyroxene and K-$feldspar_{ss}$ ascends near the surface, it would result in an assemblage comprising $forsterite_{ss}$ + $åkermanite_{ss}$ + $leucite_{ss}$ + liquid (katungite).

With reference to the study of the join Fo–Lc–Ak at 2.3 GPa, it may be observed that Kalsilite-bearing minette, pyroxene-bearing minette and olivine–pyroxene-bearing minette are heteromorphic high pressure equivalents of katungite. Likewise monticellite alnoite (Fig. 10.17a, åkermanite-rich side) is represented at high pressure by merwinite-bearing websterite.

Appearance of $merwinite_{ss}$ and $diopside_{ss}$ is related to the breakdown of åkermanite (Yoder 1973) because of reaction (4) described earlier. (Yoder 1973) showed that below 0.62 GPa and 750 °C åkermanite breaks down to monticellite + wollastonite. Yoder noted that formation of merwinite and diopside at the expense of åkermanite and forsterite is valid up to 1.05 GPa and 1,090 °C, and above this pressure diopside and merwinite may be the stable products at the expense of forsterite and åkermanite.

Arima and Edgar (1983a) studied melting relationships of a katungite composition at pressures up to 3.5 GPa and temperatures to 1,300 °C. They observed that below 2 GPa in the presence of excess water, olivine appeared as the liquids phase, but it disappeared at lower temperatures because of reaction with liquid. Their data indicate that clinopyroxene together with phlogopite has a wide *P–T* range of stability. Melilite appeared only below 1.0 GPa and 1,090 °C.

At high pressure (2.3 GPa) but low temperatures towards the leucite-rich side, instead of a breakdown product of åkermanite to form merwinite and diopside, åkermanite reacts with forsterite, leucite and vapour to produce phlogopite and diopside. Note the reaction:

$$5\,\mathrm{Fo} + 3\,\mathrm{Lc} + \mathrm{Ak} + 3\mathrm{H_2O} = 3\,\mathrm{PH} + 2\,\mathrm{Di}$$

Thus, at 2.3 GPa and low temperature towards the leucite-rich side, above reactions inhibits the presence of $merwinite_{ss}$, which disappears and $diopside_{ss}$ together with phlogopite precipitates as suggested by the above reaction. Merwinite has not been reported in the K-rich mafic rock, as it is eliminated at low temperatures because of the above reaction. Near the surface, equilibrium crystallization of a liquid lying in the Fo–Lc–Ak join has been described by Gupta (1972). He showed that a katungite magma (point 'G' in Fig. 10.17a), where liquid co-exists with forsterite$_{ss}$ + leucite$_{ss}$ + åkermanite$_{ss}$ +liq, can be derived from a monticellite alnoite, a melilite italite or an olivine italite. The assemblage at 2.3 GPa shows that katungite has heteromorphic relation to olivine-bearing madupite ('N'), minette ('O') or kalsilite minette P (Fig. 10.17b). At low pressure, the final assemblage might correspond to katungite. The univariant equilibrium involving reaction (2) (Ph = 3Fo + Lc + Ks + V) has been identified by Holmes (1942), who described lava flows containing clinopyroxene, kalsilite, olivine, melilite and leucite from the western limb of the African Rift valley system (also see Yoder 1986).

In summary the experimental data of Nag et al. demonstrated that high pressure crystal differentiation can produce various types of K-rich mafic and ultramafic igneous rocks. The data of Nag et al. suggest that olivine leucitite (ugandite) can be derived from an olivine italite. The paragenetic relationship was also inferred for phlogopite-bearing madupite, pyroxene-bearing minette and kalsilite-bearing minette. The data further suggest that katungite has heteromorphic reaction to olivine-bearing madupite, pyroxene-bearing minette and kalsilite-bearing minette. The phlogopite-bearing clinopyroxenite xenoliths commonly found in kamafugitic rocks (cf. Lloyd and Bailey 1975; Lloyd et al. 2002) would probably represent high pressure crystallization product derived from kamafugitic magmas.

Chapter 11
Phase Relations in the System Leucite-Akermanite-Albite-SiO_2

In certain ultrapotassic petrographic provinces such as Somma-Vesuvius (Savelli 1967; Cioni et al. 1995), Coli-Albani (Aurisicchio et al. 1988) and Vico, all localities from Italy (Cundari and Mattias 1974), Bufumbira region of East Africa (Holmes and Harwood 1937; Rogers et al. 1998) and Highwood Mountains, Montana, U.S.A. (O`Brien et al. 1991), leucite-bearing SiO_2-undersaturated rocks are associated with highly silicic, potassium-rich ($K_2O \geq 3$ wt%) feldspathic rocks.

Melilite and clinopyroxene-bearing leucitic rocks occur in the volcanic fields of Fort Portal and Birunga petrographic province of Uganda (Holmes 1950; Rogers et al. 1998). At Birunga, melilite- and leucite-bearing rocks are associated with more SiO_2-saturated potash trachybasalt, trachyandesite or latite series. These SiO_2-saturated varieties are apparently part of different volcanic cycles and contain alkali feldspar and clinopyroxene as essential minerals.

The system leucite-akermanite-SiO_2 was investigated to establish petrogenetic relationship between SiO_2-saturated and SiO_2-undersaturated leucite-bearing mafic rocks which crystallized under volcanic to subvolcanic conditions.

Albite has an incompatible relationship with leucite (Gupta and Edgar 1975) and akermanite (Yoder and Schairer 1969). Albite (25 and 40 wt%) was therefore, added to the leucite-akermanite-SiO_2 join (Fig. 11.1) to understand the reaction relationship inhibiting the coexistence of leucite-albite and akermanite-albite mineral pairs in more complex systems. Investigation of the system was carried out also to learn about the phase relationships involving these three minerals.

The system leucite-akermanite-albite-SiO_2 is bounded by the following joins:

1. akermanite-leucite,
2. leucite-SiO_2,
3. leucite-albite,
4. albite-akermanite,
5. albite-SiO_2 and
6. akermanite-SiO_2.

The join akermanite-leucite was studied by Gupta (1972), who found it to be pseudobinary, because diopside$_{ss}$ and wollastonite$_{ss}$ crystallize in addition to leucite$_{ss}$ and akermanite$_{ss}$ in the mixtures, Lc_5Ak_{95} and $Lc_{10}Ak_{90}$. The pseudoeutectic

A.K. Gupta, *Origin of Potassium-rich Silica-deficient Igneous Rocks*,
Springer Geology, DOI 10.1007/978-81-322-2083-1_11

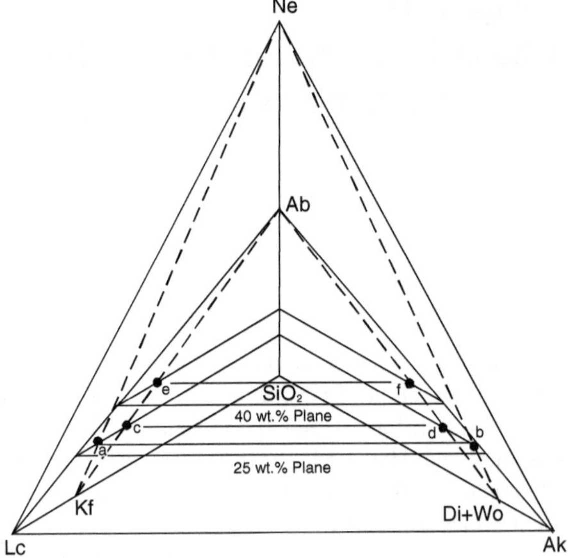

Fig. 11.1 The leucite-nepheline-akermanite-SiO_2 tetrahedron with 25 and 40 wt% albite plane (after Gupta and Gupta 1985, 1997)

point occurs at $Lc_{50.5}Ak_{49.5}$ and 1,297 ± 3 °C. The presence of liquid below 1,297 °C further confirms its pseudobinary nature.

The join leucite-SiO_2 was studied by Schairer and Bowen (1955), who found that in this join K-feldspar forms an intermediate compound. They further established that it melts incongruently to produce leucite$_{ss}$ and a SiO_2-saturated melt. The eutectic between K-feldspar and tridymite occurs at $Lc_{45.6}Q_{54.5}$ and 990 ± 20 °C.

The phase diagram of the leucite-albite join was obtained by Gupta and Edgar (1975), who found it to be pseudobinary as nepheline$_{ss}$, kalsilite$_{ss}$ and a liquid phase appear in addition to leucite$_{ss}$ and feldspar below the pseudoeutectic point ($Lc_{41}Ab_{59}$, 1,068 ± 5 °C).

The join albite-akermanite was studied by Yoder and Schairer (1969). It is pseudobinary and may be considered as a binary join of the system $NaAlSiO_4$–CaO–MgO–SiO_2. Diopside$_{ss}$, wollasto-nite$_{ss}$ and nepheline$_{ss}$ appear as additional phases with melilite or Na-feldspar. It has two piercing points:

(1) One at $Ab_{91}Ak_9$ and 1,138 °C, and
(2) The other at $Ab_{33}Ak_{67}$ and 1,274 °C.

The join albite-SiO_2 was studied by Schairer and Bowen (1935). This join is a simple binary with a eutectic point at $Ab_{32}Q_{68}$ and 1,062 ± 3 °C, where Na-feldspar and tridymite coexist with liquid.

The akermanite-SiO_2 join was investigated by Gupta and Gupta (1985), who found it to be pseudobinary with two piercing points: (1) one at $Ak_{87}Q_{13}$ and 1,355 °C, and the other at $Ak_{72}Q_{28}$ and 1,320 °C. The assemblage of diopside$_{ss}$-tridymite-liquid occurs at $Ak_{68}Q_{32}$ and 1,325 °C. The pseudobinary nature was confirmed by the presence of diopside$_{ss}$ and wollastonite$_{ss}$ in addition to akermanite$_{ss}$ in the SiO_2-undersaturated side, and tridymite in the SiO_2-saturated portion. Akermanite$_{ss}$ and polymorphs of SiO_2 do not coexist in equilibrium. It is therefore, a join of the system CaO–MgO–SiO_2. It has a thermal barrier at ($Ak_{82}Q_{18}$), which divides the system into SiO_2-saturated and SiO_2-undersaturated portions.

11.1 Phase Relations in the System Leucite-Akermanite-SiO_2

Results are summarized in Fig. 11.2, which shows that there are three piercing points:

(1) one at G ($Lc_{32}Ak_{58}Q_{10}$; Lc$_{ss}$ + Mel + Di$_{ss}$ + L) and 1,284 ± 5 °C,
(2) one at H ($Lc_{55}Ak_2Q_{43}$; Lc$_{ss}$ + Di$_{ss}$ + Kf + L) and 1,005 ± 5 °C and
(3) the other at K ($Lc_{46}Ak_2Q_{52}$; Di$_{ss}$ + Kf + Tri + L) and 970 ± 10 °C, respectively.

Subsolidus assemblages in the system, akermanite- leucite-SiO_2 is shown in Fig. 11.3. At point A, the diopside-wollastonite join crosses the unstable join akermanite-SiO_2 (akermanite and tridymite or silica polymorphs do not coexist in equilibrium) in the CaO–MgO–SiO_2 system. Beyond A, toward SiO_2-saturated side akermanite$_{ss}$ is silicated to form diopside$_{ss}$ and wollastonite$_{ss}$. Thus, beyond the line leucite-A toward SiO_2-rich side, there should not be any melilite and on the SiO_2-saturated side of the line AB, only K-feldspar, tridymite, diopside$_{ss}$ and wollastonite should be stable.

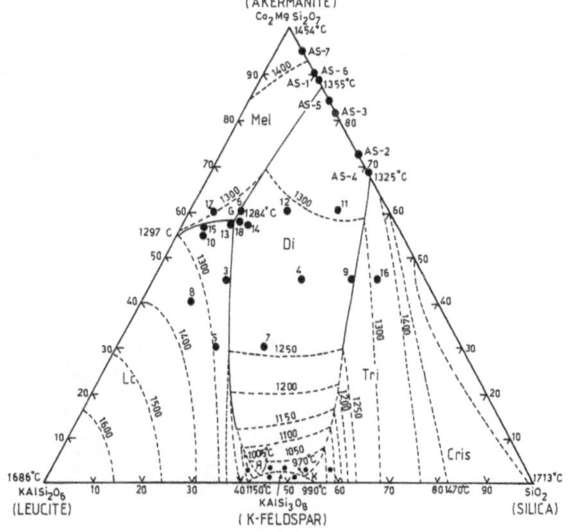

Fig. 11.2 Phase relation in the join leucite-akermanite-SiO_2 at 1 atm (after Gupta and Gupta 1997)

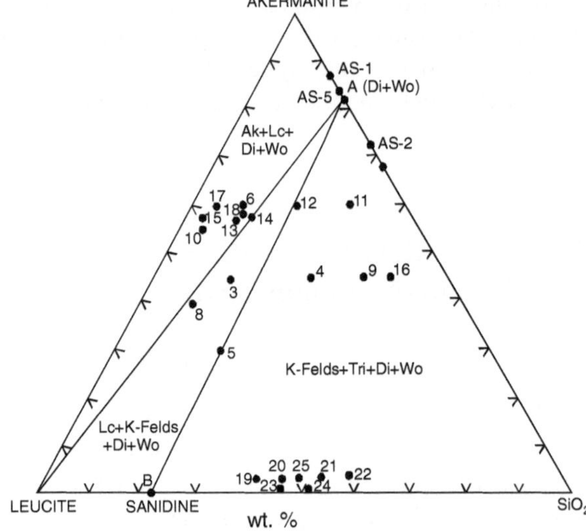

Fig. 11.3 Subsolidus phase relations in the join akermanite-leucite- SiO_2 (after Gupta and Gupta 1997)

11.2 Study of the Joins $Lc_{75}Ab_{25}$-$Ak_{75}Ab_{25}$-$Q_{75}Ab_{25}$ and $Lc_{60}Ab_{40}$-$Ak_{60}Ab_{40}$-$Q_{60}Ab_{40}$

The two joins with 25 and 40 wt% albite of the leucite-akermanite-albite-SiO_2 tetrahedron were studied to see the effect of addition of albite to the join leucite-akermanite-SiO_2. Figure 11.4 shows that the join with 25 wt% Ab cuts the phase volumes of leucite$_{ss}$, (K–Na) feldspar, tridymite, diopside$_{ss}$ and melilite. There are two piercing points:

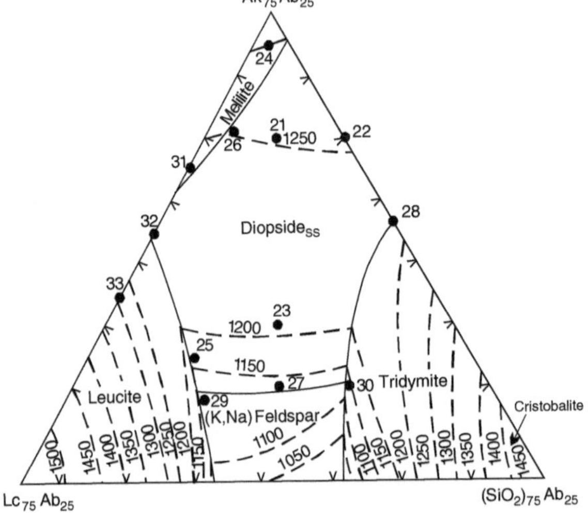

Fig. 11.4 Phase relation in the join $Ak_{75}Ab_{25}$-$Lc_{75}Ab_{25}$-$(SiO_2)_{75}Ab_{25}$ at 1 atm (after Gupta and Gupta 1997)

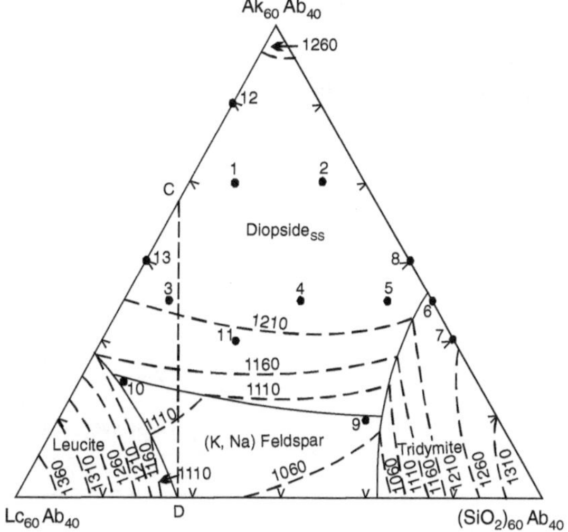

Fig. 11.5 Phase relation in the join $Ak_{60}Ab_{40}$-$Lc_{60}Ab_{40}$-$(SiO_2)_{60}Ab_{40}$ at 1 atm (after Gupta and Gupta 1997)

(1) One at $Lc_{20}Ak_{17}Q_{38}Ab_{25}$ (Di_{ss} + K–Na Felds + Tri + L) and 1,135 ± 5 °C, and
(2) the other at $Lc_{44}Ak_{15}Q_{16}Ab_{25}$ (Lc_{ss} + K–Na Felds + Di_{ss} + L) and 1,132 ± 5 °C.

Figure 11.5 shows that 40 wt% albite join cuts the phase volumes of $leucite_{ss}$, (K–Na) feldspar, $diopside_{ss}$ and tridymite. Two piercing points also occur in this join at

(1) $Lc_{13}Ak_{11.5}Q_{35.5}Ab_{40}$ (Di_{ss} + K–Na Felds + Tri + L) and 1,085 °C, and
(2) $Lc_{41}Ak_{15.5}Q_{3.5}Ab_{40}$ (Lc_{ss} + K–Na Felds + Di_{ss} + L) and 1,139 °C.

11.3 Petrological Significance of the System Leucite-Akermanite-Albite-SiO_2

If a small amount of possible alumina present in diopside (as Ca-Tschermak's molecule) and melilite (as gehlenite molecule), respectively, are ignored, the system leucite-akermanite-SiO_2 (Fig. 11.2) may be treated as ternary and the points G, H and K may be considered to have been formed by the intersection of three univariant lines and the akermanite-leucite-SiO_2 plane. The subsolidus phase relations in the join $Lc_{75}Ab_{25}$-$Ak_{75}Ab_{25}$-$Q_{75}Ab_{25}$ and $Lc_{60}Ab_{40}$-$Ak_{60}Ab_{40}$-$Q_{60}Ab_{40}$ are summarized in Figs. 11.6 and 11.7, respectively. The course of crystallisation of liquids in this system is shown in Fig. 11.8. Final assemblages in the system leucite-akermanite-SiO_2 are controlled by the joins Lc-(Di + Wo) and Sa-(Di + Wo). The invariant assemblages in different sections of the system are summarized in Fig. 11.3. It is also reflected in the flowsheet diagram (Fig. 11.8). Although $wollastonite_{ss}$ is one of the critical phases controlling subsolidus phase assemblages in

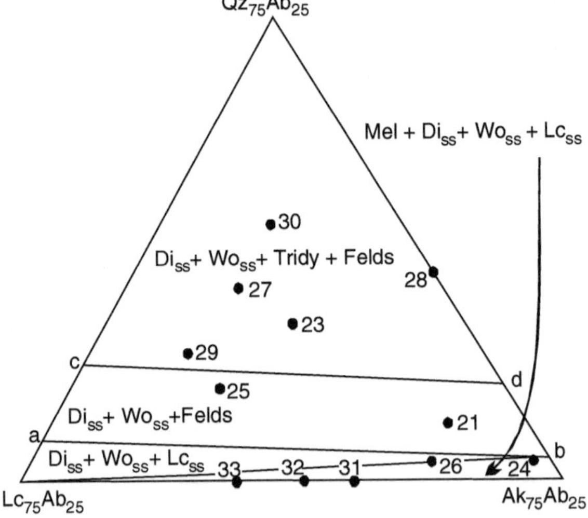

Fig. 11.6 Subsolidus phase assemblages in the join $Ak_{75}Ab_{25}$-$Lc_{75}Ab_{25}$-$(SiO_2)_{75}Ab_{25}$ at 1 atm (after Gupta and Gupta 1997)

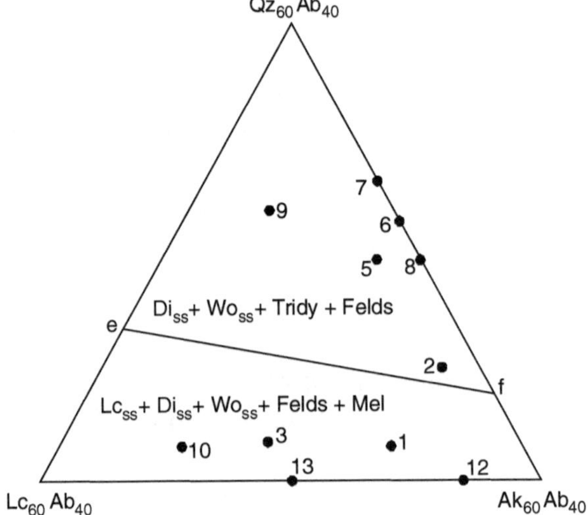

Fig. 11.7 Subsolidus phase assemblages in the join $Lc_{60}Ab_{40}$-$Ak_{60}Ab_{40}$-$Qz_{60}Ab_{40}$ at 1 atm (after Gupta and Gupta 1997)

the leucite-akermanite-SiO_2 system, its natural occurrence in K-rich volcanic rocks is extremely rare. An exceptional example is an occurrence of modal wollastonite in katungite, mafurite and uganditic rocks, in which wollastonite is associated with modal diopside, melilite and leucite (Stoppa 1988).

The flowsheet diagram (Fig. 11.8) shows that a wollastonite-melilite leucitite (i) can be derived from either a wollastonite melilitite (b–i), wollastonite-melilite-italite (a–i) or a melilite leucitite (c–i) or a wollastonite leucitite (l–i). Likewise a wollastonite-pyroxene phonolite (j) can be obtained from a wollastonite leucitite (L–j), a wollastonite-leucite phonolite (d–j), a pyroxene-leucite phonolite (e–j) or a

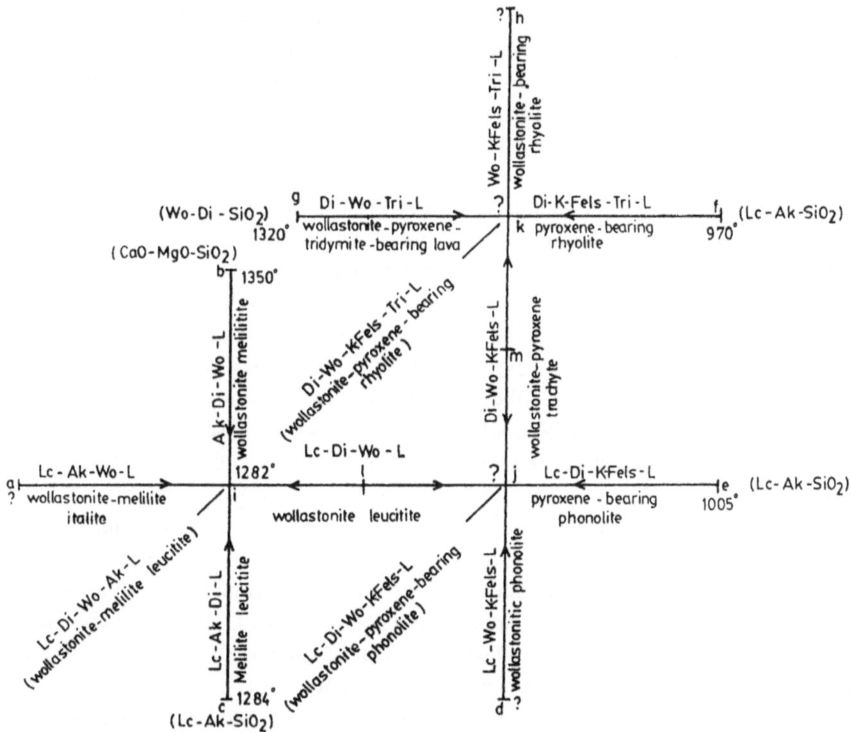

Fig. 11.8 Flow-sheet diagram summarising the course of crystallisation of liquid in the system akermanite-leucite-albite-SiO_2 at 1 atm (after Gupta and Gupta 1997)

wollastonite-pyroxene-trachyte (m–j). Association of wollastonite with pyroxene, melilite and leucite has been discussed by Stoppa (1988) from Colli Fabbri, Spoteto, Italy. The flowsheet diagram (Fig. 11.8) also shows that a wollastonite-pyroxene rhyolite can be derived from a wollastonite-pyroxene tridymite lava (g–k), a pyroxene-bearing rhyolite (f–k), a wollastonite-rhyolite (h–k) or a wollastonite-pyroxene trachyite (m–k). If a small amount of $FeSiO_3$ is present in the system, the assemblages along a–i (melilite leucitite), b–i (melilitite), c–i (melilite leucitite) and l–i (leucitite) should be represented by simplified rock types as shown within the brackets. In case of a–i, the pyroxenes are hedenbergitic but along l–i and b–i pyroxenes are diopside-hedenbergite solid solutions.

With addition of $FeSiO_3$, similarly the assemblages along m-j and e-j should be pyroxene trachyte and pyroxene phonolite. Pyroxenes in these cases are diopside-hedenbergite solid solution. In case of both e–j and c–i pyroxene are diopsidic.

If FeO is added in the system, the assemblages along g–k (pyroxene-tridymite lava), h–k (hedenbergitic rhyolite) and f–k (pyroxene rhyolite) and m–k (pyroxene trachyte) should be represented by simplified rock types as mentioned in the bracket. Along h–k pyroxene is hedenbergitic.

The assemblages at the invariant points i, j and k should be represented by wollastonite-melilite leucitite, wollastonite-pyroxene phonolite and wollastonite-pyroxene rhyolite. Once again in presence of $FeSiO_3$, the invariant assemblage should be represented by melilite leucitite (i), pyroxene phonolite (j) and pyroxene rhyolite (k). As natural rocks contain FeO, the above assumption should be valid.

Lava flows corresponding to a melilite-nepheline leucitite have been reported from Fort Portal, Uganda (Holmes and Harwood 1932). Nepheline leucitite and nepheline-melilite leucitite lava flows have been reported from Colli Albani (Alban Hills, Fornaseri et al. 1963, Aurisicchio et al. 1988). Pyroxenite, leucite and nepheline-bearing phonolites from Somma-Vesuvius are found around Naples, Italy. In this area, igneous activity began with the eruption of trachybasalt and its derivatives such as trachyandesites and phonolites (Imbo 1965). Highly potassic lavas of Rocca-monfina (Appleton 1972) are either a nepheline- or leucite-normative series or a trachybasalt or a latite series, but melilite-bearing rocks are absent in Roccamonfina. Eruption of such contrasting lavas may be controlled by such thermal divides as albite-K-feldspar- calcic pyroxene or nepheline-K-feldspar-pyroxene joins.

Melluso et al. (2003) described small outcrops of wollastonite- and melilite-bearing pyrometamorphic rocks (para-lavas), which are found along the Apennine chain in Central Italy at the localities of Colle Fabbri and Ricetto. These rocks have coarse- to fine-grained crystalline, spotted, and glassy textures. The Colle Fabbri rocks have abundant and ubiquitous wollastonite, with plagioclase ± clinopyroxene. Melilite is found only in the most Ca-rich and silica-poor samples. Federico et al. (2002) studied a suite of lithics (ejecta) collected from the latest erupted pyroclastic products of the Alban Hills volcano (Central Italy). They studied the rocks to determine their mineralogical composition and to investigate their genesis. The ejecta commonly have granular texture and consist of coarse-grained crystals often associated with a fine- to medium-grained matrix. The mineralogical composition is variable and consists of both typical igneous minerals and contact metamorphic phases. Garnet, clinopyroxene K-feldspar are almost ubiquitous, whereas leucite, wollastonite, sodalite-group minerals, phlogopite, nepheline and phillipsite are present in most of the ejecta; minor accessory phases include cuspidine, aniphibole, pyrrhotite, magnetite, apatite, uranpyrochlore, sphene, kalsilite, and melilite; anorthite, zircon and fluorine-bearing Ca, Zr silicate phases. Larnite, and baryte are found sporadically

11.4 Experimental Study of the Joins Leucite-Akermanite-Albite with or Without Anorthite in Air or Under 1 Gpa in Presence of Excess Water

Based on petrological observations of leucite-bearing rocks, Cross et al. (1902) concluded that leucite and albite do not coexist in equilibrium and they took into account this incompatibility problem in the norm calculation. Shand (1943) observed that in leucite-bearing lavas, the coexisting feldspar is either a calcium-rich plagioclase or potassium-rich alkali feldspar, or both. Yoder and Schairer

(1969) observed that plagioclase and melilite have an incompatible relationship in alkaline syenite lava. Gupta and Lidiak (1973) discussed the melilite-plagioclase incompatibility problem and concluded that melitite and plagioclase may coexist in some hybrid contact zone under water-saturated conditions. Schairer and Yoder (1970) opined that akermanite could coexist with spinel, anorthite, and diopside in the absence of forsterite. Such a calcium-rich assemblage is only found in metamorphic rocks, whereas an assemblage forsterite-akermanite-diopside-spinel has representatives among the igneous rocks. Schairer and Yoder (1970) further suggested that even in the absence of olivine, there is a limit to the anorthite content of plagioclase that can coexist with melilite. They observed that volcanic centres of the same petrographic province eject either plagioclase-bearing pyroclasts, or melilite-bearing lavas. In natural alkaline rocks, nepheline and K-feldspar are essential minerals of syenites and phonolites, but leucite and pure albite have not been observed to coexist in equilibrium.

Because of the above-mentioned incompatibility between the leucite-albite pair and akermanite-albite pair, Dwivedi et al. (2007) studied the leucite-akermanite-albite system under atmospheric pressure first, and later added anorthite, and completed the system leucite-akermanite-albite$_{50}$anorthite$_{50}$ under similar conditions to see the effect of anorthite on the incompatibility problem. Leucite also breaks down under high pressures and temperatures to form orthorhombic kalsilite and K-feldspar (Fasshauer et al. 1998). Their study showed that leucite is stable at 1 GPa at temperatures above 850 °C and akermanite breaks down to form diopside and merwinite (Yoder 1973). Therefore, the join leucite-akermanite-albite$_{50}$anorthite$_{50}$ was studied under water-saturated conditions at 1 GPa to see the effect of pressure on the leucite-plagioclase incompatibility problem in the presence or absence of melilite.

11.4.1 The Join Leucite-Akermanite-Albite Under One Atmospheric Pressure

Fourteen compositions were prepared to study this join, and the phase relations in the join leucite akermanite-albite are summarized in Fig. 11.9, which demonstrates the presence of two four-phase points: (1) at the point "A" ($Lc_{31}Ak_{47}Ab_{22}$) at 1,275 °C, where leucite$_{ss,}$ diopside$_{ss}$ and melilite are in equilibrium with liquid, and (2) at the point "B" ($Lc_{38}Ak_{22}Ab_{40}$) at 1,250 ± 5 °C, where leucite$_{ss}$ + ternary feldspar + diopside$_{ss}$ are in equilibrium with liquid (Fig. 11.9). The presence of a primary phase field of diopside$_{ss}$ and the appearance of nepheline$_{ss}$, wollastonite$_{ss}$ and alkali feldspar in the subsolidus region in the leucite- akermanite-albite join, can be explained by the following reaction.

$$\underset{\text{(akermanite)}}{Ca_2MgSi_2O_7} + \underset{\text{(leucite)}}{KAlSi_2O_6} + \underset{\text{(albite)}}{NaAlSi_3O_8} \leftrightarrow \underset{\text{(diopside)}}{CaMgSi_2O_6} + \underset{\text{(nepheline)}}{NaAlSiO_4} + \underset{\text{(wollastoniteo)}}{CaSiO_3} + \underset{\text{(orthoclase)}}{KAlSi_3O_8} \quad (11.1)$$

The experimental results of Dwivedi et al. (2007) show that sodic-plagioclase does not coexist with leucite. When leucite$_{ss}$ is in equilibrium with feldspar, it is alkali feldspar with the following compositions: $An_4Ab_{49.5}An_{46.5}$, $An_3Ab_{53.5}Or_{43.5}$, and $An_{3.5}Ab_{51.5}Or_{45.0}$.

Nepheline$_{ss}$ appearing in this system contains kalsilite and some excess silica as albite molecule. Electron microprobe analyses show that nepheline$_{ss}$ coexisting with leucite has the following compositions: $Ne_{92}Ks_7Q_1$, $Ne_{90}Ks_8Q_2$, and $Ne_{91}Ks_{7.5}Q_{1.5}$. The wollastonite$_{ss}$ appearing in the subsolidus region of the join leucite-akermanite-albite contains 1.10–1.52 wt% MgO. Melilite is present in large quantities in the akermanite-rich side of the join, and has the following compositions: $Ak_{80}Geh_{13}Sm_7$, and $Ak_{81}Geh_9Sm_{10}$, where the amount of melilite is less than that of wollastons$_{ss}$ and nepheline$_{ss}$, which appears as an additional phase. Presence of considerable amounts of soda-melilite and gehlenite componests in melilite is well known in leucite-bearing mafic and ultramafic rocks from ltaly (Meluso et al. 1996).

Le Bas (1977) described wollastonite-bearing ijolite containing significant amounts of wolastonite, pyroxene, nepheline, and small amounts of feldspar with other accessory minerals from Homa Bay. Wollastonite-bearing urtite has also been described from the same locality. In the alkaline rocks associated with alvikites from Omutageti-Nyamgurka complex of Kisingiri, Africa. Le Bas (1977) also described the occurrence of mafelsic lavas containing wollastonite$_{ss}$, nepheline$_{ss}$, clinopyroxene and alkali feldspar. In Usaki ijolite complex, LeBas described ijolite and urtite containing wollastonite, nephelin, and clino- pyroxene with trace of alkali feldspar. Le Bas (1977) observed that most of the Nyamaji phonolite dykes and many of the lavas are characterized by the presence of alkali feldspar, alkali pyroxene, wollastonite$_{ss}$, and nepheline$_{ss}$. Wollastonite-bearing nepheline syenite and ijolite (nepheline + clinopyroxene) have also been dsscribed from north Ruri by Le Bas (1977).

The occurrence of modal wollastonite in kamafugitic rocks has been described by Sahama (1974) from Uganda, Africa. In addition, the association of wollastonite along with diopside, melilite and leucite has been described by Stoppa (1988) from Colle Fabbri, Spoleto, Italy.

11.4.2 The Join Leucite-Akermanite-Albite$_{50}$ Anorthite$_{50}$ Under Atmospheric Pressure

Twenty glasses were made to study this join at atmospheric pressure (Fig. 11.10). Phase relations in the join leucite-akermanite-albite and that in the join leucite-akermanite-albite$_{50}$anorthie$_{50}$ (Fig. 11.10) are compared, it can be seen that when

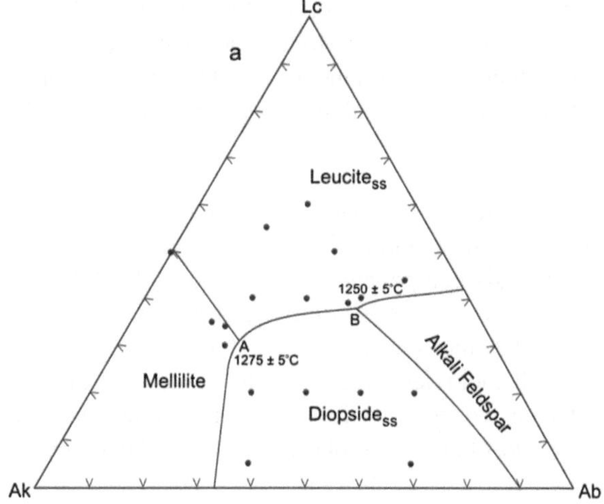

Fig. 11.9 Phase relationships in the join leucite-akermanite-albite at one atmospheric pressure (Dwivedi et al. 2007)

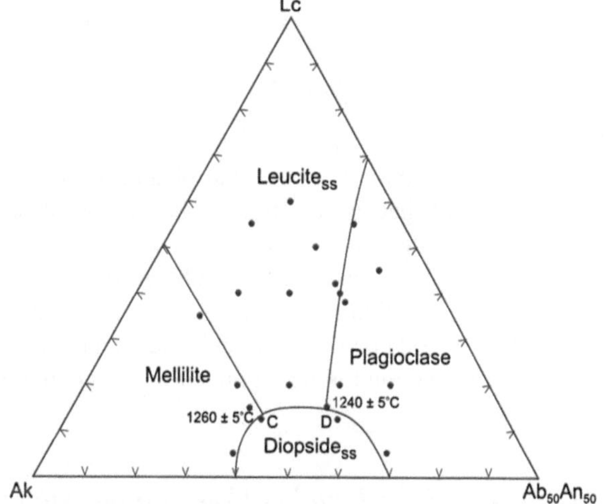

Fig. 11.10 Phase relationships in the join leucite-akermanite-albite$_{50}$anorthite$_{50}$ at one atmospheric pressure

anorthite is added, the primary phase field of diopsid$_{ss}$ shrinks considerably. In the join leucite- akermanite-albite$_{50}$An$_{50,}$ one four phase point occurs at $Lc_{14}Ak_{48}(Ab_{50}An_{50})_{38}$ at 1,260 ± 5 °C (point C) and the other occurs at $Lc_{15}Ak_{35}(Ab_{50}An_{50})_{50}$ at 1,240 ± 5 °C (point D). At C, melilite, leucite$_{ss}$ and diopside$_{ss}$ (equivalent to a rock called, melilite leucitite) are in equilibrium with liquid, whereas at D, diopside$_{ss}$ leucite$_{ss}$ and plagioclase coexist with liquid (equivalent to a rock, called leucite tephrite).

Electron microprobe analyses of plagioclase coexisting with leucite show the following compositions: An_{60} Ab_{26} Or_{14}, An_{66} $Ab_{17}Or_{17}$ and An_{64} $Ab_{21.6}$ $Or_{14.4}$. The analyses of plagioclase show that they are anorthite- rich as suggested by Shand (1943). In most of the leucite- bearing tephrites, plagioclase has been found to be anorthite- rich as observed by the Dwivedi et al. (2007). Their results on the join leucite-akermanite-albite$_{50}$anorthite$_{50}$ join indicate that when the anorthite component is absent, leucite and albite react to form nepheline and alkali feldspar with more than 45 % albite component. Over a small compositional range, melilite can coexist with calcium-rich plagioclase as observed by them.

It has been further observed that the subsolidus assemblages in this join include wollastonite$_{ss}$, and nepheline$_{ss}$, which precipitates in the akermanite-rich portion of the join. Thus, the join leucite-akermanite-albite$_{50}$anorthite$_{50}$ can be treated as a part of the six-component system $Na_2O–K_2O–CaO–MgO–Al_2O_3–SiO_2$. Electron microprobe analyses of diopside show that it contains Al_2O_3 (up to 4.37 wt%) as $CaAl_2SiO_6$ molecules in solid solution.

In the join leucite-akermanite-albite$_{50}$anorthite$_{50,}$ the subsolidus assemblages obtained at or below 1,150 °C comprise diopside$_{ss}$ + wollastonite$_{ss.}$ The presence of these subsolidus assemblages can be explained by the following reaction:

$$\begin{aligned} &\underset{\text{(akermanite)}}{8Ca_2MgSi_2O_7} + \underset{\text{(albite)}}{3NaAlSi_3O_8} + \underset{\text{(anorthite)}}{2CaAl_2Si_2O_8} \\ &\rightleftharpoons \underset{\text{(diopside)}}{8CaMgSi_2O_6} + \underset{\text{(Ca-Tschermak)}}{2CaAl_2SiO_6} + \underset{\text{(nepheline)}}{3NaAlSiO_4} \\ &+ \underset{\text{(wollastonite)}}{8CaSiO_3} \end{aligned} \tag{11.2}$$

The above reaction suggests that plagioclase coexisting with melilite is labradoritic, as established by electron probe microanalyses by Dwivedi et al. (2007). They observed that clinopyroxene contains 3.3 wt% Ca-Tschermak molecule in solid solution. The subsolidus assemblage shows the presence of nepheline$_{ss}$ and wollastonite$_{ss}$ in addition to melilite. They observed that melilite is a solid solution of akermanite, sodamelilite and gehienite. Nepheline and wollastonite combines to form sodamelilite, and is incorporated in the melilite structure.

11.4.3 The Join Leucite-Akermanite- Albite$_{50}$ Anorthite$_{50}$ at 1 GPa Under H_2O-Saturated Condition

The liquidus phase relations of this join at 1 GPa under water-saturated conditions at variable temperatures, were studied by Dwivedi et al. The experimental results are summarized in Fig. 11.11, which shows that in this join, there are only primary phase fields of diopside$_{ss,}$ plagioclase, and leucite$_{ss,}$ but melilite is absent. The join is characterized by only a single four–phase point, E, occurring at 1,060 ± 5 °C and $Lc_{43}Ak_{22}(Ab_{50}An_{50})_{35,}$ where diopside$_{ss}$ plagioclase, and leucite$_{ss}$ are in equilibrium.

Once again, nepheline$_{ss}$ appears from a reaction between leucite and an albite component of plagioclase. Elimination of melilite at 1 GPa under H_2O-saturated conditions can be explained by the following reaction:

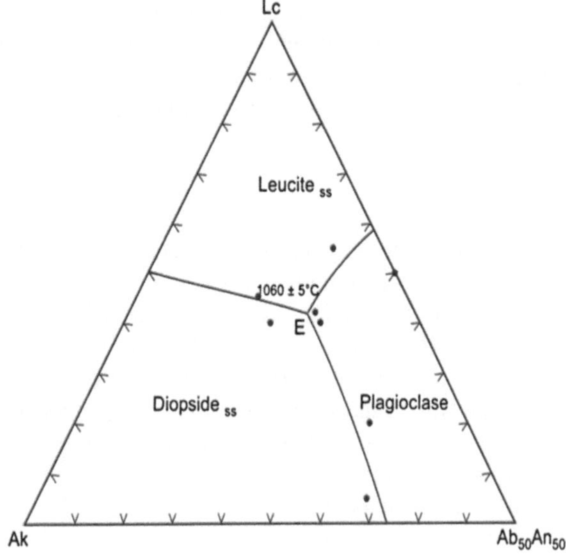

Fig. 11.11 Phase relationships in the join leucite-akermanite-albite$_{50}$anorthite$_{50}$ at 1 GPa (H_2O-saturated conditions)

$$\underset{\text{(akermanite)}}{8Ca_2MgSi_2O_7} + \underset{\text{(albite)}}{3NaAlSi_3O_8} + \underset{\text{(anorthite)}}{3CaAl_2Si_2O_8} + \underset{\text{(leucite)}}{1KAlSi_2O_6}$$
$$= \underset{\text{(grossularite)}}{Ca_3Al_2Si_3O_{12}} + \underset{\text{(diopside)}}{8CaMgSi_2O_6} + \underset{\text{(CaTs)}}{2CaAl_2SiO_6} + \underset{\text{(wollastonite)}}{6CaSiO_3}$$
$$+ \underset{\text{(nepheline)}}{3NaAlSiO_4} + \underset{\text{(K-feldspar)}}{KAlSi_3O_8}\,. \tag{11.3}$$

Nepheline$_{ss}$, alkali feldspar, wollastonite$_{ss}$, clinopyroxene, and grossular are present in the plagioclase-rich side of the join leucite-akermanite-albite$_{50}$anorthite$_{50}$ studied under 1 GPa at variable temperatures.

King and Sutherland (1976, quoted in Le Bas (1977)) described coarse- and medium-grained ijolites from Uyi located in the northeast corner of the Wasaki peninsula at Angola, located 2 km west of Homa Bay town. These rocks are similar to those at Kisingiri and Usaki, in that they include nepheline, alkali pyroxene, garnet, wollastonite, and interstitial alkali feldspar. The Usaki ijolites are also characterized by the presence of alkali feldspar, wollastonite$_{ss}$, grossular, and nepheline$_{ss}$, as observed in the study of Dwivedi et al. (2007).

It is known that leucite breaks down to form orthorhombic kalsilite and K- feldspar (Scarf et al. 1966; Fasshauer et al. 1998), which suggests that leucite$_{ss}$ will be stable at 1 GPa at temperatures above 850 °C. It is also observed that the field of plagioclase at 1 GPa is contracted, compared to the study on the same join at 1 atm, where in the anorthite-rich side of the join grossular appears slightly above the solidus. The disappearance of akermanite$_{ss}$ at 1 GPa under H_2O-saturated conditions suggests that melilite$_{ss}$ in leucite-bearing rocks crystallizes, when the ultrapotassic nepheline-bearing magma enters into the earth's crust.

The subsolidus assemblage in the join leucite-akermanite-albite$_{50}$ anortite$_{50}$ at 1 Gpa under H_2O-saturated conditions comprises wollastonite$_{ss}$, leucite$_{ss,}$ nepheline$_{ss,}$ plagioclase and grossular. In Itapirapua (Saopaulo), Wolley (1987) has described a similar garnet-wollastonite-plagioclase-Feldspathoid-bearing alkali complex.

The following conclusions are derived from the study of Dwivedi et al. (2007):

1. Leucite and sodic plagioclase do not coexist in equilibrium. In the join leucite-akermanite-albite, when leucite coexists with feldspar, it is found to be alkali feldspar with a compositional rage of $An_{3\text{-}4}Ab_{50\text{-}53}Or_{45\text{-}47.}$
2. In presence of leucite, akermanite and albite react to form diopside$_{ss,}$ nepheline$_{ss,}$ wollastonite$_{ss,}$ and alkali feldspar as shown in reaction (11.1)
3. In the join leucite-akermanite-albite$_{50}$anorthite$_{50}$, plagioclase coexisting with leucite was found to be a calcium-rich plagioclase. The An content of Plagioclase in equilibrium with leucite was found to vary between 60 and 66 mol%, the Ab content was observed to be in the range of 17–26 mol%, and the or content varied between 12 and 17 mol%, When the join leucite-akermanite-albite$_{50}$-anorthite$_{50}$was studied at 1 GPa in H_2O-enriched conditions, melilite was eliminated, as described by the reaction (11.3).

Chapter 12
P-T Stability of Phlogopite, K-Richterite and Phengite, as a Source of Potassium in the Mantle

The forsterite–kalsilite-SiO_2–H_2O plane lying in this system has significant bearing in the genesis of various potassium-rich igneous rock types. This plane has therefore, been the subject of numerous studies under variable P-T conditions in the presence or absence of a fluid component belonging to the C–H–O system (Luth 1967; Yoder and Kushiro 1969; Wendlandt and Eggler 1980a, b, c; Gupta and Green 1988; Seifert and Schreyer 1971; Foley et al. 1986). Experimental studies of these investigators are described first followed by the discussion on the P-T stability of potassium-richterite (Konzett et al. 1997) and phlogopite (Kushiro et al. 1967; Yoder and Kushiro 1969), which are considered to be the possible source materials for potassium-rich magmas. The assemblage diopside-phlogopite, found in many lamproites, have also been studied at different temperatures and pressures (3.5–17 GPa) by Luth (1997). His results have been described later, followed by the discussion on the P-T stability of phengite (Massonne 1992), which is also considered to be a source mineral for potassium in the mantle.

12.1 Phase Relations in the System Forsterite–Kalsilite-SiO_2–H_2O at Variable Temperatures up to 0.3 GPa

Luth (1967) studied the P(H_2O)-T stability of 35 univariant reactions up to a pressure of 0.3 GPa [P(H_2O) = P(Total)]. These mineral assemblages related to numerous univariant reactions have bulk compositions lying within the forsterite–kalsilite-SiO_2–H_2O join (Fig. 12.1; Table 12.1). The figure indicates the presence of nine quaternary invariant points, where six phases are in equilibrium. Figure 12.1 also shows the presence of four singular points, generated by the meeting of the following curves: 14, 15 and 16; 4, 5, and 6; 8, 9, and 11; and 1, 9, and 10.

On the basis of Fig. 12.1, Luth (1967) constructed a series of isobaric-polythermal, polybaric-isothermal diagrams (Fig. 12.2) with inferred phase relations. These diagrams are produced by projection of saturation surfaces on to the anhydrous base (Mg_2SiO_4–$KAlSiO_4$–SiO_2 plane) of the tetrahedron from the H_2O apex. The saturation surfaces are the areas, where a vapour phase is in equilibrium

A.K. Gupta, *Origin of Potassium-rich Silica-deficient Igneous Rocks*,
Springer Geology, DOI 10.1007/978-81-322-2083-1_12

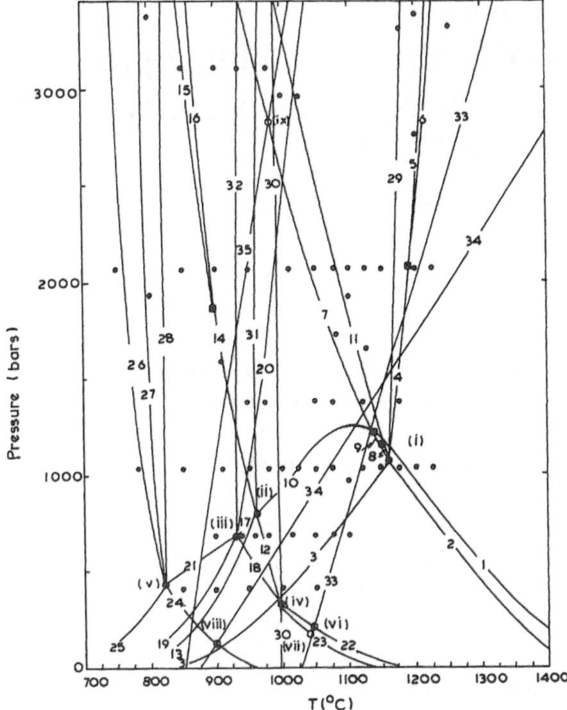

Fig. 12.1 P-T Projection of inferred invariant and univariant equilibria in the system $KAlSiO_4$–Mg_2SiO_4–H_2O. The lower case Roman numerals and Arabic numerals refer to invariant and univariant equilibria, given in Table 12.1. The positions of v is from Wones (after Luth 1967)

with a water-saturated silicate melt. The P-T condition of various inferred phase diagrams (Fig. 12.2a–i) can be obtained from Fig. 12.2j.

Figure 12.2a–i shows that at pressures at or below 0.2 GPa there are liquids of two different compositions. One is silica-undersaturated and coexists with orthorhombic kalsilite, leucite, forsterite (or phlogopite) and vapour, and the other is silica-saturated in equilibrium with K-feldspar, quartz, enstatite (or phlogopite), and vapour. According to Luth (1967), these liquids are simplified equivalents of phonolite, rhyolite, and trachyte, respectively. Comparison of Fig. 12.2a with Fig. 12.2b–j shows that at or above 0.1 GPa phlogopite appears as a phase in this system, whereas near 500 bar it completely disappears by reaction with a liquid. Bowen (1928) considered that a reaction between mica and a mafic liquid have played an important role in the genesis of potassium-rich undersaturated liquid.

Figure 12.2a–i show that forsterite is a liquidus phase in a wide range of compositions. With further variation of bulk compositions, leucite, K-feldspar, enstatite and/or phlogopite may join olivine in the course of equilibrium crystallisation. At a pressure of less than 300 bar (near the surface) the resulting crystal assemblage should consist of forsterite, leucite, and enstatite with siliceous interstitial liquid. At slightly higher water pressures, forsterite would cease to exist (depending on bulk compositions), and phlogopite and leucite, phlogopite and enstatite, or phlogopite and K-feldspar should be stable. In the case of natural

Table 12.1 Invariant points (roman numerals) and univariant curves, or reactions (arabic numerals) in the system $KAlSiO_4–Mg_2SiO_4–H_2O$ (After Luth 1967)

No.	Assemblage / Reaction
I.	Ph + Fo + Ok + Lc + L + V
1	Fo + Lc + V = L[b]
2	Fo + Lc + Ok + V = L
3	Ph = Fo + Ok + Lc + V
4	Ph = Fo + Ok + L + V[a]
5	Ph + Ok + V = Fo + L[a]
6	Ph = Fo + L + V[a]
7	Ph + Ok + Lc + V = L
8	Ph + Lc + V = Fo + L[b]
9	Ph + Lc + Fo + V = L[b]
10	Ph + L = Fo + Lc + V[b]
11	Ph + Lc + V = L[b]
29	Ph + Lc = Fo + Ok + L
II.	Ph + Fo + Or + Lc + L + V
12	Or + Fo + V = Lc + L
13	Ph + Or = Fo + Lc + V
14	Or + Ph + V = Lc + L[c]
15	Or + Ph + Lc + V = L[c]
16	Or + Ph + V = L[c]
17	Ph + L = Or + Fo + V
10	Ph + L = Fo + Lc + V
31	Ph + Or = Lc + Fo + L
III.	Ph + Fo + En + Or + L + V
18	Or + En + V = Fo + L
19	Ph + En = Or + Fo + V
20	Ph + En = Fo + L + V
21	Ph + L = Or + En + V
17	Ph + L = Or + Fo + V
32	En + Or + Ph = Fo + L
IV.	Fo + Lc + Or + En + L + V
22	Lc + En + V = Fo + L
23	Or + En + V = Lc + L
12	Or + Fo + V = Lc + L
18	Or + En + V = Fo + L
30	Or + Fo = Lc + En
V.	Ph + Or + Q + En + L + V
24	Or + En + Q + V = L
25	Ph + Q = Or + En + V
26	Ph + Or + Q + V = L
27	Ph + Q + V = En + L
21	Ph + L = Or + En + V
28	Ph + Q = Or + En + L
VI.	En + Pen + Lc + Fo + L + V
33	En = Pen
22a	Lc + En + V = Fo + L
22b	Lc + Pen + V = Fo + L
VII.	En + Pen + Lc + Or + L + V
33	En = Pen
23a	Or + En + V = Lc + L
23b	Or + Pen + V = Lc + L
VIII.	Q + Tr + Or + En + L + V
34	Q = Tr
24a	Or + En + Q + V = L
24b	Or + En + Tr + V = L
IX.	Ok + Ks + Ph + Lc + L + V
35	Ks = Ok
7a	Ph + Ok + Lc + V = L
7b	Ph + Ks + Lc + V = L

[a] Related by singular point on the univariant curve involving Ph, Fo, Ok, L, and V
[b] Related by singular points on the univariant curves involving Ph, Fo, Lc, L, and V
[c] Related by singular points on the univariant curves involving Ph, Or, Lc, and V

All reactions are written in such a way that the high temperature assemblages is on the right side of the reaction symbol, except for reaction 10. See also List of Abreviations.

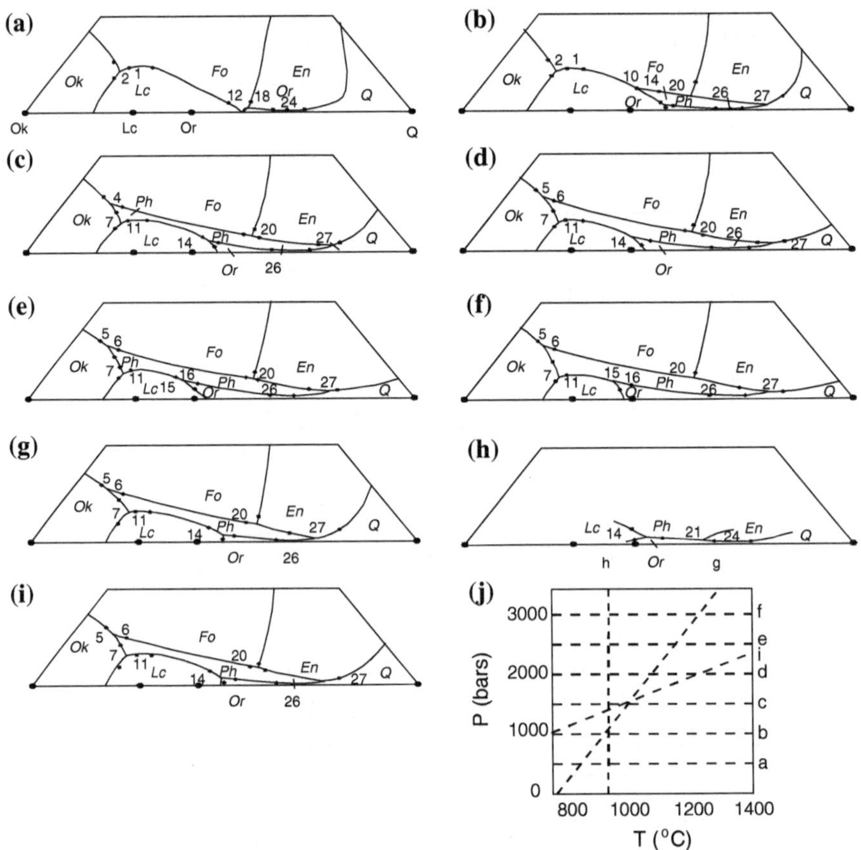

Fig. 12.2 **a–j** Inferred phase relations on the saturation surface. Projection of the saturation surface on to the anhydrous base of the tetrahedron from the H_2O apex. **a.** Isobaric-polythermal, 500 bars (a of Fig. 12.2j). **b**. Isobaric- polythermal, 1,000 bars (b of Fig. 12.2j). **c.** Isobaric-polythermal, 1,500 bars (c of Fig.122j). **d.** Isobaric-polythermal, 2,000 bars (d of Fig. 12.2). **e.** Isobaric-polythermal 2,500 bars (e of Fig. 12.2j). **f.** Isobaric-polythermal 3,000 bars (f of Fig. 12.2j). **g.** Polybaric-polythermal (g of Fig. 12.2j). **h.** Polybaric-isothermal, 900 °C (h of Fig. 12.2j). **i** Polybaric polythermal (i of Fig. 12.2j). **j** Pressure-temperature orientation for Fig. 12.2a–i. (after Luth 1967)

leucite-bearing rocks, although K-feldspar is an accompanying phase enstatite is absent.

On the basis of Fig. 12.2, Luth (1967) considered that a magma at depth initially precipitating olivine should be resorbed to form mica and then pyroxene. These early differentiates should be present as xenoliths and react with liquid. He referred to the existence of phlogopite-bearing ultramafic xenoliths in the potassium-rich lavas of the Bufumbira region (Holmes and Harwood 1937), to support his conclusion. Luth also considered that the magmas of this region cooled for a long period at moderate pressure to produce differentiates of biotite pyroxenite series, which was followed by their emplacement at shallow depth. Because of the presence of leucite (pseudoleucite), olivine and clinopyroxene and absence of phlogopite, he considered that the rocks of Shonkin Sag Laccolith were produced under low water pressures at shallow depth. He suggested that the presence of mica along with leucite, clinopyroxene, and amphibole is indicative of the fact that the rocks of West Kimberly were produced by a long cooling process of a potassic magma at a moderate depth. From the rarity of forsterite and presence of large amounts of phlogopite, he considered that the rocks of the Leucite Hills were produced at a relatively greater depth.

12.2 P-T Stability of Phlogopite

Yoder and Kushiro (1969) studied the following reactions pertaining to the stability of phlogopite up to a pressure of 3.75 GPa:

1. phlogopite + vapour $\rightleftharpoons$ forsterite + liquid,
2. phlogopite $\rightleftharpoons$ forsterite + liquid (vapour-absent condition),
3. forsterite + orthorhombic kalsilite + leucite + vapour $\rightleftharpoons$ phlogopite (Luth 1967; Wones 1967), and
4. phlogopite $\rightleftharpoons$ forsterite + leucite + kalsilite + liquid.

Their study is summarised in Fig. 12.3. Reactions (1), (2), (3) and (4) are denoted by the curves A, C, L and G in Fig. 12.3; curve B refers to the beginning of melting of the vapour-absent assemblage, forsterite + phlogopite + kalsilite (orthorhombic) + liquid. The curve D denotes minimum liquidus in the presence of a vapour phase in the phlogopite- H_2O join. The curves above the invariant point involving phlogopite, forsterite, orthorhombic kalsilite, leucite, liquid, and vapour at about 1160 °C and 0.1 GPa, indicate that the breakdown of phlogopite occurs according to the reaction, phlogopite + vapour $\rightleftharpoons$ forsterite + orthorhombic kalsilite + liquid. Under vapour-absent equilibrium conditions, phlogopite melts incongruently according to reaction, phlogopite $\rightleftharpoons$ forsterite + leucite + orthorhombic kalsilite + melt (up to 0.17 GPa), above this pressure it melts incongruently according to the reaction, phlogopite $\rightleftharpoons$ forsterite + liquid (Fig. 12.3).

The study of Yoder and Kushiro (1969) showed that the assemblage forsterite + phlogopite was stable to at least 3.75 GPa at 1,200 °C (Fig. 12.3). Kushiro et al. (1967) established that phlogopite is stable to 7.2 GPa at 1000 °C. The phlogopite + enstatite assemblage are in equilibrium up to 3.0 GPa

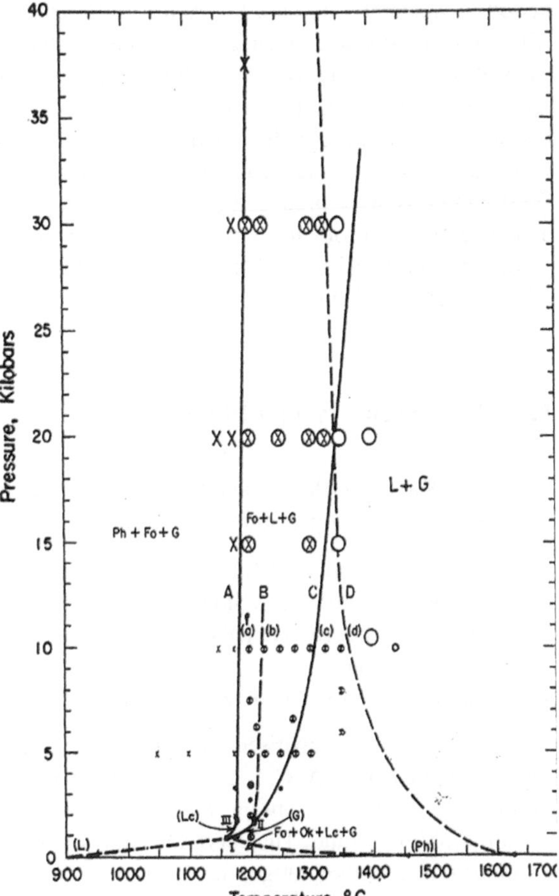

Fig. 12.3 P-T diagram for compositions in the join $K_2O.6MgO.Al_2O_3.6SiO_2$–H_2O. The *solid squares mark* an invariant point I and two singular points II and III. X = crystal + gas, *circle* with an X = crystal + liquid + gas; circle = liquid + gas. The symbols are only relevant to curves A and D. *Curve A* is the maximum stability of phlogopite in the presence of a gas phase. Curve B marks the beginning of melting of the gas-absent assemblage Ph + Fo + Ks (or Ok) + L and curve C is the maximum stability of phlogopite in the absence of a gas phase. *Black dots* are the data points of Luth (1967) (after Yoder and Kushiro 1969)

12.2.1 P-T Stability of Phlogopite up to 7 GPa in Presence of Excess Water

Tronnes (2002) made experimental investigation on phase relations of pure phlogopite and phlogopite peridotite. He observed that phlogopite broke down around 7 GPa to potassium-richterite-bearing assemblages, which in turn, yielded another assemblage comprising a hydrous phase containing high-K and low-Al.

The P-T stability of phlogopite has been determined by Tronnes (2002, Fig. 12.4), which shows that the decomposition curve is nearly coincident with that of a natural phlogopite $K_{2.1}Na_{0.1}Mg_{5.4}Fe_{0.2}Al_{2.5}Si_{5.7}O_{20}(OH_{1.4}F_{0.6})$, in the 4–8 GPa range. The phlogopite breakdown curve of Yoder and Kushiro (1969) is shown for reference in Fig. 12.4. It may be noted that the high-pressure breakdown phases of phlogopite is pyrope and forsterite + fluid. Additional breakdown products in the

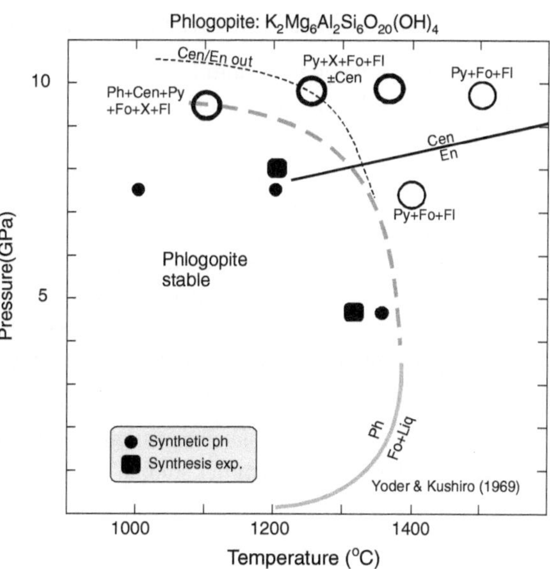

Fig. 12.4 Phase diagram of the system phlogopite $K_2Mg_6Al_2Si_6O_{20}(OH)_4$. The orthoenstatite (en)—high-p clinoenstatite (cen) boundary is from Pacalo and Gasparik (1990) and Shinmei et al. (1999) (both cited by Tronnes 2002). *Cen* clinoenstatite, *En* enstatite, *Fl* fluid, *Fo* forsterite, *Ph* phlogopite, *Py* pyrope, *X* phase X.) (after Tronnes 2002)

form of clinoenstatite and phase X are also present at 9–10 GPa. Tronnes also showed that clinoenstatite reacts out relatively close to the breakdown curve and disappears completely at lower pressure. Enstatite was not observed as a breakdown product in the natural phlogopite system in the 4–8 GPa range by Sato et al. (1997, cited by Tronnes 2002). Whereas Sato et al. (1997) noted an incipient breakdown of the natural phlogopite to pyrope and fluid at pressures of 5–8 GPa and temperatures at 1200–1300 °C, the pure phlogopite end member remains stable in this P-T interval. The amount of pyrope observed below the main phlogopite-out curve for the natural phlogopite composition however, is very low ranging from less than 5 vol% at 5 GPa to 30 % at 8 GPa. This increase in, pyrope content with increasing pressure is reflected in a systematic change in the composition of the coexisting phlogopite.

12.3 The Join $KAlSiO_4$–Mg_2SiO_4–SiO_2 up to 3.0 GPa in Presence or Absence of H_2O

Wendlandt and Eggler (1980a) studied the system forsterite–kalsilite-SiO_2 under dry condition and in presence of CO_2. They also took into account different other reactions, involving various phases, compositions of which lie in this join. These reactions include,

(1) sanidine + leucite → liquid (Lindsley 1966), and
(2) 2 leucite → kalsilite + sanidine (Scarf et al. 1966).

The reactions considered by Wendlandt and Eggler (1980) are as follows:

(3) kalsilite + leucite → liquid,
(4) kalsilite + leucite + vapour → liquid,
(5) leucite + forsterite → liquid,
(6) leucite + forsterite + sanidine → liquid,
(7) leucite + forsterite → sanidine + liquid,
(8) kalsilite + leucite + forsterite → liquid,
(9) sanidine + enstatite → leucite + liquid,
(10) leucite + enstatite → forsterite + liquid,
(11) sanidine + forsterite → leucite + 2 enstatite,
(12) plagioclase + olivine → nepheline + pyroxene, and albite + 2 forsterite → nepheline + 4 enstatite.

The reactions 12 and 13 have been suggested by Shand (1933) and Barth (1936), respectively. The other reactions studied by Wendlandt and Eggler (1980) are as follows:

(13) sanidine + enstatite → forsterite + liquid,
(14) sanidine + forsterite → leucite + liquid,
(15) sanidine + leucite + forsterite → liquid,
(16) sanidine + forsterite → liquid,
(17) sanidine + enstatite + forsterite → liquid,
(18) sanidine + enstatite → liquid,
(19) sanidine + forsterite → enstatite + liquid,
(20) leucite + forsterite → sanidine + liquid,
(21) kalsilite + forsterite + sanidine → liquid, and
(22) 4 enstatite + kalsilite → sanidine + 2 forsterite.

The P-T projection of all these reactions studied by Wendlandt and Eggler (1980), and the two reactions 1 and 2 investigated respectively by Lindsley (1966) and Scarf et al. (1966) are included in Fig. 12.5. Details of reactions associated with singular points S_3, S_4 and S_5 are shown in Fig. 12.6. The figures show that leucite and enstatite coexist in equilibrium over a limited P-T range i.e. between 980°C and one bar and the singular point I_5 (1205°C and 0.65 GPa). Wendlandt and Eggler studied the join $KAlSiO_4$–Mg_2SiO_4–SiO_2 up to 3.0 GPa in presence of CO_2. The P-T projection of the reactions investigated in this join is shown in Fig. 12.7. The details of reactions associated with singular points S_9, S_{10} and S_{11} are shown in Fig. 12.8. It is noted that phase relation in the CO_2-bearing system mimic those in the volatile-absent system, at pressures less than about 2.0 GPa, except for that reactions which occur at lower pressure. At pressures above I_{14} (Fig. 12.7), phase relations are not equivalent to the CO_2 -free system. Reactions resulting from I_{14}, pass through 3 singular points S_{18}–S_{20} (inset Fig. 12.7) and mark the rapid passage of liquid toward silica-poor magnesia-rich composition as the kalsilite-enstatite-vapour, and kalsilite-forsterite-vapour joins are pierced (Fig. 12.7). They noted that at I_8, liquid in equilibrium with forsterite + enstatite + kalsilite + vapour also attains equilibrium with magnesite; but at higher pressure forsterite does not appear in the

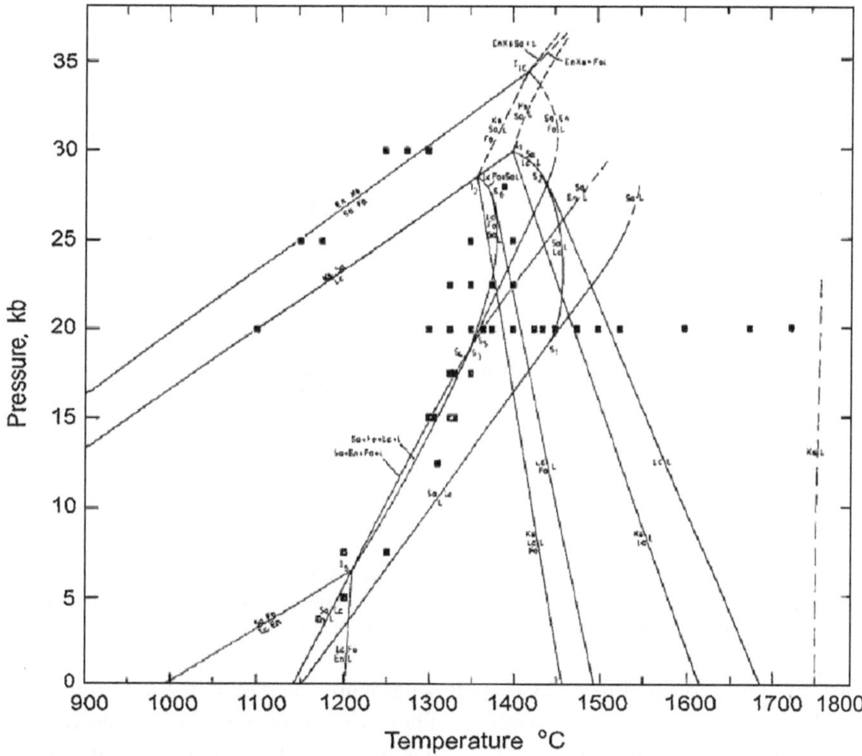

Fig. 12.5 P-T projection of the reactions investigated in the join $KAlSiO_4$–Mg_2SiO_4–SiO_2. Reactions associated with I_1 are from Lindsley (1966). The 1 atm results are from Schairer (1954) and Schairer and Bowen (1955). Because the diagram is a projection, the data points must also be projected, and there may be numerous runs involving different compositions corresponding to a pressure and temperature. *Dashed lines* are inferred. *En* enstatite, *Fo* forsterite, *Ks* kalsilite, *Lc* leucite, *Sa* sanidine, *L* liquid, *I* invariant point, *S* singular point. (after Wendlandt and Eggler 1980)

vapour-saturated assemblages because of the following subsolidus reactions that passes through I_8.

The reaction enstatite + magnesite ⇌ forsterite + vapour (24), has been studied by Newton and Sharp (1975). According to them at pressures higher than I_8 the CO_2-saturated phase assemblage melts by the following reaction:

1. enstatite + kalsilite + magnesite + vapour ⇌ liquid, and the vapour absent assemblage by the reaction,
2. enstatite + magnesite + kalsilite ⇌ forsterite + liquid.

Liquids at I_8 and higher pressures are carbonate-rich, but the exact composition of the melts are unknown.

They noted that near 1.4 GPa, for CO_2 saturated conditions and near 1.9 GPa for volatile-absent conditions, the primary phase field of enstatite crosses the sanidine-

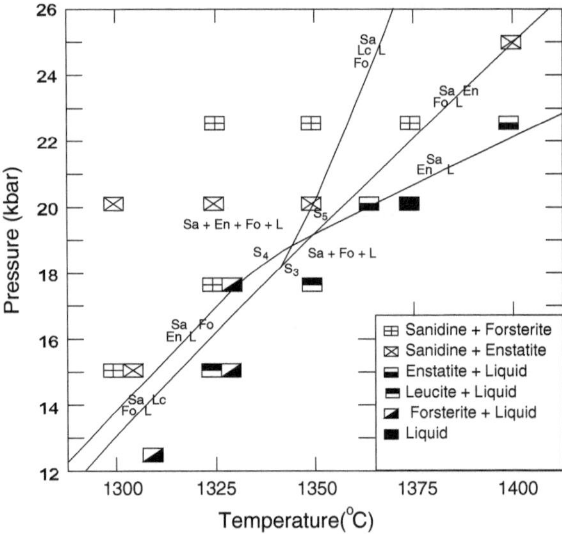

Fig. 12.6 Detail of reactions associated with singular points, S_3, S_4 and S_5 in the join $KAlSiO_4$–Mg_2SiO_4–SiO_2. The *boxes* represent runs on the compositions, $En_{16}Sa_{84}$ and $Fo_{20}Sa_{80}$. Phase abbreviations are the same as Fig. 12.5. (after Wendlandt and Eggler 1980b)

forsterite join and sanidine starts to melt congruently. The melt in equilibrium with sanidine + enstatite + forsterite with or without vapour becomes leucite-normative. They further established that at 2.75 GPa in the CO_2-saturated system and at 3.45 GPa in volatile-absent system, the solidus of the assemblage sanidine + enstatite + forsterite intersects the reaction, 4 enstatite + kalsilite $\rightleftharpoons$ sanidine + 2 forsterite, this marks the upper pressure stability limit of sanidine in the presence of forsterite. At these pressures, primary melt compositions are extremely silica-undersaturated and is kalsilite-normative. According to them at pressures above 2.9 GPa, in presence of CO_2 the primary liquids are more undersaturated and magnesite-normative.

Wendlandt and Eggler noted that at pressures below 1.8 GPa in the $KAlSiO_4$–Mg_2SiO_4–SiO_2 join and less than 1.3 GPa in the $KAlSiO_4$–Mg_2SiO_4–CO_2 join, silica undersaturated leucite-normative liquid can fractionate to silica-saturated sanidine-normative melts by crystallisation of forsterite and leucite. They further concluded that at pressures greater than 2.0 GPa in the volatile-absent join and at 1.5 GPa in the CO_2-bearing join olivine and sanidine-normative liquid can fractionate to leucite-normative liquids by crystallisation of enstatite and sanidine.

The schematic liquidus projections for the joins kalsilite–forsterite-SiO_2 and kalsilite–forsterite-SiO_2–CO_2 are shown in Fig. 12.9. In these diagrams the magnesia contents of liquids are exaggerated by Wendlandt and Eggler, except for carbonate-rich liquids for which MgO contents may be extremely low. They found that statistical characterization of 835 leucite- normative, Cenozoic volcanic rocks constitute two distinct subgroups. The rocks of first group are characterised by high

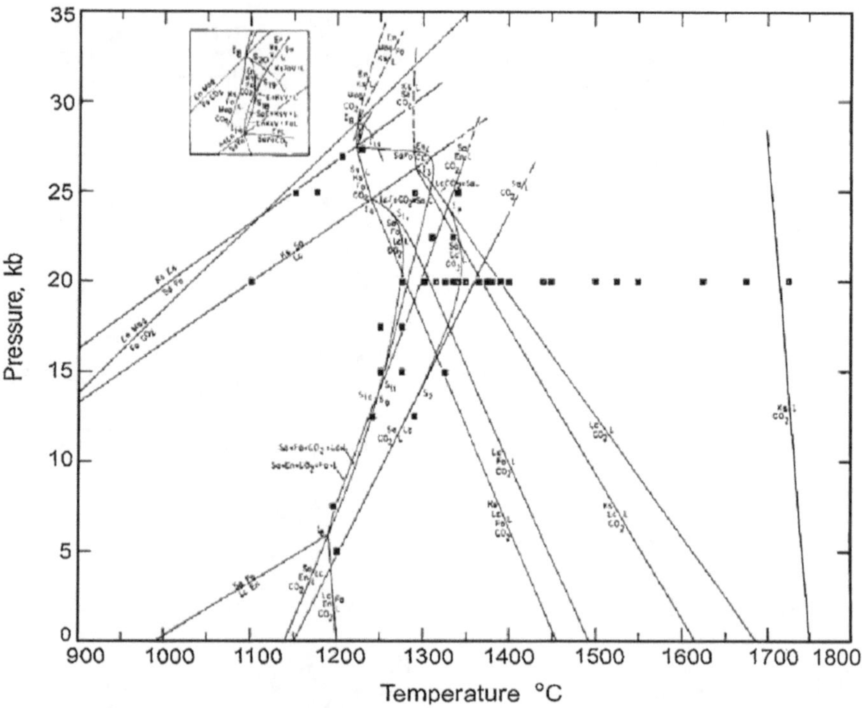

Fig. 12.7 P-T projection of the reactions investigated in the join $KAlSiO_4$–Mg_2SiO_4–SiO_2–CO_2. The 1 atm results are the same as Fig. 12.5. The reactions leucite + sanidine + kalsilite (Lindsley 1966) and enstatite + magnesite → forsterite + CO_2 (Newton and Sharp 1975) are also shown. Data points are in projections. *Dashed lines* are inferred. Phase abbreviations are the same in Fig. 12.5 and additionally *Mag* magnesite (after Wendlandt and Eggler 1980)

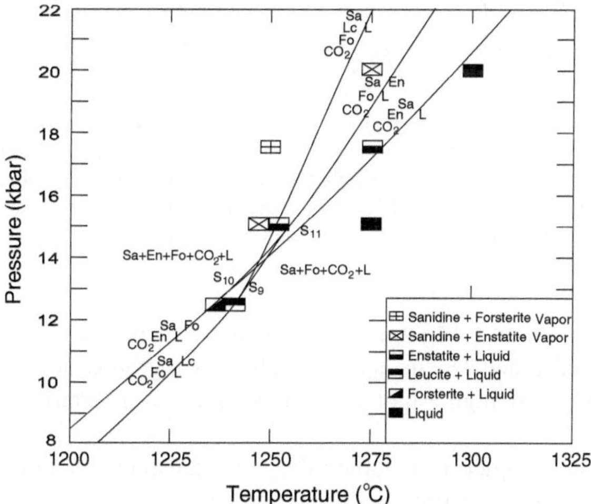

Fig. 12.8 Detail of reactions associated with singular points S_9, S_{10} and S_{11} in the join $KAlSiO_4$–MgO–SiO_2–CO_2. The boxes represent runs on the compositions $En_{16}Sa_{84}$–CO_2 and $Fo_{80}Sa_{20}$–CO_2. Phase abbreviations are same as Fig. 12.5 (after Wendlandt and Eggler 1980)

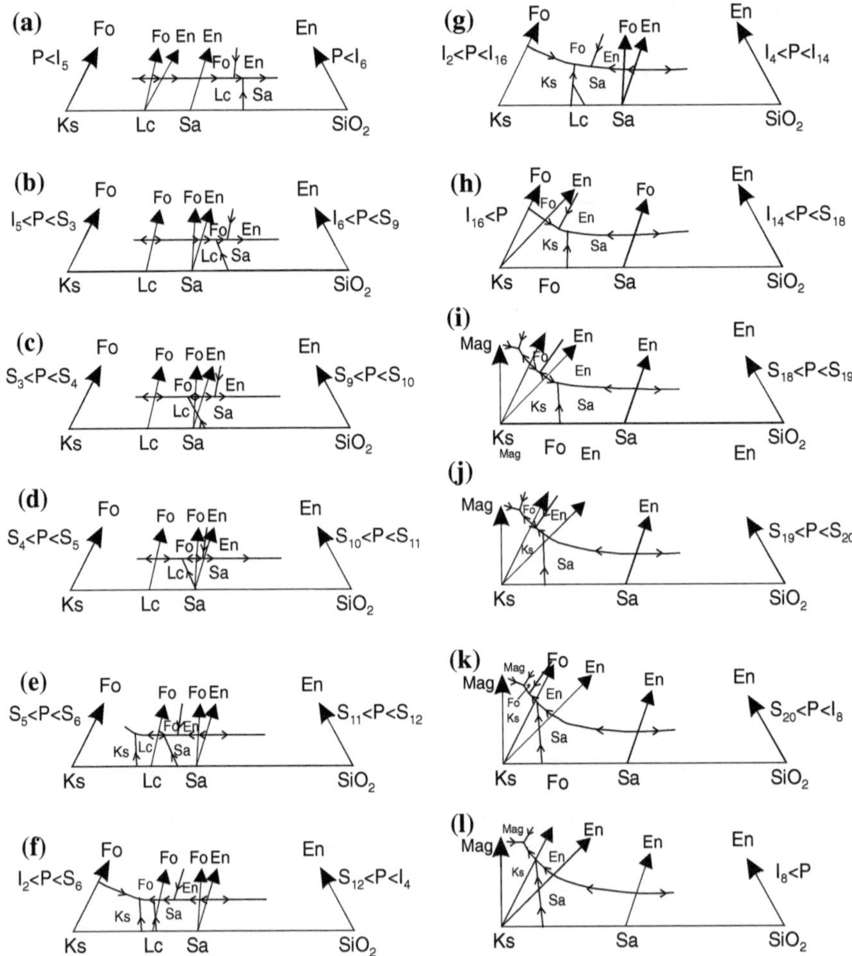

Fig. 12.9 Schematic liquidus projections for the joins $KAlSiO_4$–Mg_2SiO_4–SiO_2 and $KAlSiO_4$–MgO–SiO_2–CO_2. Magnesia contents of liquids are exaggerated, except for carbonate-rich liquids, for which the magnesia contents are probably too low. Pressures to the *left* of each figure refer to the volatile-absent join (Figs. 12.5 and 12.6): pressures to the *right* of each figure refer to the CO_2-bearing join (Figs. 12.7 and 12.8) (after Wendlandt and Eggler 1980)

K_2O and Al_2O_3 and low MgO and are similar to such rocks as leucitite, leucite-bearing nephelinite and leucite tephrite. The other groups are characterised low K_2O and Al_2O_3 and high MgO and are similar to ankaratrites, basanites, and nephelinites.

Wendlandt and Eggler also studied the joins, kalsilite-leucite (Fig. 12.10a) and leucite–forsterite (Fig. 12.10b) for fluid- absent and CO_2-saturated conditions at 2.0 GPa and different temperatures. Figure 12.10a shows that in case of the

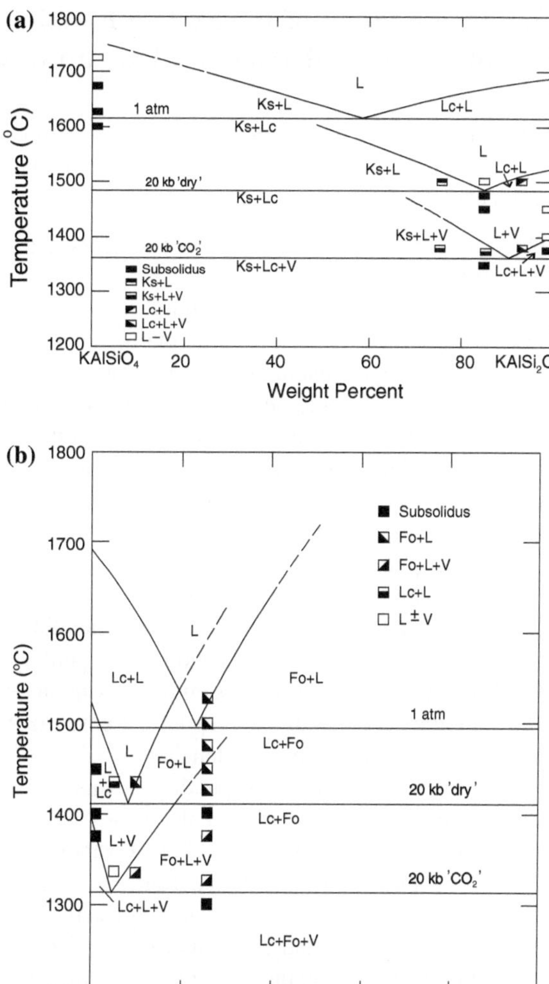

Fig. 12.10 **a** The join kalsilite-leucite at 1 atm and 2.0 GPa for volatile-absent and CO_2- saturated conditions. The results at 1 atm are from Schairer and Bowen (1955). The melting temperature of leucite at 2.0 GPa for volatile-absent conditions is from Lindsley (1966). Phase abbreviations are same as Fig. 12.6 (after Wendlandt and Eggler 1980). or the joins $KAlSiO_4$–Mg_2SiO_4–SiO_2 and) $KAlSiO_4$–MgO–SiO_2–CO_2. **b** The join leucite-forsterite at 1 atm and 2.0 GPa for volatile-absent and CO_2-saturated conditions. The results at 1 atm are from Schairer and Bowen (1954). The melting temperature of leucite at 2.0 GPa for volatile-absent conditions is from Lindsley (1966). Phase abbreviations are same as Fig. 12.4 (after Wendlandt and Eggler 1980)

kalsilite-leucite join, there is a systematic shift for the three-phase point (leucite + kalsilite + fluid) compared to one atmosphere three-phase point established by Schairer (1954). Note that since the composition of the fluid cannot be plotted in the leucite-kalsilite join, the point, where leucite, kalsilite and fluid coexist in equilibrium, is not a eutectic. In case of the leucite–forsterite join the three-phase point shifts toward leucite under both fluid-absent and fluid-present conditions, compared to the three phase point established under one atmospheric pressure by Schairer (1954). The effect of shift of the three-phase point toward leucite in case of both the joins is more in case of CO_2-saturated runs than the fluid-absent runs.

12.4 Investigation of the System $KAlSiO_4$–Mg_2SiO_4–SiO_2 in Presence of H_2O and CO_2 up to 2 GPa

Wendlandt and Eggler (1980) studied the compositional join)$KAlSiO_4$–MgO–SiO_2–H_2O–CO_2 at high pressures and high temperatures. They emphasised the significance of the following reactions:

(1) phlogopite + kalsilite + leucite ⇌ forsterite + liquid (Yoder and Kushiro 1969),
(2) phlogopite ⇌ forsterite + liquid,
(3) phlogopite + vapour ⇌ forsterite + liquid,
(4) phlogopite + kalsilite + leucite + vapour ⇌ forsterite + liquid,
(5) phlogopite + sanidine + vapour ⇌ forsterite + leucite + liquid,
(6) phlogopite + enstatite + sanidine + vapour ⇌ forsterite + liquid,
(7) phlogopite + sanidine + quartz + vapour → enstatite + liquid,
(8) phlogopite + sanidine + leucite + vapour → forsterite + liquid,
(9) phlogopite + sanidine + vapour → forsterite + liquid,
(10) phlogopite + forsterite + sanidine + vapour → liquid,
(11) phlogopite + sanidine + vapour → liquid,
(12) phlogopite + enstatite + forsterite + sanidine + vapour → liquid,
(13) phlogopite + enstatite + sanidine + vapour → liquid,
(14) phlogopite + forsterite + sanidine + vapour ⇌ enstatite + liquid,
(15) phlogopite + sanidine + vapour → enstatite + liquid,
(16) kalsilite + sanidine → 2 leucite,
(17) phlogopite + kalsilite + sanidine + vapour → forsterite + liquid,
(18) sanidine + 2 forsterite → kalsilite + 4 enstatite,
(19) phlogopite + enstatite + kalsilite + vapour → forsterite + liquid,
(20) phlogopite + enstatite + magnesite → forsterite + kalsilite + vapour,
(21) forsterite + kalsilite + vapour → enstatite + magnesite + liquid,
(22) phlogopite + enstatite + magnesite → forsterite + kalsilite + liquid,
(23) phlogopite + enstatite + magnesite + vapour → forsterite + liquid, and
(24) phlogopite + kalsilite + vapour → enstatite + magnesite + liquid.

The P-T projection of phase relations involving phlogopite, enstatite, forsterite, magnesite, sanidine, kalsilite, leucite, liquid and vapour in the compositional join $KAlSiO_4$–MgO–SiO_2–H_2O–CO_2, as determined by Wendlandt and Eggler is shown in Fig. 12.11. Dash lines at low pressure are reaction in the H_2O-saturated quaternary join. Wendlandt and Eggler obtained the results related to I_{10} are from Yoder and Kushiro (1969) and those for I_{11}, I_{12} and I_{13} are from Luth (1967). In Fig. 12.11, light solid lines are related to reactions in the CO_2-saturated quaternary joins and volatile-absent subsolidus reactions are already discussed in the preceding sections. The reaction, enstatite + magnesite → forsterite + CO_2 is from Newton and Sharp. The reaction related to the breakdown of leucite to form sanidine and kalsilite is from Lindsley (1966).

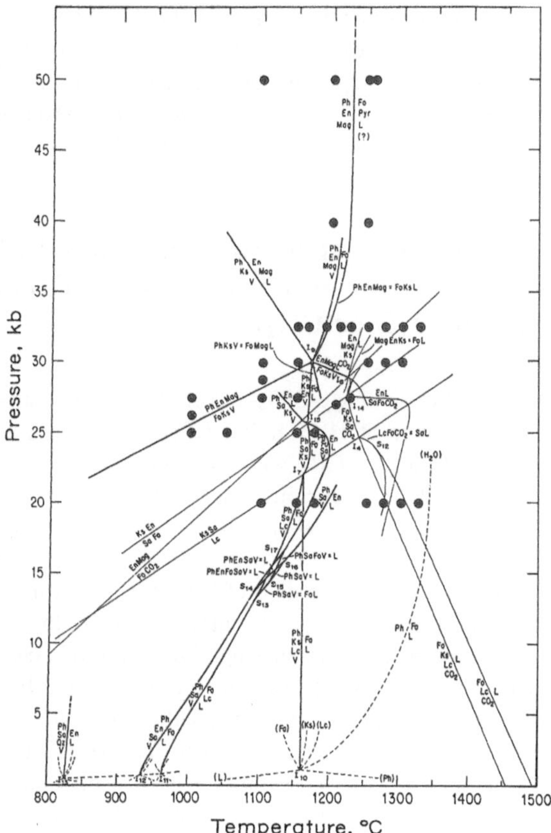

Fig. 12.11 P-T projection of phase relations involving phlogopite, enstatite, forsterite, magnesite, sanidine, kalsilite, leucite, liquid, and vapour in the system $KAlSiO_4$–MgO–SiO_2–H_2O–CO_2. *Dashed lines* at low pressure are reactions in the H_2O-saturated join (quaternary). The results for I_{10} are from Yoder and Kushiro (1969) and those for I_{11}, I_{12}, and I_{13} are from Luth (1967). *Light* and *solid lines* are reactions in the CO_2-saturated join (quaternary) and volatile-absent subsolidus reactions that are discussed in Wendlandt and Eggler (1980a). *Heavy solid lines* are univariant reactions in the H_2O–CO_2 join (quinary). The reaction leucite → sanidine + kalsilite is from Lindsley (1966) and the reaction enstatite + magnesite → forsterite + CO_2 is from Newton and Sharp (1975). Pyrope is observed as a reaction product at 4.0 GPa and higher. Although kalsilite is a predicted reaction product at pressures above I_9, it was not observed in run products. Abbreviations are same as Fig. 12.4 and additionally, En: enstatite, Mag: magnesite; Pyr: pyrope. (after Wendlandt and Eggler 1980b)

According to Wendlandt and Eggler (1980), small percentage of melting of phlogopite-bearing upper mantle between 2.0 and 3.0 GPa will result in the generation of a potassium-rich silica-undersaturated melt corresponding to a olivine leucitite or a basanite.

They concluded that at pressures in excess of 5.0 GPa, phlogopite ceases to be a solidus phase. This pressure corresponds to the breakdown of phlogopite at sub-solidus conditions and to the approximate intersection of the continental and Lesotho geotherms with the phlogopite + enstatite + magnesite solidus. Partial melting at higher pressures defined by the geotherm will not involve phlogopite.

12.5 Investigation of the System Forsterite–Kalsilite-SiO_2 at 2.8 GPa under Dry or Volatile Present Conditions, (in Presence of H_2O or CO_2)

Liquidus phase relation of the join Mg_2SiO_4–$KAlSiO_4$–SiO_2 was also studied under dry condition and in presence of H_2O and/or CO_2 by Gupta and Green (1988). Their experimental results at 2.8 GPa with or without CO_2 are summarised in Fig. 12.12, which is based on quenching data and compositions of liquids (as determined by electron microprobe analyses) in equilibrium with liquid at different temperatures. Comparison of Fig. 12.12 with that of Schairer's one atmosphere phase diagram (1954) suggests that the field of leucite is absent in 2.8 GPa isobaric polythermic section, and the eutectic point, characterised by the assemblage, leucite + forsterite + kalsilite + liquid is missing. The study of Scarfe et al. (1966) established that leucite breaks down to kalsilite and sanidine at high pressure and the study of Lindsley (1966) in the system kalsilite-SiO_2 at different pressures suggests that leucite may be stable at 2.8 GPa in the system as an incongruent melting phase of sanidine. In this study the coexistence of sanidine and kalsilite (± liquid) suggests that the stability of leucite is restricted to pressures of <2.8 GPa.

Phase equilibria study of the silica-undersaturated portion of the forsterite–kalsilite-SiO_2 join under dry conditions (Fig. 12.12) shows that there are two invariant points: (1) one occurring at $Fo_5Ks_{62}Qz_{33}$ and 1460 °C (forsterite + sanidine + orthopyroxene$_{ss}$ + liquid) and (2) the other at $Fo_6Ks_{69}Qz_{25}$ and 1440 °C (kalsilite + forsterite + sanidine + liquid). The invariant point $Fo_5Ks_{62}Qz_{33}$ lies within the forsterite- kalsilite-sanidine join and is thus a peritectic point at which the reaction, enstatite + liquid → forsterite + sanidine occurs. Enstatite crystals in runs near the peritectic point are too small (and contained quench overgrowths) for accurate electron microprobe analysis, but if the enstatite contains $MgAl_2SiO_6$ solid solution as is probable at this high temperature and pressure, then the liquids along the forsterite + enstatite cotectic would be mildly peralkaline, and would lie on the K_2O-rich side of the forsterite–kalsilite-SiO_2 plane. The peralkaline character disappears at the peritectic point as enstatite reacts out and liquids in equilibrium with forsterite- sanidine lie on the forsterite–kalsilite-SiO_2 plane.

The invariant point $Fo_6Ks_{69}Qz_{25}$ lies within the forsterite–kalsilite-sanidine field and is thus a eutectic point marking the low temperature goal of liquids crystallising in this system. It is noteworthy that the minimum melt composition lies close to the forsterite–leucite join and has an extremely undersaturated composition. The join enstatite-sanidine acts as a thermal divide so that all olivine-normative liquids

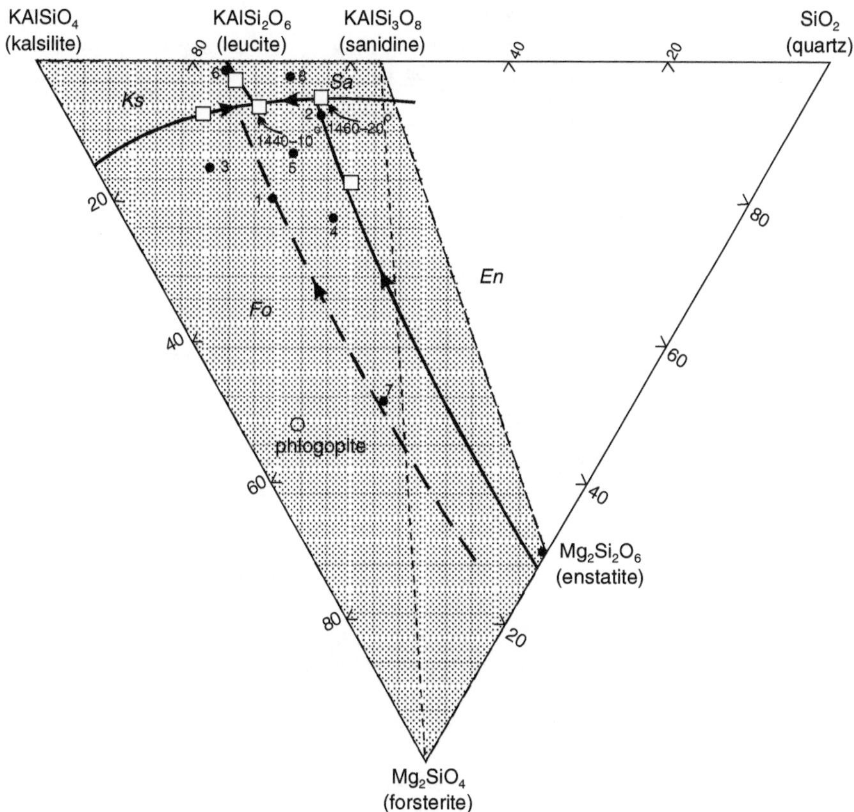

Fig. 12.12 Phase relations for liquidus surface of forsterite-kalsilite-quartz at 2.8 GPa under vapour-absent, anhydrous conditions (*solid lines*) and under CO_2-saturated conditions (*dashed line*). The *shaded area* includes all compositions which fractionate toward the silica-undersaturated minima at 2.8 GPa under both dry and CO_2-saturated conditions. *Filled circle* indicates the composition used. *Square* indicates analysed cotectic glass compositions (after Gupta and Green 1988)

fractionate toward the forsterite–kalsilite-sanidine eutectic (Wendlandt and Eggler 1980a, b).

The anhydrous joins at 2.8 GPa contrast with that at 1 atmosphere (Schairer 1954) in the contraction of the forsterite stability field in the former and the large stability field for leucite in the latter. At one atmosphere, the peritectic forsterite + liquid → enstatite + leucite, is at $Fo_1Ks_{43}Qz_{56}$ and 1200 ± 20 °C, and the eutectic lies at $Fo_{20}Ks_{65}Qz_{15}$ and 1456 ± 10 °C.

Application of phase diagram to model partial melting of a harzburgite with minor potassium-rich phase (sanidine) shows that melting at low pressure will produce silica-oversaturated liquids, but at 2.8 GPa, initial liquids are strongly silica-undersaturated (leucite-normative) but become olivine- and enstatite-normative at temperatures greater than 1560 °C. At 2.8 GPa, under dry conditions, liquids

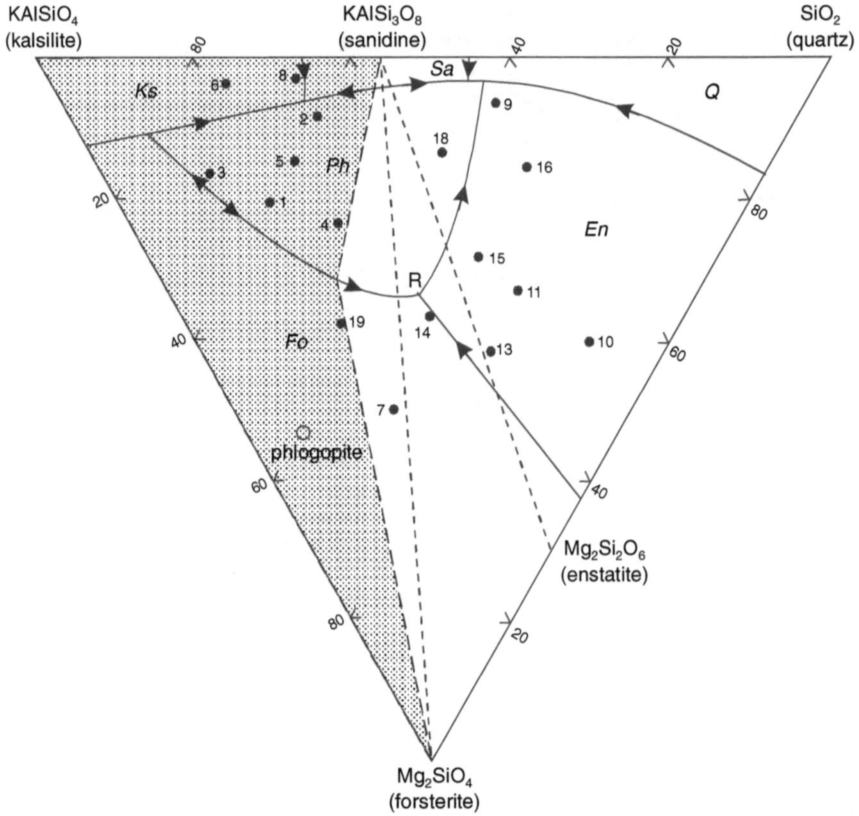

Fig. 12.13 Phase relations from the liquidus surface of forsterite- kalsilite-quartz at 2.8 GPa under water-saturated condition. Numbered points are compositions used for experiments. The reduced field of compositions fractionating to the silica-undersaturated minimum at 2.8 GPa in consequence of both the expanded field of olivine crystallisation relative to enstatite and to the role of phlogopite crystallisation (after Gupta and Green 1988)

within the forsterite–enstatite-sanidine field fractionate toward the silica -undersaturated peritectic and eutectic points.

Under water-saturated conditions, the liquidus surface is at much lower temperatures and there is a large liquidus field of phlogopite. The forsterite–enstatite cotectic moves to more SiO_2-rich conditions but this effect decreases as compositions move away from the forsterite–quartz join. The peritectic point R (forsterite + liquid $\rightleftharpoons$ enstatite + phlogopite) lies at $Fo_{33}Ks_{34}Qz_{33}$ (i.e. within the forsterite–enstatite-sanidine compositional field) and its temperature is 1160 $\pm$ 20 °C. The thermal maximum along the forsterite–phlogopite reaction boundary lies close to composition 3 (Fig. 12.13) and at a temperatures close to 1,200 °C, the liquidus surface of the phlogopite field appears to be rather flat at ~1,200 °C (Wendlandt and Eggler 1980a, b) decreasing sharply to the eutectic (kalsilite + sanidine + phlogopite) at approximately $Fo_7Ks_{62}Qz_{31}$, T < 1075 °C. The peritectic enstatite + liquid ⇌ phlogopite + sanidine, is close to $Fo_4Ks_{44}Qz_{52}$ and 1,010 °C. In Fig. 12.13, liquid compositions may be divided into those which fractionate toward the undersaturated eutectic (phlogopite + kalsilite + sanidine + liquid) and those which fractionate toward the (phlogopite + sanidine + quartz + liquid) eutectic. Comparison of Figs. 12.12 and 12.13 illustrates the strong effect of phlogopite fractionation in deriving liquids with normative olivine and leucite from fractionation toward the undersaturated minimum under dry conditions or trend to the quartz-bearing minimum in the presence of water-rich fluid.

If the data of Fig. 12.13 are applied to model melting of phlogopite-bearing harzburgite, then it may be predicted that under water-saturated (i.e. activity of water ~1) conditions, initial melts should be olivine and hypersthene-normative and would be quite rich in olivine. Crystal fractionation of phlogopite + enstatite, without olivine, would drive liquids to more silica-rich compositions. Comparison with Fig. 12.12, shows that the presence of water at these pressures plays a major role in diverting the fractionation paths of olivine or (olivine and leucite)-normative liquids away from the SiO_2-undersaturated minimum toward the silica-oversaturated minimum. This contrasts with the analogous system, forsterite–nepheline–quartz (Gupta et al. 1987), where the join enstatite-albite remains a thermal divide at 2.8 GPa under both anhydrous and water-saturated conditions and liquids formed by water-saturated melting of an albite-bearing model harzburgite will fractionate toward a forsterite–jadeite- enstatite invariant point (Gupta et al. 1987).

Composition of the phase diagram for forsterite–kalsilite–quartz under water-saturated conditions at lower pressure (Luth 1967; Wendlandt and Eggler 1980a, b) shows an expansion of the liquidus field of phlogopite with increasing pressure and a persistence of the incongruent melting of phlogopite to at least 2.8 GPa. However, the invariant point (forsterite + enstatite + phlogopite + liquid) lies at approximately $Fo_7Ks_{37}Q_{56}$ and ~1,060 °C under 0.3 GPa and marks the peritectic reaction,

$$\text{forsterite} + \text{liquid} \rightleftharpoons \text{enstatite} + \text{phlogopite},$$

i.e. olivine rather than enstatite is in reaction relation with hydrous melt. In terms of partial melting of a phlogopite-bearing harzburgite, liquids at the solidus are `rhyolitic' and strongly quartz-normative at <0.3 GPa, but are picritic and olivine and hypersthene-normative at 2.8 GPa (cf., Wendlandt and Eggler 1980a, b).

The carbonation reaction forsterite + CO_2 ⇌ enstatite + magnesite lies below 1,200 °C at 2.8 GPa (Newton and Sharp 1975). As all the experiments at 2.8 GPa are at temperatures greater than 1,250 °C, forsterite remains stable with CO_2. The results are projected from CO_2 apex on to the forsterite–kalsilite–quartz plane. It may be noted however that the depression of the liquidus surface by CO_2 is very pronounced, implying a high ($CO_3^=$) solubility in the melt phase. Although magnesite is absent at the solidus at 2.8 GPa, and high CO_2 contents (>17 wt% CO_2) were used in the capsules, it is not certain that all charges were CO_2-saturated. An investigation of CO_2 ($CO_3^=$) solubility in liquids in the forsterite–kalsilite–quartz join would have enlarged the investigation considerably but Gupta and Green noted a close approach of the experimental conditions to the carbonation reactions and marked depression of the solidus and liquidus by CO_2 near this reaction boundary (Wendlandt and Eggler 1980a, b).

In the CO_2-bearing join the forsterite–enstatite cotectic moves to much lower SiO_2 conditions such that the olivine + orthopyroxene + sanidine or leucite invariant point occurs below 1,260 °C and at compositions within the forsterite–kalsilite–leucite field (Wendlandt and Eggler 1980a, b). The experiments of Gupta and Green do not define the exact nature or position of the invariant point as it must be noted that the P-T conditions approach the carbonation reaction so that the character of the melt may change rapidly from silicate- dominated with dissolved ($CO_3^{=}$) to carbonate dominated with dissolved ($SiO_3^{=}$).

The composition of enstatite along the forsterite–enstatite cotectic is close to ideal enstatite with ≤0.5 wt% Al_2O_3 in solid solution. Liquids are therefore, not appreciably peralkaline in character in the presence of CO_2 in this join.

The forsterite–kalsilite–quartz join at 2.8 GPa is a simple system analogue for highly potassic liquids under mantle conditions and for melting of mantle harzburgite with a minor potassic phase. The study of the join at 2.8 GPa, illustrates the effect of pressure in moving phase boundaries, particularly the important forsterite-enstatite boundary, from quartz-normative to strongly leucite-normative at high pressures (Wendlandt and Eggler 1980a, b). The effects of the common volatiles H_2O and CO_2 are also large with H_2O playing a particularly important role in stabilising the low-SiO_2, high Mg phase phlogopite at liquidus temperatures. While an H_2O-rich fluid moves liquids in equilibrium with harzburgite away from leucite-normative to olivine- and hypersthene-normative compositions, a CO_2-rich fluid has the opposite effect producing liquids of extremely silica-undersaturated character from a harzburgite source. If H_2O:CO_2 ratios are variable in a relatively oxidised mantle (i.e. f(O_2) condition >MW; Taylor 1985; Foley et al. 1986b) then phlogopite stability may extend to a temperature greater than 1160 °C and the position of the peritectic point (phlogopite + enstatite → forsterite + liquid) may move to leucite-normative compositions. Analogous results were obtained for an olivine leucitite composition (Ryabchikov and Green 1978) and in the forsterite–kalsilite–quartz system by Wendlandt and Eggler (1980a, b, c). Foley et al. (1986 a, b) have shown that fluorine substituting for $(OH)^-$ in fluorphlogopite has a dramatic effect in expanding the phlogopite field to higher temperatures (1480 °C at 2.8 GPa) and moving the forsterite-enstatite cotectic to leucite-normative compositions. Natural phlogopite from high pressure xenoliths contain both fluorine (Foley et al. 1986b) and variable TiO_2 contents which will stabilisze them to higher temperatures.

The experimental study of the join forsterite–kalsilite–quartz illustrates a mechanism by which, even in this simple system, the availability of C–H-O fluid may produce diverse liquids at very low degrees of melting of a model phlogopite, harzburgite source rock. In addition, small differences in H_2O/CO_2 ratio may control the appearance of phlogopite during fractionation of highly potassic liquids in the upper mantle. The early appearance of phlogopite will drive liquids toward silica-oversaturated derivative melts, whereas the same parent magma following a P-T path avoiding the phlogopite field, will fractionate to the silica-undersaturated minimum at high pressure (kalsilite + sanidine + forsterite + liquid or leucite + sanidine + forsterite + liquid).

12.6 Phase Relations in the System $KALSiO_4$–Mg_2SiO_4–SiO_2 at 2.8 GPa in Presence of Fluorine

Fluorine concentration in potassium-rich lamproites is high and it shows positive correlation with potassium content (Aoki et al. 1981). The possible importance of fluorine in the petrogenesis of lamproites has been emphasised by Jaques et al. (1984), who suggested that at higher P-T condition, fluor-mica is more stable than hydroxy-mica. Foley et al. (1986) therefore, studied phase relationship in the join kalsilite–forsterite–quartz at 2.0 GPa with approximately 4 wt% F added to the system. Figure 12.14 shows the liquidus phase relations at 2.8 GPa with 4 wt% $(F_2O)_{-1}$.

The results presented in Fig. 12.14 for 4 % $(F_2O)_{-1}$ have been projected by Foley et al. (1986), onto the kalsilite-forsterite–quartz base from the corresponding fluorine end members, $KAlSiF_8$–Mg_2SiF_8–SiF_4. This figure therefore, represents a prismatic projection along the F_2O_{-1} exchange vector from a plane within the prism, because of the addition of fluorine by direct substitution. According to Foley

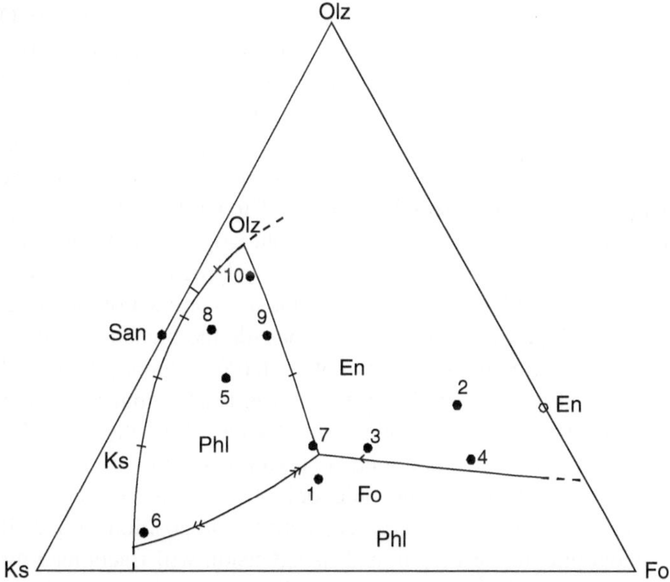

Fig. 12.14 Liquidus phase fields for the join kalsilite-forsterite-SiO_2 at 2.8 GPa in presence of 4 % F_2O_{-1}. Numbers mark starting compositions. *Ks* kalsilite, *San* Sanidine, *Phl* phlogopite, *Qtz* quartz, *En* enstatite. (after Foley et al. 1986)

et al., it is not a saturation surface projected from the apex of a tetrahedron as in the H_2O and CO_2 system, but the differences in the projection angles are not large, allowing comparisons with CO_2 and H_2O-bearing kalsilite–forsterite–quartz system at 2.8 GPa (Gupta and Green 1988).

In the dry system, kalsilite–forsterite–quartz at 2.8 GPa, MgO-rich liquid crystallises forsterite and enstatite and the composition moves toward a forsterite + enstatite + sanidine + liquid peritectic, and then to a phlogopite + sanidine + kalsilite + sanidine + liquid eutectic, as in the case of silica-undersaturated compositions (Fig. 12.12). These two four-phase points occur in the silica undersaturated portion of the forsterite-sanidine join. The crystallisation path in the kalsilite–forsterite–quartz system at 2.8 GPa is similar to those in the same join at 3.0 GPa except for the presence of leucite at low pressures. Compositions defined by enstatite-sanidine-phlogopite crystallize through a peritectic, where quartz + phlogopite + enstatite + liquid are at equilibrium at all pressures from 1 to 2.8 GPa. Phase volume of phlogopite is enlarged greatly with increasing pressure and the enstatite + phlogopite + forsterite + liquid peritectic lies at much more Mg-rich compositions at pressures equivalent to that in the mantle.

Foley et al. observed that the phase diagram for 4 % F_2O_{-1} at 2.8 GPa (Fig. 12.14) broadly resembles the water-saturated, fluorine-free system of Gupta and Green (1988) (Fig. 12.13) in having a large primary phase volume for phlogopite plus primary phase volumes for the same six minerals (enstatite, forsterite, phlogopite, kalsilite, sanidine and quartz). Foley et al. (1986, Fig. 12.15) compared the shift in the enstatite-forsterite phase boundary in presence of CH_4, CO_2 and F with respect to the system forsterite–kalsilite–SiO_2–H_2O system investigated by Gupta and Green (1988). However the fluorine-bearing system differs in that the fluorphlogopite has a much greater thermal stability (maximum 1,490–1,500 °C) than hydroxyl-phlogopite (<1,200 °C: Gupta and Green in 1988). This may be attributed to the lack of K–H repulsion in fluorphlogopite. This repulsion exists in hydroxy-phlogopite due to orientation of the O–H bond directly away from neighbouring octahedral cations and toward the interlayer potassium cations (McCauley et al. 1973). The enstatite-phlogopite phase boundary is not a peritectic reaction, despite its extension, apparently lying outside the join enstatite-phlogopite. According to Foley et al. (1986), this is an artefact of the projection due to phlogopite_{ss} and liquid composition lying outside the plane of projection. As in the H_2O-saturated system, the intersection of forsterite + phlogopite phase boundary with the extension of forsterite–phlogopite join forms a thermal maximum. Liquids with compositions to the silica-rich side of this divide will fractionate either through the enstatite + forsterite + phlogopite + liquid peritectic point or across the phlogopite phase field to either the phlogopite + sanidine + quartz + liquid or phlogopite + sanidine + kalsilite + liquid eutectics. Compositions to the silica- poor side of the phlogopite–forsterite join and its extension will fractionate through the kalsilite + forsterite + phlogopite + liquid, peritectic point or across the phlogopite phase field toward the kalsilite + sanidine + phlogopite eutectic.

In the fluorine-bearing system the primary phase field of enstatite relative to forsterite is enlarged compared to the volatile-free join, so that the

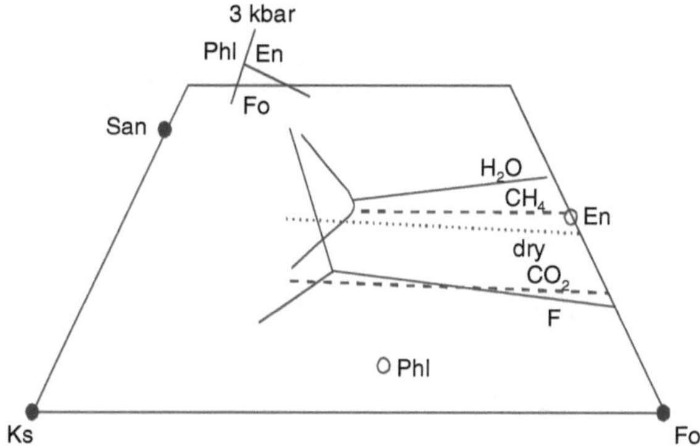

Fig. 12.15 Part of the join kalsilite-forsterite-quartz showing relative positions of the enstatite-forsterite phase boundary in the presence of various volatile species. The boundary for CH_4 is estimated from its position in the nepheline-forsterite-quartz system, taken from Gupta et al. (1987). The position with HF is uncertain, but can be expected to be close to the CH_4 position. The data for the water-saturated kalsilite-forsterite-quartz join at 0.3 GPa are taken from Luth (1967) (after Foley et al. 1986). Abbreviations as in Fig. 12.14

enstatite + forsterite phase boundary is in a similar position to that in the CO_2-saturated join (Fig. 12.12) of Gupta and Green, 1988. The position of this phase boundary is frequently taken as the degree of polymerization of the melt (Eggler 1974; Kushiro 1975). Expansion of the enstatite phase volume at the expense of forsterite with the addition of fluorine therefore, suggests that fluorine causes polymerization of the structure. The assemblage phlogopite + enstatite + forsterite + liquid in the simple join is similar to that of a phlogopite harzburgite. The position of this point may indicate that partial melting of a phlogopite harzburgite in H_2O-free conditions but in the presence of fluorine should produce a silica-undersaturated magma. Such a melt should lie to the silica-poor side of the forsterite-sanidine join at 1,480 °C. In contrast, in the water-bearing system (Gupta and Green 1988), the first melt would lie in the more silica-saturated part of the system, within the triangle, defined by forsterite, enstatite and sanidine at 1160 °C.

Foley et al. (1986) conducted several experiments with composition 1 (Fig. 12.14) with 10 % F_2O_{-1}. They observed that the near-liquidus runs contained a very minor amount of immiscible liquid phase rich in Mg and F. The immiscibility may extend above the liquidus, but this could not be ascertained because of the abundance of quench crystals in these near liquidus runs. They further observed that with 10 % F_2O_{-1}, composition 1 lies just inside the primary phase volume of phlogopite, indicating expansion of the phlogopite phase volume with increasing fluorine, They think that because of the apparent immiscibility, which occurs at high fluorine contents, it is unlikely that phlogopite will melt congruently at any fluorine content.

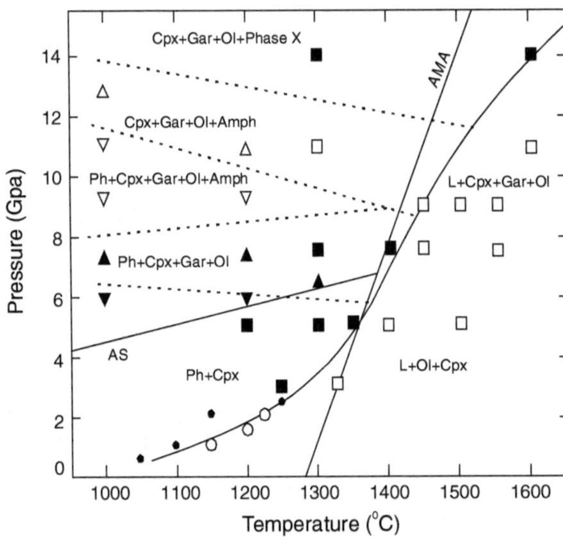

Fig. 12.16 Pressure-temperature projection of experimental results from this study (*squares*), from Modreski and Boettcher (1972) (*circles, solid symbols* are subsoildus experiments, open symbols are supersolidus experiments) and from Sudo and Tatsumi (1990) (*triangles, differnet symbols* refer to the different phase assamblages as labelled on the figure). Stable phase assamblages are noted on the diagram, and the inferred boundaries between the differing subsolidus assamblages are drawn as *dashed lines*. The solidus is shown as a *heavy solid line*. Also shown on this diagram is the average modern mantle adiabat (AMA) and a geotherm for an Archean shield (AS) (after Luth 1997)

12.7 Investigation on the Assemblage Phlogopite-Diopside up to 17 GPa

The experimental study of the system phlogopite-diopside from 3.5 to 17 GPa has been made by Luth (1997, Fig. 12.16). On the basis of both natural samples and experimental studies, clinopyroxene has been suggested as a potential reservoir for potassium in the Earth's mantle (Harlow and Veblen 1991). The amount of K partitioning into clinopyroxene depends on the phase assemblage present, the bulk composition, pressure, and temperature. To investigate some of these dependencies, subsolidus and melting phase relations in the system phlogopite-diopside was studied by Luth (1997) to 17 GPa. In this system, phlogopite becomes unstable with increasing pressure, breaking down to potassium richterite, which in turn yields potassium-bearing hydrous phase (phase X), such that a potassium-rich phase coexists with clinopyroxene to 17 GPa. Clinopyroxenes contain ≤1.3 wt% K_2O in assemblage of phlogopite + clinopyroxene ± olivine ± liquid at 3–5 GPa, phlogopite + clinopyroxene + garnet ± olivine ± liquid at 7–9 GPa, clinopyroxene + garnet + olivine ± potassium richterite ± liquid at 11 GPa, and clinopyroxene + olivine + garnet + phase X at 14 and 17 GPa.

In these assemblages, K is partitioned into hydrous phases or liquid, rather than into the clinopyroxene. By inference, phlogopite (or its higher-pressure breakdown products) is the primary host or K in the mantle (if H_2O is present), and any coexisting clinopyroxene has very low concentrations of K. Conversely, the natural occurrence of clinopyroxene with ≫1 wt% K_2O requires that phlogopite, potassium richterite, or phase X is not stable in the local source environment of such samples.

Luth (1997) observed that a pyroxene- and phlogopite-bearing assemblage yields phlogopite + diopside + garnet at 6 ± 1 GPa. Potassium-richterite joins this assemblage at 8 ± 1 GPa and at still higher pressure, phlogopite is completely replaced by potassium-richterite and the stable assemblage is diopside + garnet + forsterite + potassium-richterite above 11.5 GPa and 1,000 °C and 8 GPa and 1,400 °C (Fig. 12.16).

12.8 K-Richterite as a Source Mineral of Potassium in the Upper Mantle

12.8.1 P-T Stability of K-Richterite

Scott Smith and Skinner (1984) classified potassium-rich lamproites on the basis of the presence of following minerals: phlogopite, potassium-richterite, olivine, diopside and sanidine. In case of some of the diamond-bearing ultramafic rocks from India, which are described as kimberlites, Scott Smith classified them as lamproites because of the presence of potassium-richterite. In many lamproites potassium-richterite is quite common. This mineral is particularly an essential mineral in the mantle. It is also present as a groundmass mineral in wyomingite, cedricite, orendite and jumillite. Bergman (1987) also emphasised the presence of K–Ti- richterite, K-riebeckite and K-arfvedsonite, and, particularly the potassium-richterite as an important mineral constituent of the lamproites. The possible presence of potassium-richterite in the upper mantle and importance of this mineral as a source of potassium has been discussed by Kushiro and Erlank (1970) and Varne (1968). The P-T stability of potassium-richterite was therefore, determined by Gupta and Venkatesh. Which is shown in Fig. 12.17. The data at 0.1 GPa and below are from Charles (1974, 1975). Extension of the sodium-richterite breakdown curve as determined in Gupta and Venkatesh (1993), agrees well with low pressure stability data of Charles.

In Fig. 12.17 the univariant curves of Gilbert and Briggs (1974) related to the stabilityofpotassium-richterite [$KNaCaMg_5Si_8O_{22}(OH)_2$] and fluor-potassium richterite [$KRi(F_{100})$] are also included to compare the relative stabilities of the sodium- and potassium-richterites. It may be noted that although below 1.2 GPa, hydroxy potassium richterite breaks down at lower temperatures than richterite,

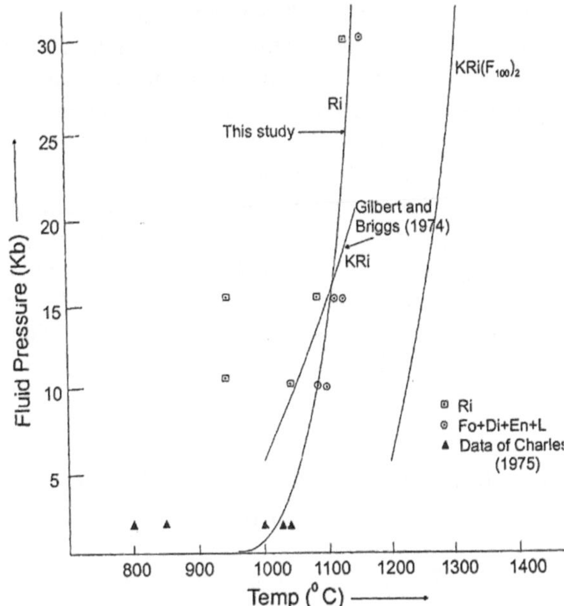

Fig. 12.17 P(H_2O)-T stability of sodium-richterite. The data below 0.1 GPa are from Charles (1975). Breakdown of richterite to various phases below 140 bar is not shown (after Gupta and Venkatesh 1993)

above this pressure however, hydroxy potassium richterite has slightly higher P-T stability. The fluor-potassium richterite breaks down under much higher P-T conditions than richterite and hydroxy potassium richterite (Fig. 12.17), whereas sodium-richterite breaks down to orthopyroxene + forsterite + diopside + liquid, potassium richterite yields diopside + forsterite + liquid. The study of Gupta and Venkatesh (1993) indicates that sodium-richterite is stable also at 3.0 GPa and 1185 °C but at higher pressure it is less stable than hydroxy—and fluorine-bearing potassium richterite. Experimental study of Gupta and Venkatesh (1993) indicates that alkali-richterite could be a source of Na_2O in the upper mantle. Experimental study of Gilbert and Briggs (1974) suggests that partial melting of a potassium-richterite should result in producing a liquid in equilibrium with wherlite—rich—fraction (clinopyroxene + forsterite-rich olivine), whereas an upper mantle with pockets of fluor-potassium richterite after partial melting, should produce a lamproitic liquid.

Konzett et al. (1997) studied the stability of potassium-richterite in a MARID (mica- amphibole-rutile-ilmenite-diopside) assemblage. They compared the results with phase relations and compositions of natural MARIDs to assess possible mechanism of formation for MARID type rocks. Their experimental study on the stability of potassium-richterite on the MARID assemblage is shown in Fig. 12.18. They found that potassium-richterite is stable in a wide range of bulk K/Na ratios in the MARID assemblage to 8.5 GPa and 1,300 °C. In this assemblage the amphibole

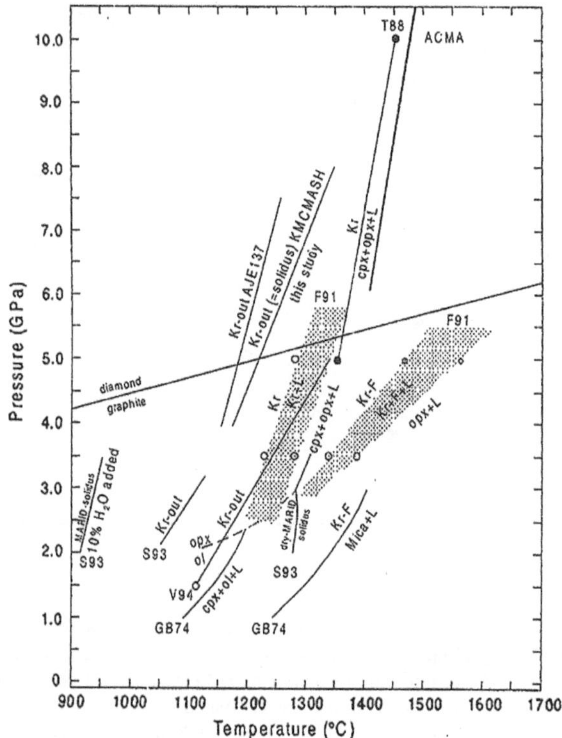

Fig. 12.18 P-T diagram summarising experimental results for a natural MARID sample (AJE137). Circles are run symbols for experiments made by Konzett et al. (1997), heptagons denote experiments by Sweeney et al. (1993). Meaning of symbols are as follows: *Cpx* clinopyroxne, *Opx* orthopyroxene, *Ga* garnet, *Ol* olivine, *Kr* potassium-richterite, *phl* phlogopite. Numbers next to run symbols are approximate percentage of melt in the charge. The *dashed line* is a 40 mW/m^2 geotherm according to Pollack and Chapman (1977);. ACM and PA are the average current mantle adiabat for a subcontinental temperature of 1280 °C and a plume adiabat as given by Thompson (1992). *Shaded areas* span the P-T range obtained from clinopyroxene, orthopyroxene-garnet geothermo-barometry (WE88, Waters and Earlank 1988), P-T for metasomatized peridotites (PH95, Phillips and Harris 1995), P-T for diamond inclusions from the Kimberley areas: northern Lesotho: P-T derived from garnet peridotite data given by Nixon and Boyd (1973). (after Konzett et al. 1997)

can accommodate significant amounts of K in the M site and shows a systematic increase in K/Na ratio with increasing pressure. At P > 7.0 GPa, potassium-richterite can coexist with garnet. This result contradicts the observation of Kushiro and Erlank (1970). Phase relations of potassium-richterite in a normal MARID composition are consistent with those in the simplified system and confirm the potential stability of potassium-richterite and potassium-richterite + garnet within the diamond stability field. The assemblage K-richterite + phlogopite + clinopyroxene is incompletely buffered in the KNCMASH system in a systematic relation between bulk and mineral compositions observed in the experiments. Such a correlation,

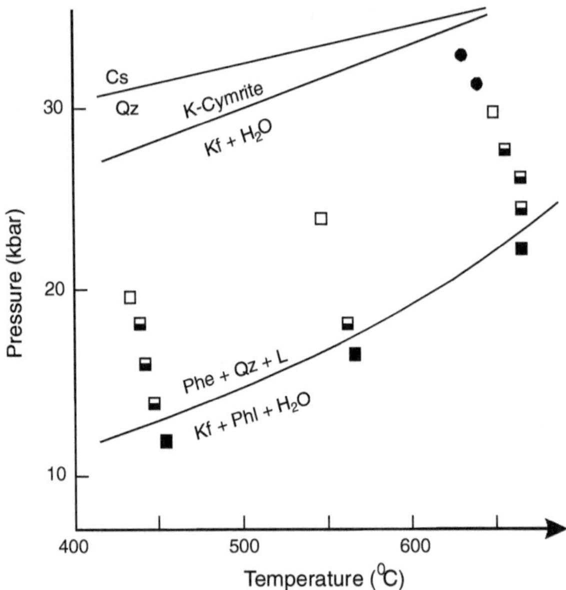

Fig. 12.19 P-T diagram showing experimental results on an assemblage comprising K-feldspar + phlogopite yielding a mixture composed of K-feldspar, phlogopite, phengite and quartz [no H_2O (*circles*) or a minimum of 3 wt% of H_2O (*boxes*) had been added]. The symbols represent experimental conditions including the estimated 2σ uncertainty; closed = no decomposition detected; half closed = partial disappearance of K-feldspar + phlogopite, in particular, when less than 11 wt% of H_2O had been added; open = total or nearly partial disappearance of this assemblage in the presence of about 3.5 wt% of H_2O. (after Massonne (1992)

however, cannot be observed in natural MARIDs. Therefore, MARID-type rocks do not represent the bulk composition from which they were formed and, hence must be products of an open system crystallisation.

12.8.2 Investigation on High Pressure Stability of Phengite

Experiments were conducted by Massonne (1992) to study following reactions in the system K_2O–MgO–Al_2O_3–SiO_2–H_2O at temperatures between 400 and 700 °C:

(1) Phengite ± quartz + potassium and magnesium-rich siliceous fluid → feldspar + phlogopite + H_2O, and
(2) phengite + talc + potassium- and magnesium-rich siliceous fluid → phlogopite + quartz/coesite + H_2O.

His experimental results are summarised in Fig. 12.19, which suggests that subduction of potassium-rich metasediments, constituting fraction of the oceanic crust, could give rise to low-temperature ultrapotassic fluid, at temperatures between 300 °C and pressures between 1.5 and 3.0 GPa. The ascending potassium-rich fluid should penetrate the overlying mantle and metasomatize it. After termination of the subduction process, heating of this mantle material, previously cooled by the subducted lithosphere, could lead to the formation of a high temperature potassium-rich magma.

Chapter 13
Experimental Studies on K-Rich Rocks

13.1 Investigations of Leucite-Bearing Rocks Under Atmospheric Pressure

Experimental studies on potassic rocks from Leucite Hills Wyoming, U.S.A. by different investigators are summarized in Table 13.1. Wyomingites and orientates reported in this table are from the same locality, but the samples are different with slightly different chemistry, hence variation in liquidus temperatures of different primary phases are expected. It may be observed that the liquidus temperatures in case of the runs conducted by Yagi and Matsumoto are higher than average tholeiitic or alkali basaltic rocks (Yoder and Tilley 1962). Although textural studies indicated that leucite is the liquidus phase in the cases of wyomingite and orendite, clinopyroxene always appeared as the first phase in their atmospheric runs. This was interpreted by them to be due to loss of water in their one atmospheric runs.

Table 13.1 shows that there are large differences between the liquidus and solidus temperatures in case of the experimental runs performed by Fyfe (quoted in Carmichael 1967). On the basis of this observation, Carmichael concluded that rocks of Leucite Hills are product of a magmatic liquid formed at great depth in the mantle.

Gupta et al. (1983) made quenching experiments on four lamproitic rock samples collected from dykes intruding Gondwana rock formations in the Damodar Valley regions (see Chap. 4). The two localities (Raniganj and Jharia Coal Fields) from where the samples were collected are respectively 200 and 250 km west of Kolkata. The powdered rock samples were heated in sealed capsules under atmospheric pressure and NNO buffer conditions. Analyses of the four rock samples from Sudamdih (S), Mohanpur (M-2 and M-3) and Poidih (P) are shown in Table 4.2. Results showed that the primary phases, liquidi and solidi are as follows: S: Leucite, 1,305 and 1,185 °C; Leucite, 1,245 and 1,150 °C; M-2: Olivine, 1,250 and 1,150 °C; M-3: Leucite, 1,242 and 1,150 °C and P: Olivine, 1,235 and 1,100 °C.

A.K. Gupta, *Origin of Potassium-rich Silica-deficient Igneous Rocks*,
Springer Geology, DOI 10.1007/978-81-322-2083-1_13

Table 13.1 One Atmosphere Melting Studies on Potassic Rocks From Leucite Hills, Wyoming, U.S.A

Sr. No.	Crystalline (assemblage shown in bracket)	Locality	Liquidus Phase	Liquidus (°C)	Solidus (°C)	Source
1.	Orendite (Lc + Sa + Cpx + Ap + Cal)	South Table Mountains	Cpx	1,302		Yagi and Matsumoto (1966)
2.	Wyomingite (Lc + Cpx + Ph + Ap + Cal + Mt)	Boars tusk	Cpx	1,322		Yagi and Matsumoto (1966)
3.	Xenolith (included within Wyomingite)	Boars tusk	Cpx	1,318		Yagi and Matsumoto (1966)
4.[a]	Wyomingite (mineralogy same as No. 2)	Boars tusk	(Ph + Cpx)	1,270 (1,250–1,150)	1,000	Sobolev et al. (1975)
5.	Wyomingite (mineralogy same as No.2)	Boars tusk	Lc	1,200	1,000	Fyfe (quoted in Carmichael 1967)
6.	Olivine orendite (Ol + Lc + Sa +Cpx + Ap + Cal)	South Table mountain	Ol	1,275	1,000	Fyfe (quoted in Carmichael 1967)
7.	Olivine orendite (mineralogy same as Sl. No. 6)	North Table Mountain	(Lc + Ol)	1,215	1,000	Fyfe (quoted in Carmichael 1967)
8.	Madupite (Ph + Cpx + Mt)	Pilot Butte	Lc	1,245	1,040	Fyfe (quoted in Carmichael 1967)

[a] Fluid inclusion study

Lc leucite, *Sa* sanidine, *Cpx* clinopyroxene, *Ap* apatite, *Cal* calcite, *Ph* phlogopite, *Ol* olivine, *Opx* orthopyroxene, *Mgt* magnetite, *Ilm* ilmenite

13.2 Investigation on a Synthetic Leucite Basanite and Melilite-Nephiline Leucitite up to 2.5 Gpa and Variable Temperatures

13.2.1 Investigation on a Natural Leucite Basanite and a Tephrite

Experiments in the presence of water vapour [$P(H_2O)$ = P(Total)] were made (between 0.1 and 0.5 Gpa) on two natural leucite-bearing rocks, collected from Kunkskopf and Hochsimmer (East Eifel, West Germany, Table 13.2). The rock from Kunkskopf contained phlogopite, plagioclase, nepheline, leucite, apatite, and magnetite in the groundmass with clinopyroxene, and small amounts of olivine as phenocrysts, thus it is a basanite. The rock from Hochsimmer is a tephrite, consisting of clinopyroxene and a small amount of phlogopite as phenocrysts with rare amounts of nepheline, plagioclase, and opaque minerals in the groundmass. At a pressure of 0.5 Gpa and temperatures of 800 and 850 °C, the powdered samples of the rocks from both the areas produced the same assemblage of clinopyroxene, amphibole, phlogopite, magnetite and liquid, although the relative proportions of amphibole and phlogopite are less in the Hochsimmer sample. At 0.5 GPa below 750 °C the tephrite from Hochsimmer produced an assemblage of phlogopite and clinopyroxene with minor amounts of amphibole and magnetite. Disappearance of leucite was noted in both cases slightly above 0.1 GPa from 600 to 800 °C. At 1 GPa below 800 °C the leucite- nepheline tephrite is represented by an assemblage containing nepheline, amphibole, phlogopite, magnetite and pyroxene. Nepheline disappears above 800 °C and the resulting assemblage at 1 GPa corresponds to a magnetite-amphibole-phlogopite pyroxenite. The amphibole is a potassium-richterite.

Experimental investigation was made on a leucite basanite ($F_{20}Di_{30}Lc_{30}An_{20}$) under different P–T conditions in presence of excess water up to 2.5 GPa by Gupta et al. (1976). Their results are summarized in Fig. 13.2, which shows that above 1 GPa the synthetic leucite basanite is converted to a garnet and phlogopite-bearing pyroxenite. Appearance of garnet may be explained by the following reaction:

$$\underset{\text{forsterite}}{Mg_2SiO_4} + \underset{\text{anorthite}}{Ca_2Al_2Si_2O_8} \rightleftharpoons \underset{(\text{pyrope} - \text{grossularite})_{ss}}{(Mg_2Ca)Al_2Si_3O_{12}} \tag{13.1}$$

It should be pointed out that in their experiments on an assemblage comprising forsterite and anorthite (present in 1:1 molar ratio), Kushiro and Yoder (1966) noted that at 1.05 GPa and 1,000 °C, the assemblage is converted to a garnet pyroxenite, which has a wide P–T stability.

Table 13.2 Analyses of k-rich mafic and ultramafic rocks used for experimental studies by various investigators

	1	2	3	4	5	6	7	8	9	10	11
SiO_2	38.05	55.14	50.20	43.56	42.33	37.09	44.65	42.60	42.02	43.70	44.4
TiO_2	3.84	2.58	2.30	2.31	5.62	6.46	6.92	5.30	4.93	2.70	2.1
ZrO_2	–	0.27	–	0.27	–	–	0.27	–	–	–	–
Al_2O_3	7.54	10.35	11.39	7.85	8.14	9.33	2.75	8.30	8.02	13.60	15.6
Cr_2O_3	–	0.04	–	0.04	–	–		0.20	0.15	–	–
Fe_2O_3	8.41	3.27	0.57	5.57	9.21[a]	13.79[a]	6.94	0.20	0.12	6.40	4.2
FeO	2.80	0.62	4.23	0.85	–	–	1.36	7.80	9.07	4.10	5.4
MnO	0.21	0.06	0.07	0.15	0.09	0.12	0.08	N.D.		0.20	0.2
MgO	13.55	6.41	7.23	11.03	11.73	10.44	14.39	15.80	16.14	9.80	7.3
CaO	13.90	3.45	6.00	11.89	6.04	9.07	3.81	10.70	10.54	12.10	11.5
SrO	–	0.26	–	0.40	–	–	–	–	–	–	–
BaO	–	0.52	–	0.66	–	–	2.62	–	–	–	–
Na_2O	1.31	1.21	0.86	0.74	0.26	0.03	0.83	0.80	0.55	3.40	3.2
K_2O	3.02	11.77	10.19	7.19	6.22	2.87	6.80	7.10	7.26	3.40	4.1
H_2O^+	2.27	1.23	3.59	2.89	4.92	8.24	3.22	–	–	0.20	0.6
H_2O^-	3.00	0.61	–	2.09	3.01	–	2.18	–	–	–	–
CO_2	0.53	–	–	–	–	–	0.93	–	–	–	–
P_2O_5	0.95	1.40	1.89	1.50	2.40	2.53	1.41	–	–	0.50	0.2
F	0.27	–	–	–	–	–	0.59	–	–	0.65	–
Cl	0.02	–	–	–	–	–	0.12	–	–	–	–
Total	99.67	99.19	98.52	98.99	99.97	99.97	99.87	98.80	98.8		–

(continued)

Table 13.2 (continued)

	12	13	14	15	16	17	18	19	20	21	22	23
	12	13	14	15	16	17	18	19	20	21	22	23
SiO_2	52.94	56.30	49.13	32.37	30.31	30.30	54.91	55.41	47.96	42.6	44.4	55.41
TiO_2	5.64	2.49	20.02	9.32	7.33	3.60	0.31	0.26	0.94	21.53	1.25	1.41
ZrO_2	–	–	–	–	–	–						
Al_2O_3	8.55	10.50	10.51	9.84	6.61	9.33	22.30	21.97	14.72	8.3	12.71	8.83
Cr_2O_3	–	–	–	0.053	0.52	–				0.2	0.08	–
Fe_2O_3	5.65	3.55	3.82	8.68[a]	11.51	8.58				0.2	0.08	1.17
FeO	–	–	4.30	–	–	–	2.16	2.90	7.77	8.70	8.7	3.40
MnO	0.07	0.88	0.12	0.12	0.17	0.15	0.24	0.24	0.14	–	0.17	0.08
MgO	8.38	6.35	9.87	8.09	11.20	19.89	0.23	0.65	6.38	15.8	13.93	13.59
CaO	4.51	3.11	9.07	8.26	11.76	10.25	3.11	3.69	11.32	10.7	10.68	3.97
SrO	–	–	–	–	–	–			0.22			
BaO	–	–	–	–	–	–			0.15			
Na_2O	0.96	1.25	2.06	0.13	0.12	0.15	6.22	5.32	1.67	0.8	2.17	1.13
K_2O	8.84	11.25	4.86	7.09	4.82	2.89	9.89	9.17	6.97	7.1	4.20	8.63
H_2O^+	0.87	–	2.36	2.35	2.41	2.86				–	–	0.59
H_2O^-	–	–	0.38	2.12	2.22	–				–	–	0.23
CO_2	–	0.10	1.35	6.58	5.42	15.01						
P_2O_5	1.26	1.35	0.97	2.93	4.84	1.20	0.09	0.12	0.50	–	1.34	1.35
F	–	–	–	–	–	–			0.22	–	–	0.36

(continued)

Table 13.2 (continued)

	12	13	14	15	16	17	18	19	20	21	22	23
Cl	–	–	–	–	–	–	0.54	0.26		–	–	–
Total	–	–	–	98.93	98.72	99.20		99.76	98.96	98.8	100.00	100.15

1. Analysis of the katungite used by Arima and Edgar (1983). Analyst: Ramlal and Kuehner
2. Orendite. North Table Mountain, Leucite Hills, Carmichael (1967)
3. Wyomingite, Boars Tusk, Lucite Hills. Analyst: Jension. Manchester University
4. Madupite. Pilot Butte, Leucite Hills, Carmichael (1967)
5. Lamproite from Mohanpur, Damodar Valley, India, Analyst: Gupta et al
6. Lamproite from Satyanarayanpur, Damodar Valley, India. (Gupta et al. 2002)
7. Wolgidite from West Kimberley Australia. Analyst: Ramlal and Arima
8. Glass composition used as starting material by Edgar et al. (1976). Analyst: Kiss
9. Starting glass composition used by Ryabchikov and Green (1978). Analyst: Kiss
10. Leucite basanite from Kunks Kopf, East Eifel, Germany (Duda and Schminke 1978)
11. Leucite tephrite from Hochsimmer, East Eifel, Germany (Duda and Schminke 1978)
12. Armalcolite, Phlogopite lamproite from Smoky Butte, Montana (U.S.A.) (Edgar et al. 1992)
13. Sanidine Phlogopite lamproite from Leucite Hills, USA (Mitchell 1995)
14. Minette from Buell Park, Arizona, USA, Analyst: Roden and Smith (1979). (See Esparanca and Holloway 1987)
15. Lamproite from Sudamdih, Jharia Coal field (Gupta et al. 1983). The amount of H_2O^+, H_2O^- and CO_2 was determined by Gupta et al
16. Lamproite from Poidih, Raniganj Coal field (Gupta et al. 1983). The amount of H_2O^+, H_2O^- and CO_2 was determined by Gupta et al. (2002)
17. Analysis of Jarangdih lamproite (Gupta et al. 2002)
18. Major element chemistry of a 79 AD Vesuvian lava flow from Italy
19. Another sample of a 79 AD Vesuvian lava flow. (Shea et al. 2009)
20. Analysis of a phono-tephrite from Mount Melone lava flow, Alban hills Central Italy (Freda et al. 2008)
21. Glass analysis of a synthetic biotite-mafurite, (Edger et al. 1976)
22. Glass analysis of a synthetic olivine leucitite from Sierra Nevada (W-1) (Elkins-Tanton and Grove 2003)
23. Chemical analysis of Cancarix lamproite, Albacete (Spain), (Wagner and Velde 1986)

[a] Total iron oxide. FeO content was not determined

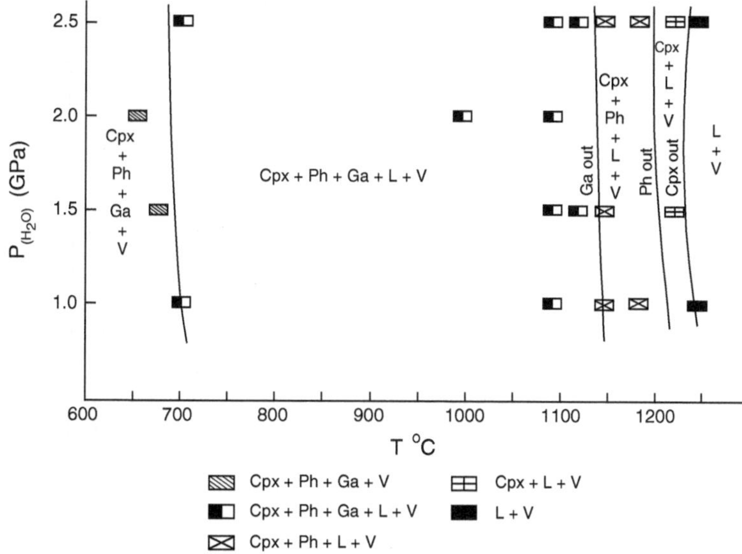

Fig. 13.1 P(H_2O)-T diagram of a synthetic leucite basanite ($Fo_{20}Di_{30}Lc_{30}An_{20}$) (after Gupta et al. 1976) Abbreviation same as in Table 13.1; Ga: garnet, L: liquid, V: vapour

13.2.2 Investigation on a Synthetic Melilite Nepheline Leucitite in Presence of Excess Water

A synthetic melilite-nepheline leucitite ($Di_{28}Ne_{29}Lc_{43}$) was heated at different temperatures (650–850 °C) and pressures (0.1–0.5 GPa in presence of excess water) by Gupta et al. (1976). The starting material was crystallized from a glass of its composition at 900 °C under 1 atmospheric pressure. Appearance of melilite and olivine in this composition is related to the reaction between nepheline and diopside (Bowen 1922a; Schairer et al. 1962). With the mixture $Di_{28}Ne_{29}Lc_{43}$, studied under 1 atm, leucite was the liquidus phase at 1,197 °C under atmospheric pressure and was followed by the appearance of forsterite$_{ss}$ (1,194 °C), nepheline$_{ss}$ (1,185 °C), melilite (1,178 °C) and finally diopside$_{ss}$ (1,100 °C). Forsterite$_{ss}$ disappeared at 1,160 ± 10 °C, and the final assemblage corresponds to a melililite nepheline leucitite. The composition of melilites and nepheline$_{ss}$ are as follows: $Ak_{70}Sm_{30}$ and $Ne_{68}Ks_{29.8}Si_{2.2}$, respectively.

Experimental study of Gupta et al. (1976) on this composition is summarized in Fig. 13.2, which shows that melilite disappears at or above 0.3 GPa (AB). Breakdown of leucite occurs at slightly higher pressure (CD, Fig. 13.2). Phlogopite and pyroxene coexist with liquid above DEF, below which phlogopite disappears at low P(H_2O) by reaction with a liquid; thus, supporting the findings of Luth (1967). The reaction, leucite → kalsilite + K-feldspar (CE, Fig. 13.2), has been studied by Scarfe et al. (1966; see Fig. 6.1). The results of this investigation suggest that leucite and melilite-bearing rocks are strictly confined to volcanic and subvolcanic conditions.

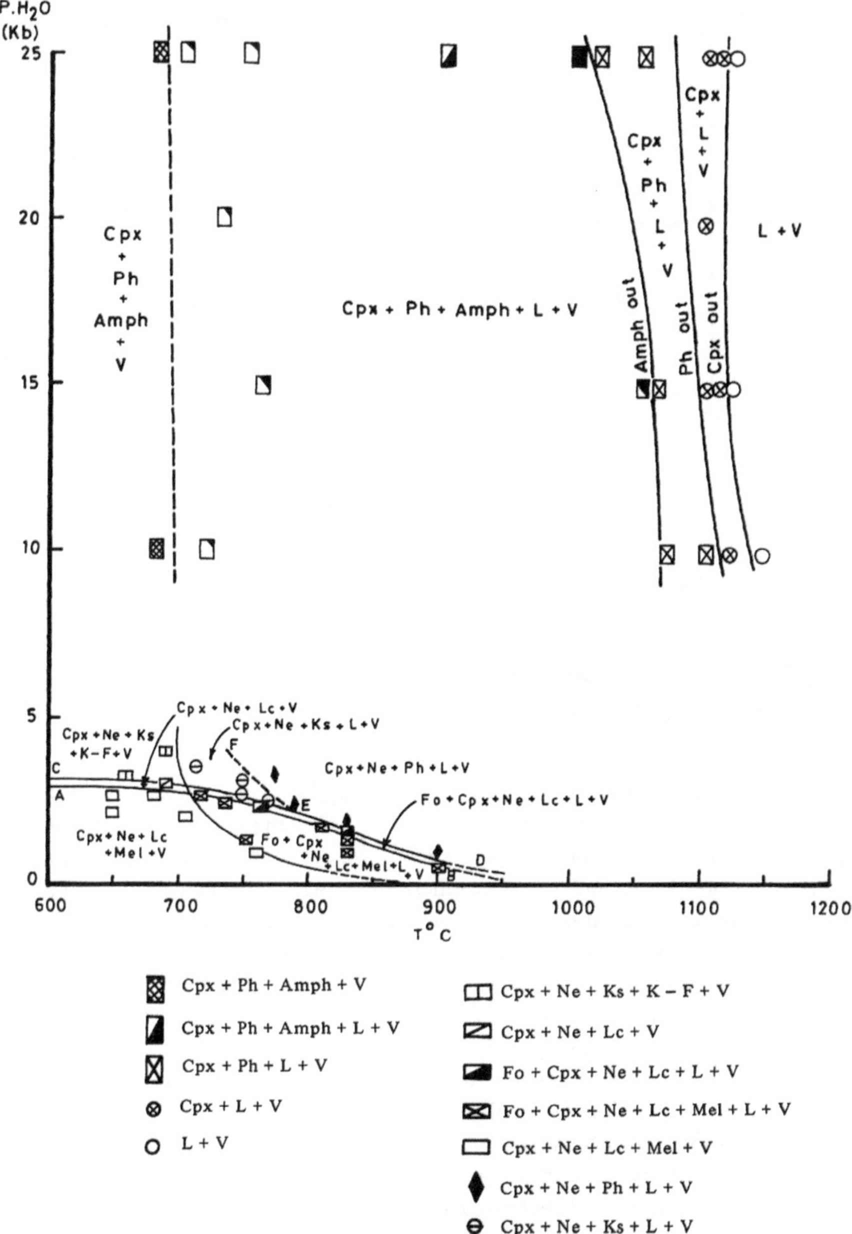

Fig. 13.2 $P(H_2O)$-T diagram of a synthetic melilite-nepheline leucitite ($Di_{28}Ne_{29}Lc_{43}$). (after Gupta et al. 1976). Abbreviation: Amph: amphibole (richterite), Ne: nepheline, Ks: kalsilite, K–F: K-feldspar, Mel: melilite. Other abbreviations are same as in Table 13.1

Figure 13.2 further shows that a melilite-nepheline leucitite and a olivine-melilite-nepheline leucitite are represented at 1 and 2.5 GPa by an assemblage corresponding to amphibole-phlogopite pyroxenite. Although clinopyroxene was the last phase to crystallize in the one atmospheric run, it was the first phase to crystallize in the high pressure runs. The results of the investigation at atmospheric pressure showed that the liquidus temperature of 1,197 °C is lowered by 82° and 64 ± 5 °C at 2.5 and 1 GPa, respectively. Microprobe studies show that the amphibole is a K–Na richterite.

13.2.3 Experimental Study on a 79 AD Vesuvian Lava Flow

Shea et al. (2009) made experimental investigation on two samples of a Vesuvian lava flow from Italy. The compositions of the two samples are included in Table 13.2 (Analysis 18 and 19). Their experimental results are summarized in the Fig. 13.3a and b. It is observed that pyroxene is the liquidus phase in both the samples. Amphibole is an accompany phase both the samples. Leucite appears after both sanidine and plagioclase. The results indicate that the two rocks are leucite-bearing tephri-phonolite.

Freda et al. (2008) studied a phono-tephrite from Alban hills (Central Italy). Their study was made in presence of volatiles under 0.5 GPa at variable temperatures, the content and nature of volatiles are given in the Fig. 13.4a–d. The figures depict only paragenetic relationships of various phases.

13.2.4 Phase Relations on Katungites

13.2.4.1 Investigation on a Synthetic Katungite (up to 1 GPa)

Experimental study was made by Yoder (1973) on a synthetic katungite (akermanite + leucite + forsterite under fluid pressures of H_2O and CO_2 (mixed in different proportions) up to a pressure of 1 GPa. He found that at high pressure the assemblage consisted of phlogopite, olivine, and calcite, which is similar to the assemblage of kimberlitic groundmass. A synthetic katungite ($Fo_{20}Ak_{40}Lc_{40}$) consisting of forsterite, akermanite and leucite was studied between 1.0 and 2.5 GPa $P(H_2O)$, (by Gupta and Yagi 1980) below 1,100 °C and was found to convert to an assemblage consisting of phlogopite, pyroxene, and liquid. At these pressures liquid appeared at 750 ± 25 °C. Breakdown of diopside$_{ss}$ + phlogopite$_{ss}$ at low pressures can be explained by the following reaction, if the small amounts of solid solutions in these phases are ignored:

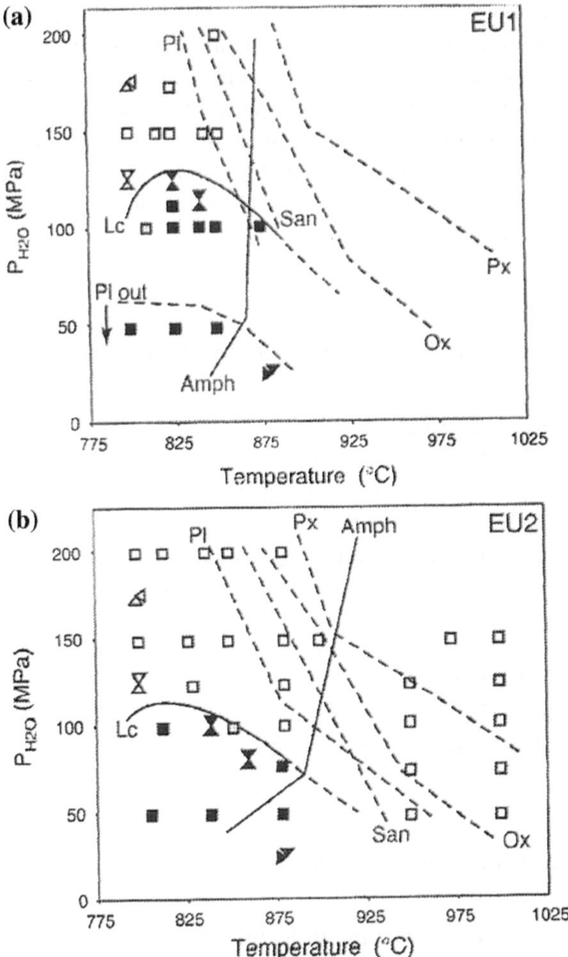

Fig. 13.3 **a** and **b** phase relations on two samples of a 79 AD Vesuvian lava flow. Px: cllinopyroxene, San: sanidine, Pl: Plagoclase, Amph: amphibole, Lc: leucite (after Shea et al. 2009)

$$\underset{\text{(phlogopite)}}{6KMg_3AlSi_3O_{10}(OH)_2} + \underset{\text{(diopside)}}{4CaMgSi_2O_6} \rightleftharpoons \underset{\text{(forsterite)}}{10Mg_2SiO_4} + \underset{\text{(akermanite)}}{2Ca_2MgSi_2O_7} + \underset{\text{(leucite)}}{6KAlSi_2O_6} + 6H_2O$$

This experimental study suggests that an assemblage consisting of forsterite, akermanite, and leucite should be stable only at high temperatures and low P (H_2O), but would be converted to a phlogopite and pyroxene-bearing assemblage at high water pressures. The above study, suggests that katungites were formed under volcanic to subvolcanic conditions, and they are represented at greater depths by phlogopite pyroxenite, closest equivalent of which is a madupite.

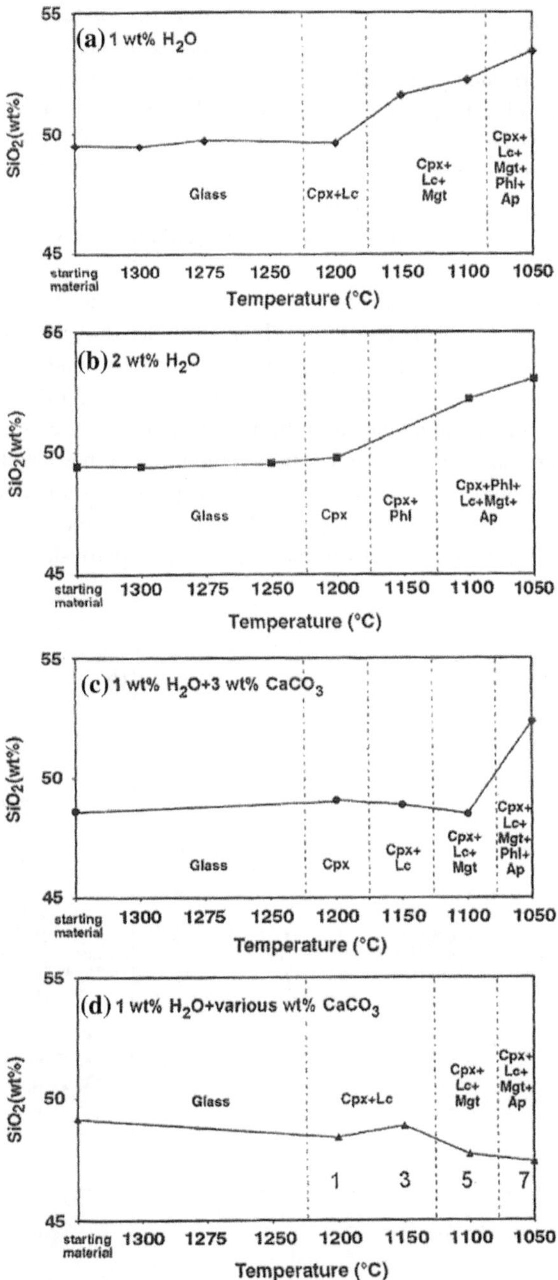

Fig. 13.4 **a**, **b**, **c** and **d** phase relations on a Alban Hills lavas under 0.5 GPa and variable temperatures in presence of fluid. Cpx: cllinopyroxene, Lc: leucite, Mgt: magnetite, Phl: phlogopite: phlogopite, Ap: apatite (after Freda et al. 2008)

Arima and Edgar (1983) made experimental study on a katungite (for analysis of the rock see Table 13.2) collected from the western branch of the African Rift Valley. They determined phase relations in that rock between 0.8 and 4 GPa under “dry” and wet conditions (in presence of 5 and 15 wt% H_2O) Fig. 13.5a, b. They also carried out experiments in presence of both H_2O and CO_2 (corresponding to $CO_2/(CO_2 + H_2O) \sim 0.5$ and 0.75 molar ratio) under NNO buffer condition. They observed that under H_2O-undersaturated conditions, the liquidus temperature for ‘dry’ runs (Fig. 13.5a) range from 1,375 °C at 1 GPa to nearly 1,510 °C at 3 GPa (Fig. 13.5b).

In case of experiments with 5 % H_2O (Fig. 13.5a) at 1 GPa, the liquidus temperature is 225 °C lower than those obtained under dry condition at the same pressure. Likewise the liquidus temperature for the H_2O-undersaturated runs at 3 GPa is 220 °C lower. It may be observed that above 2 GPa the phases obtained with lowering of temperature are as follows: (1) liquid, (2) clinopyroxene + Ti-magnetite + liquid and (3) clinopyroxene + phlogopite + perovskite + Ti-magnetite + apatite + liquid. At 1 GPa however, the phase assemblages observed at successively lower temperatures are, (1) liquid, (2) olivine + liquid, (3) olivine + chromite + liquid and (4) olivine + melilite + perovskite + apatite + Ti-magnetite + liquid.

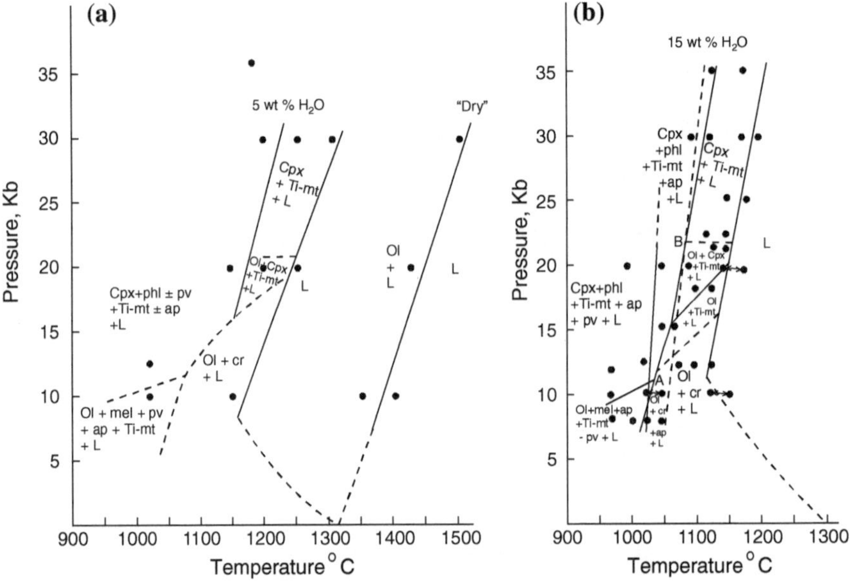

Fig. 13.5 Pressure-temperature diagram for phase relations in katungite composition under (**a**) ‘dry’ and 5 wt% H_2O-added experiments, and under (**b**) 15 wt% H_2O added experiments. *Arrows* joining experimental points indicate reversal runs. *Full lines* represent experimentally determined boundary curves, dashed lines inferred boundary curves. *Line A–B* represents reaction relationship of olivine with liquid to produce phlogopite (after Arima and Edgar 1983). TiMt: Titanomagnetite, Pv: perovskite, Cr: chromite, Mel: melilite

13.2.4.2 Investigation on an Orendite, Wyomingite and a Madupite from Leucite Hills

Water-saturated melting relations to 0.5 GPa on three Leucite Hill lavas were carried out by Barton and Hamilton (1978). Composition of these three lavas are given in Table 13.2. The wyomingite is characterized by the presence of microphenocryst of phlogopite and diopside in a fine-grained groundmass comprising leucite, diopside, K-richterite, apatite and glass. The orendite was constituted of microphenocrysts of phlogopite set in a groundmass containing leucite, diopside, K-richterite, apatite, sanidine and sometimes rare amount of glass, which in some samples are absent. The madupite had microphenocryst of diopside (often enclosed by poikilitic crystals of phlogopite) set in a groundmass containing leucite and diopside.

The water-saturated phase relationship of orendite, wyomingite and madupite is shown in Fig. 13.6a–c. These figures show that leucite is a primary liquidus phase in case of both orendite and madupite at low pressures (<0.05 GPa). In the wyomingite, leucite olivine and clinopyroxene co-precipitate within a very small temperature interval. Olivine and clinopyroxene are the primary liquidus phases from 0.05 to 0.5 GPa and with increase of pressure the proportion of near liquidus olivine decreases relative to clinopyroxene. Phologopite is the primary liquidus phase at a pressure close to 0.1 GPa in both the orendite and wyomingite. In case of wyomingite, clinopyroxene is the liquidus phase at all pressures, whereas in the orendite, clinopyroxene is the liquidus phase at pressure close to 0.5 GPa. Barton and Hamilton noted that like the madupite, the relative proportion of near liquidus olivine decreases with increasing pressure in both the wyomingite and orendite, compared to one atmospheric liquidus temperature, there is a drop of 100–140 °C at 0.1 GPa (Fig. 13.6).

Barton and Hamilton noted that experimentally determined order of phase appearance is in agreement with the inferred crystallization sequences for natural lavas. They found that phlogopite is the major primary liquidus phase of orendite at pressures above ~0.1 GPa and joined by clinopyroxene only at pressures close to 0.5 GPa. The orendite contains microphenocrysts of phlogopite and clinopyroxene as groundmass phases. Initial crystallization of these lavas might have started at pressures between 0.1 and 0.5 GPa. Phologopite and clinopyroxene co-precipitate at the liquidus of wyomingite at 0.1 GPa . They further concluded that absence of olivine in some wyomingites and orendites may be related to reaction relationship between olivine and liquid, resulting in the resorption of olivine with crystallization of phlogopite.

Experimental results on the madupite indicate that the clinopyroxene crystallizes with phlogopite and olivine above 0.3 GPa. Olivine + liquid appear as a liquidus phase at or above 0.05 GPa, whereas phlogopite + olivine coprecipitate as primary phases above 0.1 but below 0.3 GPa. The presence of leucite suggests that the groundmass of the lavas and their hypabyssal equivalents crystallize at low pressures. The abundance of amphibole at low temperature in the near solidus temperature is in agreement with the observation that this phase is a late crystallization

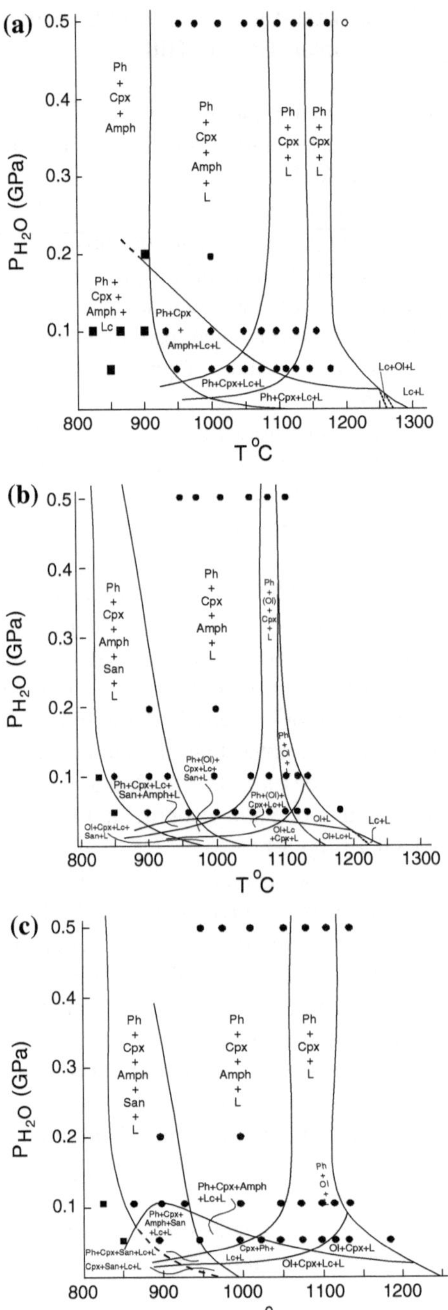

Fig. 13.6 **a** The water-saturated phase relationships of an orendite (A1770). Symbols: open circles: above liquidus; solid circles: melt + crystals; solid squares: subsolidus (after Barton and Hamilton 1978). Abbreviations are same as in Table 13.1 and Fig. 13.2. **b** The water-saturated phase relationships of a madupite (A1805) (after Barton and Hamilton 1978). San: sanidine. Abbreviations are same as in Table 13.1 and Fig. 13.2. **c** The water saturated phase relationships of a wyomingite (A1745) (after Barton and Hamilton 1978). Abbreviations are same as in Table 13.1 and Fig. 13.2

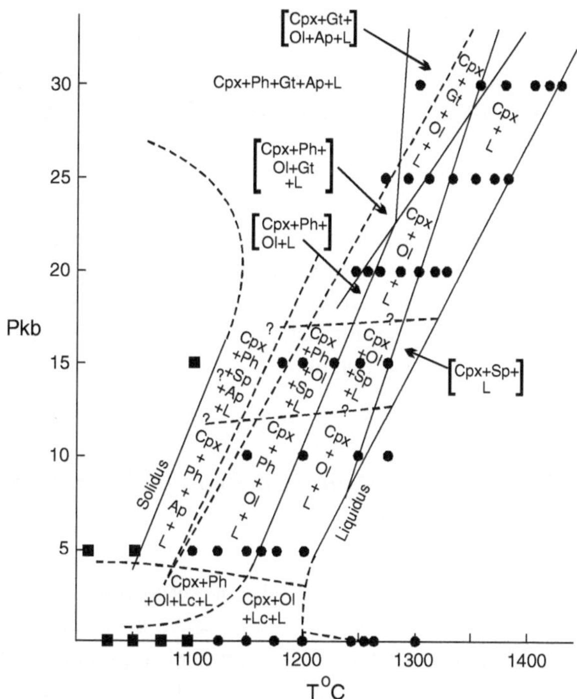

Fig. 13.7 Interpretation of water-undersaturated melting experiments on madupite +2.9 % H_2O. The less-well determined boundaries are shown as *dashed line* for clarity. Some of the smaller fields are not labeled (after Barton and Hamilton 1979). Sp: spinel, Gt: garnet. Other abbreviations are same as in Table 13.1

product. The experimental results further indicate that in lavas quenched at high temperatures and low pressures, leucite and clinopyroxene should be abundant groundmass phases but in lavas quenched at temperature near the solidus, sanidine and amphibole should be present in the groundmass in large amount. They further concluded that absence of sanidine in wyomingite, may suggest that the lava was quenched at temperatures above those of the reaction between leucite and liquid to produce sanidine.

13.2.4.3 Phase Relations on a Madupite (up to 3 GPa)

Barton and Hamilton (1979) studied a madupite from Leucite Hills. Analysis of the rock is given in Table 13.2. They studied the powdered sample of the rock up to a pressure of 3 GPa at variable temperatures. Their results are summarized at Fig. 13.7, which demonstrate that at low pressures (<0.5 GPa), leucite is the dominant liquidus phase, but is replaced at high pressure (0.5–0.7 GPa) by clinopyroxene and olivine and only clinopyroxene (0.7–1.25 GPa). At still higher pressure, clinopyroxene and spinel (1.25–1.75 GPa) and only clinopyroxene (1.75–3 GPa) are the liquidus phases. Olivine reacts with liquid and probably with clinopyroxene to yield phlogopite. They further observed the presence of apatite just above the solidus at pressures lower than 1.5 GPa, or as the fourth phase before the precipitation of phlogopite (above 2.5 GPa). They established that clinopyroxene is diopsidic with low alumina and titanium content and phlogopite is alumina-deficient (up to 12 wt%). Figure 13.7 demonstrates that olivine is not a liquidus phase at pressures above 0.7 GPa, and orthopyroxene is not stable at any pressure up to 3.0 GPa, which suggests that madupite was not probably produced by melting of a hydrous lherzolite or a garnet lherzolite in the upper mantle. They concluded that madupitic melt might have been derived by partial melting of a phlogopite pyroxenite or a phlogopite peridotite. Barton and Hamilton pointed out that the chemistry of some of the potassium-rich volcanic rocks might have been affected by volatile transfer and other such processes during eruption and the experimental studies on materials affected this way may have little bearing upon the genesis of potassic magmas.

13.2.5 Investigation on a Leucite Lamproite from Gaussberg, Antarctica

Experimental study on a olivine leucitite from Gaussberg was studied under buffered (low $f(O_2)$) conditions in presence of an aqueous fluid having minor amount of CH_4 (Foley 1989, Fig. 13.8a). Spinels present in the lamproite suggested reduced mantle conditions for the formation of this lamproitic magma. He therefore, added CH_4 to the aqueous fluid. His experimental results are summarized in Fig. 13.8a, which shows the presence of low temperature liquidus field for olivine. Figure 13.8b shows high pressure liquidus for orthopyroxene. This is expected, if the source rock is mica harzburgite but the intervening field of phlogopite does not explain this concept. Foley considers that the likely explanation is related to addition of excess amount of water as an experimental convenience so that $f(O_2)$ can be suitably buffered. If the amount of water present was less, the phlogopite field should have shrunk and olivine could have appeared as an incongruent melting phase of phlogopite. The experimental study of Foley (1985) may be compared with that of Barton and Hamilton (1982). The chemistry of the two lamproites is similar, but the two starting materials have different silica content and mg-number, and the experiments were conducted in case of the Leucite Hills lamproite in presence of water-deficient condition, whereas the Gaussberg lamproite was studied in presence of excess water.

It is observed that orendite with higher silica concentration has a four-phase point at the liquidus at 2.7 GPa (Fig. 13.8a). The assemblage at this point is similar to a lherzolite. The Gaussberg lamproite does not show the presence of olivine at pressures greater than 0.5 GPa. These two studies are comparable if allowance is made for the uncertainty about the presence of olivine. Barton and Hamilton (1982) in the study of lamproites had difficulty in recognizing olivine in runs above 1.2 GPa, because the grain size of olivine was small and microprobe analyses could not be conducted. Thus if the assumption is made that olivine is stable only below

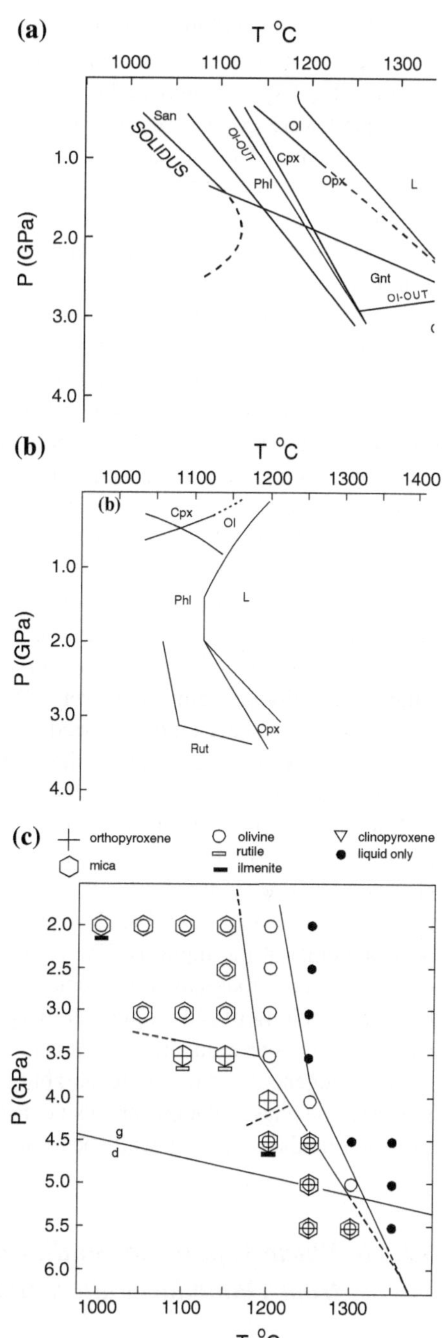

Fig. 13.8 **a** Water-undersaturated phase relationships for a olivine leucitite from Gaussberg. First appearance of minerals labeled in **a** and **b**. The four-phase lherzolite multiple saturation point is labeled at 2.7 GPa (after Foley 1989). L: liquid, Ol: olivine, Opx: orthopyroxene, Cpx: Clinopyroxene, Phl: phlogopite, San: Sanidine. **b** Phase relation in presence of volatile (H_2O >> CH_4) (after Foley 1989a). Abbreviations are same as Fig. 13.8a. **c** Liquidus phase relationships for an olivine lamproite composition from West Kimberley under fluid present (H_2O >> CH_4) conditions. Phase relationships were determined to 4.0 GPa (Foley 1989a), and are extrapolated here to 6.0 GPa to show that they are consistent with phlogopite-harzburgite partial melting at approximately 5.5 GPa (after Foley 1990)

1.2 GPa, then the difference in the liquidus relationship between the two sets of study can be explained. The Gauss-berg lamproites were studied in presence of excess water. Experimental studies on orendites (Barton and Hamilton 1982) on the other hand were conducted in H_2O-deficient condition. Foley (1990) considered that in presence of intermediate H_2O contents the liquidus temperatures should be comparable. He thought that if more H_2O was added in the orendite system, there should have been expansion of the phlogopite field towards the liquidus, due to depression in the liquidus temperature and increase in the phlogopite phase. This would not allow olivine to be stable at higher pressure in the H_2O-rich system. In such a case there should be a four-phase field of mica + olivine + orthopyroxene + liquid in the liquidus of the orendite at a pressure of less than 2.7 GPa. Berton and Hamilton (1982) doubted that orendite with 55 wt% SiO_2 could be derived from a lherzolite source, but suggested that olivine orendites, which are olivine-normative, despite having 53 wt% SiO_2 (Kuehner et al. 1981), might have originated by melting of a mica and garnet-bearing lherzolite at a pressure of about 2.7 GPa. According to Foley (1990), the effect of water on the phase relation should expand the phase volume of olivine and mica; and garent should appear at much higher pressure (Fig. 13.8a). The assemblage olivine + mica + orthopyroxene + liquid could be near liquidus phases in the pressure range of 1–2 GPa. According to Foley, the two sets of experiments are compatible with a mica harzburgite or mica lherzolite source, which is free from garnet. Foley (1990) studied an olivine lamproite from West Kimberley in presence of a fluid ($H_2O >> CH_4$). This study showed the presence of following liquidus phases: olivine at lower pressure, phlogopite under intermediate pressure and orthopyroxene at higher pressure.

Liquidus experiments on an olivine lamproite (Fig. 13.8c) has been performed at pressures between 4.5 and 5.5 GPa in the presence of a reduced fluid (with $H_2O > CH_4$) (Foley 1990). Results show that olivine is the liquidus phase up to a pressure in excess of 5.0 GPa, above which olivine, orthopyroxene, and phlogopite occur together at the liquidus. The results are consistent with the hypothesis that a range of primary magmas from leucite lamproite to olivine lamproite can be derived by partial melting of phlogopite harzburgite as a function of pressure between <1.5 and >6 GPa under reducing conditions. Simple interpretation of the results in terms of a homogeneous phlogopite harzburgite, however, belies, a complex melting process. The phase relations are also consistent with melting of a veined source rock in which neither vein nor wall-rock consists of phlogopite harzburgite.

13.2.6 Phase Equilibria Studies on (Lamproites from Damodar Valley, India

Experimental study was made by Gupta et al. (2002) on powdered samples of three lamproites from Gondwana basins occurring in the Damodar valley region of east India. Whereas one of the lamproites is from the Jarangdih mine of Bokaro basin, the other two are from Mohanpur and Satyanarayanpur, respectively. Both are from the Raniganj Basin. The analyses of these rocks are given in Table 13.2. Experimental studies on these lamproites are summarized below.

13.2.6.1 P–T Study on a Phase Relation in a Jarangdih Lamproite

Experimental results on the Jarangdih sample are summarized in Fig. 13.9 (for chemical analyses of the lamproites from Jarangdih, Mohanpur and Satyanarayanpur see Table 13.2), which shows that olivine is the liquidus phase in the P–T range of this study. With lowering of temperature following assemblages successively, appear:

$$\begin{aligned}&\mathrm{Ol+Liq+Vap\rightarrow Ol+Phl+Liq+Vap\rightarrow Ol}\\&\mathrm{+Phl+Opx+Liq+Vap\rightarrow Ol+Phl+Opx+Il+Liq+Vap\rightarrow Ol}\\&\mathrm{+Phl+Opx+Ap+Il+Liq+Vap\rightarrow Ol}\\&\mathrm{+Phl+Opx+Ap+Il+Cb+Liq+Vap\rightarrow Ol}\\&\mathrm{+Phl+Opx+Ap+Il+Cb+Vap.}\end{aligned}$$

Below the solidus, the assemblage corresponds to a Mg-ilmenite, apatite and carbonate-bearing phlogopite harzburgite. It is interesting to note that although

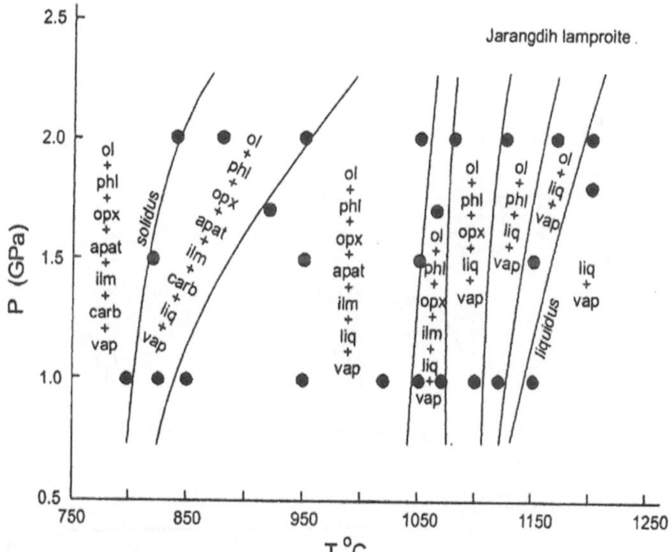

Fig. 13.9 Phase relations of a olivine lamproite from Jarangdih coal mine (Bokaro Basin). For abbreviations, see Appendix I (after Gupta et al. 2002). apat: apatite, opx: orthopyroxene, ilm: ilmenite, liq: liquid, phl: phlogopite, carb: carbonatite, liq: liquid, vap: vapour

orthopyroxene is not present in the Jarangdih lamproite, it is observed as a phase in the high pressure runs. Ryabchikov and Green (1978) studied a mafurite in presence of variable CO_2 and H_2O. They noted that in presence of high CO_2 content orthopyroxene appeared as a phase even through the mafurite did not contain orthopyroxene. In their study of an armalcolite-bearing lamproite, Edgar et al. (1992) also observed the presence of orthopyroxene at pressure above 1.0 GPa although it is not present in the natural rock. The presence of orthopyroxene at high pressure may be related to the following reaction:

$$Mg_2SiO_4 + CO_2 \Leftrightarrow MgSiO_3 + MgCO_3$$

The carbonate-bearing phase appearing as the last phase may be enriched in this component. Electron microprobe studies showed that La_2O_3 (0.35–0.42) and Ce_2O_3 (O.78–O.89) contents of the apatites appearing at 800 °C and 1.0 GPa are high. It suggests that this accessory apatite in the mantle harzburgite may be a possible source of LREE in these rocks.

13.2.6.2 P–T Study on a Phase Relations in a Mohanpur Lamproite (Raniganj Basin)

Experimental studies on Mohanpur Lamproite are summarized in Fig. 13.10, which shows the appearance of following assemblages with progressive lowering of temperature:

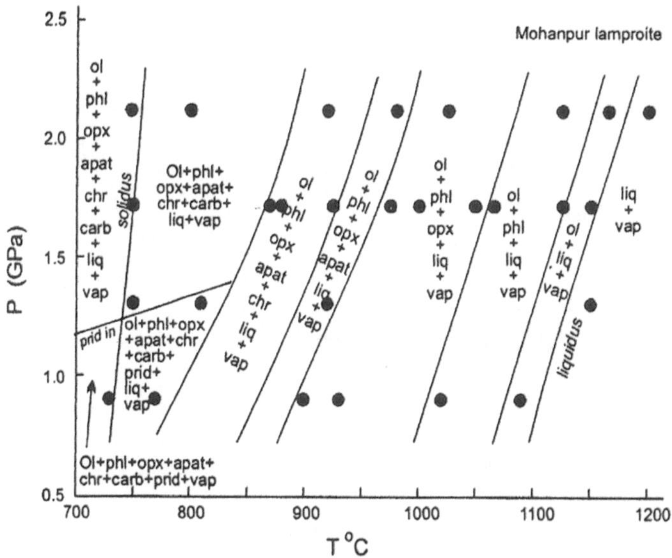

Fig. 13.10 Phase relations of a lamproite from Mohanpur (Raniganj Basin) (after Gupta et al. 2002). For abbreviations, see Fig. 13.7. prid: priderite, chr: chromite. Other abbreviations are the same as in Fig. 13.7

$$\mathrm{Ol + Lig + Vap \rightarrow Phl + Ol + Liq + Vap \rightarrow Phl}$$
$$\mathrm{+ Ol + Opx + Liq + Vap \rightarrow Phl + Ap + Ol + Opx + Liq + Vap \rightarrow Phl + Chr}$$
$$\mathrm{+ Ol + Opx + Ap + Liq + Vap \rightarrow Phl + Ap + Opx + Chr}$$
$$\mathrm{+ Ol + Cb + Liq + Vap \rightarrow Phl + Opx}$$
$$\mathrm{+ Ol + Ap + Chr + Carb + Vap}$$

High pressure-low temperature assemblage in case of Mohanpur lamproite is also an apatite- and phlogopite-bearing harzburgite containing minor amount of priderite and carbonates. The La (0.41) and Ce (0.64) contents of apatites crystallized from the Mohanpur sample at 975 °C and 2.5 GPa are also high. Apatite may therefore, be host mineral for these elements in the mantle rocks. Foley et al. (1994) studied the stability of priderite and concluded that it is stable up to 5 GPa and 1,500 °C; however, in this particular case priderite appears to be a low pressure phase at 1.0 GPa. It is possible that the P–T stability of priderite is restricted to a lower pressure in presence of various phases in lamproite.

13.2.6.3 High P–T Study on a Satyanarayanpur Lamproite

The phase equilibria study of the lamproite sample from Satyanarayanpur is summarized in Fig. 13.11, which shows that with the decrease of temperature following assemblages appear:

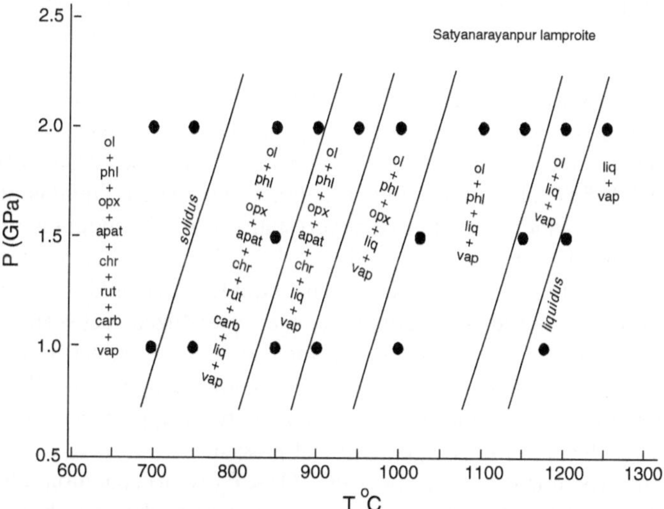

Fig. 13.11 Phase relations of a lamproite from Satyanarayanpur (Raniganj Basin) (after Gupta et al. 2002). For abbreviations, see Fig. 13.7. rut: rutile

$$\begin{aligned}
&Ol + Liq + Vap \rightarrow Ol\\
&\quad + Phl + Liq + Vap \rightarrow Ol\\
&\quad + Phl + Opx + Liq + Vap \rightarrow Ol\\
&\quad + Phl + Opx + Ap + Chr + Liq + Vap \rightarrow Ol\\
&\quad + Phl + Opx + Ap + Chr + Carb + Rut + Liq + Vap \rightarrow Ol\\
&\quad + Opx + Phl + Ap + Chr + Rut + Carb + Vap.
\end{aligned}$$

Subsolidus assemblage once again is a phlogopite-bearing harzburgite containing chromium spinel, appatite, rutile and a carbonate phase. Orthopyroxene appears once again as an important phase because of high CO_2 fugacity in the volatiles.

Priderite does not appear in the case of Satyanarayanpur lamproite. Absence of K-feldspar, which is present as a minor phase in all the three lamproites can be explained by the reaction with (Mg, Fe)CO_3 to produce phlogopite (Rice 1977).

The La_2O_3 (0.40) and Ce_2O_3 (0.83) contents of apatite crystallized at 950 °C from the Satyanarayanpur lamproite are once again high.

13.2.7 Experimental Investigation on a Natural Wolgidite

Arima and Edgar (1983) made experimental studies at variable temperatures and pressures (up to 4.0 GPa) on a wolgidite collected from Mt. North, West Kimberley, Australia. Analysis of the rock is given in Table 13.2. They made two sets of experiments; in the first set of experiments 13 wt% H_2O and 0.84 wt% CO_2 [X (CO_2) = 0.03] were added. The second set of experiments were conducted in absence of excess water [3.22 wt% H_2O, 0.93 wt% CO_2, X(CO_2) = 0.11]. Results of the experiments are summarized in Fig. 13.12a, b. In their study under P (H_2O) = P(Total) condition, liquidus temperature increases from 1,065 °C at 1.0 GPa to 1,165 °C at 4.0 GPa (Fig. 13.12a). Olivine is the liquidus phase up to 2.0 GPa, above which orthopyroxene crystallizes in the liquidus. Phlogopite appeared about 50 °C below the liquidus at 1.0 GPa and its temperature of crystallization is ~1,150 °C at 4.0 GPa. The liquidus increased with increase in pressure. They observed that clinopyroxene crystallized between 1,000 and 1,025 °C at 1.0 to 3.0 GPa. Rutile, armalcolite and priderite crystallized 30 to 40 °C below the liquidus. Below 1.7 GPa, rutile is predominant at above 2.0 GPa also. Crystallization of rutile is close to the temperature of appearance of phlogopite. Priderite crystallized at low temperature and pressure.

In case of experimental study under P(H_2O) < P (Total) condition, olivine is the liquidus phase at 1.0 GPa, but orthopyroxene appears as the first phase at 2.0 and 3.0 GPa (Fig. 13.12b). When experimental results under vapour-absent conditions are compared with those studied in presence of excess water, it is observed that the primary liquidus volume of orthopyroxene decreases relative to that of olivine. Phlogopite is the second phase appearing 30 °C below the liquidus at 3.0 GPa,

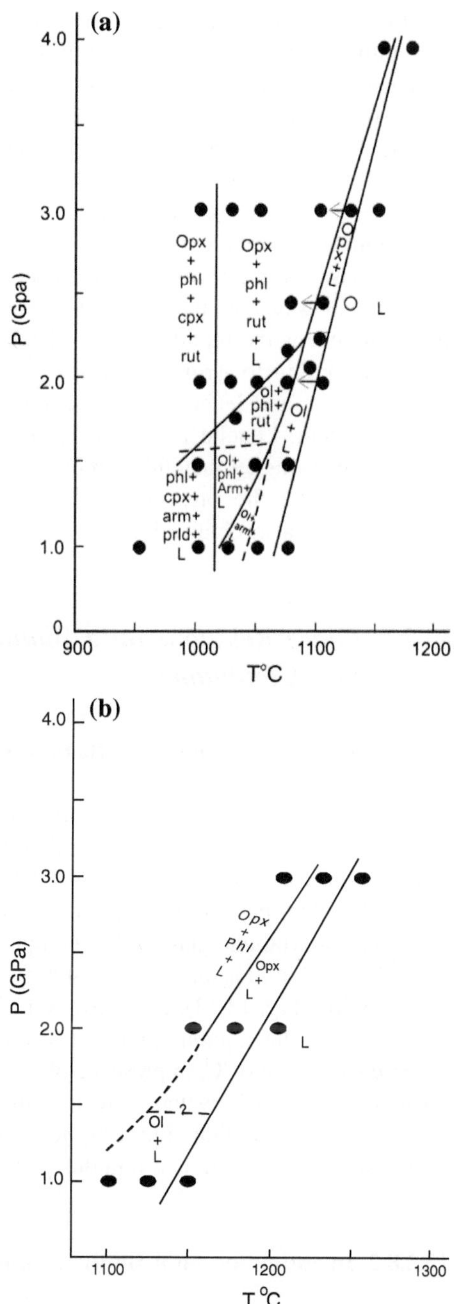

Fig. 13.12 **a** Pressure-temperature diagram for phase relation in wolgidite composition "H_2O added" experiments. Arrows joining experimental points indicate reversal runs. Full lines represent experimentally determined boundary curves; *dashed lines*: inferred boundary curves. Arm: armalcolite, prid: priderite. For abbreviations see Table 13.1 **b** Pressure-temperature diagram for phase relation in wolgidite composition "No H_2O added" experiments. (after Arima and Edgar 1983). Abbreviations are same as in Table 13.1

when no excess water is added to the system. The assemblages obtained from the same bulk composition with lowering of temperature between 2.0 and 3.0 GPa (in presence of excess water, Fig. 13.12a are as follows: orthopyroxene + phlogopite + rutile + liquid and othopyroxene + phlogopite + clinopyroxene + liquid. At pressure below 1.5 GPa, assemblages obtained in presence of excess water are: (1) olivine + liquid; (2) olivine + armalcolite + liquid; (3) olivine + phlogopite + armalcolite + liquid; (4) phlogopite + clinopyroxene + armalcolite + priderite + liquid. Since the experiments were conducted in presence of excess water, a vapour phase should have been present in association with the above assemblages. It is interesting to note that under both vapour-absent and excess water condition, orthopyroxene is the liquidus phase whereas below 1.5 GPa olivine is the liquidus phase. According to Arima and Edgar except for the presence of orthopyroxene, Ti–K-richterite and leucite, the mineralogy of natural wolgidite is similar to that found by them. Leucite is absent because the P–T range of study of these two investigators is higher than the stability field of leucite. Barton and Hamilton (1982) did not observe the presence of leucite from an orendite composition above 0.4 GPa. They also obtained Ti–K-richterite below the subsolidus.

13.2.8 Phase Relations on a Biotite Mafurites under High P–T Conditions

13.2.8.1 Investigation on a Synthetic Biotite Mafurite Glass

A synthetic biotite mafurite glass was studied by Edgar et al. (1976) between 1.0 and 3.0 GPa under dry and wet conditions. A natural biotite mafurite of similar composition, was studied by Holmes (1942), who found that the rock consisted of olivine, kalsilite, perovskite, and biotite with xenoliths of glimmerite. Edgar et al. 1976 made a synthetic glass of biotite mafurite compositions see Table 13.2. They noted that the liquidus range from 1,490 °C at 1.0 GPa to 1,625 °C at 3.0 GPa under dry condition (Fig. 13.13). Clino-pyroxene is the primary phase at higher pressures, and olivine is the liquidus phase at lower pressures. The liquidus, however was lowered by 200–300 °C, in presence of 5 % water, olivine being the liquidus phase. Their runs with 15 % water, show a further lowering of temperatures, olivine still remaining as the liquidus phase up to a pressure of slightly lower than 3.0 GPa, above which phlogopite is a liquidus phase.

13.2.8.2 Investigation on a Biotite Mafurite by Ryabchikov and Green

A synthetic glass of a biotite mafurite composition very similar to that used by Edgar et al. (1976) was investigated by Ryabchikov and Green (1978) in presence of variable amounts of H_2O, and different concentration of CO_2 and H_2O. The composition of the glass used by them is same as that used by Edger et. al. (1976),

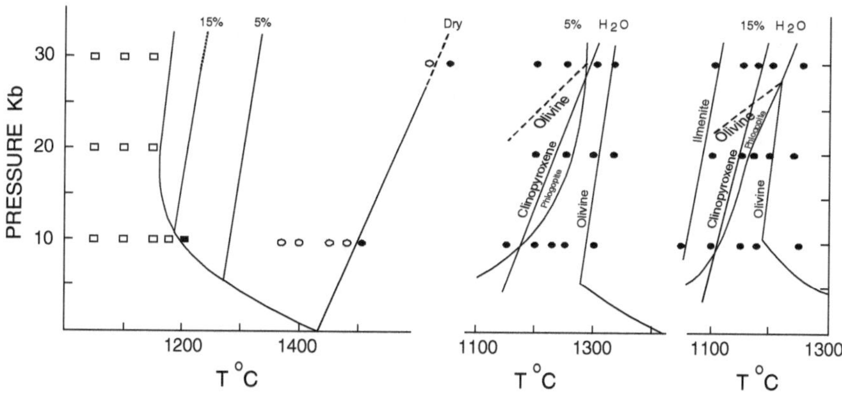

Fig. 13.13 P(H_2O)-T diagram of a synthetic biotite mafurite. The left hand diagram shows subsolidus (*open symbols*) and above-solidus runs for water-saturated and anhydrous conditions and the positions of the liquid for 5 and 15 % H_2O. Olivine is the liquidus phase, shown as *solid lines*. The centre diagram shows experimental points and incoming of phases for 5 % H_2O and the right diagram shows similar data for 15 % H_2O. The olivine-out boundary at high pressure, due to reaction with liquid and other solid phases with decreasing temperature, is shown as a *dashed* boundary. Below 2 GPa olivine crystallization is followed by phlogopite and clinopyroxene. In case of 15 % H_2O ilmenite appears as the last phase (after Edgar et al. 1976)

Table 13.2. The results of "anhydrous" experiments and runs with various amounts of H_2O and/or CO_2 added are summarized in Fig. 13.14a, b. The phase relations on the system biotite mafurite-H_2O at 3.0 GPa is shown in Fig. 13.14a, b. In this figure, they took the data on 5, 15 and 40 % H_2O from Edgar et al. (1976). They had two sets of runs on the biotite-mafurite at 3.0 GPa. In one set of runs H_2O or CO_2 was not added and in case of another set 30 wt% CO_2 was added. Comparison of both sets of run products, show that in both cases olivine is the first phase crystallizing from the melt. However, the liquidus temperature for the CO_2-saturated composition is approximately 120° lower than that for volatile-free biotite mafurite, suggesting substantial solubility of carbon dioxide in the silicate liquid. They analysed quenched glasses from biotite mafurite + 30 wt% CO_2 runs, and found that it contains 10.4 wt% CO_2 at 1,380 °C and 3.0 GPa and 10.9 wt% CO_2 at 1,370 °C and 3.0 GPa giving an average of 10.6 wt% CO_2. This value is only slightly higher than the CO_2 solubility in olivine melilite melts (Brey and Green 1976; Brey 1976). Both magmas contain high Mg and Ca have but very different K and Na contents.

Ryabchikov and Green established that olivine was the liquidus phase for both volatile-free and CO_2-saturated compositions, but the second crystallizing phase was clinopyroxene in the first case and orthopyroxene in the second. Similar observation was noted by them in the runs with both CO_2 and H_2O. They further established that at sufficiently high concentration of CO_2, clinopyroxene is replaced by orthopyroxene. They concluded from their experiments that approximately 6 wt% CO_2 is necessary to bring orthopyroxene to the liquidus in biotite mafurite composition at

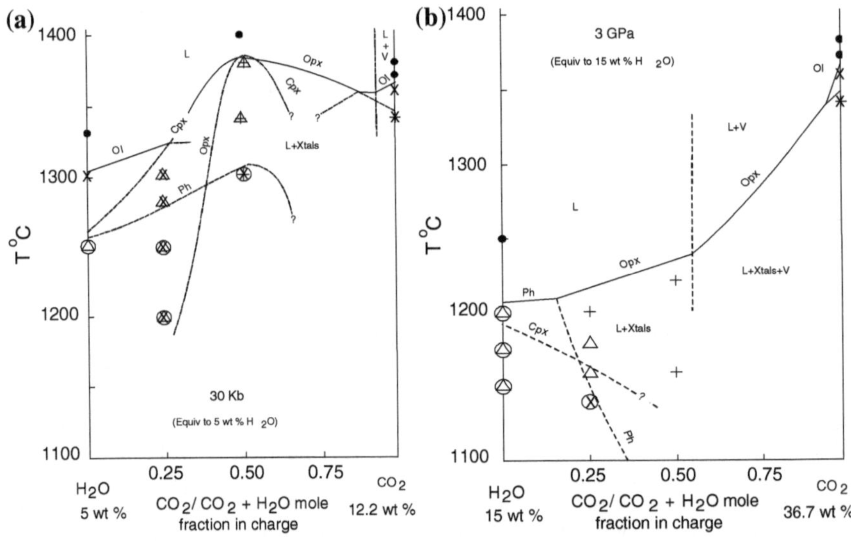

Fig. 13.14 **a** Phase relations for biotite mafurite at 3.0 GPa as a function $CO_2/(CO_2 + H_2O)$ values in the charge in the presence of an external MH-buffer. Data for compositional joins equivalent to 5 wt% H_2O. (after Ryabchikov and Green 1978). For abbreviations see Table 13.1, Xtals: crystals. **b** Data for the biotite mafurite- H_2O side were taken from Edgar et al. (1976). Note that at low $CO_2/(CO_2 + H_2O)$, experiments were carried out under vapour- undersaturated condition, but were vapour-oversaturated for high $CO_2/(CO_2 + H_2O)$ values. Xtals: crystals. For other abbreviations see Table 13.1

3.0 GPa. This result is similar to that obtained by Brey and Green (1975) for olivine melilitite.

Ryabchikov and Green conducted a series of runs at various pressures with biotite mafurite containing 6.8 wt% H_2O and 16.5 wt% CO_2 (H_2O: CO_2 = 1:1 molar ratio, mixture equivalent to 15 wt% H_2O). They observed that orthopyroxene persists at the liquidus with decreasing pressure down to at least 2.0 GPa. They did not find phlogopite as a liquidus phase at 3.0 GPa in their experiments containing CO_2. They however, speculated that it should appear as the first crystallizing phase in H_2O-rich compositions as is evident from Fig. 13.14a, b, which combine their experimental data with those for the biotite mafurite + H_2O composition (Edgar et al. 1976). From Fig. 13.14, it is apparent that at 10–11 wt% H_2O, phlogopite should start to crystallize at 1,240–1,250 °C, whereas in their experiments with 10.7 wt% H_2O and 8.7 wt% CO_2, primary phlogopite precipitates from the melt between 1,160 and 1,140 °C. This observation suggests that large amount of CO_2 strongly suppresses the stability of phlogopite. They however noted that this effect is, not so evident in the series of runs with lower CO_2 concentration. In the mixture with 3.7 wt% H_2O and 3.0 wt% CO_2 and 2.4 wt% H_2O and 5.9 wt% CO_2, they found that crystallization temperatures of phlogopite are higher than that observed for composition in the biotite mafurite-H_2O join with equivalent water content

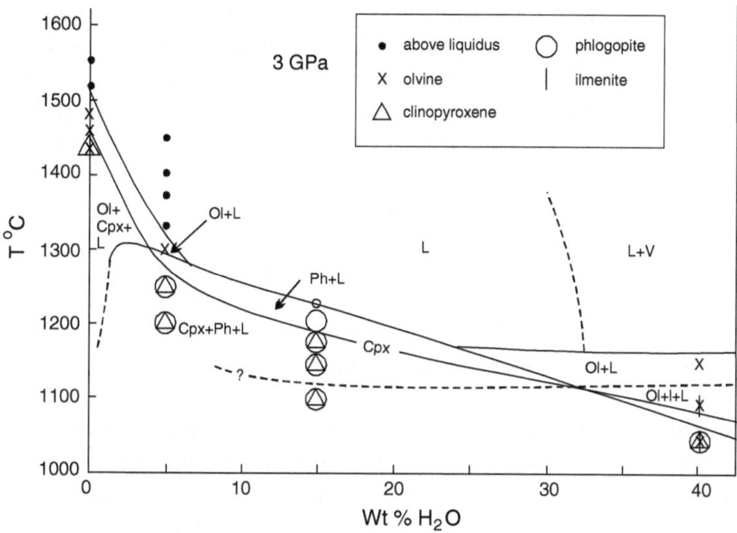

Fig. 13.15 Phase relations for biotite mafurite at 3.0 GPa as a function of water content. Date for 5 wt% H_2O, 15 wt% H_2O and 40 wt% H_2O are taken from Edgar et al. (1976) and for anhydrous biotite mafurite from Ryabchikov and Green (1978). For abbreviations see Table 13.1

(5 wt% H_2O, Fig. 13.15). They found that in case of the composition, biotite mafurite with 2.4 wt% H_2O and 5.9 wt% CO_2, the highest temperature of crystallization for primary phlogopite (above 1,300 °C at 3.0 GPa) was observed in experimental runs. Phlogopite is not a liquidus phase in these runs and it crystallizes from liquid enriched in potassium compared with biotite mafurite due to the preceding crystallization of olivine or pyroxene.

In their runs liquidus phlogopites are characterized by slightly lower Mg/(Mg + Fe) values than coexisting olivines or orthopyroxenes, whereas previously published experimental results on biotite mafurite with H_2O (Edgar et al. 1976) show the opposite relation. Ryabchikov and Green (1978) concluded that these differences are possibly due to the fact that their experimental runs were carried out under much more oxidizing conditions than the previous study, and part of iron in the runs of Ryabchikov and Green may be present in trivalent state. They therefore, concluded that $Mg^{2+}/(Mg^{2+} + Fe^{2+})$ values should be higher than the measured $Mg/(Mg + Fe^{2+} + Fe^{3+})$ values, obtained from microprobe analyses. In case of runs of Edgar et al. (1976), the lower $f(O_2)$ values should cause little difference between $Mg^{2+}/(Mg + Fe^{2+})$ and $Mg/(Mg + Fe^{2+} + Fe^{3+})$ values.

The TiO_2 contents of liquidus phlogopites in case of runs of Ryabchikov and Green (1978), are somewhat higher than in phlogopites crystallizing from biotite mafurite with H_2O (Edgar et al. 1976) crystallized under similar P, T and $P(H_2O)$ conditions. This may be related to either the difference in oxidation states or to the effect of CO_2 dissolved in the melt or the activity of titanium. They carried out a run without CO_2, but with 15 % H_2O at 3.0 GPa, 1,160 °C with oxygen fugacity controlled by hematite-magnetite buffer. The run yielded phlogopite with relatively high TiO_2 content (5 wt% TiO_2). This result supports the first of the above explanations. Their experiments confirmed the previously-observed (Edgar et al. 1976) positive correlation between TiO_2 contents of phlogopite and temperatures of equilibration of melt or negative correlation with H_2O content in the system. Ryabchikov and Green found that primary phlogopites crystallized at 3.0 GPa and 1,300 °C from compositions with 2.4 wt% H_2O and 5.9 wt% CO_2, are characterized by the highest titanium concentrations (8.1 wt% TiO_2). Their study demonstrates that TiO_2 content of phologpite positively correlates with increased temperature, increased Fe^{2+}/Mg ratio and increased Fe^{3+} content [higher f(O_2)] inversely correlates with increased pressure. Of these controls, the temperature and Fe^{2+} contents may be the most important.

Experimental studies of Edgar et al. (1976) demonstrated that over a wide P–T range (in presence of water), orthopyroxene did not crystallize within the melting interval of biotite mafurite. This suggests that the melt of biotite mafurite composition was not derived by partial melting of a mantle mineral assemblage containing olivine, clinopyroxene and orthopyroxene and water as the only volatile component. In contrast, the study of Ryabchikov and Green demonstrates that, in the presence of CO_2 and H_2O, biotite mafurite melt may become saturated with respect to orthopyroxene thereby allowing the possibility of generation of silicate liquids similar to biotite mafurite in the course of the partial melting of an upper mantle peridotite. According to Ryabchkov and Green, the appearance of orthopyroxene in the melting range of this potassium-rich silica-undersaturated highly calcic rock results from drastic reduction in the activities of CaO and alkalis due to the interaction of these components with $(CO_3)^-$ complexes. Similar observation was made in a number of other silicate systems containing H_2O and CO_2 (Eggler 1974; Brey and Green 1975). Ryabchikov and Green also anticipated this effect on the basis of general consideration of the interaction of basic cations with acid volatile compounds (Kushiro 1975). Ryabchikov and Green constructed a schematic diagram showing the position of the liquidus fields of various minerals in biotite mafurite as a function of H_2O and CO_2 content in the melt (Fig. 13.16). They drew the vapour-saturation boundary tentatively in the compositional triangle by analogy with the data of Brey and Green (1977) on olivine melilitite composition at variable P–T conditions. They drew the boundaries between the liquidus fields of the various crystalline phases as lines, but they are more probably bands in which two neighbouring phases coexist with liquid (cf. Brey and Green 1977). This diagram demonstrates that there are three 'points,' where three crystalline phases may simultaneously coexist with the liquid of biotite mafurite composition: A—phlogopite + orthopyroxene + clinopyroxene + liquid; B—orthopyroxene + olivine + clinopyroxene + liquid; and C-phlogopite + olivine + liquid. In their diagram, point A corresponds to $CO_2/(CO_2 + H_2O) \sim 0.15$. The crystalline assemblage at this point does not include olivine, but has phlogopite, orthopyroxene and clinopyroxene. On the basis of which, they concluded that if the $CO_2/(CO_2 + H_2O)$ ratio in the melt is approximately 0.1–0.2, then the silicate liquid coexisting with

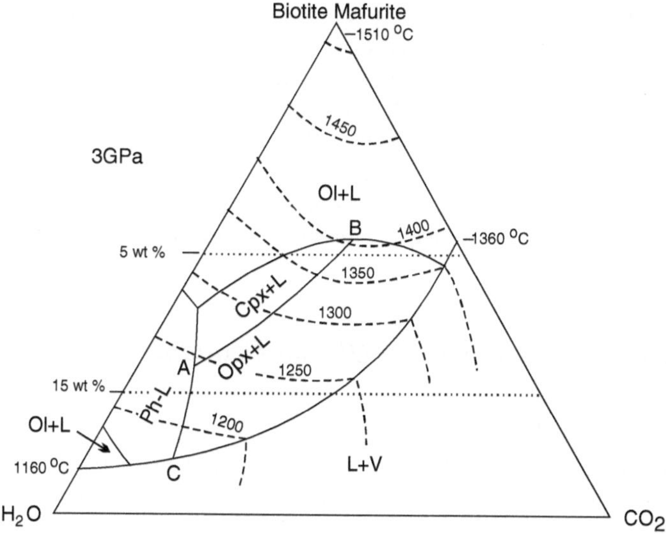

Fig. 13.16 Schematic diagram showing approximate position of liquidus phases in the system biotite mafurite-H_2O–CO_2 at 3.0 GPa (after Ryabchikov and Green 1978)

phlogopite + olivine + orthopyroxene + clinopyroxene assemblage should be characterized by approximately the same K_2O content (~7 wt%) as the investigated biotite mafurite, but it would be impoverished in silica. On the other hand clinopyroxenes in their experiments contain less Na_2O than subsolidus clinopyroxenes in pyrolite under comparable P–T conditions (Green 1973). They therefore, concluded that partially melted phlogopite-bearing pyroxene pyrolite should produce a silicate liquid having higher Na_2O content than biotite mafurite. The near-liquidus clinopyroxenes in the olivine melilitite composition containing H_2O and CO_2 at 3.0 GPa are characterized by approximately the same Na_2O concentrations as subsolidus pyrolite (Brey and Green 1975); and the melt equilibrated therefore, with phlogopite-bearing mantle mineral assemblage should contain Na_2O in similar amounts to olivine melilitites (3–4 wt% Na_2O); and the K/Na ratio in such melts may be estimated as equal to 1.2–1.5. Olivine, othopyroxene, clinopyroxene and liquid are in equilibrium at point B with $CO_2/(CO_2 + H_2O) \sim 0.7$. It may be noted that phlogopite is absent at this point. It may be concluded therefore, that silicate melts with $CO_2/(CO_2 + H_2O) > 0.5$ and in equilibrium with phlogopite + olivine + orthopyroxene + clinopyroxene, should contain more than 7 wt% K_2O, and thus have K/Na ratio >1.5 (Fig. 13.16).

The phase assemblage with volatiles having $CO_2/(CO_2 + H_2O)$ ratio ~ 0.05 does not include orthopyroxene, they therefore, concluded that liquids with low $CO_2/(CO_2 + H_2O)$ ratios and equilibrated with phlogopite + olivine + orthopyroxene + clinopyroxene, should contain about 7 wt% K_2O, but they should be more silicic and probably less calcic and less alkaline than liquids with higher CO_2 contents in equilibrium with the same assemblage (Green 1973).

Garnet did not appear as the near-liquidus phase at 3.0 GPa in the biotite mafurite composition. This mineral, however, was found in the melting interval of another highly potassic olivine leucitite (Cundari and O'Hara 1976) at pressures above 3.0 GPa and therefore, the above conclusions should be also qualitatively valid for the melting of an assemblage containing, phlogopite + olivine + orthopyroxene + clinopyroxene + garnet, at pressures higher than 3.0 GPa.

13.2.9 Phase Relations in an Olivine Ugandite under High P–T Conditions

An olivine ugandite from South West Uganda was studied at pressures between 1.0 and 4.0 GPa under volatile-present conditions with 5, 15, 25 and 40 wt% added H_2O under an oxygen fugacity, which is slightly greater than the NNO buffer condition. The same lamproite was studied by Edgar et al. (1984) in presence of volatiles $[H_2O + CO_2,\ X(CO_2) = 0.75]$ under similar oxygen fugacity condition. Results of their experiments in presence of variable H_2O contents are shown in Fig. 13.17a–c, whereas run products related to their experiments in presence of $[(H_2O + CO_2,\ X(CO_2) = 0.75)]$ are summarized in Fig. 13.18. Figure 13.17a displays change in the slope of the liquidi as a function of pressure in presence of

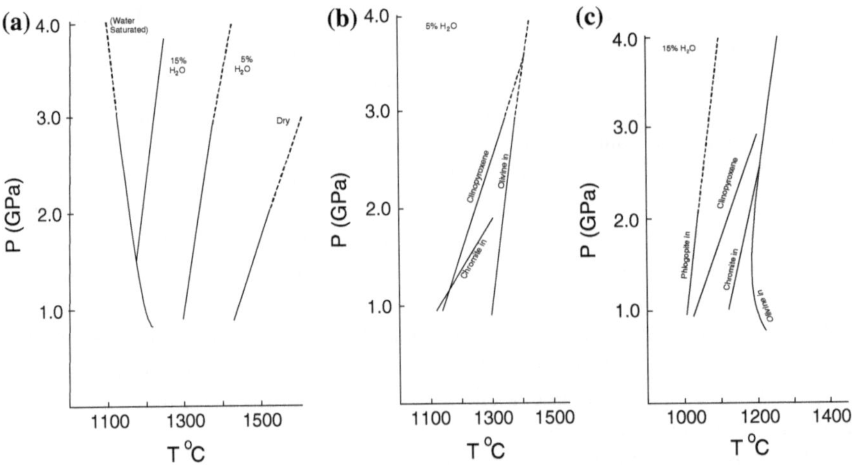

Fig. 13.17 **a**, **b**, **c** Results of experiments on a olivine ugandite with H_2O only. All H_2O added by weight. *Open symbols*: above liquidus, close symbols: below liquidus. (after Edgar et al. 1980).S. **a** Liquidus relation under 'dry' 5 and 15 wt% added H_2O and water-saturated (25 and 40 wt% H_2O) condition. **b** Detailed phase relations with 5 wt% H_2O. **c** Detailed phase relation with 15 wt% H_2O

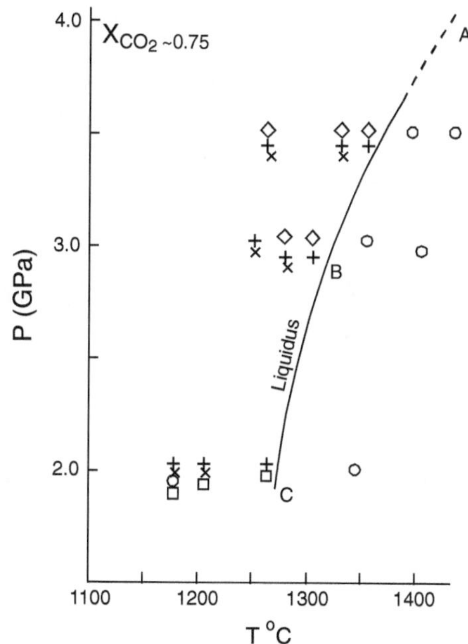

Fig. 13.18 Results of experiments with $H_2O + CO_2$ at $X(CO_2) \simeq 0.75$ Symbols: ◇- garnet, + orthopyroxene. *Open circle*: all glass, *Open square*: olivine, Cross: chromite. (after Edgar et al. 1980)

variable H_2O contents. It is apparent that under dry condition there is an increase in the liquidus temperature and the curve has a positive slope, whereas with the increase of the H_2O content there is a gradual change in the slope of the liquidi, and under H_2O-saturated condition (>25 wt% H_2O), there is a decrease in the liquidus temperature with the increase in pressure.

Figure 13.17b shows that in presence of 5 wt% H_2O, chromite is in the liquidus phase up to 2.0 GPa, above which olivine is the primary phase. Above 3.6 GPa clinopyroxene appears as the first phase. Figure 13.17c shows the appearance of olivine as the primary phase up to 3.2 GPa followed by chromite. Clinopyroxene is the liquidus phase above 3.2 GPa, and phlogopite precipitates as the fourth phase with the progressive decrease of temperature. In their experiments with added H_2O and $CO_2[X(CO_2) \sim 0.75]$at 2.0, 3.0 and 3.5 GPa (Fig. 13.18), Edgar et al. noted that at 3.0 and 3.5 GPa, garnet and orthopyroxene appear as the liquidus phase followed by the appearance of clinopyroxene as the third phase. The curve A–B–C (Fig. 13.18) is the liquidus curve for the ugandite established in presence of volatile $[X(CO_2) = 0.75]$ under variable P–T conditions. Garnet does not appear at all in the 2.0 GPa runs but chromite and orthopyroxene are liquidus phases at this pressure. Clinopyroxene appears as the third phase with decrease in temperature. They did not encounter phlogopite and olivine at all in runs equilibrated in presence of fluid with high $X(CO_2)$ content. Their runs at lower $X(CO_2)$ show that at 2.0 GPa and $X(CO_2) = 0.5$, stability of orthopyroxene increases but that of clinopyroxene decreases. With further lowering of $[X(CO_2) = 0.23]$ at 1,200 °C and 2.0 GPa both pyroxenes are eliminated and olivine becomes the liquidus phase. They did not observe the presence of phlogopite in any of their runs, suggesting that presence of high CO_2 content in the volatile suppresses the phologopite stability as suggested by Ryabchikov and Green (1978). The clinopyroxenes are found to be calcic [Ca/Ca + Mg) > 0.46] and olivines are forsteritic [Mg/(Mg + Ca) = 0.83 – 0.89]. Orthopyroxene is rich in $MgSiO_3$ molecule [Mg/(Mg + Fe) = 0.76 – 0.85] and garnets are pyrope-rich as expected. Mica is phlogopitic [Mg/(Mg + Fe) = 0.85] and appears to be the main source of K_2O in the mantle.

13.2.10 Experimental Investigation on a Phlogopite-Bearing Minette

Experimental study was made on a phlogopite minette from Buell Park, Alizona by Esperanca and Holloway (1987). Chemical analysis of the rock is given in Table 13.2. The minette was studied under oxygen fugacities equivalent to that maintained by iron-wustite-graphite (IWG) and quartz-magnatite-fayalite (FMQ) buffers at pressures between 1.0 and 2.0 GPa. Their results are summarized in Fig. 13.19, where comparison between experimental products obtained under these two buffer conditions are shown. Their results demonstrate that olivine together with clinopyroxene appears as the primary phase followed by precipitation of phlogopite.

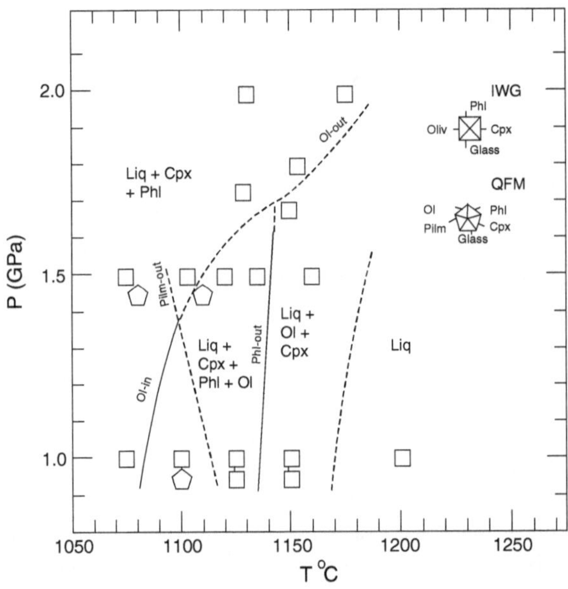

Fig. 13.19 Liquidus phase diagram for Buell Park minette sample. Phase boundaries are drawn based on models for IGW (iron-wustite-graphie experiments), "Picroilmenite out" curve (Pilm out) extrapolated from QFM (quartz-fayalite-magnetite) buffer experiment. Filled parts represent coexisting near liquidus phase counter clockwise from the bottom, glass, clinopyroxene, phlogopite, olivine, picroilmenite. Boundaries were approximate (after Esperanca and Holloway 1987)

Picroilmenite coexists with olivine, clinopyroxene and phlogopite at 1.0 GPa under QFM buffer condition below 1,100 °C. This phase shows lower TiO_2 and FeO content at 1.5 GPa where it coexists with clinopyroxene and phlogopite. They suggested that the phlogopite-bearing minette magma might have originated from a metasomatized garnet-bearing peridotite source at deeper level in the mantle ($P \geq 2.0$ GPa) but can also be in equilibrium with a phlogopite-bearing wherlite (± orthopyroxene) source at 1.7–2.0 GPa under oxidizing as well as reducing mantle condition. Because of their rapid crystallization rate and high liquidus temperatures, a series of potassic daughter melts like K-rich latite and felsic minettes may originate by segregation of mafic minette magmas during their ascent to the cooler continental crustal environment. They thought that preservation of olivine in equilibrium with phlogopite phenocrysts in primitive minettes suggest that assimilation was not involved in their genesis. Their study supports the model that the phlogopite minette magmas are brought to near surface condition at temperatures between 1,000 and 1,200 °C and they got chilled rapidly during their emplacement.

13.2.11 Experimental Studies on a Phlogopite-Pyroxenite Nodule from South-West Uganda

Melting experiments on a mantle-derived nodule consisting of clinopyroxene, phlogopite and minor titanomagnetite, sphene and apatite, was conducted by Lloyd et al. (1985) at 2.0 and 3.0 GPa between 1,175 and 1,300 °C (Fig. 13.20). The nodule composition was selected on the basis of modal and chemical analyses of 84 mantle-derived nodules with metasomatic textures from the Katwe-Kikorango and Bunyaruguru volcanic fields of south west Uganda. At 3.0 GPa, 1,225 and 1,250 °C, representing 20 to 30 % partial melting, the composition of glasses compare favourably to those of the average composition of 26 highly potassic mafic lavas from the same region. Glasses produced by sufficiently low degrees of partial melting at 2.0 GPa could not be analysed. Glass compositions obtained by 20–30 % melting at 3.0 GPa have high K_2O (3.07–5.05 wt%), low SiO_2 (35.0–39.2 wt%), high K/(K + Na) (0.54–0.71), (K + Na)/Al (0.99–1.08) and Mg/(Mg + Fe_t) of 0.59–0.62. These results support the suggestion of Lloyd and Bailey (1975) that the nodules represent the source material for the high K-rich lavas of south-west Uganda. If this conclusion is correct it implies that anomalies in the mantle source rock of phlogopite clinopyroxenite composition could produce the Uganda lavas by relatively higher degrees of partial melting than normally considered. Highly alkaline mafic magmas could have been derived also from a pyrolitic mantle source. Higher degree of melting is considered likely from different source regions, rich in alkalis rather than radioactive elements. Steeper geotherms and increased fluxing of sub-rift mantle by degassing would also reduce higher degrees of partial melting.

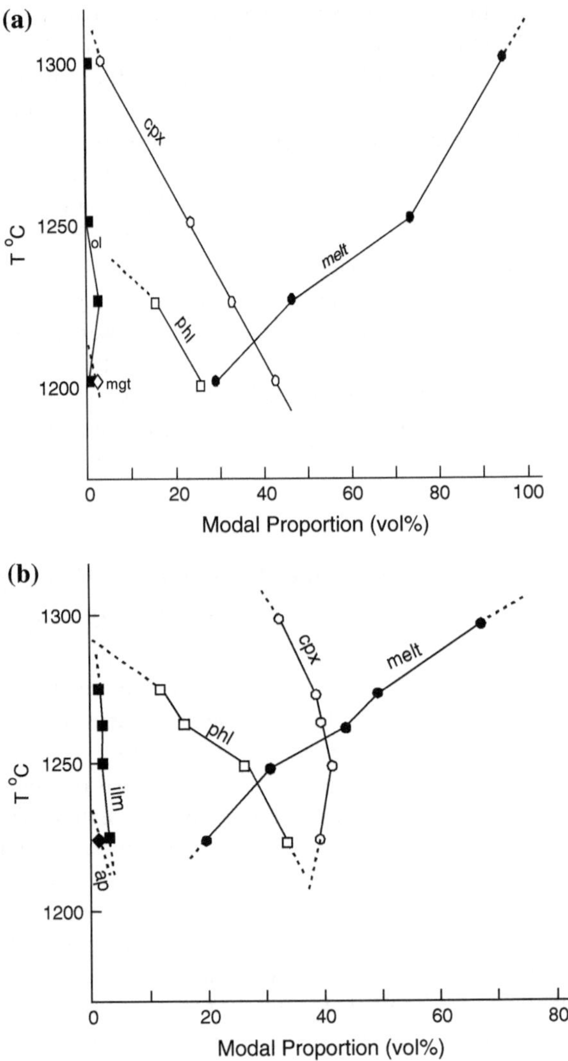

Fig. 13.20 **a** Modal proportions of phases in experiments at 3.0 GPa. The modes were calculated by the least squares method using microprobe analyses of all phases (after Lloyd et al. 1985) **b** Modal proportions of phases at 2.0 GPa. The modes were calculated by the least square method using microprobe analyses (after Lloyd et al. 1985). For abbreviations see Table 13.1

13.2.12 High P–T Investigation on an Armalcolite-Phlogopite Lamproite from Smoky Butte, Montana

Suprasolidus phase relations on an armalcolite-phlogopite lamproite (Table 13.2) was determined at pressures between 0.8 and 3.0 GPa and temperatures from 950 to 1,380 °C by Edgar et al. (1992). Velde (1975) earlier described this armalcolite-bearing phlogopite-diopside lamproite from Smokey Butte, Montana. The starting material for the study was a glassy armalcolite-phlogopite lamproite (Fig. 13.21) from the chilled margin of a medium-grained lamproite from Smoky Butte, Montana. For abbreviations in Fig. 13.21 see Table 13.1.

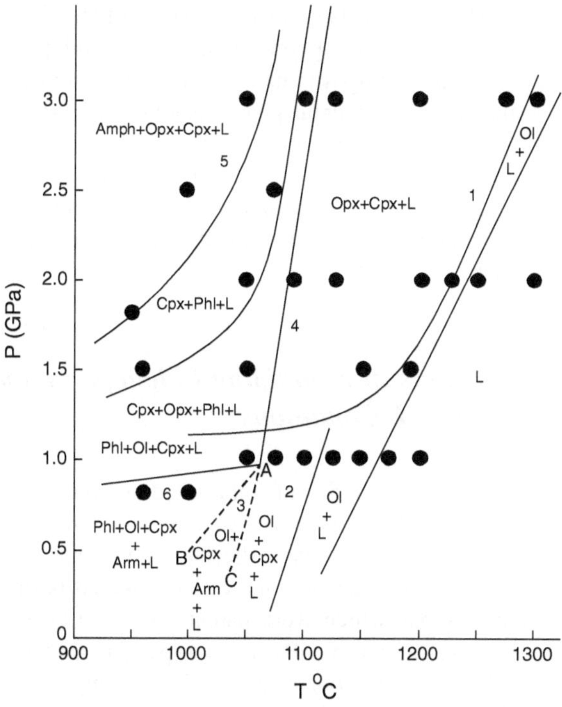

Fig. 13.21 Pressure-temperature plot of the phase relations of an armalcolite-phlogopite lamproite from Smoky Butte, Montana. *Heavy lines* indicate well established data boundaries, *light lines* are less well-established., *dashed lines* are inferred. A narrow field of olivine + pyroxene (orthopyroxene or clinopyroxene) probably coexist with liquid between the olivine + liquid and orthopyroxene + clinopyroxene + liquid assemblages (after Edgar et al. 1992). Arm: armalcolite

The lines AB and AC have been drawn arbitrarily and there are no data points to substantiate the claim of two different assemblages, Ol + Cpx + Arm and Ol + Cpx + L.

The Armalcolite-phlogopite lamproite has microphenocrysts of olivine in a groundmass of phlogopite, sanidine, armalcolite, clinopyroxene, chromite, priderite, apatite and abundant glass. The lamproite is SiO_2-rich and has high F/H_2O ratio relative to lamproites that have been investigated in previous experimental studies. Their data show that with decreasing temperature from the liquidus at pressure above ~1.2 GPa, a melt coexists successively with olivine; orthopyroxene + clinopyroxene; orthopyroxene + clinopyroxene + phlogopite; clinopyroxene + phlogopite; and clinopyroxene + K-richterite. Below 1.2 GPa, the assemblage succession is olivine; olivine + clinopyroxene; olivine + clinopyroxene + phlogopite; and olivine + clinopyroxene + phlogopite + armalcolite. The main difference from the natural paragenesis is that the rock does not contain any orthopyroxene,—a feature, which is rather remarkable as it has ~16 % normative hypersthene—and the rock differs also in that it contains sanidine and priderite. In the experiments, sanidine is observed only as ghost-like domains in some of the glass fragments, and appears to have been formed during quenching. The solid phases crystallized experimentally are generally compositionally similar to the minerals in the rock. These similarities and the experimental phase relations support the concept of a rapid initial magma ascent with only a small temperature drop and crystallization of olivine, but not of orthopyroxene. At lower pressure, less than ~1.2 GPa, it appears that the magma ascended more slowly with a larger temperature drop suggested by the similarity of the experimentally determined sequence of assemblage to the paragenesis of the rock.

13.2.13 Phase Relations in a Sanidine Phlogopite Lamproite under High P–T Conditions

Mitchell (1995) made melting experiments on a sanidine phlogopite lamproite (for analysis see Table 13.2) at 4–7 GPa and temperatures from 1,000 to 1,700 °C (Fig. 13.22). This particular sample was collected from Table Mountain, Leucite Hills, Wyoming. The lamproite is a silica-rich variety, and has been postulated to be representative of the magmas which were parental to the Leucite Hills volcanic field. Near-liquidus phases above 5 GPa are pyrope-rich garnet and jadeite-rich pyroxene. Below 5 GPa, jadeite-poor pyroxene is the only near-liquidus phase. Near-solidus assemblage consists of clinopyroxene, titanian potassium richterite and titanian phlogopite with either potassium titanian silicate above 5 GPa or potassium feldspar below 5 GPa. The potassium titanian silicate is a high-pressure phase ranging in composition form $K_4Ti_2Si_7O_{20}$ to $K_4TiSi_8O_{20}$. It coexists at pressures above 6 GPa at 1,100–1,400 °C. A previously unrecognized K–Ba-phosphate is a common near-solidus phase. The phase relationship is interpreted to suggest that lamproite cannot be derived by partial melting of simple lherzolite sources. However, it is proposed that sanidine phlogopite lamproites are derived by high degrees of partial melting of ancient metasomatic veins within a harzburgitic-lherzolitic lithospheric substrate mantle. The veins are considered to coexist with phlogopite, K-richterite, K–Ba-phosphate and K–Ti silicates.

The change in slope of phase boundaries at A, B and C would denote change in mineralogical assemblages, but the phase diagram of Michell (1995, Fig. 13.22) does not show any new phase assemblage in the right hand side of the line BC. If this change in slope is correct careful study of the run products might reveal the presence of a new assemblage in that region.

Mitchell and Edgar (2002) studied phase relations on a leucite-phlogopite lamproite from Oscar plug (West Kimberely, Australia, Fig. 13.23) at pressures between 2 to 6.4 GPa and temperature ranging between 800 and 1,700 °C.

Their experimental study on a madupitic lampriotes from Middle Table Mountain, Wyoming, (U.S.A) is summarized in Fig. 13.23, whereas phase relationships of a sanidine-richterite lamproite from Cancarix, Murcia-Almeria, Spain are shown in Fig. 13.24.

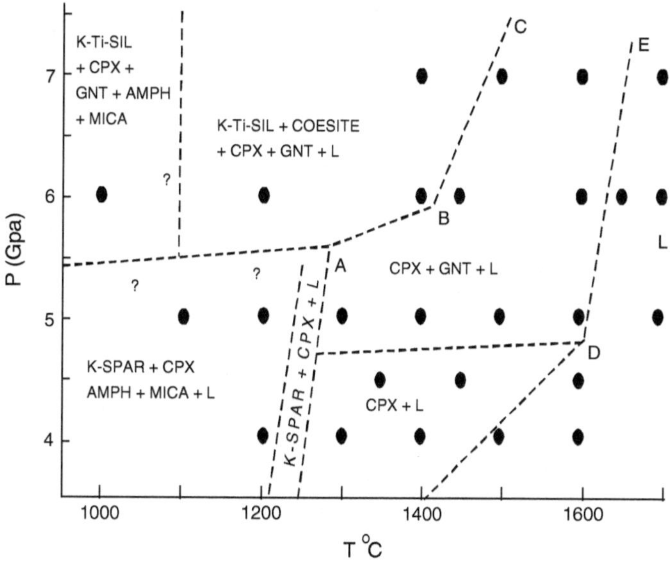

Fig. 13.22 Experimentally determined phase relationships of the sanidine phlogopite lamproite (after Mitchell 1995). Abbreviations Cpx: clinopyroxene, Gnt: garnet, K-spar: K-felspar, K–Ti–SIL: potassium-titanium-silicate, AMPH: amphebol

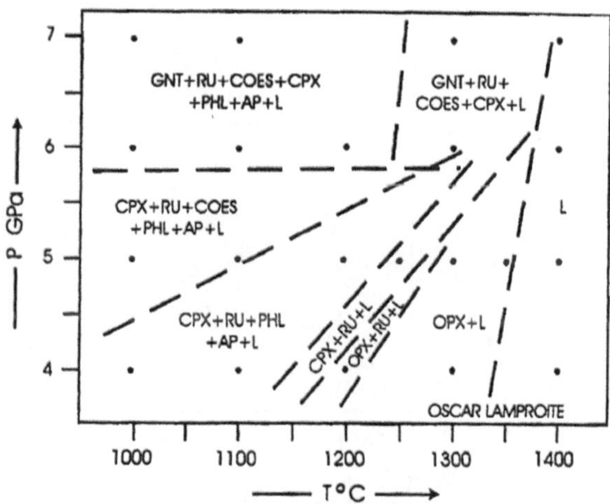

Fig. 13.23 Experimentally determined phase relationship of the Oscar lamproite. L liquid, OPX orthopyroxene. PHL phlogopite, COES coesite, GNT garnet, RU rutile. (after Mitchell and Edger 2002)

They showed that bulk composition has a significant control on the nature of the initial liquidus phases, with orthopyroxene occurring at low pressure (<4 GPa) in the relatively calcium-poor Oscar and Cancarix lamproites. At higher pressure (>6 GPa) orthopyroxene is replaced by garnet plus clinopyroxene as near-liquidus phases in the Oscar lamproite and by orthopyroxene plus clinopyroxene in the

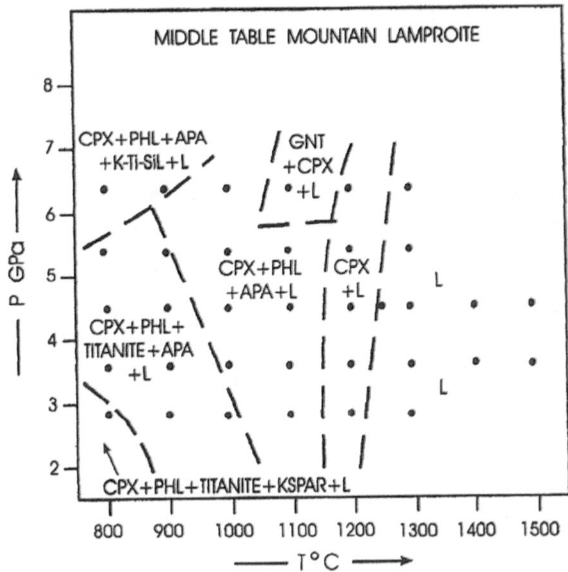

Fig. 13.24 Experimentally determined phase relationships of the Middle Table Mountain lamproite. L liquid, CPX clinopyroxene, PHL phlogopite, GNT garnet, KTiSiL potassium titanium silicate, APA apatite, KSPAR potassium feldspar (after Mitchell and Edger 2002)

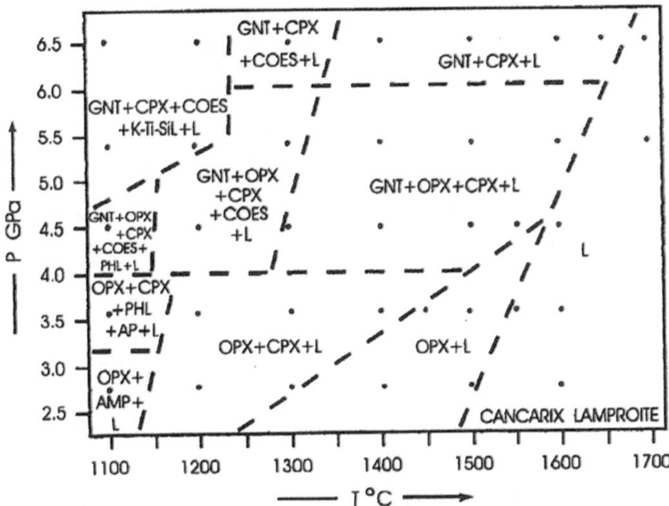

Fig. 13.25 Experimentally determined phase relationships of the Cancarix Lamproite. L Liquid, Opx Orthopyroxene, Cpx Clinopyroxene, Phl Phlogopite, Coes Coesite, Gnt Garnet. (after Mitchell and Edger 2002)

Cancrix sample. Clinopyroxene is a near-liquidus phase at all pressures in the Middle Table Mountain lamproite. Near –solidus phase assemblages at high pressure (>5 GPa) are: clinopyroxene + phlogopite + coesite + rutile + garnet (Oscar); clinopyroxene + garnet + coesite + K–Ti-silicate (Canccarix); clinopyroxene + phlogopite + apatite + K–Ti-silicate (Middle Table Mountain). In all compositions olivine is never found as a liquidus phase at any of the temperatures or pressures studied here. The phase relationships are interpreted to suggest that silica-rich lampriotes cannot be derived by the partial melting of lherzolitic sources. According to them their genesis was involved with high degree of partial melting of ancient metasomatic veins within a harzburgitic-lherzolitic lithospheric substrate mantle. They considered the veins to haver mineralogy similar to that of an experimentally-observed, high pressure, near-solidus phase assemblages.

13.2.14 Experimental Study on an Olivine Leucitite up to 3.5 GPa at Variable Temperatures

Elkines-Tanton and Grove (2003) made experimental investigation on a Pliocene olivine leucitite (WC-1) from the central Sierra-Navada, California. The chemical composition of the rock has been given in Table 13.2:

Phase relation on this composition (Wc-1) is shown in Fig. 13.26.

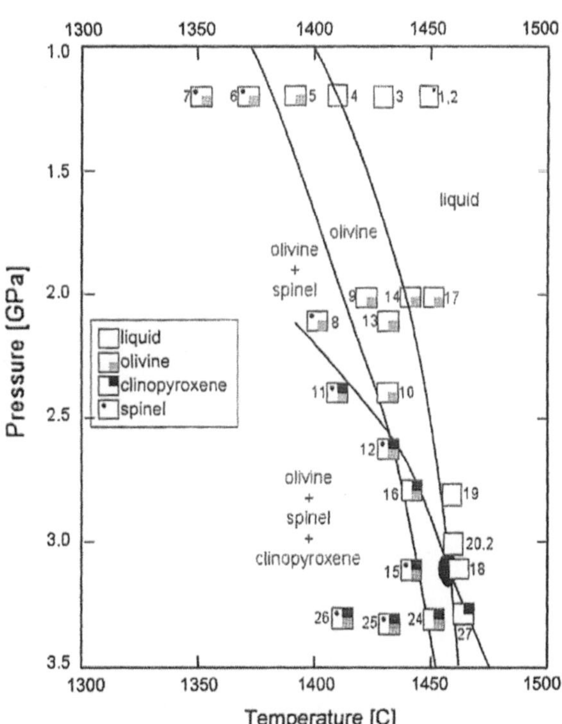

Fig. 13.26 Phase equilibrium experiments on synthetic analogue of Sierra Navada olivine leucitite WC-1. Experiments numbers are marked. The solid oval marks the multiple saturation point (after Elkines-Tanton and Grove 2003)

They observed that clinopyroxene crystallizes over a temperature interval of 50 °C, at pressures between 3.1 and 3.3 GPa below which spinel is the accompanying phase. At lower pressures (below 2.3 GPa) olivine is the liquidus phase followed by the appearance of spinel. They observed that with 6 % water in the coexisting melt phlogopitef is stable. These results indicate that the magma was derived from a hydrous source at greater than 100 km depth.

Xenoliths carried by other young Pliocene lavas in the vicinity of WC-1 have yielded temperatures of equilibrium from 700 to 900 °C, with one outlier at 1,060 °C. These xenoliths are consistent with the hypothesis that the lower lithosphere under the Sierra Navada lamproites the mantle got metasomatized prior to the Pliocene, then got partially melted to produce the highly K-rich lavas. They suggested that subduction-derived fluids were derived by a reaction that consumed garnet + orthopyroxene to create clinopyroxene + phlogopite, and that the high-potassium Sierran magmas were generated by melting phlogopite-clinopyroxene metasomatized peridotite.

Chapter 14
Structural and Tectonic Evolution of K-Rich Silica-Deficient Volcanic Provinces of Different Continents

Genesis of an ultrapotassic magma in the central part of east Africa (Ruwenzori and Birunga volcanic province), the Rhine Graben region of Germany, Damodar Valley Coal Field (east India) and Rio Grande region of the USA is considered to have been associated with the development of rift systems (see Chap. 4). According to modern concept, development of rift valleys and associated large scale volcanism, are related to plume activities. Potassic volcanism in Indonesia, Ulleung Island, Spain, Turkey, and Highwood Mountains of the United States is considered to be subduction-related. Ultrapotassic volcanism in different parts of Italy (the Vulsini complex, Vico, Phlegrean Fields, Vesuvius and Aeolian arc region) is either related to partial melting of the subducted lithosphere underneath the European plate (Edgar 1980) or related to rifting (Cundari 1980). Bianchini et al. (2008) supports post collisional and intraplate Cenozoic volcanism in the rifted Apennines/Adriatic domain. The K-rich volcanism of Alto Paranaiba province is considered to be related to plume-related activity. Different tectonic processes pertaining to the genesis of potassium-rich mafic and ultramafic rocks are described below.

14.1 Tectonism in European Volcanic Provinces

14.1.1 Development of the Rhine Rift Valley

Ziegler (1992) studied the Cenozoic rift system of western and central Europe. This rift system is extended from the shores of the North Sea to the Atlantic coast of North Africa over a distance of about 3,000 km. It comprises the Spanish Valencia Trough, the Gulf of Lions, Saone, Limagne and Bresse grabens of southeastern France, the Rhine, Ruhr Valley and Leine Grabens (Fig. 14.1).

The north–northeast-trending graben bifurcates near Frankfurt. The Leine graben maintains the same strike direction of the Rhine Graben and the northwest-trending Ruhr Valley graben. According to them reactivation of tectonic activity during Permo-Carboniferous and Mesozoic resulted in the formation of fracture systems

A.K. Gupta, *Origin of Potassium-rich Silica-deficient Igneous Rocks*,
Springer Geology, DOI 10.1007/978-81-322-2083-1_14

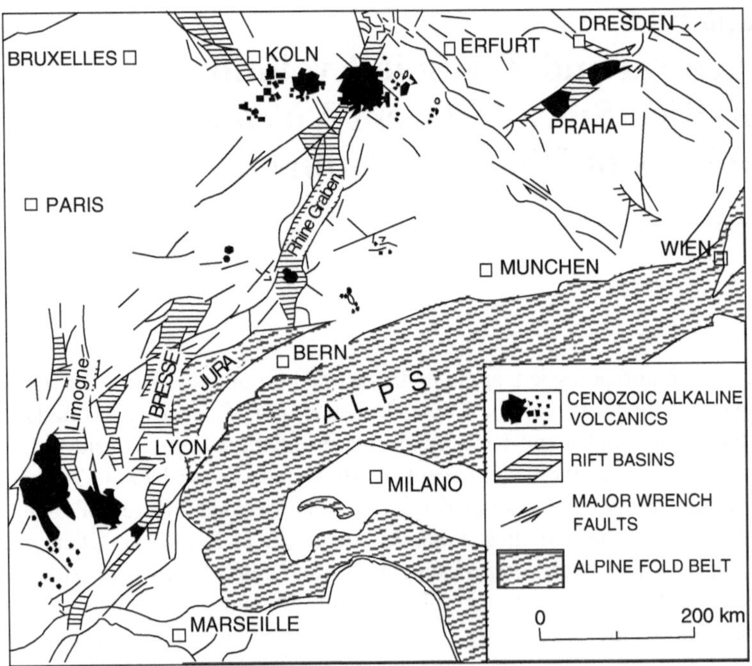

Fig. 14.1 The Cenozoic rift system of western and central Europe (after Ziegler 1992)

associated with Leine graben. This had major role in the localization of these grabens. The Ruhr Valley graben dies out in the central areas of the Netherlands.

According to Ziegler (1992), the subsidence of Rhine Graben system started during late Eocene and volcanism was more or less continuous throughout the Cenozoic, but it was maximum during Quaternary. He further emphasized that the area of Rhine–Leine–Ruhr Valley triple junction was the main centre of volcanic activity. There are however, smaller volcanic centres, which are located 100 km east of the Rhine Graben. These centres include Rhon in the north and the Urach and Hegau areas in the south. Neogene volcanism was extended over a broad area on the Rhenish Shield. This area straddles the triple junction of the Rhine, Ruhr Valley and Leine grabens. This zone of Neogene volcanism spreads over an east–west-trending zone, and is 250 km wide. The Quaternary volcanism was concentrated over the east Eifel area, where the last volcanism took place 11,000 years ago (Mertes and Schminke 1983). The rate of up lift of the western part of the Rhenish shield is estimated to be nearly 0.5 mm/year (Ziegler 1992).

Marine connection between Alpine foreland basin, the Paris and NW European basins were severed during early Miocene, because of thermal uplift of the Massif Central and the Rhenish Massif. Geological evidence suggests that uplift around the southern Rhine Graben, Vosges–Black Forest rift dome started during the late middle Miocene. The Vosges and Black Forest areas were further uplifted during

Pliocene and Pleistocene. This was paralleled by slow intermittent subsidence of the southern part of the Rhine Graben, in which Quaternary sediments were deposited.

The Ruhr Valley graben started its subsidence during the Miocene to Recent times and its major strike direction is parallel to the present northwest-directed maximum horizontal compressional stress field. The Rhenish shield progressively domed upward during Miocene till Recent times. The area was affected by volcanic activity, but the Quaternary volcanic activity shifted to the Eifel region located, west of Rhine, where Laacher See volcano erupted 11,000 years ago. According to Zigler precision survey indicates that the Rhenish shield straddling the Rhine river was uplifted at present at a rate of 0.4–0.6 mm per year, whereas in the northern part of the Rhine Graben, subsidence rates reached a maximum of 1 mm/year between Karlsruhe and Frankfurt. The southern parts of the Rhine Graben was however, stable.

Ziegler (1992) concluded that the asthenosphere–lithosphere boundary below the thermally uplifted region of Massif Central (France), the southern Rhine Graben and the Rhine–Leine–Ruhr Valley graben triple junction, domed up to the level of 50–60 km below the surface and laterally descended to a depth of 80–100 km. According to Zeigler (1992), the refraction and reflection seismic and gravity data could be interpreted either in terms of a tensional failure of lithospheric extension model (Fig. 14.2, top) or in terms of mantle plume model (Fig. 14.2, bottom), both of which should develop crustal doming and the associated gravity anomalies. According to the mantle plume model, the crustal doming was caused by lithospheric extension and thermal thinning of the subcrustal lithosphere over upwelling mantle diapirs and plumes, involving small scale mantle convection. The other model involved tensional failure and assumed that the melts ascended diapirically from the base of the lithosphere to M-discontinuity, where they spread out laterally, maintaining density equilibrium which resulted in the updoming of the overlying crust.

Telesiesmic survey however, suggested that under the southern parts of Rhine Graben, where a low density body was suspected in the upper mantle, no important uplift of the asthenosphere/lithosphere boundary occurred. Presence of weak upper mantle velocity anomalies, which were aligned with the graben axis, could be related to the intrusion of the partial melts, during the middle and late Miocene doming stage of the southern Rhine Graben. Geophysical data of the Rhenish Shield Dome suggested an uplift of the asthenosphere/lithosphere boundary almost to the crust/mantle boundary. Petrological data suggested that volcanism both within the rift zone and on the flanks were derived from a source lying at a depth of 70–100 km, corresponding to the base of lithosphere of the Rhine Graben area. This was in accordance with geophysical configuration of the southern Rhine Graben region. Deep reflection seismic data showed that thinning of the lower crust started along the flanks of the Vosges–Black Forest rift dome at a considerable distance away from the zone of upper crustal extension into the Rhine Graben. The reflectivity of the lower crust increased under the flanks towards the graben margin. Below the Rhine Graben the reflectivity below the lower crust decreased possibly as a result of energy absorption of the sedimentary field of the graben. The top of the

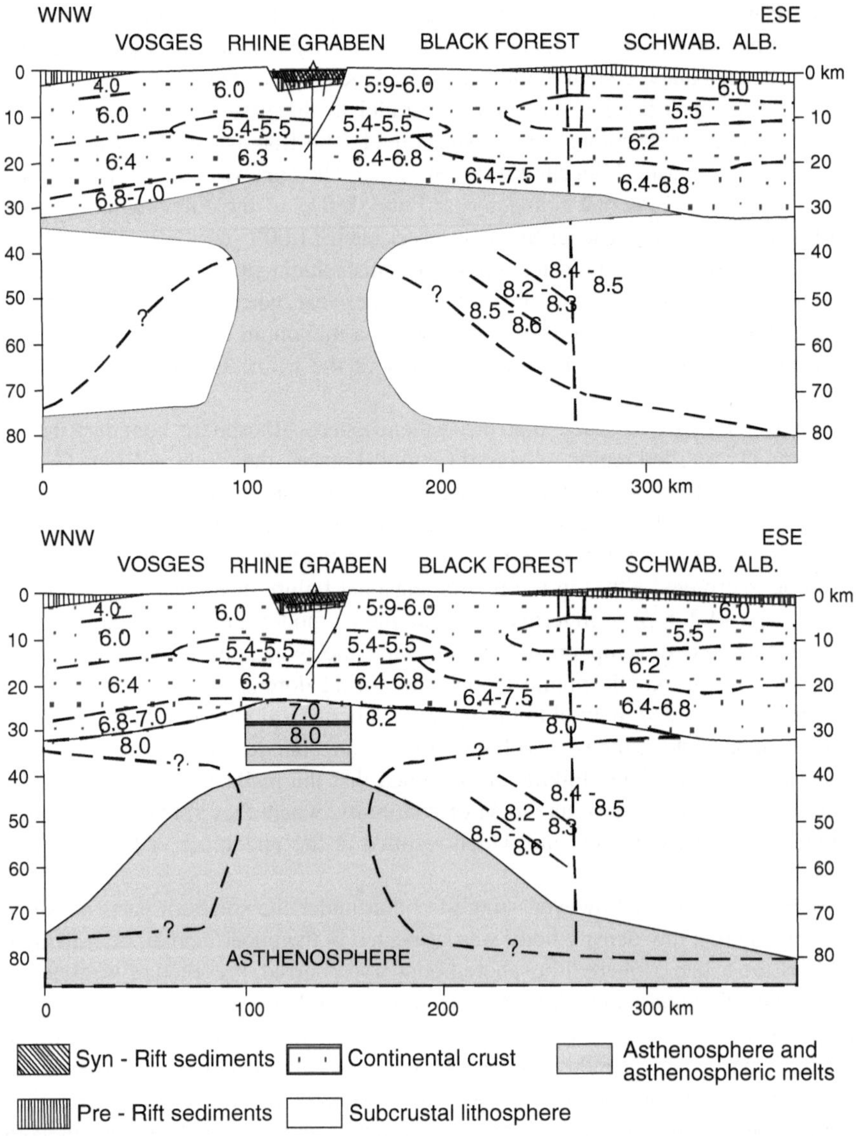

Fig. 14.2 Central configuration of southern Rhine Graben, interpretations along tensional failure (*top*) and mantle plume (*bottom*) models (modified after Edel et al. 1975; Prodehl 1989)

highly reflective lower crust coincided with the base of a highly conductive low velocity layer, which itself had low reflectivity. He thought that this high conductivity–low velocity zone was possibly related to the presence of fluids, which were released from the laminated lower crust.

Reflection seismic data suggested that extension of the upper crustal section of the Rhine Graben ranging between 5 and 7 km (Sittler 1969) and the reflection profiles indicated that crustal configuration of the southern Rhine Graben had an extension value of 17 km for the lower crust. It was estimated that pre-rift crust had a uniform thickness and its volume was preserved during the rifting process.

The doming of the Rhenish Shield, which commenced at the beginning of Miocene was controlled by progressive thinning of the subcrustal lithosphere, but uplift of the Vosges–Black Forest arch was probably related to middle and late Miocene emplacement of melts in the upper mantle near the crust/mantle boundary. According to Ziegler (1992), both processes could cause broad arching of a rift zone, and if rates of thermal uplift exceeded the rate of tectonic subsidence of the axial graben in response to lithospheric extension, the subsidence of the rift could be reversed.

Two models were advanced in an attempt to explain the evolution of the Cenozoic rifts in the Alpine foreland of western and central Europe. One model involved a hotspot driven mechanism (Cloos 1939) through continent–continent collision (Molnar and Tapponnier 1975; Dewey and Windley 1988) to a progressive assertion of a new cycle of plate reorganization during late Alpine orogeny that probably resulted in the break-up of the present continent assembly (Ziegler 1988). The hotspot model suggested that the evolution of the Saone–Limagne–Bresse-Rhine rift system, was a result of crustal extension, accompanied by some volcanic activity, which was preceded by the development of discrete thermal domes 20–40 Ma ago. The formation of these discrete thermal domes could be related to the presence of deep-seated mantle plumes.

The development of the Rhine, Bresse, Limagne, Saone and Gulf of Lions grabens during Eocene to Oligocene were considered to be collision-related. The Gulf of Lions, Saone and Bresse grabens in particular have strike, nearly normal to the Pyrenean thrust front, but have strike sub-parallel to the West Alpine orogenic front.

This model however, is incompatible with the propagation of the Rhine–Rhone rift systems towards south into the west Mediterranean area during Oligocene. This rift system was superimposed from the off-shore parts of the Pyrenees and the compositionally different Celtiberian and Catalonian Coast Ranges as well as the Corsica–Sardinia block.

The evolution of the Cenozoic rift systems of Western and Central Europe was paralleled by the development of East African-Red Sea-Gulf of Suez and Libyan rift systems and the Dead Sea wrench zones. Rifting propagated northward onto the Pelagian Shelf through the Messinian Strait onto the internides of the Apennine during Plio-Pleistocene, when late phases of Alpine orogeny was in operation (Ziegler 1992). During this period the Rhine–Rhone rift propagated southward across the West Mediterranean area and through northwest Africa to the Cape Verde Islands.

According to Ziegler (1992), the coincidence related to the development of mega rift and the late Alpine orogenic phases could not be explained only in terms of the Africa–Arabia and European collision which interacted with tensional forces. This controlled the development of these new rift systems.

14.1.2 Structure and Tectonic History Associated with Potassic Volcanism in Italy

Ninkovitch and Hess (1972) considered that intermediate depth earthquakes in the Mediterranean are distributed beneath the two arcs at the Tyrrhenian and southern margin of the Aegean sea. The active volcanoes in both the areas are located above these intermediate depth earthquakes. They thought that shallow focus earthquakes are associated with axis of orogenic belt passing through Sicily and the Apennine peninsula.

According to Ellam et al. (1989), the Benioff zone is continuous to a depth of 500 km dipping at an angle of 70° to around 250 km depth, and then flattening to around 45°. The regions of active volcanoes in Italy are Tyrrhenian Sea, Sicily and the Eolian islands and the Neopolitan regions. Ellam et al. (1989) considered that the subduction zone in the southern part of Italy, and is probably in the late stage of its evolution and the arc is located on a thin continental basement of about 16 km thickness. There is no oceanic crust remaining in the fore-arc i.e. in the southern part of the archipelago. Anderson and Jackson (1987) suggested that the process of subduction has ceased. Ellam et al. (1989) considered that the subduction beneath the Roman Province is highly speculative. They thought that the deep seismicity beneath the Aeolian and Naples area was associated with different portions of a bifurcating slab.

Alverez (1972) thought that the subduction beneath the Roman Province is not directly related to Benioff zone associated with the Aeolian arc, but an earlier episode of subduction. He presented palaeomagnetic evidence that the Corsica–Sardinia microplate was rotated anticlockwise between 11.5 and 6 Ma ago by tectonic activities. According to him the rotation was accompanied by Eastward subduction in the Tyrrhenian Sea, located close to the modern west coast of main land Italy. He thus believed that both Roman and Aeolian magmatism may be related to subduction processes and the timing of subduction was probably different in the two areas. According to Elam et al. (1989) the magmatic activities of the Roman Province and Aeolian region might have been coeval. The age of eruption range from 0.9 to 0.15 Ma (Ferrara et al. 1986) and the age of K-rich volcanic rocks of the Roman province is 1 Ma. The oldest volcanic rock in the Aeolian arc lavas are 0.5 Ma old, The dredged samples from the seamounts related to Aeolian volcanism to the west of the exposed arc is 1.3 $\pm$ 0.2 Ma old (Beccaluva et al. 1982). The Aeolian arc and Roman Province magmatism was essentially contemporaneous but mostly post date the activity of the Tuscan province (0.7–0.43 Ma) and Mt. Etna has been active during the past 0.1 Ma.

Generation of potassic volcanism in the Roman province evoked a question whether it could be linked to present day subduction beneath the Calabrian Arc or whether it was related to a continental rift system very similar to those in East Africa (Cundari 1980); resolution of the problem is ambiguous (Beccaluva et al. 1991). While most of the Italian volcanoes erupting K-rich silica-undersaturated rocks (except for Somma–Vesuvius and Ischia—Phlegraean Fields), lie on the lithospheric segments not associated with the present day active Calabrian–Aeolian Benioff zone, their geochemical characteristics are very different from those of continental rift lavas and similar to potassic volcanics from island arcs (Foden and Varne 1980).

Affinity of the potassic rocks of central and southern Italy and its possible association with subduction-related metasomatized mantle have been suggested by many authors (Thompson 1977; Civetta et al. 1981; Rogers et al. 1985). The discovery of two-pyroxene highly potassic calc-alkalic basalts, basaltic andesite and andesites from the north-western margin of the Phlegraean Fields, buried under highly potassic alkalic rocks (Di Girolamo 1977) threw new light on subduction-related volcanism in this area. On the basis of this observation, Beccaluva et al. (1991; Fig. 14.3) proposed that the mantle wedge overlying the subducted slab was hybridized by melts produced by partial fusion of subducted material derived from the continental crustal sediments. This process might have played a dominant role in the generation of the mantle sources from which the high-K-calc-alkaline, shoshonitic, leucite basanitic and leucititic magmas of the Roman Province, were derived.

14.1.3 Neogene Tectonics of Southern Spain

Lopez Ruiz and Badiola (1980) discussed about Neogene volcanic activity in Spain. The volcanism in this region was associated with calc-alkalic, high K-calc-alkalic, shoshonitic and ultrapotassic lamproitic rocks and alkali basalt (Fig. 4.27, Chap. 4). The volcanic episode started during Middle-Miocene and ended at the beginning of Pliocene, 17 and 6 Ma ago. During this period, the volcanism was associated with progressive enrichment of K_2O and incompatible trace elements (Rb, Ba, Pb, Sr and Zr) and ferromagnesian components towards north, and their age also progressively decreases in the northward direction. Thus, the calc-alkalic rocks are oldest and they erupted in the southern part at Cabo da Grata followed by slightly younger K-rich calc-alkalic and shoshonitic rocks towards north and highly potassic silica-undersaturated lamproites in the northernmost sector. According to them the last volcanic episode associated with alkali volcanism took place 2–4 Ma ago.

They considered that the mineralogy and geochemistry of calc-alkalic, high K-calc-alkalic and shoshonitic and K-rich rocks suggest that their genesis is associated with subduction of the oceanic crust, which started in the Alboran Sea at the end of Oligocene. The subduction of oceanic crust resulted in the genesis of andesites at a depth of 100 km, and high K-calc-alkalic shoshonitic lavas at a depth of 150 km. During the ascent of the fluid, they might have picked up high quantity of incompatible elements and also assimilated crustal material in varying degrees. Apparently the ultrapotassic magmas were produced due to melting at greater depth. According to Lopez Ruiz and Badiola (1980), the alkali basaltic magmatism was apparently of subcrustal origin produced at a depth of 35–40 km. They considered this magmatism to be related to distensive phase, which started developing during tortonian. Instead of assimilation with crust during ascent of the magmas, this author prefers partial melting of metasomatised upper mantle after subduction of crustle material.

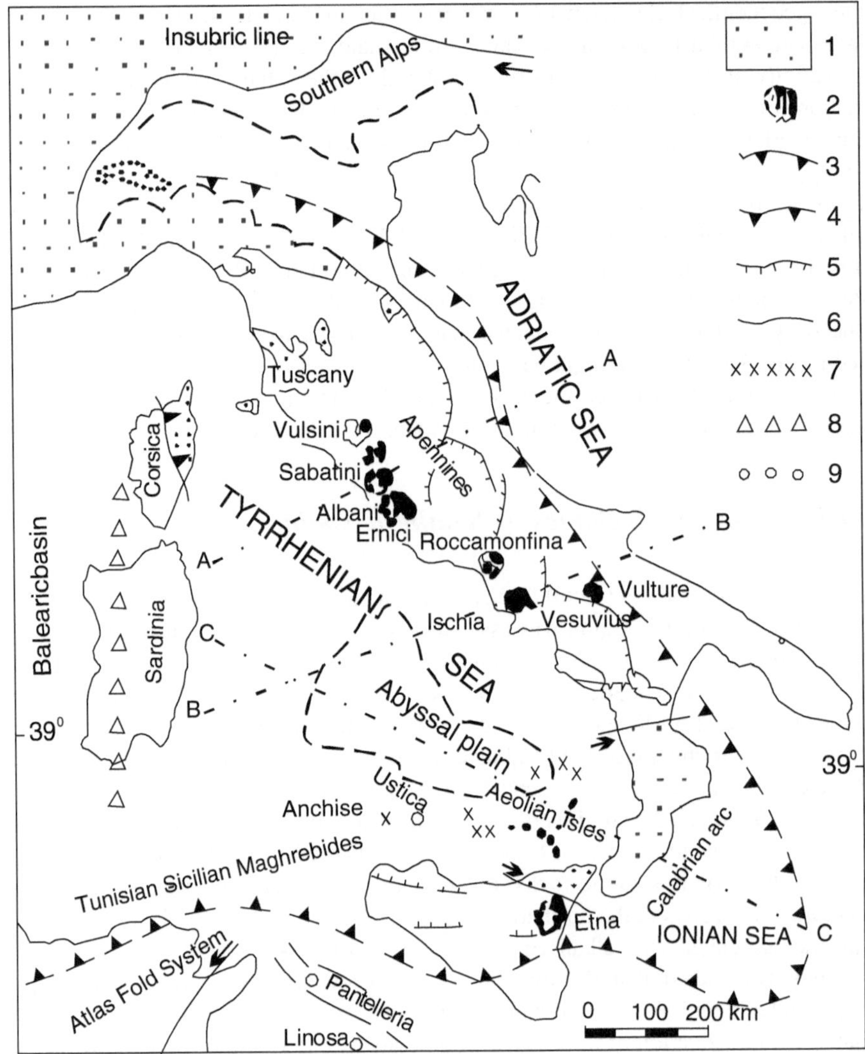

Fig. 14.3 Structural sketch map of Italy: (*1*) Palaeogene–Neogene Alpine chain, (*2*) main Quaternary volcanic complexes in central and southern Italy, (*3*) compressional front of the Alpine chain, (*4*) Neogene compressional front of the Apennine–Maghrebian chain, (*5*) front of carbonate units in the Apennines and Sicily, (*6*) some major faults, (*7*) Quaternary arc volcanism of the Aeolian Islands and adjacent seamounts, (*8*) Oligocene–Miocene arc volcanism of Sardinia, (*9*) Quaternary transitional to alkalic (sodic) volcanic complexes. These also include the Etna tholeiitic and alkalic series (after Beccaluva et al. 1991)

14.1.4 Mantle Upwelling Beneath Eastern Atlantic and Western and Central Europe

A detailed seismic tomography of the upper mantle below the eastern Atlantic volcanic province, the western Mediterranean and the central European rift valley region was made by Hoernle et al. (1995). They established that the eastern Atlantic volcanic province forms an elongated belt of islands and seamounts, which parallels the coast of north-west Africa and Liberian peninsula. This province extends for approximately 2,500 km towards north–northeast from the Saharan seamounts and the Canary Islands in the south to Madeira Island in the centre to the Biscay abyssal plain in the north. Subaerial volcanism in the Canary and Madeira Islands began in the early Miocene, but submarine volcanism might have started 60 Ma ago.

According to them the central European volcanic province extends from France through southwest Poland to Hungary. The province includes the Massif Central volcanic province of France, where volcanism took place during 0 and 20 Ma and 35–65 Ma ago. There is a systematic record of recent volcanism in Rhenish massif (Eifel) and Rhine Graben in Germany (0–20 Ma), in Lower Silesia, Poland (15–14 Ma) and in the western Pannoian basin in Czechoslovakia and Hungary (Eocene-Miocene, and 1–12 Ma). According to Hoernle et al. (1995) most of the volcanism in the central Europe is related to a NNE-trending rift belt formed in the early Cenozoic Era. This continental rift belt includes the Rhone, Limagne, Bresse, Rhine, Ruhr and Leine Grabens. They think that in contrast to other central European volcanism, the Pannonian basin volcanism was associated with back-arc spreading.

They further suggested that the Western Mediterranean volcanic province comprises the Roman province of Italy (Quaternary), eastern Sicily (Late Cretaceous–Quaternary) and Sardinia and Corsica (Oligocene–Pleistocene). They observed that each of these volcanic region is underlain by continental crust, which has been undergoing extension as evidenced by intensive block faulting and rift zone.

On the basis of seismo-tomographic study they established that there is a low S-wave velocity anomaly (LVA) with a width of about 2,500 km to NNE direction and elongated about 4,000 km in the ESE direction, which underlies the eastern Atlantic, northern Africa, western Mediterranean and central Europe. The LVA zone in this model is a planar feature with a NNE strike and westward dip.

According to Hoernle et al. (1995) the local S-wave velocity model shows the presence of a LVA zone at depths shallower than 200 km beneath western and central Europe and the Mediterranean segment. Local P-wave velocity model also indicates the presence of LVA at depths of 150–250 km beneath the same region. They interpreted this low velocity zone to be a relatively hot (possibly volatile-enriched) region of mantle upwelling. The LVA zone could be a single feature or may reflect an averaging of spatially isolated low velocity region such as cluster of plumes. The local S-wave models however, shows the LVA to be continuous and does not show any evidence of deep roots of the LVA beneath central Europe or the western Mediterranean region in agreement with general global model.

The question is whether the LVA zone is continuous over a large area. Some degree of lateral discontinuity in the LVA zone is to be expected. It could result from the localized up- and down-wellings near the lithosphere–asthenosphere interface or rising blobs and pipes of hot volatile-rich mantle surrounded by cooler asthenosphere entrained at shallower depth.

Hoernle et al. (1995) compiled the isotope data from Cenozoic volcanic rocks overlying the LVA zone (Fig. 14.4). They subdivided the data into four groups based on regional domain:

1. eastern Atlantic oceanic domain (western Canary and Madeira islands) erupted on oceanic lithosphere,
2. eastern Atlantic continental margin domain (eastern Canary Islands) erupted on lithosphere, which is transitional in character between oceanic and continental lavas,
3. central European domain, erupted on continental lithosphere and apparently affected to some degree by ancient subduction associated with the Hercynian orogenic and
4. western Mediterranean domain, a region influenced by recent subduction.

The eastern Atlantic oceanic domain, includes western Canary and Madeira islands. The eastern Atlantic continental margin domain, consists of eastern Canary islands. The central European continental domain, includes K-rich mafic volcanic rocks and carbonatites from the Massif Central, Rhenish massif, Rhine Graben, Eger graben and Pannonian basin. The western Mediterranean domain, contains Italy, Sicily, the Aeolian Islands and islands in the Straits of Sicily. According to Hoernle et al. (1995) the data for each group is crudely consistent with binary mixing. The arrays for each region has a restricted compositional range in the Low Velocity Anomaly region (LVC), which has been interpreted by them as having similar isotopic composition of the seismic low velocity anomaly; and is referred to by them as low velocity composition (LVC).

The volcanic rocks from the western Mediterranean domain can be divided into two groups based on isotopic composition and tectonic setting. The sodic lavas ($Na_2O/K_2O > 0.7$ from Mt. Etna, Sicily, Straits of Sicily and Pietre Nere, south-eastern Italy have Sr–Nd and Pb isotope ratios that overlap the LVC. Potassic and calc-alkalic lavas from the regions are influenced by recent subduction of the African plate (Western Italy and Aeolian Islands), as evidenced by very high $^{87}Sr/^{86}Sr$ and low $^{143}Nd/^{144}Nd$ and $^{206}Pb/^{204}Pb$ ratios. According to them, lavas from western Italy have the most extreme isotopic composition with values similar to deep sea sediments, whereas lavas from Aeolian Islands fall in the middle of the array. Oxygen isotope and positive Eu anomalies in samples from western Italy provide additional evidence related to contamination by sedimentary components probably subducted to shallow depths in the upper mantle. They concluded that a minor amount of a third, MORB-like component could also be seen in the lavas from the Aeolian Islands, and in ancient tholeiites from Mt. Etna.

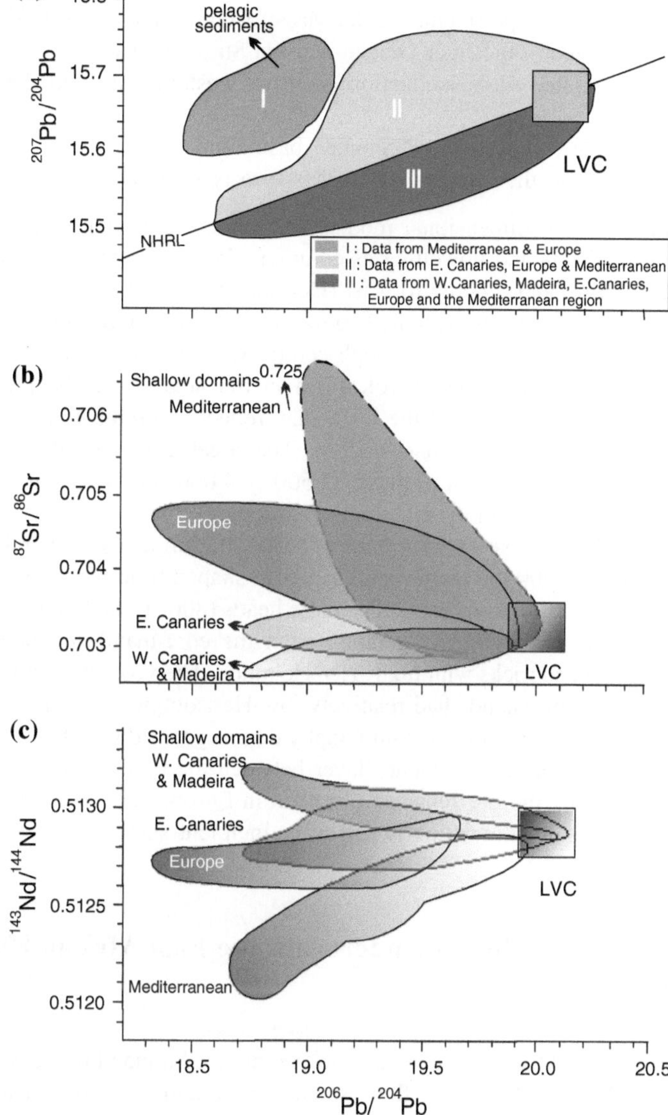

Fig. 14.4 A plot of ^{207}Pb/^{204}Pb (**a**), ^{87}Sr/^{86}Sr (**b**) and ^{143}Nd/^{144}Nd (**c**) isotopic ratios against ^{206}Pb/^{204}Pb for different K-rich rocks from eastern Atlantic, western and central Europe. *NHRL* The Northern Hemisphere Reference Line, *LVC* low velocity component (modified after Hoernle et al. 1995)

The tomographic study raised an important question concerning the structure of the LVA. Does the region represent a large-scale upwelling or a multiple smaller, discrete upwelling? They concluded that evidence for the presence of LVC is present in all tectonic environments above the LVA. This includes:

(1) regions associated with rifting and extension such as the NNE rift belt running though central Europe (including the Massif Central, the Rhenish Massif and the Rhine Graben), the Eger Graben and the Straits of Sicily,
(2) regions associated with subduction, such as western Italy and the Aeolian Islands,
(3) back-arc basins, such as the Pannonian basin, and
(4) hotspots, such as the Canary Islands, Madeira Islands and possibly Mt. Etna.

In contrast to three dimensional models of mantle convection, which showed downwelling in sheets and upwelling as columns, the LVA appeared to be sheet-like stalling to NNE and dipping to west (Hoernle et al. 1995). They thought that planar upwelling is consistent with the large-scale tectonic features, in particular the NNE-trending rift belt extending through central Europe and Western Mediterranean and with the distribution of volcanism in the area above the LVA. They thought that the westward dip of the LVA may reflect a combination of eastward asthenospheric flow and lithospheric drag. The areal extent of this upwelling sheet along the base of the lithosphere (2,500 × 4,000 km) is about twice that estimated for plume heads in the mantle. For example, White and McKenzie (1989) proposed that plume consisted of a narrow (150–200 km across) central column, and a wide (1,000–2,000 km diameter) mushroom-shaped head of anomalously hot mantle (100–200 °C above ambient). The large heads (800–1,200 km diameter) had been predicted for the plumes originating from the core-mantle boundary. Such plume heads produced rocks with high $^{3}He/^{4}He$ ratios. The volcanic rocks from Mt. Etna and the Canarian Islands had relatively low-He isotopic ratios. In case of the LVA from this region, the seismic tomography data suggested that this was a upper mantle feature fed from the boundary layer between the upper and lower mantle, which explained why more extensive volcanism in Europe and Mediterranean was absent and why continental rifting is still in its incipient stage.

14.2 Deep-Seated Plumes Underneath the East African Rift Valleys

Rift system of East Africa is extended from the interior highland of Kenya through Ethiopia and Afar to the ocean floor spreading system of Red Sea and Aden (Fig. 14.5). The second rift system is finally connected to mid-oceanic ridge system of the Indian Ocean.

The uplift of continental segments of Africa is more pronounced than the rifting. Sometimes there are uplifts of thousands of metres along certain zones (500–600 km in width and several thousand km in length). The erosional surfaces have their vertical separation along the flanks of the rifts. These erosional surfaces converge and overlap if they are traced into the neighbouring basinal areas.

Davies (1998) discussed about plume activity in relation to volcanism in East Africa. According to him East Africa has the following interesting characteristics:

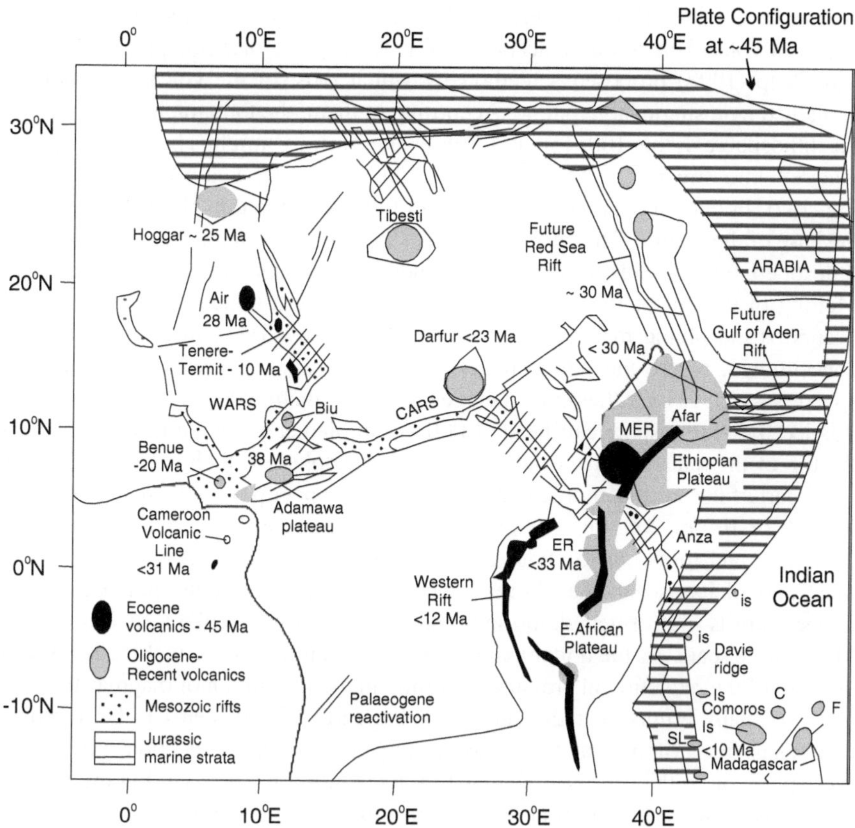

Fig. 14.5 Reconstruction of the African plate at ~45 Ma ago, before the separation of Africa and Arabia. Shown are Cenozoic plateaux, swells, Eastern Rift (ER); Main Ethiopian Rift, MER) and magmatism (maximum ages in each province indicated). The distribution of Mesozoic–Palaeocene rifts, and marine strata along passive margins (after Bosworth et al. 1992; cited by Ebinger and Sleep 1998)

(a) it has highest mean elevation of all the continents,
(b) it hosts the great East African Rift system,
(c) there is young and active sea-floor spreading in the Red Sea and Gulf of Aden, and
(d) presence of great volumes of volcanic rocks in the northeast and minor volcanism in association with rift valley extended across a large area of the eastern, central and northern parts of the continents.

These features suggest upwelling of hot mantle under Africa associated with mantle plumes. A maximum number of 40 hot-spots was proposed by Burke and Willson (1976) in Africa. Geophysicists working on mantle dynamics have debated whether so many plumes could rise beneath one continent. Duncan and Richards (1991, cited by Davies 1998) advocated only one hot-spot within the African continent. Davies (1998) disagreed with this idea and wondered how so many scattered volcanism could be accounted for by only one or a few plumes. Ebinger and Sleep (1998) proposed a method by which most of the volcanism could be explained by a single large plume. According to them, the topography at the base of the lithosphere lying 100–150 km below the surface, might channel the plume material into streams and pools. The model of Ebinger and Sleep is shown in Fig. 14.6. In that model a large plume head followed by a thin plume tail rise from the base of the mantle. The head flattens and spreads as it approaches the base of the lithosphere, and it tends to flow up and under the thinnest parts of the lithosphere. Melting of the hot plume material only begins above a critical depth, and it only occurs as the plume material rises and decompresses. Thus, melting tends to be concentrated, where the plume material flows up a slope towards thinner lithosphere, as well as over the plume tail. According to them, when a new plume starts, it is led by a large spherical `head' 1,000 km in diameter. As the head comes near the top of the mantle it `pancakes' under the lithosphere, which is the cool, strong outer part of the earth comprising the crust and the upper part of the mantle down to a depth of 100 or 200 km. The plume head is then spread to a diameter of about 2,000 km and thins out to about 200 km vertically. According to Davies during most part of its ascent, the plume material is in the solid state, but because the temperature is high it gets deformed gradually and behaves like a fluid on a geological time scale. As the hotter material reaches shallow depths, melting is initiated because of the reduction in pressure, resulting in the generation of magmas that rise, and eventually erupts near the surface of the earth. There seems to be a general consensus that eruption of flood basalts (covering millions of cubic kilometres) takes place on to the earth's surface within a few million years. During the last 250 million years there had been at least a dozen of such eruptions.

Ebinger and Sleep suggested that the Afar plume was probably responsible for not only the Ethiopian flood basalt, but might also have been instrumental in promoting rift valleys that radiate from the region. They further envisaged that this plume was also responsible for rifting of the African continental offshore and south to the Comoros Islands near Madagaskar, west of the Darfur uplift and probably further to the Adamawa plateau and Atlantic coast (Fig. 14.5). This explains the scattered nature of the associated volcanism due to variation in the thickness of the lithosphere through this region. The model of Ebinger and Sleep comprises two mechanisms: (1) the plume tends to rise into the regions of thinner lithosphere because of its buoyancy, (2) melting is initiated at a critical depth during the ascent of the plume.

According to this model if the plume rises under a thick lithosphere it may melt very little or fusion may not occur at all. If the material flows horizontally it will not melt any more. If later, the plume reaches thinner lithosphere and can flow upward again, it can melt further as a consequent of this a rifting takes place. Subsequently it may trigger more melting and may promote the flow of plume material over larger distance.

George et al. (1998, cited by Davies 1998) however, were of the opinion that there are actually two plumes in this region. Eruption on the southern Ethiopian plateau, which took place 45 million years ago may be related to an older plume, which lies under Lake Victoria. According to them the later eruptions in southern Ethiopia (19–12 millions year ago) might be related to the spreading influence of the more northerly Afar plume.

The volcanism of alkalic rocks without rifting and associated with topographic swells and ocean island chains of Africa was considered to be related to separate plumes (Wilson and Giraud 1993; Emerick and Duncan 1982). Burke (quoted in Ebinger and Sleep 1998) considered that the Cenozoic volcanism and high elevation in this region is related to stationary position of the African plate relative to mantle circulation since 35 Ma ago. This resulted in the penetration of 40 mantle plumes without creation of hot-spot tracks. The swells at Hoggar, Tibesti, Darfur and Adamawa and eruptive centres of the Cenozoic period lie on a Mesozoic and a Mesozoic-Palaeocene rift systems between East African and Ethiopian plateaux (Fig. 14.5). Wilson and Giraud (1993) suggested preferential melting of mantle, which was metasomatized during the breakup of Gondwana in the presence of many plumes during the Mesozoic. Ebinger and Sleep (1998) suggested that seismic velocity and geochemical anomalies below Red Sea, Afar and Eastern Rift (Prodehl and Mechi 1991; Green et al. 1991). According to Ebinger and Sleep, tomographic model suggested that the lithosphere beneath the cratonic core was more than 200 km thick.

Very little is known of the upper mantle structure beneath the Ethiopian plateau away from the rifts. The high $^3He/^4He$ ratios throughout a 2,000-km-wide region centred on the Ethiopian plateau indicated the presence of a broad mantle thermal anomaly. According to Ebinger and Sleep (1998) the largest geoid anomalies coincided with the Ethiopian and east African plateau, where more than 8×10^5 km^3 of primary basaltic material erupted since 45 Ma ago (Latin et al. 1993). Cenozoic flood basaltic magmatism took place in the Ethiopian plateau

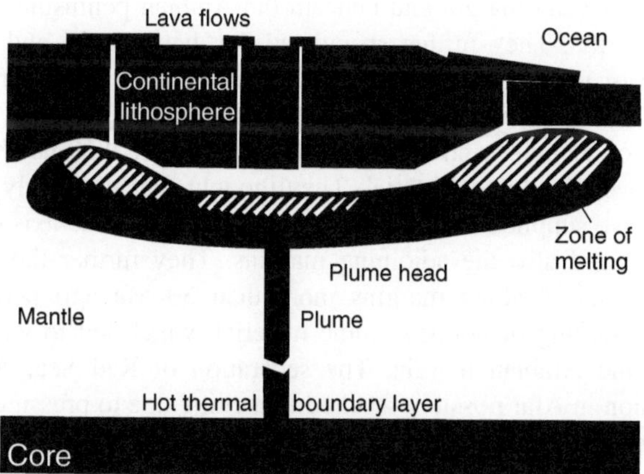

Fig. 14.6 Ebinger and Sleep's proposed model showing how a mantle plume interacting with topography on the base of the lithosphere, might explain scattered volcanism across a very broad region of east Africa (see also Davies 1998)

region although the Cameroon chain onshore witnessed discontinuous activity since Late Cretaceous (Lee et al. 1994). Incomplete thermal equilibration of thinned lithosphere beneath Late Cretaceous-Palaeocene rifts and Mesozoic margins might have caused variation in depth to the 1,300 °C isotherm at the base of the lithosphere beneath Africa 45 Ma ago and the basin became active during Palaeocene. Sleep (1996) thought that the topographic relief at the lithosphere–asthenosphere boundary before the arrival of the plume head probably resulted in the deflection of plume material away from its centre, as is expected from cratonic roots. In the numerical models of Sleep, topographic relief at the lithosphere–asthenosphere boundary disrupts the radial spread of buoyant plume material, which resulted in ponding below the thinner lithosphere. According to Sleep et al. melting is enhanced, where plume material cascades over steep relief at the lithosphere–asthenosphere boundary.

The distribution of plume material and melt beneath the central and northern Africa was determined by Ebinger and Sleep (1998, Fig. 14.7a, b). Assuming a linear depth versus melting parameterization they predicted the final thickness of the liquid layer above the plume. In their model they did not distinguish between plume head and tail material during slow movement of the African plate since 45 Ma ago. According to them the plume was initially centred below a site, where earliest basalts were found and it moved to a spot, the projection of which on the surface occurs at 36°E, 4°N. They further thought that 45 Ma ago, a circular plume head (750 km radius) arrived at the base of lithosphere producing a thick low density layer beneath the Ethiopian plate. They speculated a net upward movement of linearly flowing plume material with steep to vertical gradient having increasing vertical velocity. This was associated with fusion related to pressure release (Fig. 14.7b). They further envisaged that pre-existing structural features might have controlled location of eruptive centres. About 45 Ma ago, plume materials moved an additional 500–800 km southward and eastward to a region of thin lithosphere along the Indian Ocean margin and beneath the Arabian peninsula after Mesozoic and Palaeogene rifts. They further speculated that between 35 and 10 Ma ago, a narrow tongue of plume material moved upward beneath the central African rift zone reaching the western Adamawa plateau 5 Ma ago.

The model of Ebinger and Sleep took into account thermal thinning of the lithosphere above the plume material. They thought that the newly thinned lithosphere might have supplied melt from the eastern limb into the axis of the Red Sea and Aden Rifts and also the adjoining margins. They further thought that these extension along the Arabian margins more than 30 Ma ago perhaps triggered decompression melting of ponded plume material, which led to widespread magmatism along the Arabian margin. The separation of Red Sea, Aden and East African extension in Afar possibly resulted in melting due to pressure release rather than at the centre of the plume configuration. Lateral density variation within the upper mantle should generate extensional stresses, which might have been large enough to initiate rifting. They further established that adiabatic decompression melting beneath the rift should enhance flow of plume material and rift propagation. Ebinger and Sleep (1998) thought that these mechanism might explain extension after magmatism and uplift and propagation of the rift towards south in the east African region.

Episodic extension in Red Sea and Gulf of Aden captured ponded melt and resulted in pulses of basaltic magmatism in east Africa (Ebinger and Sleep 1998). Reactivation of pre-plume basins and initial extensions in the northern Kenyan Rift about 25 Ma ago probably created a conduit for the distribution of the plume material towards south beneath the eastern and western rift systems (Figs. 14.5 and 14.7a, b). They thought that this additional extension perhaps caused outward flow as far south as Madagascar and the western rift, where magmatic centres are situated near the Mesozoic rifts.

Ebinger and Sleep (1998) thought that the steep gradient at the lithosphere–asthenosphere boundary along the Tanzanian Craton and mobile belt boundary perhaps resulted in flow, which explains the generation of more than 30 Ma old carbonatitic and kimberlititc magmas obtained from small melt volume at greater depths (Boyd and Gurney 1986; Dawson and Smith 1988). Ebinger and Sleep considered that the reliefs present in the lithosphere and asthenosphere boundary before the emplacement of the plume materials probably diverted it and helped decompression melting along the craton margins preserving their thick strong cores (Fig. 14.7b).

The scheme of Ebinger and Sleep implied the rejuvenation and thermal subsidence of continental rifts. They further thought that ponded plume material contributed to regional uplift, but the sills intruding into sedimentary strata, which underplated at the base of the previously thin crust could locally augment or inhibit the uplift. This would however, depend on density contrast. Genik (1992) thought that many Mesozoic–Palaeozoic rift basins lying along the paths of plume material show late Cenozoic reactivation and igneous intrusions. Coffin et al. (1986) opined the presence of anomalous upper mantle velocity of 7.8 km per second below the western Indian Ocean basin, which suggested elevated mantle temperature above the south-eastern lobe (Fig. 14.7).

The studies of Ebinger and Sleep suggested that the location of continental flood basalts perhaps did not coincide with the centre of plume head and the presence of several discrete magmatic provinces could be related to a single plume. Their model of Ethiopian plume also supported a plume origin for the east African rift systems.

Rogers et al. (1998) studied volcanism in the Eastern Birunga province. They noted that the magmas were derived from the mantle lithosphere with the source age of 1 Ga and 0.5 Ga and the youngest ages corresponded to the deepest magma sources. The magma production rate in Birunga was low (0.04 km^3 per year), reflecting prolong heating (10–15 Ma) of the lithosphere by the East African plume. They suggested that conductive heating and melting of the lithosphere at the volatile-enriched mantle solidus resulted in the production of lava. According to the model of Turner et al. (1996), the amount of melt produced, is represented by the melt thickness, which depended on the duration of heating, the potential temperature of the underlying mantle plume (T_p) and the thickness of the lithosphere. A melt thickness of 1–2 km can be calculated for the Birunga province from the above total volume estimates integrated over the outcrop area of the whole province ($\sim$3,500 km^3). If it is assumed that East Africa arrived over the plume during the

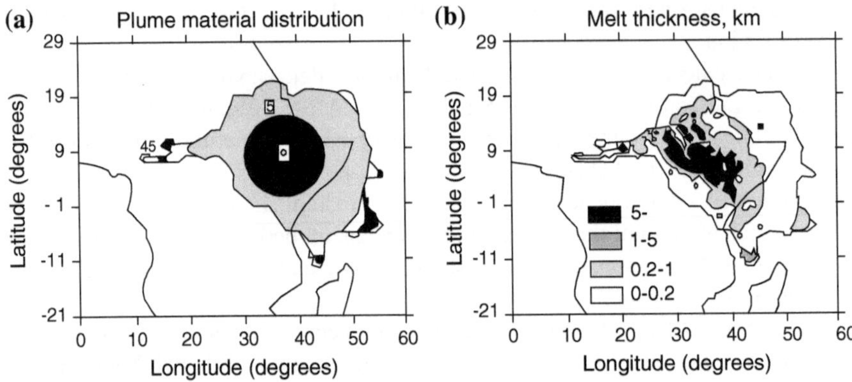

Fig. 14.7 **a** Plume material distribution: contours of the maximum horizontal extent of plume material between 45 Ma ago (contour O) and present (contour 45) predicted by the interactions of plume with the lithospheric lid (modified after Ebinger and Sleep 1998). **b** Melt thickness in km: the equivalent thickness of the melt produced from 30 Ma ago to the present. The calculation includes pressure-release melting for lateral flow of material and lithospheric thinning, caused by the interaction of plume material and the base of the lithosphere. Plume material was assumed to melt by 0.1 % per km of ascent, independent of depth, melting was enhanced due to steep gradients (modified after Ebinger and Sleep 1998)

early Miocene, then the lithosphere had been heated by the plume during the period between 10 and 15 Ma.

Applying these values to the model of conductivity heating described by Turner et al. (1996) and assuming a lithosphere thickness of 150 km gives a maximum T_p for the subjacent plume of 1,430 °C. The calculated melt thickness, which assumes 50 % melt trapped at depth, is almost certainly an overestimate, and so 1,430 °C should be regarded as a maximum temperature for the underlying plume. The estimated temperature should be considerably lower than that often assumed for mantle plume (T_p > 1,500 °C), but it is consistent with the location of the Birunga province at the periphery of the East African plateau, the present-day topographic and geophysical expression of the East African plume the presence of small volumes of magmatism in and around the Birunga province dates back to 12 Ma, which suggested that contact with the plume occurred between 10 and 15 Ma ago, if the duration of conduction heating had been <10 Ma, then a melt thickness of 2 km implies higher plume temperatures. However, the duration of heating should have to be less than ~7 Ma to necessitate plume temperatures >1,500 °C.

In comparison with models of melt generation for the Kenyan Rift, which Latin et al. (1993) related to the decompression of the East African plume as a consequence of extension across Kenyan Rift, it should be emphasized that the duration of magmatism in the Birunga province was shorter. Its volume was much smaller, and the time of onset was generally later than that of the Kenyan Rift. However, as the relative contribution from lithospheric and sub-lithospheric sources to Kenyan basalts was not determined, the similarity between melt generation rates between the Birunga province and the Gregory Rift could be regarded as being largely fortuitous.

Finally, the rates of melt generation in both the Birunga province and the Kenyan Rift are at least an order of magnitude less than those of flood basalt provinces such as those in Parana (~ 0.4 km^3/year, Stewart et al. 1996) and almost two orders of magnitudes less than that for the Deccan volcanic region (Courtillot et al. 1988; Ducan and Pyle 1988). Although this observation is not unexpected for Birunga, it is surprising that the Kenyan Rift has such a low production rate, considering the β factor of 2.5 proposed by Latin et al. (1993) against that inferred for the Parana (1.5). It is possible that amounts of extension across the Kenyan Rift are much less than estimated and or/the underlying plume is significantly cooler than that involved in the generation of continental flood basalt provinces.

14.3 Tectonic Evolution of Silica-Deficient Potassic Rocks from Brazil with Reference to Trinidade Plume

The significance of the Trinidade Plume in relation to the genesis of potassic rocks of Alto Paranaiba Igneous province was discussed by Gibson et al. (1994). They thought that one of the major phases of plume activity and a major foci of this upwelling was probably along the future site of the South Atlantic Rift (Willson 1992). White and McKenzie (1989) considered that the voluminous early Creteceous magmatism of Parana-Etendeka and NE Brazil-Central Africa was related to Tristan and St. Helana Plumes. According to Gibson et al. (1995), the recent geoid anomaly maps of south Atlantic suggested that Trinidade (or Martin Vaz) plume and other small-scale unnamed mantle plumes were present during Late Cretaceous. Herz (1977) first suggested that the alkali magmatism of this age in Brazil was related to plume activity. He also realized the significance of alkaline complexes in the reconstruction of the movements of the South Atlantic plates during the breakup of Gondwana supercontinent. Later Gough et al. (1980) suggested that plume-related to magmatism in Brazil was traceable as early as 120 Ma ago on the South American Plate. This observation was based on age dating of kimberlites form Brazil, and the idea was however, debatable. The zircons form kimberlites occurring both west (Pimenta Bueno, Rondonia) and east (Poco Verde, Minas Gerais) yielded ages between 80 and 90 Ma. Much recent volcanic activity associated with this plume occurred on the islands of Trinidade and Martine Vaz, when volcanism took place 0.7 and 0.2–3 Ma ago, respectively. These islands lie 1,300 km off the coast of Brazil and are linked by Trinidade–Vitoria seamount chain. The chain of islands associated with the seamounts may represent the track of the Trinidade mantle plume (Herz 1977; Griffith and Campbell 1991; Gough et al. 1980). The Eocene basaltic eruption (42–52 Ma; Cordiani 1970) in the continental shelf,

200 km north of Trinidade–Vitoria seamount chain, which form the Aboralhos platform perhaps was related to the plume. The Pb-isotope ratio of these basalts are similar to that of the Island of Trinidade. According to Morgan (1983) the present day centre of the Trinidade hotspot perhaps were located beneath the Alto Paranaiba province ~90 Ma ago. If the Trinidade–Vitoria seamount chain represented the surface expression of the same plume that was responsible for the alkali magmatism in the northeast margin of the Parana basin, then the width of the hotspot track has narrowed significantly with time. This view is in agreement with the suggestion of O'Conner and Duncan (1990), who suggested additional seamounts (Hotspur, Rodgers, Minerve and Almirante Saldanha) located to the north and south of Trinidade–Vitoria sea mount chain, which represented the presence of a hotspot with large diameter, when the South American continental margin passed over the plume 50 Ma ago.

Fluid dynamic studies by Richards et al. (1989) showed that a plume generated from the mantle at a depth does not displace the overlying mantle. It ascended quickly until the near superficial plume head attained a large amount of buoyancy. A narrow tail is formed behind the part of the plume head and this led to rapid ascent of hot low viscosity materials. The plume head gradually flattened and was dissipated, but the tail possibly persisted long after this.

Various types of plume-related igneous provinces could be recognized depending on whether the lithosphere moved over a plume head or a tail. Continental flood basalts or oceanic plateau basalts now cover an equant area (1,500–2,000 km across) and are considered to have been generated by melting, as the lithosphere passed over the plume head, whereas linear belts, which formed hotspot tracks (<300 km) were formed, whenever the overlying lithosphere passed over the plume tail. According to Gibson et al. (1994), the 100 km wide zone of Late Cretaceous (80 Ma) mafic potassic magmatism in the Alto Paranaiba Igneous Province may be the surface expression of a starting plume head; and the 150 km wide Trinidade–Vitoria seamount chain was a long-lived (post 50 Ma) hotspot track of the plume tail. No effect of melting related to the plume could be observed and the plume might have been turned off during the early. Tertiary, when the old thick lithosphere of the San Francisco Craton (South America) passed over it. "The plume activity reemerged from beneath the westward drifting of South American continent and became the magma source for the ocean island volcanism of the Atlantic Ocean and the seamounts (Gibson et al. 1994)."

According to Courtney and White (1986) plume-related volcanism is usually associated with uplift, and this was probably related to (1) positive buoyancy of the plume, (2) due to thermal expansion of the lithosphere and (3) possibly because of magmatic underplating (Cox 1989). According to Gibson et al. the temporally and spatially associated uplift and magmatism in the Alto Paranaiba Igneous Province (Fig. 14.8) were caused by the Trinidade mantle plume. Pressure estimates from a garnet lherzolite xenolith (Leonardos et al. 1993) and analogies with experimental

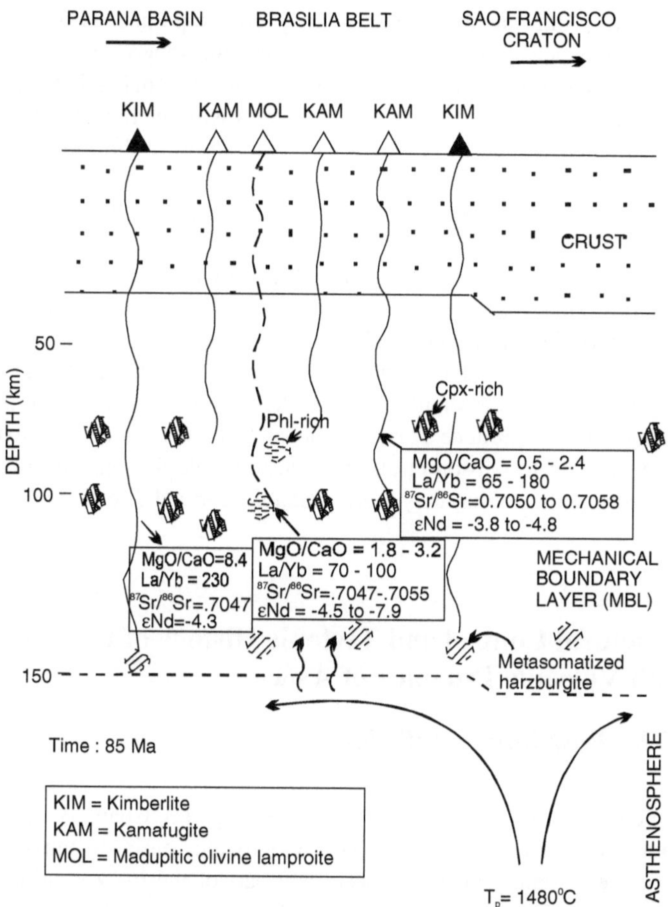

Fig. 14.8 Schematic diagram to illustrate variations in composition and depth of the mantle source regions of kamafugitic rocks, madupitic olivine lamproites and kimberlites in the Alto Paranaiba Igneous Province. *Cpx* clinopyroxene, *Phl* phlogopite, *Kim* kimberlite, *Kam* kamafugitic rocks (after Gibson et al. 1994)

studies (Eggler and Wendlant 1979), suggested that the Tres Ranchos Kimberlite originated from a depth of >150 km. Low CaO/Al_2O_3 and high La/Yb ratios suggest that the kimberlite magma was derived from a metasomatized mantle harzburgite source material. Kamafugitic lavas and intrusions have high CaO/Al_2O_3 and appear to have been derived from a relatively clinopyroxene-rich (wehrlitic?) source. The madupitic olivine lamproites have intermediate CaO/Al_2O_3 ratios and perhaps was derived from a more clinopyroxene-rich source than the kimberlite. Their high K_2O content suggested that phlogopite had an important contribution in their melting regime. This was distinct from the kamafugitic magmas and the kimberlites, where phlogopite probably was residual in their mantle source. Similar

Nd-isotope ratios indicate that approximately contemporaneous and widespread LREE-enrichment occurred in the subcontinental lithospheric mantle. Hill (1991) considered that the uplift taking place in the Alto Paranaiba Igneous Province during pre- and postmagmatic episodes agreed well with the time constraints for the plume impact. Gibson et al. further considered that uplift of the cool lithosphere up to 20 Ma before magmatism probably reflect the ascent of the plume through the asthenosphere, and 20 Ma post-magmatic uplift may reflect heating and displacement of the underlying lithosphere.

Gibson et al. further added that transfer of heat by conduction and advection by asthenospheric source melts into the subcontinental lithosphere probably coincided with the impact of the plume. The melts produced due to the plume activity should mobilise into subcontinental lithosphere enriched in volatile-rich segments. The enriched nature of horizons should enhance their predominance in the final melts, which consolidated to produce K-rich mafic rocks of the Alto Paranaiba, Igneous Province. According to Gibson et al. the enriched isotopic signature of these volcanic rocks may indicate that they were probably derived due to Trinidade plume activity.

14.4 Structural Control and Tectonic History of Potassium-Rich Volcanic Province of Asia

14.4.1 The East Indian Rift Zone

The tectonic history of the Jharia basin (also see Fig. 4.3, Chap. 4) was studied by Ghosh and Mukhopadhyay (1985). This Gondwana basin is intracratonic and located in eastern India. The Jharia basin is a part of the E–W-trending Damodar Valley Coal Fields. The eastern fringe of the basin lies about 175 km northwest of Kolkata. The Gondwana Supergroup in the Jharia basin is characterized by the Talchir formation (upper Carboniferous to Permian) at the basement followed by fluviatile Permian sediments comprising the Barakar, the Barren Measure and the Raniganj formations. According to Ghosh and Mukhopadhyay, the basin has an oval shape and the dip of the sedimentary beds is around 5–10°. The sediments are intruded by numerous dykes and sills of dolerite and lamproitic rocks. The dolerites are post-tectonics with reference to different movements in the basin. The lamproites are found to be displaced by many faults. For example, towards northeast of Mohuda, the lamproite dykes show evidence of shearing, where they were emplaced along fault planes. Transverse geological sections of the western part of the Jharia basin is shown in Fig. 14.9a–e.

In rare instances the dykes cut across pre-existing faults. On the basis of above observations they considered that the time of emplacement of lamproites overlapped with that of late stage faulting. The movement was related to reactivation of early contemporaneous basement faults and the emplacement of lamproitic rocks took place after the deposition of at least the major part of Gondwana sediments in

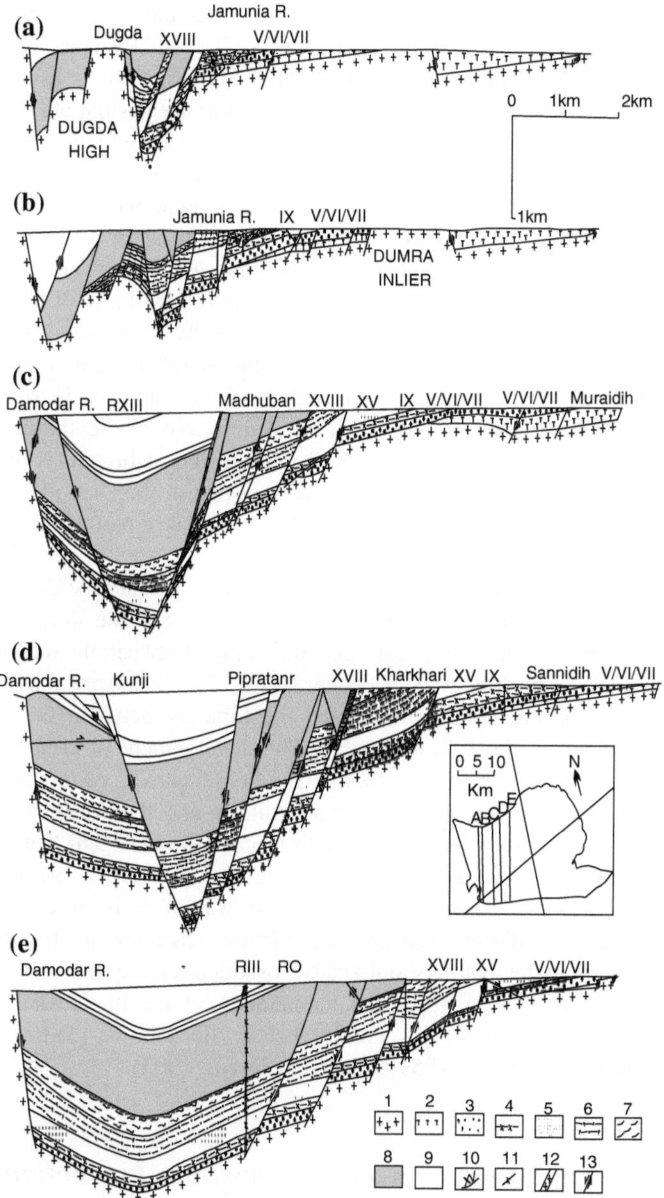

Fig. 14.9 Transverse geological sections of the western part of Jharia basin. *1* Archaean, *2* Talchir Formation, *3* Lower Barakar sediment, *4* combined V/VI/VII–IX interseam sediments, *5* IX–XV interseam sediments, *6* XV–XVIII interseam sediments, *7* XVIII supraseam sediments, *8* Barren Measure Formation, *9* Raniganj formation, *10* coal seam, *11* mica peridotite, *12* dolerite, *13* faults (after Ghosh and Mukhopadhyay 1985)

the Damodar Valley coal fields. The lamproitic dykes cut across Upper Permian sedimentary beds in the Jharia basin and the middle Triassic Panchet beds occurring in the adjoining Raniganj basin. Ghosh and Mukhopadhyay found that the emplacement of the lamprotic rocks were restricted to the basinal area only. These dykes and sills occur extensively within the Gondwana sediments and are usually confined to the Precambrian framework of the basin. There is only one example of the occurrence of lamproitic dykes in the Dughda high, which is very close to basinal border. According to them the subsurface data show that the maximum concentration of the lamprotic intrusives coincides with the zones of maximum subsidence in the central part of the Jharia basin and the volume of the lamproites increase with depth. They found that in the other Gondwana basins of the Damodar Valley, the volume of lamproites is larger in the Barakar formation.than in the higher formations. The magnetic anomaly data of the Jharia basin suggested the presence of large intrusive bodies within the central axis of the Jharia basin. The occurrence of lamproitic intrusives within the Gondwana sediments is maximum in the Jharia and Raniganj basins towards the eastern part of the Damodar lineaments. The volume of the intrusives decreases towards the western part of the basin and is minimum in the north Karanpura basin.

Although the emplacement of the lamproites was later than the development of the basin, the spatial correlation between the total mass of the intrusives and the amount of subsidence suggests a close genetic relation between the development of the basin and the igneous activity at deeper levels. The close spatial association of the basin and lamproite emplacement and the time-lag between the two events could be explained in the following way. The development of contemporaneous normal faults and the intrabasinal graben no doubt indicates a crustal tension during the development of the basin. The normal faulting, and the crustal tension associated with it, continued even after the deposition of the Raniganj beds, which are the youngest beds of the Jharia basin. It is suggested that during the late phase of such crustal tensions, when the Jharia basin was already filled with lower Gondwana sediments, did the resulting late fractures reach the mantle. Propagation of the fractures up to the mantle should cause a sudden release of pressure, which in turn would lead to a lowering of the temperature of the mantle solidus. If the existing geothermal gradient was not far below the solidus, melting would be initiated (Yoder 1952; Uffen 1959; Uffen and Jessop 1963).

14.4.2 Tectonic Setting of K-Rich Rocks from Indonesian Archipelago

The Indonesian Archipelago (for tectonic setting of Indonesian archipelago and Benioff zone contours see Fig. 4.5a) has been evolving since the late Palaeozoic, by an evolving system of island arcs (Whitford et al. 1979; Edwards et al. 1993). The fossil arcs are now represented by a series of linear volcano-plutonic igneous complexes with associated sedimentary deposits now representing the Indonesian

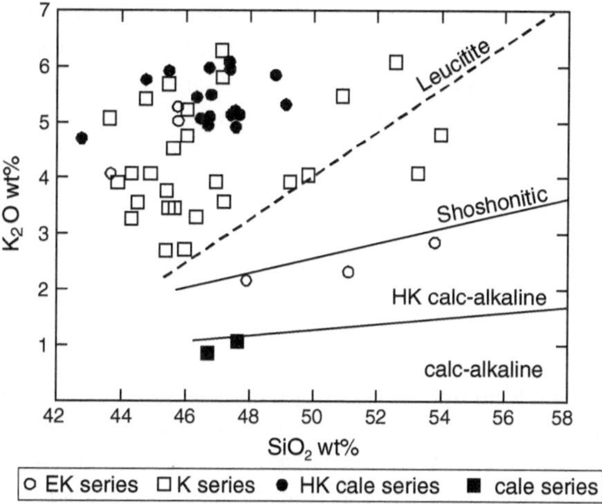

Fig. 14.10 The K_2O versus SiO_2 for the calc, HK-calc, K and EK (highly evolved K-series) rocks of Ringgit–Besser. The boundaries are drawn according to the suggestion of Peccerillo and Taylor (1976) [after Edwards et al. 1994]

Archipelago. The age of these complexes decreases from Sumatra (late Palaeozoic) to eastern end of the system in the Banda arc (late Tertiary). According to Audley-Charles (1975), the present arc became a single entity in the mid-Tertiary. The pre-mid-Tertiary geological evolution of the arc west of Sumbawa was distinct from that to the east. McElhinny et al. (1974) thought that the pre-mid-Tertiary evolution of the eastern Sunda and Banda arcs was related to that of the neighbouring Australian continent while the western Sunda arc might be unrelated (McElhinny et al. 1974). Figure 14.10 shows the presence of calc-series, highly potassic calc-series, shoshonites and leucititic rocks from Ringgit–Beser complex in east Java. Occurrence of these rocks are related to their melting from a Benioff zone and the leucititic rocks coming from a deeper source.

This evolutionary pattern reflects the tectonic environment along Sunda arc. The crust is relatively thick (~25 km) in Sumatra and the presence of Palaeozoic granites and pelitic schists (Katili 1973) suggests a mature continental crust. The young eastern Sunda and Banda arcs (Tertiary) present many immature features on oceanic crust (Purdy et al. 1977). The 20 km thick crust beneath Java and Bali, has a velocity structure intermediate between that of continental and oceanic and the oldest exposed rocks are of Mesozoic age (Katili 1973).

The Sunda arc presents a regular spatial distribution of earthquake epicentres (Fitch and Molnar 1970). According to Fitch (1970) shallow focus earthquakes (<70 km) are confined to a less than 200 km wide narrow zone along the entire length of the arc, located between the trench and the volcanic arc. He also described the presence of intermediate and deep focus earthquakes (>70 km) occurring in a parallel zone further from the trench. Hatherton and Dickinson (1969) discussed andesitic volcanism and seismicity in Indonesia, the lesser Antilles and other island arcs, and Cawthorn (1977) showed a systematic correlation between the depth of the Earthquake foci with respect to the potash content of the orogenic magmas (also see Beswick 1976).

There is a poorly defined Benioff zone occurring west of Sunda Strait (Sumatra). It dips to the northeast, and has not been observed at depths greater than 250 km. Earthquakes at depths between 500 and 650 km are relatively common towards east of the strait. Although there is a considerable gap in seismic activity at depths between 300 and 500 km. The deep focus events define a Benioff zone dipping at 60°, although it is less steep at depths of less than 100 km.

The oldest rocks in Java and Bali are composed of lower Tertiary shallow marine sediments which were intruded and overlain by plutonic and related volcanic rocks. They are observed in a zone only slightly south of the present-day volcanic arc. According to Van Bemmelen (1949), the igneous activity represents first cycle of andesite volcanism, which are relatively common in the Southern Mountains of Java and in the Barisan Range of Sumatra. The andesite volcanism culminated in a period of uplift and intrusion in the mid-Miocene.

A second cycle of volcanism marked by the appearance of more alkaline lavas took place during late Tertiary. This cycle is represented by the high-K calc-alkaline suite, which represents the second cycle, and is located further northeast from the Miocene 'Old Andesites'. These rocks are found together with the Quaternary 'third cycle of volcanism'. The distinction between the products of these latter cycles is rarely clear.

According to Fyfe and McBirney (1975), phlogopite having higher P–T stability field than other commonly-occurring hydrous phases, should survive up to a greater depth during the underthrusting of the crust into the mantle and eventually at considerable depths, dehydration of phlogopite should produce potassic magmas (also see Beswick 1976). Sodium-rich amphibole, having a lower P–T stability field than phlogopite, should break down at shallower depths and the magmas generated at such depths should thus be relatively richer in sodium than potassium.

14.5 Plate Tectonic Model for Potassic Volcanism in the USA

14.5.1 Tectonic History of Potassic Volcanism in the Highwood Mountains Region

The plate tectonic model for potassic magmatism in the Highwood Mountains region was discussed by O'Brien et al. (1991). According to them, the Montana alkalic province in general and the K-rich mafic volcanic rocks in particular constitute a broad zone of Eocene magmatism, which stretches form the southern British Columbia and part of South Dakota. According to them, the calc-alkalic volcanism, which was contemporaneous with the Highwood potassic volcanism (54–50 Ma) is observed in the Castle Mountain of Central Montana, the Lowland Creek-Pioneer volcanics of the Southwest Montana and the Gallatin-Absaroka volcanics of southern Montana and northwest Wyoming.

Voluminous calc-alkalic rocks (46–42 Ma) also cover central Idaho, southwest Montana and further west in Oregon and Washington. Close association of calc-alkalic and K-rich silica-undersaturated rocks in Sunda arc and Mexico, provided indirect support for the model that suggested that the Eocene volcanism of the northwestern United States to subduction-related process. O'Brien et al. considered that the recent minette lava in western Mexico was erupted in an extensional environment, and the volcanism was similar to the eruption of such rocks in the northwestern United States during the Eocene (Coney 1980). Lipman (1983) and Heller et al. (1987) reconstructed the tectonic history of the region for this period. According to their hypothesis, there was a NW–SE-trending convergent plate boundary throughout the western United States. Lipman et al. (1971) and Dickinson (1979) thought that there was a single low dipping slab of Farallon plate subducting under the lithosphere in this region.

O'Brien et al. (1991) pointed out the presence of calc-alkalic rocks in the southwest within the Larmide fold and thrust orogenic belt to potassic rocks within the foreland, is consistent with the deeper and mineralogically distinct source region below the Wyoming cratonic knell. They therefore, considered that the genesis of parent magma for the Highwood Mountains region is related to partial melting of asthenospheric mantle wedge, which resulted in the infiltration of fluids from a low-angle subducted slab (Fig. 14.11). The relationship between the subduction slab with respect to the position of different volcanic fields at Challis, Rocky Mountains, Highwood Mountain (HM), Haystack Butte (HB), Bearpaw Mountains (BM) and

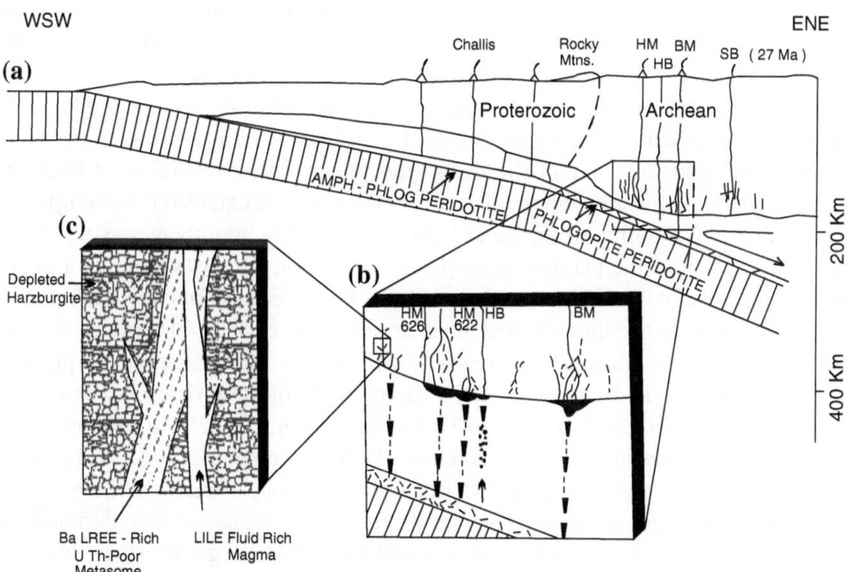

Fig. 14.11 Schematic cross section illustrating tectonic model for the formation of the Highwood Mountains magma (no vertical exaggeration) (after O'Brien et al. 1991)

Smoky Butte (SB) are shown in Fig. 14.11. During the period (55–50 Ma, Fig. 14.11a), relatively rapid convergence of the Farallon plate resulted in low-angle subduction, with a trench approximately 750 km southwest of the Highwood Mountains. Fluids enriched in LILEs released at relatively shallow depths from the subducted slab reacted with the overlying asthenospheric mantle to form a phlogopite—and amphibole-bearing peridotite carapace that was dragged down by induced flow. The Challis Volcanics probably formed as a result of amphibole breakdown in this metasomatised carapace.

Figure 14.11b shows that as the slab descended to greater depth below the Wyoming Province craton, breakdown of residual phlogopite in the peridotite carapace induced melting to form the proto-Highwood magmas. In an alternative, two-stage model the carapace components probably had been added to the wedge during the late Cretaceous or Palaeocene, and reactivated in the Eocene by back-arc asthenospheric flow. In either case, these small-volume LILE and fluid-rich magmas ascended to the lower lithosphere interface, but only those magmas (solid lines) that migrated along pre-existing pathways (dashed lines) where they retain sufficient heat to rise very far into the thick lithosphere. Figure 14.11c demonstrates that this preferential channelling facilitated the movement of alkali magma, which reacted with ancient crystallization and metasomatic products present as veins within the depleted harzburgite mantle lithosphere. Assimilation of these Ba-rich metasomes overprinted the Highwood magmas with the distinctive ancient U-Th-HFSE-depleted Ba-LREE-enriched geochemical/isotopic signatures. The degree of overprinting was variable depending on residence time (e.g. more for HM-622, considerably less for HM-626b). At a slightly later time (48 Ma), CO_2 fluid or carbonatite melt fluxing in the asthenospheric mantle included melting to from the alnoitic suite of magmas represented by Haystack Butte and the Missouri Breaks diatremes. These magmas ascended explosively through the lithosphere and acquired very little of the ancient metasome signature. In Oligocene (27 Ma), very small volume of lamproitic melts containing a dominant ancient metasome component were generated within the deep lithosphere (either by influx of back arc asthenospheric melts or by unspecified processes linked to extension) and erupted at Smoky Butte. According to the model, about 55–50 Ma ago, there was rapid low angle subduction of the Farallon plate (Dickinson 1979; Bird 1984), along a trench located approximately 750 km southwest of the Highwood Mountains (Heller et al. 1987). O'Brien et al., thought that fluid enriched in LILEs, released at shallow depths from the subducted slab reacted with the overlying mantle to produce phlogopite-amphibole-bearing peridotite carapace. With the progressive descent of the slab to a greater depth below the Wyoming Craton, breakdown of amphibole-bearing segments resulted in the production of Chalis Volcanics. This was followed by breakdown of phlogopite in the peridotite carapace. Partial melting of the phlogopite-rich segment resulted in the production of magmas that intruded the Highwood Mountains localities. They considered that magmatism in that area would be related to a two-stage process (Fig. 14.11b). Addition of carapace components perhaps were added during the Cretaceous or Palaeocene and became reactivated during Eocene due to back-arc asthenosperic flow. A very small volume of lamproitic melt was generated within the deep lithosphere possibly due to back-arc asthenospheric melting as a result of some process linked to extension and eruption 27 Ma ago in Smoky Butte.

14.5.2 Generation of Potassic Rocks Associated with Rio Grande Rift

Gibson et al. (1993) discussed about ultrapotassic magmatism associated with Rio Grande Rift. According to them the rift development (Fig. 14.12) was connected with complex geological events in the western USA during Oligocene. During this period there was extension and widespread magmatism in the Basin and Range Province, as discussed in the preceding section.

Cenozoic magmatism was associated with plate motion. It is believed that there was convergence between the North American and Farallon plate with the subduction of the lithosphere. There was rapid fragmentation of the slab and during its assimilation by the asthenosphere a slab `gap' was produced around 35 Ma ago beneath northern Mexico and New Mexico. This was associated with a E–W-spreading zone of calc-alkalic magmatism (Coney and Raynolds 1971), which is extended up to the western margin of the Great Plains, and covers volcanic fields of Mogollon–Datil (39–24 Ma), San Juan and later volcanic fields. The Rio Grande Rift is located close to the eastern edge of the Tertiary magmatic arc and the development of the rift was initiated immediately after this volcanic episode. There

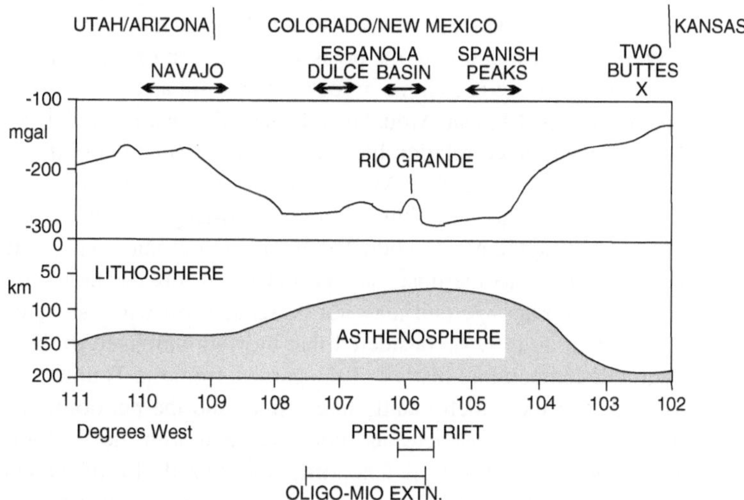

Fig. 14.12 Gravity versus distance profile along west–east direction for the Reo Grande Rift region. The axis of the rift is delineated by a broad uplift on both sides of an axial graben, which is 540 km wide and 15 km deep (after Gibson et al. 1993)

was probably a change in the east-dipping subduction 30 to 40 Ma ago to a strike-slip margin near the coast of Pacific because of the development of the San Andreas transform system (Gibson et al. 1993). Lipman (1980) considered that the Rio Grande depression was caused due to decreased rate of convergence and subduction resulting in the back floor of the asthenosphere and interaction with the overlying lithosphere. Wernicke et al. (1987, cited in Gibson et al. 1993) however, presented a different hypothesis, according to which extension related to thermal relaxation and gravity-driven spreading of materials resulted in thickening of the crust during Larmide compression. They thought that the Mid-Tertiary extension was confined to the Rio Grande Rift. Due to thrusting during the Mesozoic period, there was crustal shortening. Gans et al. (1989) however, believed that the spatial and temporal distribution of Cenozoic magmatism has nothing to do with zones of Mesozoic shortening or Cenozoic plate motions. According to them, the earlier magmatism in the Basin and Range province, predates the beginning of crustal extension and is related to upwelling of the asthenosphere but not due to stretching of the lithosphere.

The region of potassic magmatism is far inland within the continental margin at a distance of about 1,500 km and has been interpreted due to a shallow dipping subduction zone beneath the area. Increase in the K_2O content with respect to SiO_2 and the K_2O/SiO_2 ratio of the Cenozoic volcanic field is supposed to have good correlation with the distance from the trench axis of the Pacific coast. This is related to steepening dip of the slab 40 to 20 Ma ago after truncation and later detachment of the slab by the San Andreas fault. Rowell and Edgar (1983) observed that except for the Leucite Hills lamproites, there is in general NE–SW younging trend of the K-rich Cenozoic magmatism, which coincides with the younging trend of the outer limit of magmatism. On the basis of this observation, they suggested that potassic magmatism in the western USA is related to subduction rather than intraplate tectonic activity. Age determination of the potassic rocks, using K/Ar technique and geochemical data of Cenozoic K-rich mafic rock localities in an around Rio Grande rift such as Spanish Peaks, Elkhead Mountains, Dulce dike swarms and Two Butte, show that there is no simple correlation between age and spatial distribution of the K-rich mafic rocks in Colorado, New Mexico and Arizona. Helmstead and Doig (1975) suggested that eclogite, xenoliths within Navajo-Hopi minettes were originally considered to be fragments of subducted Mesozoic oceanic crust, which was perhaps genetically related to the host magma, but recent observation shows that eclogites are of Precambrian age and may not be related to Cenozoic subduction-related event. Roden et al. (1979) considered that they are unrelated to the genesis of minettes. Variable temperature history of the garnets from the Thumb have been considered to be due to local melt-mantle interaction, and the peridotite xenoliths are part of the continental lithosphere and not associated with the Farallon plate. According to Gibson et al. (1993) the Chondrite normalized plot of incompatible element concentration in the K-rich mafic magmas of the Rio Grande rift zone has inter-elemental ratios similar to the plots of rocks produced due to subduction-relatedvolcanism. The contents of LILE and LREE are more in the K-rich mafic rocks from the Rio Grande rift zone than in subduction-related volcanic rocks.

Gibson et al. therefore, thought that the cause of LILE enrichment of the Cenozoic volcanic rocks in the western USA is thus, quite controversial. The Cretaceous to Pleistocene magmas from the Wyoming Province, are considered to have been derived from a very ancient LREE-rich source (Dudas 1991; Eggler et al. 1988). They considered that genesis of the compositionally similar K-rich mafic magmas throughout the Cenozoic era in the western USA (Leucite Hills, Crazy Mountains) was independent of prevailing tectonic setting. This view is thus, quite contrary to that suggested by O'Brien et al. (1991).

There is a rigid conductive lid overlying the convecting interior (Thompson et al. 1989). They recognized a mechanical boundary layer constituting the lithosphere immediately below the continental crust. This crustal segment is considered to have remained isolated for geologically long period from the underlaying mantle, which had been convecting below. Heat passes through the crustal segment by the process of conduction and there is inherent buoyancy resulting from earlier chemical differentiation. The mechanical boundary layer below the continent is separated from the highly convecting asthenosphere by a thermal boundary layer, which transmits heat upward by convection at its base and conduction at the top. Thompson et al. considered that as the thermal boundary layer overturned during convection on a geological time scale, it remains chemically inherent part of the underlying asthenosphere. If there is change in composition after a shift in tectonic regime, there should be a period of several million years, when the overlying thermal boundary layer retains the previous asthenospheric composition. McKenzie (1989) proposed a petrogenetic model, according to which, there is continuous or semi-continuous leakage of extremely small melt fractions from the convective mantle into the overlying ridge in the subcontinental lithosphere. McKenzie (1989) proposed that the continental thermal structure is usually stable for a long period, the small batches of volatile-rich ultrapotassic liquid gets frozen within small depths. Subsequent heating or stretching related to decompression of this region, results in remelting of the accumulated frozen ultrapotassic liquid and their associated metasomatic materials.

Chapter 15
Genesis of Ultrapotassic Rocks

Different hypotheses, invoked during the first half of the last century, involved, (1) assimilation between different magma types and rocks (Daly 1910, 1933; Shand 1931), (2) Subtraction of Eclogite from a Picrite or Peridotite Magma (Holmes 1932; O'Hara and Yoder 1967), (3) gaseous transport (Kennedy 1955), (4) zone refining process (Harris 1957), and (5) partial melting of phlogopite–richterite-bearing peridotitic mantle (Gupta and Yagi 1980).

Different types of assimilation hypotheses as invoked by different earth scientists during the first half of the last century are described below. This would show that most petrologists recognized that some processes of mixing between mantle components and crustal segments were required to explain the genesis of ultrapotassic rocks. Before the view of this author on the genesis of ultrapotassic magmas is discussed, different hypotheses are described briefly.

15.1 Assimilation Processes

Any hypothesis related to the genesis of K-rich magmas must explain the following:

(1) extreme enrichment of potassium over sodium,
(2) high concentration of LREE,
(3) high abundance of HREE (6–20 times more than that of chondrites),
(4) high Ni, Co and Cr concentration in these rocks particularly with high mg-number,
(5) high $^{87}Sr/^{86}Sr$ ratio (sometimes as high as 0.725–0.730) and their negative ε_{Nd} content.

Because of these peculiarities, most serious investigators (see Chap. 1), who studied these rocks thought that some mixing processes between a mantle-derived magma and crustal materials were involved in the generation of ultrapotassic magmas. Following the suggestion of Shand (1910, 1931), the process of assimilation of some primary magmas with different types of crustal components were invoked by many. A brief critical evaluation of these assimilation hypotheses are given below:

A.K. Gupta, *Origin of Potassium-rich Silica-deficient Igneous Rocks*,
Springer Geology, DOI 10.1007/978-81-322-2083-1_15

(1) a basaltic magma and granitic rocks (Turner and Verhoogan 1960),
(2) a basaltic magma and peridotitic rocks (Holmes and Harwood 1932, 1937),
(3) a carbonatite magma and granitic rocks (assimilation)(Holmes 1965),
(4) a monchiquite magma with granitic rocks (Williams 1936),
(5) a trachytic magma and carbonate-bearing sediments (Rittmann 1933).

In case of all assimilation hypotheses, Bowen's argument (1928) seems to be very valid. He pointed out that at the time, any magma punctures through the earth's crust it is not superheated enough to assimilate foreign materials and avoid concomitant crystallization. It seems highly unlikely that after assimilation, they should be completely modified to a new magma type and then should be still hot enough to undergo fractionation so that different daughter products of a K-rich magma could be generated (see also Wyllie 1974).

Regarding the first assimilation hypothesis, it should be further mentioned that this hypothesis does not explain the high Ni, Co, Cr, Sc content of many K-rich ultramafic rocks, as neither granites nor basalts are enriched in these elements. Experimental study with four compositions involving a mixture of dolerite and granite in following variable proportions were made: 80:20, 60:40. 40:60 and 20:80. Experiments were conducted in one atmosphere between 800 and 1,000 °C (Yagi and Gupta 1977) and at 750 °C under 0.1 GPa. The run products failed to produce any feldspathoid-bearing assemblage.

Holmes (1950, 1965) argued that assimilation reaction between a carbonatitic magma and granitic rocks could produce highly potassic silica-undersaturated magma like katungite (Bell and Powell 1969). Carbonatites are enriched in incompatible elements and LREE, and thus such an assimilation hypothesis could explain observed trace and minor element distribution pattern. The P–T stability of carbonates and other minerals in carbonatites, however suggest their low temperature of formation (Koster van Groos and Wyllie 1973). The eruption temperature of carbonatitic liquids at Oldoinyolengai, Tanzania was only around ~500 °C (Dawson 1966). Thus, a magma with such a low temperature of eruption, before incorporating any other rock type should undergo concomitant crystallization before being transformed into a potassic magma type. In many regions such as Mundwara (Rajasthan, India), Ambadongar, Gujarat (India), West Eifel (Duda and Schminke (1985) and Kaiserstuhl (Wimmenauer 1974), both the places are from Germany), carbonatites invading granitic country rocks are found to be in close association with fenites and Na-rich undersaturated rocks; and in these localities potassic rocks are not found. Besides, in most localities ultrapotassic rocks are found to occur independent of carbonatites.

In cases of assimilation hypotheses involving carbonate sediments and trachyte or monchiquite and granitic rocks, it may be noted that trachytes and monchiquites themselves are of secondary in nature, produced by fractionation of other primary magmas and therefore, are not hot enough to assimilate other rock types, and then undergo further fractionation to produce ultrapotassic rocks. Such a process should not also explain enrichment of mantle compatible elements like Ni, Co. Cr, etc. found in many K-rich ultramafic rocks.

In order to explain variable as well as high $^{87}Sr/^{86}Sr$ and $^{18}O/^{16}O$ ratios in potassium-rich volcanic rocks of Vulsinian lavas, Turi and Taylor (1976, also see Ferrara et al. 1986) concluded that the primary magma was contaminated by high ^{18}O and ^{87}Sr crustal materials.

According to Holm et al. the two samples of Turi and Taylor with high $\delta^{18}O$ values (+11), are in fact quartz-normative trachytic lavas, which are members of the silica-saturated group, and do not belong to a undersaturated variety. The $\delta^{18}O$ values of the primary Vulsinian lava may be 2–3 % higher than typical values for primary mantle-derived magmas. According to Holm et al. (1982), it is difficult to explain high $\delta^{18}O$ in the mantle, other than mixing of normal $\delta^{18}O$ primary mantle melt with a fluid of high $\delta^{18}O$ content as suggested by Muelenbachs and Kushiro (1974). Holm et al. (1982) concluded that such a fluid could have been derived by interaction with or melting of materials which were in equilibrium with surface water at low temperatures, which is with sediments or meta-sediments. They further considered that such a fluid was derived from sediments and/or hydrated oceanic basalt of the subducted crustal slab.

They further observed that the $^{87}Sr/^{86}Sr$ ratio is nearly constant in the entire SiO_2-undersaturated series with the values falling in the range of 0.7105–0.7108. The olivine–leucite melilitite of San Venanzo, which may be a part of the Vulsini volcanic province has similar values: 0.71037 and some silica-saturated ones have the value between 0.7098 and 0.7100. The available $^{87}Sr/^{86}Sr$ ratio of the silica-saturated and under-saturated lavas cannot be distinguished (Holm et al. 1982), and they suggested a narrow range of $^{87}Sr/^{86}Sr$ ratio (0.7104–0.7108) for the entire suite. They found that their $^{87}Sr/^{86}Sr$ values and those of Vollmer (1976) are not intermediate between silica-saturated high $^{87}Sr/^{86}Sr$ ratio of Tuscan lavas and low $^{87}Sr/^{86}Sr$ ratio of K-rich lavas of Roman district. Their observation thus, contradicts the conclusion of Turi and Taylor (1976). Holm et al. thought that the lavas of Vulisini district have a primitive composition and they represent rather uniform magmatism extending from Alban Hills and northward.

15.2 Subtraction of Eclogite from a Picrite Magma

Experimental studies at high pressures (up to 3 GPa) on a nepheline-normative and a hypersthenes-normative picritic lava were made by Gupta and Yagi (1979). In both the cases above the solidus, omphacitic pyroxene together with pyrope-rich garnet (similar to eclogite fraction) crystallized just above the solidus from the both types of picritic liquid, but when the composition of the melt minus the eclogite fraction was determined they were not ultra potassic in nature as suggested by O'Hara and Yoder (1967).

Holmes (1932) observed the presence of eclogite xenoliths included in K-rich lavas at Ruwenzori (Africa). Holmes and Paneth (1936) however, observed that eclogite xenoliths included in Ruwenzori lavas were much older in age than these lavas and therefore himself, could not be genetically related. Holmes therefore, disproved his own hypothesis.

Further, with reference to the second hypothesis, even if sufficient heat is provided to a region having a gabbro and peridotite interface by a mantle plume, the liquid thus, produced should not be enriched in Ba, Rb, Sr, Zr etc., as either of the two rock types are impoverished in these elements.

15.3 Zone Refining Hypothesis

Zone refining hypothesis of Harris (1957) suggests that incompatible elements have very low K_D values [X_{tr} (liquid/crystal $\leq$ 0.01–0.02)], hence during partial melting, such elements as K, Rb, Sr, Ba, etc. should be saturated in the first liquid after low degree of melting, but this mechanism would have produced K-rich rocks as an exception rather than a rule.

15.4 Genesis of Potassic Rocks by Volatile Transport

Lindgren (1933) considered that differentiation largely through gaseous transport produced trachytes, phonolites, basanites and many other volcanic rocks. Rittmann (1933) also suggested this mechanism together with assimilation between limestone and trachyte to be an effective way of producing leucite-bearing rocks of Vesuvius. Morey and Hesselgesser (1952) showed that 40 % by weigh of alkali silicates could be dissolved by aqueous vapour phase at 400 °C and 2,200 bar. From these observations, Kennedy (1955) considered that high solubility of alkali silicates in gaseous water indicates that they may strongly associate with water molecules and their solubility is very much dependent on pressure and temperature. Thus, when there is a release of pressure, resulting from fracturing during escape of volatiles, alkali silicates should be left behind because of decrease of solubility. Concentration of alkalis at the top of an igneous stock may result from such a process, as in the case of potash enrichments in igneous rocks of Jumbo basin, Alaska (Kennedy 1955) or incase of formation of pseudoleucite crystals in the contact zone of shoshonitic dyke and country rocks in the Highwood Mountains region (Larsen et al. 1941).

In case of this gaseous transport hypothesis alkaline rocks should have been emplaced always at the roof tops in isolated cupolas during the last stage of volcanism. Such a process could lead to accumulation of alkaline rocks locally. This process cannot explain large scale lava flows found in the surface in many localities throughout the world.

15.5 Phlogopite–Richterite-Bearing Peridotitic Mantle

A brief review of all the hypotheses (Chap. 14) shows that all the petrologists realized that the generation of these ultrapotassic magmas could not be derived by a simple process of partial melting from a chondritic mantle. A magma produced from a peridotitic upper mantle needed to undergo some kind of mixing process to account for the extreme enrichment of LIL elements and the LREE over HREE in the mantle compatible elements. Before discussing further on the genesis of potassium-rich magmas it is important to know first what should be the appropriate parent materials in the mantle for these magmas.

In Chap. 13, experimental studies by various petrologists on different types of K-rich rocks have been discussed. The subsolidus assemblages obtained by different investigators are summarized in Table 15.1.

The above table suggests that different types of K-rich rocks under mantle P–T conditions are represented either by a phlogopite pyroxenite, a phlogopite-bearing websterite, a phlogopite-bearing harzburgite, a phlogopite-bearing lherzolite or a phlogopite-bearing wherlite. Garnet is an accompanying phase in the high pressure subsolidus assemblage, if plagioclase is present in the starting composition. Experimental study of Luth (1997) and Tronnes (2002) on the stability of phlogopite shows that at pressures above 7 Gpa, richterite and a K-bearing X-phase appear as a break down product. The X-phase has a structure similar to that of wadeite. Appearance of K-richterite at moderate pressure has also been noted in many lamproitic rocks by different petrologists. In these high pressure subsolidus assemblages accessory phases present, are various combination of the following phases: richterite, apatite, chromian spinel, rutile, priderite, armalcolite, wadeite and different types of carbonates.

Presence of phlogopite-bearing lherzolites, harzburgites and pyroxenites as mantle xenoliths in ultrapotassic lavas (Menzies and Hawkesworth 1987) suggests that different types of phlogopite-bearing ultramafic rocks, containing various accessory minerals like K-richterite, apatite, crichtonite (Haggerty et al. 1983), wadeite, priderite, rutile, perovskite, ilmenite and armalcolite) should represent the mantle source material for ultrapotassic magma.

Crichtonite group of minerals are characterized by a variety of end members rich in K, Ba, Ca, Pb, Sr, U, and REE (Hagerty et al. 1983; Hagerty 1983). Such minerals may coexist in equilibrium with Nb and Cr-bearing rutile, Cr- and Mg-bearing ilmenite and armalcolite, rich in such elements as Nb, Zr, Ca. These minerals are stable under mantle P–T condition in peridotite, which is rich in Mg and Cr and low in alumina content (Haggerty 1983, also see Foley et al. 1987).

Tanton and McKenzie (1994) thought that the source rock for K-rich magmas was initially a garnet peridotite containing 0.6–6.7 % garnet and 0.8–5 % clinopyroxene. In the second stage, partial melting (about 20 %) of such a source rock in the garnet stability field, should result in the depletion of such elements as Na, Ca, Al etc. In the third stage the depleted mantle source is metasomatised by addition of

Table 15.1 Subsolidus assemblages as obtained by different petrologists in their experiments on various K-rich rock types

Sr No.	Rock type investigated	Subsolidus assemblage (phases shown in brackets are accessory minerals)	References
1.	Alban hills leucite basanite (synthetic)	Garnet–phlogopite pyroxenite	Gupta et al. (1976)
2.	Melilite–nepheline leucitite (synthetic)	K-richterite–phlogopite pyroxenite	Gupta et al. (1976)
3.	Katungite	Phlogopite pyroxenite (perovskite, Ti-magnetite)	Arima and Edgar (1983)
4.	Madupite	Garnet–apatite–phlogopite pyroxenite as a source rock	Barton and Hamilton (1979)
5.	Leucite lamproite	Phlogopite lherzolite	Foley (1989)
6.	Jarangdih lamproite	Phlogopite harzburgite (apatite + ilmanite + carbonate)	Gupta et al. (2002)
7.	Raniganj lamproite	Phlogopite harzburgite (apatite + chromite + carbonate)	Gupta et al. (2002)
8.	Mohanpur lamproite	Phlogopite–harzburgite (apatite + chromite + rutile + carbonate)	Gupta et al. (2002)
9.	Wolgidite	Phlogopite websterite (rutile)	Arima and Edgar (1983)
10.	Biotite mafurite	Phlogopite wehrlite (ilmanite)	Ryabchikov and Green (1978)
11.	Olivine ugandite	Phlogopite wehrlite	Edgar et al. (1980)
12.	Phlogoite minette	Phlogopite wehrlite (±orthopyroxene)	Edgar et al. (1980)
13.	Armalcolite–phlogopite lamproite	Phlogopite websterite (±richterite, armalcolite)	Edgar et al. (1992)
14.	Olivine lamproite	Phlogopite harzburgite	Foley (1993)
15.	Sanidite–phlogopite lamproite.	Phlogopite–richterite pyroxenite	Mitchell (1995)

4–10 % metsomatic fluid. This metasomatic fluid is considered to have been formed by less than 0.5 % melt of the MORB source rock.

Arima and Edgar (1980) and Dubeau and Edgar (1985) showed that wadeite and priderite are stable at pressures up to at least 2.5 GPa in the mantle. Whereas wadeite can be a source of Zr, and K, priderite may supply K, Ba, and Ti in the mantle in an alumina-deficient environment. They are therefore, not expected in the mantle source rock for tephritic and basanitic liquid. Experimental study of armalcolite-bearing K-rich rock shows that it may be stable under mantle pressure up to 2.0 GPa, and at still higher P–T condition, it breaks down to rutile and other phases.

Ionov et al. (1937) observed that mica is the dominant host mineral for Rb and Ba. This phase also has high but variable Rb/Sr (0.13–60) and Ba/Sr ratios. Amphibole and mica are major hosts for Nb in the peridotitic upper mantle and their presence strongly affects its Zr/Nb ratio. Accessory wadeite can be a source of Zr, and K and priderite may supply K, Ba, and Ti in the mantle in an alumina-deficient environment. Ionov et al. further observed that apatite is a major host for Th, U, Cl, Br, LREE and Sr, and clinopyroxene contains significant amount of Sr, Zr, Y, REE. The presence of both these minerals in fertile lherzolites with disseminated amphibole and mica can account for significant enrichments of Th, U, Pb.

Edgar and Pizzolato (1995) observed the presence of a F-rich phase in their experiments on the system hydroxy-fluor phlogopite and hydroxy fluor richterite at 2.0 GPa and 900–1,000 °C. This amphibole has a F content of 4.2 wt%. Such a phase may be present in the mantle, and enhance F-enrichment in the melt.

15.6 Production of Fertile Source Rocks by Mantle Metasomatism

The low REE content of komatiites suggests that primitive mantle had concentration of these elements similar to those of chondrites. Modern tholeiitic basalts have the REE contents 10–30 times the abundance of chondrites, and the alkali basalts have even higher REE enrichment than the tholeiites. Ultrapotassic rocks show high enrichment in LREE/HREE ratio. There seems to be a general agreement that on the basis of currently known partition coefficients, such an enrichment cannot easily be accounted for using simple partial melting models from depleted mantle source regions with low $^{87}Sr/^{86}Sr$ and high $^{143}Nd/^{144}Nd$ ratios. Geochemists argue that there has been recent enrichment within the mantle by metasomatic events during last 0.2 billion years.

Lloyd (1987) studied spinel lherzolites and alkali pyroxenite xenoliths from K-rich lavas of west Eifel (Germany) and Southern Uganda. The K-rich mafic rocks from both the Eifel and S. Ugandan lavas are LILE-enriched. Lloyd (1987) showed that concentration of K and associated LILE elements in these rocks would require enrichment of K by nearly a factor of 150 and more than 200 in the mantle below Eifel and Uganda, respectively. The regional K-enrichment is most satisfactorily explained in terms of pervasive K-metasomatism in the underlying mantle. Selvedges, veins and interstial crystals of K-richterite and phlogopite (Erlank et al. 1987; Mitchell et al. 1987; Tanton and McKenzie 1994) in lherzolites are considered to be evidence for the influx of magmas and fluids into the mantle (Kempton 1987).

The metasomatic agents, which are responsible for influx into the mantle include a variety of melts or fluids either rich in H_2O or CO_2 (Roden et al. 1984, cited by Kempton 1987). Incompatible trace element enrichment in anhydrous lherzolite may be attributed to fluxing of incompatible element-laden with CO_2-rich fluid into the mantle. Menzies et al. (1987) have recorded effects of such silicate melts and hydrous fluids in mantle xenoliths.

Experiments on partitioning of REE in H_2O-rich fluid on peridotite minerals at pressure ranging between 1.0 and 3.0 GPa have been conducted by Mysen (1983). On the basis of these data fluid—melt partition coefficients could be determined. In this determination Ce/Sm and Se/Tm ratio is assumed to be equal to 1. Wendlandt and Harrison (1979, cited by Eggler 1987) also studied REE partitioning between coexisting CO_2-rich fluid and two immiscible melts (one silica-rich and the other carbonate) in the system K_2O–Al_2O_3–SiO_2–CO_2. In the investigation by both the groups, fluid compositions were calculated by mass balance from analyses of crystals and melts. Their data have been summarized by Eggler (1987), which show that at 1.0 GPa, the REE is compatible with H_2O and with CO_2 at any pressure. Fluid-melt partitioning data show LREE enrichment. Analyses of the C-1 carbonaceous chondrites show the presence of 5–10 wt% of total C, H and S.

15.7 Crust–Mantle Mixing

It has been estimated by Richter et al. (1992) that Sr fluxes of present day rivers worldwide is as high as 0.710, but those contributed by the trans-Himalayan rivers (the Ganges, the Brahmaputra and Tsangpo) during early Miocene are on an average usually high (0.7236). Thus, ocean sediments should be extremely enriched in this ratio; and subduction of ocean sediments should result in the increase in this ratio in the mantle wedge above the lithosphere.

The presence of "high ^{3}He" and "low ^{4}He" in lavas from Herd islands (Indian Ocean) has been reported by Hilton et al. (1995). According to them whereas the high ^{3}He lavas provide unambiguous evidence for deep-seated plume in their origin, they also thought that the low ^{3}He/^{4}He ratios in other lavas result from shallow-level contamination by radiogenic helium before eruption.

Fairly (1995) reported rapid cycling of subducted sediments into the Samoan mantle plume. He found that harzburgite xenoliths from the Samoan hot spot carry extraordinarily enriched Sr, Nd, and Pb isotopic ratios derived from recycled sediments. He described gas composition and micro-thermometric properties of fluid inclusions, which are dominantly pure CO_2 and were apparently formed at the base of the crust. They have high ^{3}He/^{4}He ratios (12 times the atmospheric ratio) inconsistent with the radiogenic values expected for recycled sediments. Moreover, C/^{3}He ratios of inclusions have higher values of -3×10^9, which are indistinguishable from typical mantle values and far lower than those in sediments ($>10^{11}$). The metasomatic fluid is probably a product of mixing with Samoan hot spot of volatiles high in ^{3}He/^{4}He ratios in the plume melt and recycled sediments. The retention of high ^{3}He/^{4}He requires very rapid cycling of sedimentary component through the mantle (probably $< 10^7$ years) than billions of years envisaged. The time scale indicates involvement of materials recently returned to the mantle at the nearby Tonga trench, and pelagic sediments near the trench have approximately the requisite signature. The unique isotopic characteristics of Samoan plume likely arise from its tectonic setting at the northern terminus of subduction in the Tonga trench.

Addition of subducted materials to mantle wedge might have caused variation in Pb isotopic data of K-rich lavas of Ringgit Beser complex (Indonesia, Edwards et al. 1994) and that both highly and moderately enriched lavas are derived from heterogeneous mantle source.

15.8 Metasomatic Fluid Source

Nelson (1992) divided the potassic rock suite (with molar $K_2O/Na_2O > 1$) into an "orogenic" subgroup that occur in the subduction—related tectonic settings and an "anorogenic" sub-group that are confined to stable continental settings. Representatives of both sub-groups possess trace element and isotopic signature consistent with the contamination of their magma sources by isotopically evolved "metasomatic" components enriched in incompatible elements. It is generally argued that these metasomatic components are principally derived from subducted lithosphere, containing sediments. Most examples of orogenic potassic magmatism (e.g. Italian potassic rocks, Spanish lamproites, Sunda arc leucitites) have trace-element and Sr and Nd and Pb isotopic characteristics consistent with the contamination of their mantle sources by a component derived from marine sediments. Nelson (1992) argued that potassic magmas belonging to anorogenic subgroup, have generally similar incompatible trace elements and Sr and Nd isotopic characteristics to those of orogenic potassic magmas, but many examples have unusual Pb isotopic compositions with unradiogenic $^{206}Pb/^{204}Pb$. Modern marine sediments characteristically have low U/Pb ratios and the unradiogenic $^{206}Pb/^{204}Pb$ of anorogenic potassic magmas may have evolved during long-term storage of subducted sediments (or components derived from them) within the subcontinental lithosphere. These unusual Pb isotopic compositions require substantial time period (>1 Ga) to have elapsed between the fractionation event and subsequent potassic magmatism. The fractionation events might have lowered the U/Pb ratio due to erosion and sedimentation at the Earth's surface. It is therefore, not surprising that most examples of anorogenic potassic magmatism are not associated with recent subduction processes. According to Nelson, although the eruption of potassic magmas is commonly related to rifting or hotspot activity, these processes do not necessarily play an important role in the genesis of unusual sources from which potassic magmas are derived.

Ultrapotassic volcanism related to subduction has been reported from southern Spain, Italy, Highwood Mountains and Rio Grande Rift, U.S.A., Tibet, Indonesia, Manchuria and many other localities. Thus, there may be arguments, whether some or all strongly potassic and ultrapotassic mafic magmas are directly related to subduction and whether their immediate sources are within lithospheric upper mantle (also see Thompson et al. 1990).

Sharda et al. (1999) conducted step-heating analyses for Mid-Atlantic Ridge glass samples to show that maximum $^{40}Ar/^{36}Ar$ values correlate with $^{206}Pb/^{207}Pb$ and $^{208}Pb/^{204}Pb$ ratios. These correlations hold for the whole Atlantic Ocean and therefore, are unlikely to result from shallow-level contamination processes. Instead, they are taken as mixing between the degassed-depleted upper mantle and a recycled component characterized by high $^{206}Pb/^{204}Pb$ ratio (19–21) and low $^{40}Ar/^{36}Ar$ ratio (300–1,000). These relations imply that argon may also be a tracer of mantle recycling.

Petrological experiments on oceanic crust samples characterize recycling of potassium from mid-oceanic ridge basalts and sediments. Metasomatism could develop directly and continuously from subducted potassium-bearing crust from shallow levels to a maximum depth of 300 km.

15.9 Crustal Contamination

There are numerous lines of evidence, which rule out any significant crustal contamination during ascent of a K-rich magma to the surface, and therefore, influence of crustal nitrogen on the lamproites. From the petrological and mineralogical studies on these Gondwana ultra-potassic rocks, Jia et al. (2003) observed the following: (1) the dominant primary mineralogy of olivine, phlogopite, Cr-spinel, and priderite is typical of ultrapotassic mantle rocks, but not typical crust; (2) As a corollary, no xenocrysts of abundant crustal minerals such as quartz or plagioclase have been observed; (3) xenoliths of harzburgitic composition in these lavas confirm their mantle source; (4) REE patterns are much more fractionated than upper continental crust (La/Yb_{cn} values of 68–191 in ultrapotassic rocks versus 9.4 in continental crust), and all fractionated HREE data are indicative of residual garnet in the source, unlike continental crust (Table 5.2 and Fig. 5.4); (5) the REE patterns all feature an inflection at Sm–Nd, and zero or minor negative Eu anomalies (Eu/Eu* = 0.646–0.982), whereas upper continental crust has a systematic negative Eu anomaly of 0.661 (Hall 1999) both Mg# of 90–75, and Cr, Ni, Ba, Th, LREE contents are high relative to continental crust (Table 5.2).

The results are also consistent with previous studies in some of these Gondwana ultrapotassic suites. According to Middlemost et al. (1988), Rock et al. (1992) and Paul and Sarkar (1986), the K-rich rocks in the Gondwana lamproites from Raniganj, Jharia, and Bokaro basins are all of mantle origin, given the presence of mantle-derived nodules and xenoliths of harzburgitic composition, with undetectable crustal contamination. Jia et al. also noted their primitive mafic–ultramafic major element chemistry; enrichment of refractory trace elements like Ni, Cr, Co, and Sc; high Mg-value of the phlogopite xenocrysts or phenocrysts associated with them and high concentrations of REE (up to 1,139 ppm) and pronounced REE fractionation (La/Lu_{cn} of 16–560) all rule out crustal contamination.

Strontium isotopic compositions of these rocks also support an enriched mantle source. The $^{87}Sr/^{86}Sr$ ratios in the Bokaro suite range from 0.70394 to 0.71544

(Basu et al. 1997). Three Raniganj K-rich rocks range between 0.70493 and 0.71067 (Middlemost et al. 1988), and in seven Jharia ultrapotassic ultamafic rocks $^{87}Sr/^{86}Sr$ ratios range between 0.70526 and 0.70595 (Rock et al. 1992). The Bokaro samples thus encompass a much wider range of $^{87}Sr/^{86}Sr$ ratios. These enriched Sr-isotope data of the Jharia, Raniganj and Bokaro rocks are similar to those of lamproites world wide (Bergman 1987). For a given suite $^{87}Sr/^{86}Sr$ ratios do not co-vary with SiO_2. A plot of $^{87}Sr/^{86}Sr$ versus 1/ Sr of the Raniganj and Jharia ultra-potassic rocks do not generate a straight line for mixing between crustal and mantle sources (Fig. 5.4). Consequently, these K-rich rocks have not been affected by crustal contamination, and are believed to reflect mantle source characteristics (Bergman 1987; Nelson et al. 1986; McCulloch et al. 1983).

Hall (1999) measured NH_4^+, contents of up to 54 ppm in hydrothermally altered oceanic basalts, and suggested that subduction of such rocks provides a mechanism for recycling N into the mantle. Fyfe (1997) addressed recycling of volatiles, including N, from surface reservoirs into the mantle. The subducted flux of N is estimated at 2×10^{12} g/yr, or five times the mantle outgassing flux Jia et al. (2001). However, according to Marty and Humbert (1997), N is quantitatively released from the slab at convergent margins by arc magmatism, consistent with the high N/ Ar ratios of volcanic gases in arcs. In contrast, Javoy (1998) argues that N may accumulate in the mantle from progressive recycling, stored in a stable form such as Osbornite (TiN).

Four leucite samples from tephritic rocks of Roccamonfina (Italy) reported by Gupta et al. (2002) also show enriched $\delta^{15}N$ values ranging from 2.7 to 4.6 % and nitrogen contents ranging from 60 to 270 ppm (Fig. 5.1). These results are also consistent with nitrogen being added to the mantle by subduction of sediments or continental crust.

Recycling of subducted sedimentary/ crustal N into the deep mantle has also been found by Dauphas and Marty (1999), who analysed samples of the 370 Ma ultramafic–alkaline–carbonatite complex in the Kola peninsula, which have a range of $\delta^{15}N$ from +0.7 to +6.5 % (Fig. 5.2; one outlier). Based on noble gas (He, Ne, and Ar) isotopic compositions, and coupled Sr- and Nd-isotopic compositions, which provide no evidence for crustal contamination, they interpreted the ^{15}N-enriched signature of a Deep mantle material to be recycled sedimentary N through subduction processes.

A recent study on N-rich fluids in the mantle provides evidence that molecular nitrogen, as a major component in fluids, was trapped by minerals in the upper mantle xenoliths during a fluid (melt) rock interaction process within the mantle (Andersen et al. 1995). Ammonium-bearing phlogopite and amphibole breakdown by an oxidation–dehydration reaction such as:

$$\begin{aligned} &2\{(NH_4)Mg_3AlSi_3O_{10}(OH)_2\}_{phl}+(Mg_2SiO_4)_{Fo}+3/20_2 \\ &= (MgAl_2O_{4})_{Sp} + 7(MgSiO_3)_{En}+N_2 + 6H_2O \end{aligned} \tag{15.1}$$

and

$$\begin{aligned}&2\{(NH_4)Ca_2Mg_5AlSi_7O_{22}(OH)_2\}_{amph}+(Mg_2SiO_4)_{Fo}+3/2O_2\\&=(MgAl_2O_4)_{Sp}+7(MgSiO_3)_{En}+4\ CaMgSi_2O_6N_2+6H_2O\end{aligned} \tag{15.2}$$

These ammonium-bearing minerals may be stable under low oxygen fugacity and elevated water activity (Andersen et al. 1995).

Conticelli et al. (2007) investigated the cause of eruption of ultrapotassic magmas during Neogene and Quaternary in the Mediterranean area. According to them the Roman Magmatic Province is dominated by leucite-bearing rocks with variable degrees of silica saturation, from undersaturated (leucitites and plagioleucitites) to strongly undersaturated (kamafugites), but minor amounts of shoshonitic to high-potassium calc-alkalic rocks are still present. The Lucanian Magmatic Province, located at the southeasternmost edge of the volcanic belt, is dominated by foiditic (hauynites and leucitites) and kamafugitic (melilitites) members, all strongly undersaturated in silica. In spite of these petrologic differences, the Neogene Italian potassic and ultrapotassic rocks display similar trace-element patterns. Depletion in high field strength elements with respect to large ion lithophile elements is a common feature. Sr, Nd, and Pb isotopic compositions of mafic high-MgO rocks range widely, relating mainly to geographic location of eruption and to enrichment in alkalies. The Os isotopic composition of these samples, however, does not clearly correlate with eruption location, but is dependent on the amount of "continental crust component" added to the magmas. Some of the studied samples are compatible with crustal contamination en route to the surface. In most cases, however, there are several lines of evidence suggesting the possibility that crustal components were added directly to the mantle source prior to partial melting. Large amounts (many tens of weight percent) of "crustal component" must be added to the peridotitic mantle in order to obtain the $^{187}Os/^{188}Os$ of the lamproites in Tuscany. These large amounts of crustal components have been recycled into the mantle in the form of either melts or fluids. The recycling can be reconciled with a veined mantle in which the crustal component is concentrated. Partial melting of veins would then produce the high-silica and high-magnesium lamproitic magmas from Tuscany. Dilution of the crustal components by increasing partial melting of surrounding mantle peridotite, or alternatively, a reduction of metasomatic veins, could then produce shoshonitic and high-K calc-alkalic mafic magmas.

Southeastward geochemical and isotopic variations are reconciled with decreasing direct contributions from crustal components introduced into the mantle by subduction, but an increasing role of subducted fluids from dehydration of CO_2^- rich sediments. The coupled isotopic and chemical characteristics of Italian magmas cannot be reconciled with an ocean-island basalt (OIB)-like primary magma composition due to the substantial overprinting by crustal- and or subduction-related components.

While discussing potassic and ultrapotassic magmatism circum-tyrrhenian region Avanzinelli et al. (2009) suggested pelitic sediment recycling at destructive plate margin. They thought the central-western Mediterranean is one of the most important areas on Earth for studying subduction-related potassic and ultrapotassic magmatism. In the circum-Tyrrhenian area leucite-free (i.e., lamproite) and leucite-bearing (i.e., kamafugite, leucitite, and plagioleucitite) ultrapotassic rocks have been emplaced and are associated with shoshonites and high-K calc-alkaline volcanic rocks. Four different magmatic provinces are recognised from this area. Eastward and then south-eastward migration of magmatism with time occurred following roll-back of the subducting plate. Leucite-free silica-rich lamproites are restricted to the early stages of magmatism, associated with shoshonites and high-K calc-alkaline volcanic rocks. Present day volcanic activity is restricted to the Neapolitan district where ultrapotassic rocks with variable geochemical and isotopic characteristics occur. Ultrapotassic rocks are strongly enriched in incompatible trace elements with variable fractionation of Ta, Nb, and Ti with respect to Th and large ion lithophile elements. Mafic ultrapotassic rocks are also variably enriched in radiogenic Sr and Pb and unradiogenic Nd. The main geochemical and isotopic signatures result from sediment recycling within the upper mantle via subduction. Selected trace element ratios suggest that high temperatures are required to generate sediment-derived melts. Recycling of carbonated pelites play an important role in the Roman province controlling the genesis of leucite-bearing magmas. Large volumes of metasomatic components are predicted to be accommodated within a vein network in the sub-continental lithospheric mantle. Partial melting of the pure vein mineralogy is likely to generate ultrapotassic magmas of either lamproitic or kamafugitic nature. Over time, increased interaction between the metasomatic vein lithology and the surrounding mantle dilutes the alkaline component producing shoshonites and high-K calc-alkaline rocks. The addition of a further subduction-related component shortly before magma generation is required to explain the isotopic composition of rocks from the Neapolitan district. In the last phases of circum-Tyrrhenian evolution, a within-plate component appears within south-eastern Italy. This component is evident at Vulture volcano, in the Lucanian Magmatic province (SE Italy). Boari et al. (2009) supported such a view related to cluster recycling of sedimentary materials for the genesis of K-rich magma of the Middle Latin Valley Volcanic field of the Roman Magmatic Province.

15.10 Recycling of Nitrogen from Crust into the Mantle

Most of the ocean floor crust is returned to the mantle at subduction zone. The process is one of the most deep cycling events in the planet (Fyfe 1997). The overall significance of the subduction process is that oceanic crust including pelagic sediments (Gilluly 1971) carries with it the memory of interactions with the hydrosphere (H_2O, CO_2, S, Cl, Na, U, K, Sr, O_2 and N_2). Fyfe considered that nitrogen content in subduction-related volcanism may provide important clue as to the crust mantle mixing.

Lunar surface provides information on influx of different nitrogen isotopes due to solar wind as it implants itself on individual soil grains to a depth of tens of nanometer (Sanno et al. 2001). Deciphering pre-solar and planetary nitrogen contribution, they record that the lunar soil has an abundance of this element between 100 and 1000 ppm. The nitrogen concentration decreases with depth from several thousands to 100 ppm at ~150 nm depth in the grains.

Sano et al. (1998, also see Itihara and Honma 1979) determined the global flux of nitrogen from the earth. They found that nitrogen flux from island arcs is 6.4×10^8 mol/year. This is related to large flux of subducted sedimentary nitrogen, whereas nitrogen flux from back-arc basins (5.6×10^8 mol/year) have a relatively smaller but measurable subducted component. They observed that volcanic gas and hydrothermal samples show a concentration of mantle-derived nitrogen of 9–30 %, which is in agreement with those of phenocrysts in subareal volcanic rocks and fluid inclusions in hydrothermal minerals (Sano et al. 1998, also see Itihara and Tainosho 1989). They considered that in general nitrogen content in island arc samples is dominated by atmospheric and/or sedimentary components (also see Williams et al. 1989). They thought that sedimentary nitrogen in the sample may be attributed to subducted marine sediments. Their concentration in volcanic gas and fluids in island arc region can be indeed very high: 40–52 % (Papandayan), 47–57 % (Galeras crator), 47 % (Purace) and 65 % (Colima).

Figure 15.1 of Sano et al. (1998, also see Itihara and Honma 1983) showed the schematic diagram of global influx of nitrogen into the mantle of the earth. It may be noted that back-arc basin and island arc indicate significant mantle flux of nitrogen from the earth. This figure also demonstrates that in the island arcs regions, there is a large flux of subducted sedimentary nitrogen, whereas back-arc basin has a relatively smaller concentration, this is because the total length of subduction zone (continental island arc) is 4×10^4 km, whereas about 40 % of the subduction is not accompanied by back-arc basin volcanism, such as South and Middle America (Uyeda and Kanamori 1979).

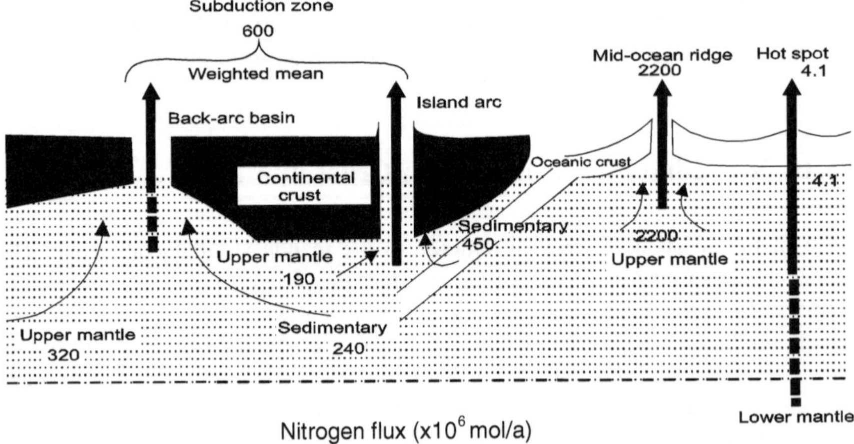

Fig. 15.1 A schematic diagram of nitrogen flux from the solid Earth. Note that both back-arc basin and island arc indicate significant mantle flux of nitrogen. Contribution of hot spot region is significantly smaller (after Sano et al. 2001)

15.11 Recycling of Potassium from Subducted Oceanic Crust

It has been argued by Schmidt (1996) that metasomatism should occur directly and continuously from subducted potassium-bearing crust from shallow levels to a depth of 300 km. He therefore, conducted experiments on mid-ocean ridge basalts (MORB) and andesite bulk compositions in order to explain the mechanism for transport Metasomatic Fluid of K_2O to mantle depths and to understand the generation of transport of K_2O rich metasomatic fluids into the mantle depth. His results are summarized in Fig. 15.2, which show that the principal K-bearing phase observed at subsolidus condition was phengite, a white mica. This phase formed at low pressures (<1.5 Gpa) and remain present until 9.5–10.0 Gpa (750–1,050 °C, Fig. 15.2). At pressures representing depths >110 km, omphacite (a sodium and calcium-rich clinopyroxene) became an important host for potassium in addition to phengite. K-clinopyroxene component in omphacite increases from <0.5 mol% (2.5 GPa) to as much as 6 mol% at 10.5 GPa (900 °C); thus omphacite can incorporate significant amounts of K also, at great depths in subduction zones.

Schmidt (1996) observed that when the weight solidus of MORB is achieved with increasing temperature ~1050 °C at 9.0 GPa, then phengite disappeared rapidly resulting in the generation of 3–5 volume per cent of a K-rich liquid. Small euhedral garnets were observed in the interstitial melt patches. He found that the weight solidus for K-bearing MORB is located at temperatures 100–200 °C lower than the weight solidus for synthetic K-free MORB (970°–1,160 °C at 4.5–6.5 GPa). The experimental results of Schmidt (1996) suggest that the solidus

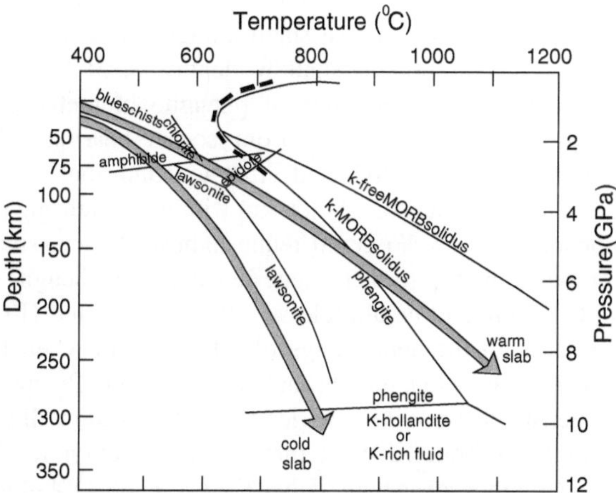

Fig. 15.2 Experimental studies on a MORB rock, showing the P–T stability of chlorite, lawsonite, epidote and phengite. The solidi for K-free MORB and K-bearing MORB are also shown. The breakdown curve of phengite to produce K-hollandite and fluid around 300 km is also shown (after Schmidt 1996)

of K-bearing MORB is defined by incongruent melting of phengite through a reaction, phengite + coesite + omphacite = garnet + melt.

Schmidt (1996) considered that the subduction history of highly soluble elements such as potassium is directly related to fluid content. The quantity of free fluid phase present in the descending oceanic crust is expected to be small and most of the fluid should ascend. In his experiments with 1 wt% H_2O, Schmidt observed that in case of the MORB bulk composition, almost pure $KAlSi_3O_8$ (K-hollandite structure) instead of phengite appears. In experiment with >1 wt% water K-hollandite or other K-bearing crystalline phase was not observed as a breakdown product of phengite instead K-bearing silicate phases (submicroscopic to 10 μm appear between grain boundaries between garnet and clinopyroxene. These phases probably precipitated from the fluid phase and is interpreted as the breakdown product of phengite to omphacite and garnet and a K-rich fluid.

The experiments of Schmidt suggest that K was strongly partitioned in the fluid phase and may be interpreted as evidence of breakdown of phengite to omphacite + garnet and a K-rich fluid. These experiments suggest that phengites (at low pressures), and K-hollandite and also omphacite, can carry K to great depths in the subduction zones, but that any fluid resulting from phengite breakdown will be strongly enriched in K and other such mobile elements as Rb, U, Pb and B. Thus, fluids passing through or arising from subducted oceanic crust at great depths may leach some of the K from the crust during metamorphism.

Tatsumi (1989, cited by Schmidt 1996) proposed that a descending oceanic crust should release all its potassium when amphiboles break down at 2.5 Gpa. This should lead to subsequent metasomatism of the overlying mantle wedge and transportation of K_2O through such mantle phases as phlogopite or K-richterite. The experiments of Schmidt suggest an alternative mechanism for storing K_2O in the subduction zones. Minor amounts of K-rich sediments and abundant K-poor basaltic and gabbroic crust will loose a small amount of K into fluids, produced during initial partial dehydration at shallow depths (<1.0 GPa). According to Schmidt the presence of minor amounts of phengite in metabasalts and major amounts of phengites in metasediments from coesite-bearing eclogite terrains, documents that K is stored in subducted crust to conditions beyond depths of 90 km. He suggests that at depths of 75–300 km with the increasing pressure during subduction, omphacite (40–55 %) could retain almost all potassium present in a MORB bulk composition. He further suggested that the phengites finally break down at depths of 300 km and would release only minor K and H_2O from MORB composition. However, in the depth range of 100–300 km when K-omphacite is formed continuously from phengite through the reaction, Phengite- → K-feldspar + pyroxene + enstatite + coesite + K-rich fluid. Potassium will then be released continuously from the subducted crust through the solution in fluids. Schmidt (1996) suggested that greywackes with higher K_2O contents (1.4–4 wt%) and lower omphacite abundances (20–30 vol%) might still have significant amount of phengite (10–35 vol%) at 300 km depth (10 GPa). He thought that a pulse of K-rich fluid may be generated in sediments at a depth of ~300 km, when K-hollandite + K-rich fluid is formed from phengite. At still greater depths the descending crust should almost completely dehydrate and should not cause any metamorphism. Schmidt therefore, concluded that phengite is the principle potassium host at sub-solidus conditions. It transports potassium and water to depths of up to 300 km and could yield over the entire depth range K-rich fluids or melts (depending on the specific geotherm), which are likely to constitute one of the primary metasomatic agents.

15.12 Metasomatic Fluid Transport

Navon and Stolper (1987) used the analogy between mantle processes and chromatographic fractionation in the laboratory to explain mantle metasomatism, and envisaged large scale percolation. They thought that as the fluid moves upward it interacts and equilibrate with the matrix composition. As the column is used up, the next batch of fluid passes through the column without any interaction with the matrix. Each element behaves differently and mass fractionation takes place. According to Navon and Stolper, the factors governing fractionation include equilibrium partition coefficients (K_D, matrix to melt) and the equilibrium mass fraction of an element in the liquid (X_f). The compatible element reacts with the matrix before incompatible elements move through the column at a slower rate. The compatible elements have a high K_D value relative to peridotite and lower X_f for a specific melt composition.

Navon and Stolper thought that a refractory harzburgite or lherzolite, which lost basalt fraction from previous melting, might represent the lithospheric mantle. Silicate melts produced in the asthenosphere or fluid produced from melting of a subducted slab, might be enriched in LREE relative to chondrite (also see Bergman 1987). Theoretical values of K_D and X_f used by them predict that relatively compatible middle or heavy REE (HREE) will be retained in the matrix and incompatible LREE will move to a greater distance. They concluded that small quantities of melt pass through the column and the fronts of the more incompatible LREE reach the top of the mantle column ahead of the more compatible REE. Thus, relative to other REE, the LREE should increase in the infiltrating melt.

Nielson and Wilshire (1993) studied xenoliths from Lherz and Horoman massif. According to them if their model is correct, even modest LREE enrichments in the xenoliths of Lherz massif would require melt percolation for as long as 25,000 years, an unrealistic life span for mantle dyke conduits, that supply metasomatic fluid. Effects of porous medium flow have been inferred in mantle peridotite samples on scales of tens of meters at the most, but are well-documented only on scales of centimetres or decimeters. In all these hypotheses porous flow is controlled by proximity to magma-filled fractures. Theoretical and experimental studies suggest that the factors, which control the degree of permeability in partially fused peridotites include, 1) melt fraction and dihedral (wetting) angle and the possible presence of C–O–H fluids in the mantle (Watson et al. 1990). Low dihedral angles are observed in mafic and ultramafic melt compositions in peridotite at upper mantle temperature and pressure, which favour inter-connection of inter-granular melts and indicate that this mechanism may be the most likely condition for porous-medium flow. There are driving forces that may direct the flow of inter-granular buoyancy and deformation in upwelling mantle and rising diapirs. Numerous field-oriented studies of alpine massif, ophiolites and xenoliths show that a common mode of transport in the upper mantle is in conduits formed by hydraulic fracturing.

Such fracturing occurs even in rocks with large melt fraction; the process is generally accompanied by plastic deformation. The process of hydraulic fracturing in the mantle, requires melt pressures in excess of the peridotite yield stress, which is estimated by Nicolas (1986) to be <50 MPa. Davies (1999) considered that during subduction of a lithospheric slab along a continental margin, as the temperature and pressure are increased, fluids are released because of the breakdown of fluid-bearing phases. The presence of non-percolating fluid should increase by pore pressure, which may approach the lithostatic pressure. A greater amount of de-fluidization may result in higher pore-pressure. This may ultimately lead to the generation of intermediate earthquake. He further considers that ductile creep can also increase pore-pressure even in cool systems with finite permeability and negative volume change due to de-volatilization. Faulting may be initiated along well-healed faults that were present earlier prior to subduction. Isolated pockets of fluid should nucleate micro-cracks and under compression fracture should form due to coalescence of dilating tensile micro-cracks. The fluid should flow locally to open under higher confining pressure. With increase in strain micro-cracks would interact to produce echelon crack arrays, which should extend and coalesce further to form fracture system.

Foley (1992) discussed about vein-plus-wall-rock melting mechanism in the lithosphere and the origin of potassic magmas. He suggested remobilization of several generations of veins, which should explain the occurrence within a restricted space and time of rocks bearing chemical characteristics which are generally thought to indicate contrasting tectonic setting (e.g. central Italy). According to him the generation of ultra-potassic rocks is explained by melting of vein-rich segments (i.e. with high vein/wall-rock ratio). Further dilution of the vein components by wall-rock, supplemented by asthenospheric melt in advanced cases, leads to the production of more voluminous lavas. These lavas bear incompatible element signatures reminiscent of the ultrapotassic rocks.

15.13 Potassic Volcanism Associated with Rift and Tectonic Processes

Formation of rift systems in the earth's crust may be related to ascent of mantle plumes. With the rise of such a plume and presence of two opposite convective limbs, tensional forces are created. This causes thinning and stretching of the crust. As a consequence of the emplacement of the plume, asthenoshere upwelling and decompression melting takes place. First, there is generation of basaltic magma by partial melting of the ultramafic mantle, and formation of lava pool below the mechanical boundary layer (MBL) immediately below the region of rifting. Thompson et al. (1990) suggested that potassic mafic magmas are generated in the sub-continental MBL. They think that eruption of the liquids along the axis of continental rift is much more consistent with pure-shear than simple-shear models of continental extension. The emplacement of strongly potassic magmas along major strike-slip faults is an evidence in support of the view that the latter is generated at the base of continental lithospheric plates rather than terminating in the lower crustal detachment zones. This model suggests the cause of magmatism is probably a combination of both extension of the continental lithosphere (McKenzie 1989) and heating from below by progressively hotter asthenosphere. McKenzie proposed that the sub-continental lithospheric mantle is semi-continuously being infiltrated by small partial melts from the asthenosphere, which introduced volatiles and incompatible elements.

While describing the genesis of ultrapotassic lava from Tibetan Plateau, Turner et al. (1996) did not share the view of McKenzie and argued that this mechanism should equally enrich elements of similar incompatibility and therefore, does not provide a mechanism for producing the observed fractionation of Nb, Ta and Ti relative to rare earth. Thus, this mechanism does not explain the negative Nb–Ta and Ti anomalies as observed in case of ultrapotassic rocks in the Tibetan Plateau and elsewhere. According to Turner et al. such anomalies are characteristics of upper mantle sediments into the source region during a subduction event (e.g. Hergt et al. 1989), which might also explain Archaean Pb model ages. They further add that Rb/Ba is probably strongly fractionated during sediment subduction resulting in having low ratio of these two elements. Otherwise, such a low Rb/Ba ratio of the majority of mafic lavas could not be explained due to the subduction of upper crustal sediments. As negative Nb/Ta and Ti anomalies are characteristics of island arc magmas, they suggested that ancient subduction processes were responsible for the metasomatism of the Tibetan sub-continental lithospheric mantle. While discussing the genesis of post-collision potassic volcanism on the Tibetan plateau, Turner et al. (1996) concluded that convective thinning of the lithospheric mantle brought cooler and insulated zones of the lithospheric mantle into direct contact with asthenospheric temperatures such that any metasomatized peridotite, would undergo partial fusion, as its solidus should be lowered due to addition of volatiles. They think that the thermal time constant of the lithosphere is such that gradual conductive heating will continue over a period of $\sim$10–15 Ma. Thus, their model agrees with the observation that eruptions continue from $\sim$13 Ma into this century and there is seismic evidence for the presence of nearly 7 % partial melt over the 80 km thick slab of the upper mantle beneath Tibet.

Voluminous potassic magmatism in the Alto Paranaiba Igneous Province (Chap. 4) was described by Gibson et al. (1995) from the margin of the San Fransiscan Craton of Brazil during the Late Cretaceous period. This produced wide variety of rock types including madupite, kamafugitic rocks, olivine-rich madupitic lamproite and kimberlite. There is also occurrence of minor amounts of melilitite and several carbonatite intrusive complexes. This indicates that these different types of rocks were produced from compositionally varied ranges of mantle sources. According to Gibson et al. low $^{143}Nd/^{144}Nd$ and high $^{87}Sr/^{86}Sr$ ratio along with high concentration of incompatible trace elements and La/Yb ratios of different rocks types indicate that the source rocks were in the sub-continental lithospheric mantle. They thought that variable K_2O content in the liquid and modal phlogopite content was related to different carbonate and hydrous phases in the mantle. This resulted in different CO_2/H_2O ratios in the source rocks during melting. The presence of garent lherzolitic xenolith in Tres Ranchos kimberlite suggests that they originated at great depth within the mantle. The kamafugitic rocks were probably derived at relatively shallower mantle source, which might have undergone metasomatism.

Gibson et al. considered that the mantle source rocks for the olivine lamproites in Alto Paranaiba Igneous Province might have been metasomatized by introduction of subducted materials. According to Gibson et al. low $^{143}Nd/^{144}Nd$ and high $^{87}Sr/^{86}Sr$ ratio along with high concentration of incompatible trace elements and La/Yb ratios of different rock types of the Alto-Paranaiba Province indicate that the subcontinental lithospheric mantle constituted the source rock.

Reference to different experimental studies (Chap. 13) suggest that at least high temperature was required to generate the kimberlitic melts (Tres Ranchos) and later the kamafugitic melts at relatively lower temperature. The heat was provided by a mantle plume. They thought that heat transfer by conduction and advection during ascending asthenospheric melt might have been responsible for melting of volatile-rich fusible component of the lithosphere.

Rogers et al. (1998) discussed volcanism of the Birunga complex comprising four volcanoes. They estimated that these volcanoes yielded 2,000 km^3 of potassic lava during the last 100 ky. This leads to an eruption rate of $\sim$0.02 km^3 per year. They suggest that even if 50 % of the melt is trapped at depth as cumulates then the production rate could be as high as 0.04 km^3.

Rogers et al. (1998) concluded that the production rate of lava (0.04 km^3) at Birunga is low compared to the production of lava in the flood basalt area of Parana (0.4 km^3 per year, Stewart et al. 1996). According to them, significantly low rate of lava production and the lack of significant plume contribution to the magmatism are consistent with both the thick lithosphere inferred from the depth of magma source and the remoteness of the Birunga province from the axis of the putative East African mantle plume. They further thought that the control exerted by the lithosphere on the location of both extension and magmatism and on the composition of the magmatism in the early stage of continental extension is dominant. They thought that the timing of the magmatism is consistent with a model of conducting heating of the lithosphere by the cooler outer part of the East African mantle plume.

15.14 Possible Causes for the Frequent Occurrence of K-Rich Silica-Poor Volcanic Rocks in the Recent Evolutionary History of the Earth

Eruption of K-rich lavas are recent (Table 15.1); but it has been recorded by most petrologists (See Table 15.2) that there had been emplacement of very few alkaline silica-undersaturated rocks during Precambrian (e.g. some shoshonitic rocks and minor amounts of calc-alkalic lamprophyres), which might indicate the presence of rather a few regions of subduction-related magmatism and presence of ancient convective cycles within the Archaean mantle.

It has been argued that scarcity of alkaline rocks during the Precambrian could be due to one of the following reasons (Blichert-Toft et al. 1996):

1. Preferential destruction of alkaline rocks as the feldspathoids present are rather facile (Gupta and Fyfe 1975),
2. The majority of alkaline rocks are present in tectonic settings that have little chance of survival,
3. Rift-related volcanic rocks are easily eroded away,
4. Those formed in the oceanic islands might have been subducted and therefore, lost.

Although the above mentioned arguments could be correct, judging by the abundance of K-rich volcanic rocks in the modern geological history, and occurrence of komatiites almost exclusively during Archaean, it is most likely that the thermal-tectonic condition of the mantle during Precambrian was different. On the basis of experimental studies the estimated temperatures of eruption of komatiites (Green et al. 1975) were quite high, and according to Campbell et al. (1990) they represent ancient analogues of modern plume-related magmas. Such high temperatures and extensive melting might indicate that alkaline rocks could not have been formed because they need low degree of melting (Frey et al. 1978). It is possible that even during ancient time low degree of melts might have been produced at the cooler margins of the plumes, but Blichert-Toft et al. (1996) consider that they might have been swamped by lavas of high degree melts from the central hot zone. This would be similar to absence of alkalic suite from Mid-Oceanic plateau.

According to Green and Taylor (1987), experimental studies indicate that in presence of CO_2, partial melting of peridotite results in the production of silica-undersaturated magmas. It is possible that carbon was concentrated in the atmosphere during Archaean, and CO_2 was not introduced to the mantle. For example, Lowe (1980), stated that carbonates are relatively rare in the Archaean sedimentary formation and oceanic crust was probably not introduced to the mantle as frequently as today.

Modern view states that genesis of K-rich magma is related to partial melting of fertile mantle, because the original mantle has been metasomatized by introduction of crustal segments. This also involves introduction of hydrous fluid to the asthenosphere. This mechanism happens more frequently in the modern time

Table 15.2 Archean potassic rocks (after Blichert-Toft et al. 1996)

Rock type	Geologic unit	Location	Age (Ga)
The Superior province,	Timiskaming group	Ontario	2.68
North America shoshonites	Bijou point	Ontario	2.70
	Oxford lake group	Manitoba	2.65
	Opemisca group	Quebec	2.70
	Rainy lake)	Quebec	7.70
Lamprophyre dikes	Abitibi greenstone belt, OntarioAbitibi greenstone belt	Ontario	2.70
	Kamiskotia–Montcalm suite	Ontario	2.70
	Bristol township suite	Ontario	2.69
	Roaring river complex	Ontario	2.70
	Vermilion complex	Minnesota	2.70
	Yollowknife greenstone belt	Northwest territories	2.70
Syenite and nepheline	Poohbah lake	Ontario	2.70
Syenite stocks	Garrison stock	Ontario	2.68
	Otto stock	Ontario	2.68
	Murdock creek	Ontario	2.68
	Herman lake	Ontario	2.67
	Clericy, Aldermac, Nora lake	Quebee	2.68
	Icarus and Linden Plutons	Ontario/ Minnesota	2.70
Lamprophyre dikes		Tanzania lamprophyre dikesTanzania/ Zimbabwe	2.70
Lamproite dikes	Caroline–Amalie, across, sound	Skjoldungen, Southeast Greenland	2.70
Syenite and nepheline syenite stocks	Ruinnaesset, sphinxen, blabaersaraent	Skjoldungen, Southeast Greenland	2.70

particularly in the margin of the continents by the subduction processes, which was not efficient during the Archaean because of the lack of old and thick lithosphere during Archaean. This also might explain why rift-related magmas formed by partial melting of metasomatized lithosphere could not have produced K-rich magma during the ancient geological history of the earth.

Chapter 16
Petrologic Conclusions

Silica-deficient potassic rocks are characterized by high K_2O content (>3), high K_2O/Na_2O ratio (>1) and low Si_2O/K_2O ratio (<15). They have high LREE concentration, sometimes 1,000–1,400 higher than that of chondrites. The LILE/HREE ratios are usually low, but $^{87}Sr/^{86}Sr$ ratios are usually high (0.707–0.730). The ε_{Nd} content of these rocks is usually negative compared to average basaltic rocks. If we exclude some exceptional cases, these rocks are relatively young (<25–30 Ma) in the evolutionary history of the earth.

Potassic rocks can be classified into three series: (1) low-K, (2) medium-K and (3) high-K. All kinds of tephrites, tephriphonolites, phonotephrites and trachytes can be classified with respect to A (alkali feldspar)–P (plagioclase)–F (feldspathoid) contents. The classification with respect to total alkali–SiO_2 (TAS) is inadequate. Classification with respect to modal minerals is preferable.

On the basis of mineralogy, K-rich lavas can be classified into (1) plagioclase-free melilite-bearing varieties and (2) plagioclase-bearing types. Tephrites, phonolites, phonolitic tephrites and tephri phonolites are the most abundant rock types. The plagioclase-free ultrapotassic rocks can be plotted in a kalsilite–forsterite–larnite–SiO_2 tetrahedron (Yoder 1986). When H_2O is added, bulk composition of many of the lamproitic rocks can be plotted in this system.

The K-rich rocks are associated with either rift valleys formed by the ascent of the mantle plumes or subduction related volcanism. Those well-known localities, where K-rich rocks were produced by rift-related systems include, (1) East and equatorial Africa, Damodar Valley coal fields, Aldan province (Russia), the Eifel region (Germany), Gaussbergh and Alto-Paranaiba province (Brazil). Those K-rich rocks, which were generated by the process of subduction include Aeolean Arc region (Italy), Southern Spain, Indonesia, Manchuria (China) and the Western part of the USA. There is a debate about Roman province volcanism. There are some, who believed that Roman Province volcanism is rift-related (Beccaluva et al. 1991), but there are others (Alverez 1972; Ellam et al. 1989), who thought that the volcanism to be subduction related.

A.K. Gupta, *Origin of Potassium-rich Silica-deficient Igneous Rocks*,
Springer Geology, DOI 10.1007/978-81-322-2083-1_16

The *P*–*T* stability of leucite and sodalite group of minerals with or without melilite (present in plagioclase-free rocks) suggest that these rocks crystallized under volcanic to subvolcanic conditions. The study of the nepheline–kalsilite–SiO_2 system (Gupta et al. 2010) also supports this conclusion.

According to some petrologists these rocks have been formed under FMQ–NNO buffer conditions (Turbeville 1993), but there are others suggesting their formations between IW–WM buffer (Foley et al. 1986). On the basis of the chemistry of oxide minerals, the log $f(O)_2$–T condition of formation of these rocks from different localities have been suggested to be as follows: −6.5 (1,300 °C, Foley 1985), −10.4 (815 °C, Roden and Smith 1979), −13.5 (827 °C) and 10^{-5} (755 °C, Ghiorso and Carmichael 1981).

In his study of the system forsterite–kalsilite–SiO_2–H_2O, Yoder showed that orthopyroxene and phlogopite coexist together but by the time orthopyroxene appears, leucite is silicated to form sanidine, which can be in equilibrium with phlogopite and enstatite.

The study of the system diopside–nepheline–sanidine at 1 and 2 GPa under *P* (H_2O) = *P*(total) by Gupta et al. (2006) suggests that a pyroxene-bearing nepheline phonolite can be generated either from a trachyte or a nephelinite or a pyroxene-free nepheline phonolite. It is interesting to note that the field of $forsterite_{ss}$ appearing as a reaction product of diopside and nepheline, is eliminated at 1 and 2 GPa. Study of the same system at 0.1 GPa shows that a nepheline leucitite can be generated from either a leucitite, an olivine–nephelinite, or a pyroxene-free nepheline-leucite phonolite.

Experimental study of the system pressure, forsterite–diopside–leucite–anorthite under atmospheric shows that a leucite basanite).can be derived from an olivine leucitite, a leucite tephrite, an olivine italite and a murambite. The flow sheet diagram of the system leucite–akermanite–albite–SiO_2 also studied under atmospheric pressure, shows that a shoshonitic liquid can move either to a quartz latite or towards tephri–phonolite. A tephri–phonolite in turn can be derived from either (1) a leucite tephrite, (2) a plagioclase-bearing phonolite, (3) a pyroxene-bearing phonolite or a shoshonite.

The paragenetic relationship amongst many leucite, nepheline and melilite-bearing kamafujitic rocks can be adequately described from experimental study of the system forsterite–akermanite–diopside–leucite and diopside–nepheline–akermanite–leucite (Nag et al. 2007) under one atmospheric pressure.

Experimental study of the system $KAlSiO_4$–Mg_2SiO_4–SiO_2–H_2O at pressures between 0.05 and 0.3 GPa (Luth 1967), at 2 GPa (Wendlendt and Eggler 1980a, b, c) and Gupta and Green (1988) suggests that at low *P*–*T* condition (0.05 GPa) the system cuts through the phase volumes of kalsilite, leucite, forsterite, enstatite and quartz. Leucite-quartz, kalsilite-quartz and forsterite-quartz do not coexist. With increase in water pressure phlogopite appears at the expense of feldspathoid and olivine. At low pressure (<0.3 GPa) there are 35 univariant curves and nine invariant points. At 2.0 GPa Wendtland and Eggler (1980a, b) have described 20 univariant reactions with or without CO_2 and 24 univariant reactions in presence of CO_2. Various combinations of the following minerals appear in the univariant curves and quaternary invariant points: kalsilites, leucite, phlogopite, forsterite, orthopyroxene, quartz, different types of carbonates, liquid and vapour. These univariant reactions and various singular points represent assemblage similar to different types of K-rich rocks. These different assemblages are found in various localities, where ultrapotassic rocks have been reported. Because of reaction relationship between feldspathoids and silica, there are thermal barriers, which inhibit crystallization of silica-undersaturated and silica-saturated rock types.

Phase equilibria study of the silica-undersaturated portion of the forsterite–kalsilite–SiO_2 system under 2.8 GPa (Gupta and Green 1988) illustrates mechanism by which, even in this simple system, the availability of C–H–O fluid may produce diverse liquids during very low degrees of melting of a model phlogopite-bearing harzburgitic source rock. In addition, small differences in H_2O/CO_2 ratio may control the appearance of phlogopite during fractionation of highly potassic liquids in the upper mantle. The early appearance of phlogopite will drive liquids towards silica-oversaturated derivative melts, whereas the same parent magma following a *P–T* path avoiding the phlogopite field, will fractionate to the silica-undersaturated minimum at high pressure (kalsilite + sanidine + forsterite or leucite + sanidine + forsterite). Luth (1997) observed that a pyroxene and phlogopite-bearing assemblage yields phlogopite + diopside + garnet + quartz at 6 ± 1 GPa. K-richterite joins this assemblage at 8 ± 1 GPa, and at still higher pressure, phlogopite is completely replaced by K-richterite. The stable assemblage is diopside + garnet + forsterite + K-richterite above the line defined in the *P–T* space by the points 11.5 GPa and 1,000 °C and 8 GPa and 1,400 °C.

Experimental study of a melilite–nepheline–leucitite, a katungite, a madupite, a leucite lamproite, a phlogopite mafurite, a olivine ugandite and a phlogopite-bearing minette, shows that under mantle *P–T* conditions all these rock types are represented by a phlogopite–pyroxinite (±richterite, ±apatite ± spinel ± rutile ± priderite ± wadeite ± paragonite). Only a wolgidite under mantle *P–T* conditions, is represented by a rutile and phlogopite-bearing websterite.

Presence of both compatible (Ni, Co, Cr, and Sc) and incompatible elements (K, Rb, Ba, Sr and U, F, P, etc.) in these rocks and great variability in $^{87}Sr/^{86}Sr$ and $^{18}O/^{16}O$ ratios suggest that these rocks are product of geosphere mixing processes. Jia et al. (2007) have determined nitrogen concentration for four leucite-bearing tephrites) and lamproites. They found that the N-content of these rocks varies between 60 and 270 ppm, a maximum being 270 ppm and possibly this high concentration is related to nitrogen components added to the mantle by sedimentary subduction processes. The global flux of nitrogen in rocks of island arcs is 6.4×10^8 mol/year. This is related to large flux of subducted sedimentary nitrogen, whereas nitrogen flux from back-arc basins (5.6×10^8 mol/year) has a relatively smaller but measurable subducted components. The volcanic gas and hydrothermal components have a concentration of mantle-derived nitrogen of up to 9–30 %, which is in agreement with those of phenocrysts in subareal volcanic rocks and fluid inclusions. In general nitrogen in island arc samples is dominated by atmospheric and/or sedimentary components.

The Mg, Cr and Ni concentrations of many K-rich mafic and ultramafic rocks signify their mantle origin. However the N-isotope and the N contents are too high. Accordingly, a multi-stage process may be suggested. Subduction of crustal rocks first into asthenospheric mantle, then melting of the rocks to generate a K-rich melt and subsequent hybridization of the melt with lithospheric mantle. Subsequently, during rifting, the metasomatized mantle underwent decompression melting.

Petrological experiments on oceanic crust samples characterize recycling of potassium from mid-ocean ridge basalts and sediments (with high Sr isotopic ratio). Metasomatism could develop directly and continuously from subducted potassium-bearing crust from shallow levels to a maximum depth of 300 km. The analogy between mantle processes and chromatographic fractionation in the laboratory has been suggested by Navon and Stolper (1987) to explain large scale mantle metasomatism. The method of mantle metasomatism by the process of Navon and Stolper should induce metasomatism within a few centimetres at the most. Numerous field-oriented studies of Alpine massif, ophiolites and xenoliths show that a common mode of transport in the upper mantle is through the conduits formed by hydraulic fracturing. Most probable method of fluid migration may be related to the presence of non-percolating fluid, which should increase pore pressure, approaching lithostatic pressure. Isolated pockets of fluid should nucleate microcracks and with the increase in strain microcracks would interact to produce echelon of crack arrays, which should extend and coalesc further to form fracture system.

In case of potassic volcanism in the regions associated with continental rifting, the following sequence of events should take place: (1) impinging of a mantle plume below a lithospheric plate, (2) continental rifting associated with lithospheric heating, which causes continental extension and (3) generation of K-rich magma from the lithospheric mantle boundary layer. Emplacement of K-rich lava along the regions of continental rift axis is associated with pure shear rather than simple-sheer model. The eruption of K-rich liquids along major strike-slip faults suggest their genesis at the base of the continental lithospheric plate.

References

Abovyan SB, Agamalyan VA, Aslanyan AT, Magalkyan IG (eds) (1981) Magmatic and metamorphic formations of the Arnenian SSR. Akadem. Nauk Armyanskoi SSR, Erevan, 331 pp

Abraham K, Gebert W, Medenbach O, Schreyer W, Hentschel G (1983) Eifelite, KNa_3Mg_4-$Si_{12}O_{30}$, a new mineral of the osumilite group with octahedral sodium. Contrib Mineral Petrol 82:252–258

Ageeva OA, Borutzky BY (2004) Kalsilite in the rocks of Khibiny massif: morphology, paragenesis, genetic conditions. New Data Min 39:40–49

Allan JF, Carmicheal ISE (1984) Lamprophyric lavas in the Colima graben, SW Mexico. Contrib Min Petrol 88:203–216

Allen DG, Panteleyer A, Armstrong AT (1976) Galore Creek. In Porphyry deposits of the Canadian Cordillera. Can Inst Min Metall (Special volume) 15:402–414

Alsopp HL, Bristow JW, Skinner EMW, Scott Smith BH, Danchin RV (1985) Rb-Sr geochronology of some Miocene West Australian Lamproites. Trans Geol Soc S Africa 88:341–345

Altherr R, Meyer HP, Holl A, Volker F, Alibert C, McCulloch MT, Majer V (2004) Geochemical and Sr-Nd-Pb isotopic characteristics of Late Cenozoic leucite lamproites from the East European Alpine belt (Macedonia and Yugoslavia). Contrib Min Petrol 147:58–73

Altherr R, Topuz G, Siebel W, Sen C, Meyer HP, Satir M, Lahaye Y (2008) Geochemical and Sr-Nd-Pb isotopic characteristics of paleocene plagioleucitites from the Eastern Pontides (NE Turkey). Lithos 105:149–161

Alverez E (1972) Rotation of the Corsica-Sardinia microplate. Nature 235:103–105

Amato A, Biella G, De Franco R (1991) Velocity structure of the Vulsinian volcanic complex (Latium, Italy) from seismic refraction data and three-dimensional inversion of travel times. J Geophys Res 96:517–535

Ambrosetti P, Azzaroli A, Banadona FP, Follieri M (1972) A scheme of Pleistocene chronology for the Tyrrhenian side of central Italy. Boll Soc Gesl Ital 91:169–184

Anderson HJ, Jackson JA (1987) The deep seismicity of the Tyrrhenian Sea. J Geophys Res Astron Soc 91:613–637

Anderson O (1915) The system anorthite-forsterite-silica. Am J Sci 39(4):407–454

Anderson T, Burke EAJ, Neumann ER (1995) Nitrogen-rich fluid in the upper mantle: fluid inclusion in spinel dunite from Lanzarote, Canary Islands. Contrib Miner Petrol 120:20–28

Aoki K, Ishikawa K, Kanisawa S (1981) Fluroine geochemistry of basaltic rocks from continental and oceanic region and petrogenetic application. Contrib Miner Petrol 76:53–59

Appleton JD (1972) Petrogenesis of potassium-rich lavas from the Roccamonfina volcano, Roman region. Italy J Petrol 13:425–456

Appleyard EC (1969) Nepheline gneisses of the Wolfe Belt, Lyndoch Township, Ontario. II. Textures and mineral paragenesis. Can J Earth Sci 6:689–717

A.K. Gupta, *Origin of Potassium-rich Silica-deficient Igneous Rocks*,
Springer Geology, DOI 10.1007/978-81-322-2083-1

Araujo ALN, Carlson RW, Gaspar JC, Bizzi LA (2001) Petrology of kamafugites and kimberlites from the Alto Paranaiba alkaline province, Minas Gerais, Brazil. Contrib Min Petrol 142:163–177

Arima M, Edgar AD (1980) Stability of Wadeite ($Zr_2K_4Si_6P_{18}$) under upper mantle conditions, petrological implications. Contrib Min Petrol 72:191–195

Arima M, Edgar AD (1981) Substitution mechanisms and solubility of titanium in phlogopites from rocks of probable mantle oringin. Contrib Min Petrol 77:288–295

Arima M, Edgar AD (1983) A high pressure experimental study on a magnesium-rich leucite lamproite from the west kimberlite area, Australia: petrological implications. Contrib Min Petrol 84:228–234

Arima M, Edgar AD (1983) High pressure experimental studies on a katungite and their bearing on the genesis of some potassium-rich magmas of the west branch of the African-rift. J Petrol 24 (2):166–187

Armienti P, Barberi F, Bizourad H, Clocchiatti R, Innocenti F, Metrich N, Rosi M, Sbrana A (1983) The Phlegrean fields: magma evolution within a shallow chamber. J Volcanol Geotherm Res 17:289–311

Atkinson WJ, Hughes FE, Smith CB (1982) A review of the kimberlitic rocks of Western Australia. Trans Cognita 2:204

Atkinson WJ, Hughes FE, Smith CB (1984) A review of the kimberlitic rocks of Western Australia. In: Kornprobst J (ed) Kimberlites I: kinberlites and related rocks. Elsevier, Amsterdom, pp 195–224

Audley-Charles MG (1975) The sumba fracture: a major discontinuity between eastern and western Indonesia. Tectonophysics 26:213–228

Aurisicchio C, Federico M, Gianfagna A (1988) Clinopyroxene chemistry of the high-potassium suit from the Alban Hills. Italy Min Petrol 39:1–19

Avanzinelli R, Lustrino M, Mattei M, Melluso L, Conticelli S (2009) Potassic and ultrapotassic magmatism in the circum-Tyrrhenian region: significance of carbonated pelitic vs. pelitic sediment recycling at destructive plate margins. Lithos 113:213–227

Azizberkov SA, Bagirov AE, Viliev MM, Ismail Zada AD, Nizhseradze NS, Emelyanova EN, Mamedov MN (1979) Geology and volcanism of Talish. Akadem. Nauk, Azerbaijanskoi, SSR Baku, 246 pp

Badasarayan GP (1966) Intrusive rocks of the Basumo-Pambaksky region. Geology of the Armenia SSR. Petrology of intrusive rocks, vol 3. Armenian Academy of Science, Erevan, pp 256–308

Bagdasarov EA, Orlova MP, Kozyerv VI (1974) Rare and trace elements of alkaline gabbroids and basalts of the north Tyarn-Shan. Zapiski Vsesoyuznogo Mineralogy, Ichaskogo Obshchasta 6:682–694

Bailey DK (1964) Crustal warping: a possible tectonic control of alkali magmatism. J Geophys Res 69:1103–1111

Bailey DK (1974) Continental rifting and alkaline magmatism. In: Sorensen H (ed) The alkaline rock. Wiley, London, pp 148–159

Baker BH (1987) Outline of the petrology of the Kenya Rift alkaline province. In: Fitton JG, Upton BGJ (eds) Alkaline igneous rocks, vol 30. Geological Society, London, pp 273–291

Baker PE, Gass IG, Harris PG, Le Maitre RW (1964) The volcanological report of the Royal Society expedition to Tristan da Cunha. Phil Trans Royal Soc London 256(A):439–578

Baldridge S, Carmicheal ISE, Albee AL (1981) Crystallization of leucite-bearing lavas: examples from Italy. Contrib Min Petrol 76:321–335

Bannister PA, Sahama ThG, Wiik HB (1953) Kalsilite and venanzite from San Venanzo, Umbria. Italy Min Mag 30:46–58

Barberi F, Innocenti F (1967) Le rocce selagitiche di Orciatico e Montecatini in Val di Cecina. Atti Soc Tosc Sci Nat Mem 74(A):139–180

Barberi PA, Innocenti F, Ferra G, Kellar J, Villari L (1974) Evolution of Eolian arc volcanism (southern Tyrrhenian Sea). Earth Planet Sci Lett 21:269–276

Barker DS (1974) Alkaline rocks of North America. In: Sorensen H (ed) The alkaline rocks. Wiley, London, pp 160–171

Barth TFW (1936) The crystallization process of basalt. Am J Sci 31:321–351

Barton M (1979) A comparative study of some minerals occurring in the potassium-rich alkaline rocks of Leucit-Hills, Wyoming, the Vico Volcano, Western Italy, and the Toro-Ankole region. Uganda Neues Jahrb Min Abh 137:113–134

Barton M, Hamilton DL (1978) Water-saturated melting relation to 5 kb of three Leucite Hills lavas. Contrib Min Petrol 66:41–49

Barton M, Hamilton DL (1979) The meling relationship of a madupite from Leucite Hills Wyoming, to 30 kbar. Contrib Min Petrol 66:41–49

Barton M, Hamilton DL (1982) Water-saturated melting experiments bearing upon the origin of K-rich magma. Min Mag 45:267–278

Barton M, Van Bergen MJ (1981) Green clinopyroxene and associated phases in a potassium-rich lava from Leucite Hills. Wyoming Contrib Min Petrol 77:101–114

Barton M, Varekamp KC, van Bergen MJ (1982) Complex zoing of clinopyroxenes in the lavas of Vulsini, Latium, Italy: evidence for magma mixing. J Volcanol Geortherm Res 14:361–388

Baruah A, Gupta AK, Mandal N, Singh RN (2013) Rapid ascent conditions of diamond-bearing kimberlitic magmas: findings from high pressure-temperature experiments and finite element modeling tectonophysics. J Tectonophys 594:13–26

Basu A, Bhattacharya AK, Paul DK (1997) Petrology and geochemistry of lamprophyric rocks from the Bokaro coal field, Bihar and their economic potential. J Geol Soc Ind 50:255–266

Basu AR, Goodwin AM, Tatsumoto M (1984) Sm-Nd study of Archean alkalic rocks from the Superior province of the Canadian shield. Earth Planet Sci Lett 70:40–46

Bauer M (1896) Jadeit von Tibet. Neues Jahrb Min 1:89–95

Bazarova TYu, Kazaryan GA (1977) Peculiaritis of crystallization of the leucite phonolites of the Azent-Vedi river. Doklad Akadem Nauk SSR 6:137–142

Bazarova TY, Kazaryan GA (1986) Neogene alkaline effusives south Armenia and condition for their formation. Volcanol Seismol 2:34–45

Bazarova TYu, Kostuk VP, Khmelnkova OA (1981) Features of the formation of alkali basalts of Great Aniui (a tributary of the Kolyma). Doklad Akadem Nauk SSSR 259:1192–1194

Beard BL, Glazner AF (1998) Petrogenesis of isotopically unusual Pliocene olivine leucitites from Deep Springs Valey. Calif Contrib Min Petrol 133:402–417

Beccaluva L, Di Girolamo P, Serri G (1991) Petrogenesis and tectonic setting of the Roman volcanic province, Italy. Lithos 26:191–221

Beccaluva L, Di Girolamo P, Vincerzo M, Franca S (1990) Phlegrean fields volcanism revisited: a critical reexamination and deep eruptive system and magma evolutionary processes. N Jh Geo Palaont Mh H.5:257–271

Beccaluva L, Rossi PL, Serri G (1982) Neogene to Recent volcanism of the southern Tyrrhenian-Sicilian area: implication for the geodynamic evolution of the Calabrian Arc. Earth Evol Sci 3:222–238

Becker H, Wenzel T, Volker F (1999) Geochemistry of glimmerite veins in peridotites from lower Austria-implications for the origin of K-rich magmas in collision zones. J Petrol 40(2):315–337

Bell K, Powell JL (1969) Strontium isotopic studies of alkalic rocks: the potassium-rich lavas of the Birunga and Toro-Ankole regions, east and central equatorial Africa. J Petrol 10:536–572

Bergman SC (1987) Lamproites and other potassic rocks: a review of their occurrence, mineralogy and geochemsitry. In: Fitton JC, Upton BGJ (eds) Allkaline igneous rocks, vol 30. Geological Society, London, pp 103–190

Beswick AE (1976) K and Rb relations in basalts and other mantle-drived minerals: Is phlogopite the key? Geochim Cosmochim Acta 40:1167–1193

Bhowmik SK (2000) Ultrapotassic rocks along late ductile shear zones from the Eastern Ghat Belt, India. Gondwana Res 3(1):55–63

Bianchini G, Beccaluva L, Siena F (2008) Post-collisional and intraplate cenozoic volcanism in the rifted Apennines/Adriatic domain. Lithos 101:125–140

Bird P (1984) Laramide crustal thickening event in the Rocky mountainig foreland and Great basin. Tectonophysics 3:741–758

Black LP, James PR (1983) Geological history of the Archean Napier complex of Enderby land. In: Plover RL, James PR, Jago JB (eds) Proceedings of the 4th international symposium on Antarctic earth sciences, pp 11–15

Blichert-Toft J, Arndt NT, Ludden JN (1996) Precambrian alkaline magmatism. Lithos 37:97–111

Boari E, Tommasini S, Laurenzi MA, Conticelli S (2009) Transition from ultrapotassic kamafugitic to sub-alkaline magmas: Sr, Nd, and Pb isotope, trace element and Ar^{40}/Ar^{39} age data from the Middle Latin Valley volcanic field, Roman magmatic province central Italy. J Petrol 50:1327–1357

Boettcher AL, Wyllie PJ (1969) Phase relationship in the system nepheline-silica-water at 35 kbar pressure. Am J Sci 267:875–969

Bogatikov OA, Ryabchikov ID, Kononova VA (1991) Lamproite. Nauka, Moscow, pp 307–320

Borley GD (1967) Potassium-rich volcanic rocks from southern Spain. Mineral Mag 36:364–379

Borley GD (1974) Oceanic islands. In: Sorensen H (ed) The alkaline rocks. Wiley, Toronto, pp 311–330

Borodin LS (1974) The principle provinces and formations of alkaline rocks. Izdatel'stvo, Nauka, Moscow, 376 pp

Bowen NL (1912) The binary system $Na_2Al_2Si_2O_8$ (nepheline, carnegieite)-$CaAl_2Si_2O_8$ (anorthite). Am J Sci 33:551–573

Bowen NL (1922) Genetic features of alnoite rocks at Isle Cadieux. Quebec Am J Sci 3:1–34

Bowen NL (1928) The evolution of igneous rocks. Princeton University Press, Princeton, 334 pp

Bowen NL, Ellestad RB (1937) Leucite and pseudoleucite. Am Mineral 22:409–415

Bowen NL, Schairer JF (1935) The system MgO-FeO-SiO_2. Am J Sci 29:151–217

Boyd FR, Gurney J (1986) Diamonds in the African lithosphere. Science 323:272–477

Boyd SR, Mattey DP, Pillinger CT, Milledge HL, Mendelessohn M, Seal M (1987) Multiple growth events during diamond genesis: an integral study of carbon and nitrogen isotopes and nitrogen aggregation state in coated stone. Earth Planet Sci Lett 86:341–353

Bragg L, Claringbull GF, Taylor WH (1965) Crystal structures of minerals. G. Bell.and Sons, London, 409 pp

Brenna M, Price R, Cronin SJ, Smith IEM, Sohn YK, Kim GB, Maas R (2014) Final Magma storage depth modulation of explosivity and Trachyte-Phonolitic genesis at an interplate volcano: a case study from Ulleung Iseland, South Korea. J Petrol 1–39

Brey GP (1976) CO_2 solubility and solubility mechanism in the silicate melts at high pressures. Contrib Mineral Petrol 57:215–221

Brey GP, Green DH (1975) The role of CO_2 in the genesis of olivine and melilite. Contrib Mineral Petrol 49:93–103

Brey GP, Green DH (1976) Solubility of CO_2 in olivine melilitite at high pressures and role of CO_2 in the earth's upper mantle. Contrib Mineral Petrol 55:217–230

Brey GP, Green DH (1977) Petrogenesis of olivine melilitite. Contrib Mineral Petrol 34:301–319

Briot D (1988) Genese d'une serie alcaline intraplaque continentale. Ph.D. Thesis, University of Clermont-Ferramd

Briot D, Cantagrel JM, Dupy C, Harmon RS (1991) Geochemical evolution in crystal magma reservoirs: trace element and Sr-Nd-O isotopic variations in two continental intraplate series at Monts Dore, Massif Central France. Chem Geol 89:281–303

Brooks CK, Fawcett JJ, Gittins J, Rucklidge JC (1981) The Batbjerg complex, east Greenland: a unique ultrapotassic Caledonian intrusion. Can J Earth Sci 18:274–285

Brotzu P, Beccaluva L, Conte A, Fonseca M, Garbarino C, Gomes CB, Leong R, Macciotta G, Mansur RL, Melluso L, Morbidelli L, Ruberti E, Sigolo JB, Traversa G, Valenca JG (1989) Petrological and geochemical studies of alkaline rocks from continental Brazil. The syenitic intrusion of Morro Redondo, RJ. Geochemmical 3:63–80 Brasil

Brown FH (1971) Volcanic petrology of Toro-Ankole region, Western Uganda. Ph.D. Thesis, University of California, Berkeley

Brown FH, Carmicheal ISE (1969) Quaternary volcanoes of the Lake Rudolf Region. Pt. 1. The basanite-tephrite series of the Korath range. Lithos 2:239–260

Buddington AF (1922) On some natural and synthetic melilites. Am J Sci 35(5):87

Buddington AF, Lindsley DH (1964) Iron-Titanium oxide minerals and synthetic equivalents. J Petrol 5:310–357

Bullard EC (1936) Gravity measurements in east Africa. Philos Trans Royal Soc London 235:445–531

Bundy FP, Bovenkirk HP, Strong HM, Wentorf RH (1961) Diamond-bearing graphite equilibrium lisr from growth and crystallization of diamonds. J Chem Phys 35:383–391

Burke KC, Wilson JT (1976) Hot spots on earth's interior. In: Decker R, Dacker B (eds) Volcanoes and earth's interior. W.H. Freeman and Co., New York, pp 31–42

Campbell IH, Hill RET, Griffiths RW (1990) Melting in an Archaean mantle plume: heads its basalts, tails its komatiites. Nature 339:697–699

Carmichael ISE (1967) The mineralogy and petrology of the volcanic rocks from the Leucite Hills, Wyoming. Contrib Mineral Petrol 15:24–66

Carmichael ISE, Nicholls J, Smith AL (1970) Silica activity in igneous rocks. Am Mineral 55:246–263

Carmichael ISE, Turner F, Vernoogen J (1974) Igneous petrology. McGraw-Hill, New York, 740 pp

Castor SB (2008) The mountain pass rare-earth carbonatite and associated ultrapotassic rocks, California. Can Mineral 46:779–806

Cattigny P, Boyd SR, Harris JW, Javoy M (1997) Nitrogen isotopes in peridotite diamonds from Fuxian, China: the mantle signature. Terra Nova 9:175–179

Cawthorn GR (1977) Petrological aspects of the correlation between potash content of orogenic magmas and earthquake depths. Mineral Mag 41:173–182

Charles RW (1974) The physical properties of hydrous Mg-Fe richterites. Am Mineral 59:518–528

Charles RW (1975) The phase equilibria of richterite and ferrorichterite. Am. Mineral 60:367–374

Chatterjee, S.C. (1974): Petrology of Igneous and Metamorphic Rocks of India, The Macmillan Co. India Ltd., Madras, 559 pp

Chattopadhyay S, Chattopadhyay B, Gupta AK (1999) Preliminary phase equilibria study of the system nepheline-diopside-sanidine under 1 kbar in presence of excess water. Ind J Geol 71(1 and 2):81–91

Chihara K (1971) Mineralogy and paragenesis of jadeite from the Omi-Kataki area, central Japan. Mineral Soc Japan 1:147–156 Special Paper

Christiansen RL, Lipman PW (1972) Cenozoic volcanism and plate-tectonic evolution of the western United States: II, Late Cenozoic. Phil Trans Royal Soc London A-271:249–284

Cioni R, Civetta L, Marianelli P, Metrich N (1995) Compositional layering and syn-eruptive mixing of a periodically refilled shallow magma chamber: the AD 79 Plinian eruption of Vesuvius. J Petrol 36:739–776

Civetta L, Innocenti F, Manetti P, Peccerillo A, Poli G (1981) Geochemical characteristic of potassic volcanic from Mts. Ernici (Southern Latium, Itali). Contrib Mineral Petrol 78:37–47

Clayton RN (1981) Isotopic variations in primitive meteorites. Trans R Soc London 303:339–349

Cloos H (1939) Hebung-Speltung-Vulkanismus element einer geometrischen analyse indischer gross formen. Geol Rund Sch 30:401–527

Coban H, Flower MFJ (2006) Mineral phase compositions in silica-undersaturated 'Leucite' lamproites from the Bucak Area, Isparta, SW Turkey. Lithos 89:275–299

Coffin M, Rabinowidtz P, Houtz RE (1986) Crustal structure in the western Somali basin. J Geophy Res Astron Soc 86:331–365

Collerson KD, McCulloch MT (1983) Nd and Sr isotope geochemistry of leucite-bearing lavas from Gaussberg, East Antarctica. In: Oliver RL, James PR, Jago JB (eds) Proceedings of the 4th international symposium on Antarctica earth sciences, pp 676–680

Comin-Chiarmonti P, Meriani P, Mosca R, Singigor S (1997) On the occurrence of analcime in the northeastern Azarbaizan volcanics (northwestern Iran). Lithos 12:187–198

Compagnoni R, Maffeo B (1973) Jadeite-bearing metagranites IS and related rocks in the Mount Mucrone area (Sesia—Lanzo zone, Western Italian Alps). Schweiz Mineral Petrogr Mitt 53:355–378

Conceicaocao RV, Green DH (2004) Derivation of potassic (shoshonitic) magma by decompression melting of phlogopite + pargasite lherzolite. Lithos 72:209–229

Coney PJ (1980) Mesozoic Cenozoic cordilleran plate techonics. In: Smith RB, Eaton GP (eds) Cenozoic Tectonics and Regional Geophysics of the Western Cordillera, vol 152. Memoir of Geological Society of America, Cordillera, pp 33–50

Coney PJ, Reynolds SJ (1977) Cordilleran Bemioff zones. Nature 270:403–406

Conticelli S (1989) Genesi del magmatismo alcalino-potassico dell'Italia centrale: evidenze petrologiche, geochimiche e petrologico Sperimentali. Doctoral theses, Dipartimento di Scienze della Terra, Univ. Studi Firenze, 404 pp

Conticelli S, Carlson RW, Widom E, Serri G, Beccaluva L, Bianchini G, Wilson M (2007) Chemical and isotopic composition (Os, Pb, Nd, and Sr) of neogene to quaternary calc-alkalic, shoshonitic, and ultrapotassic mafic rocks from the Italian Peninsula: inferences on the nature of their mantle sources. Cenozoic volcanism in the Mediterranean and geological society of America special papers, vol 418, pp 171–202

Conticelli S, D'Antonio M, Pinarelli L, Civetta L (2002) Source contamination and mantle heterogeneity in the genesis of Italian potassic and ultrapotassic volcanic rocks: Sr-Nd-Pb isotope data from Roman Province and Southern Tuscany. Mineral Petrol 74:189–222

Conticelli S, Manetti P, Menichetti S (1992) Mineralogy, geochemistry and Si-isotopes in orendites from South Tuscany, Italy: constraints on their genesis and evolution. Eur J Mineral 4:1359–1375

Conticelli S, Marchionni S, Rosa D, Giordano G, Boari E, Avanzinelli R (2009) Shoshonite and sub-alkaline magmas from an ultrapotassic volcano: Sr-Nd-Pb isotope data on the Roccamonfina volcanic rocks, Roman magmatic province, Southern Italy. Contrib Mineral Petrol 157:41–63

Conticelli S, Melluso L, Perini G, Avanzinelli R, Boari E (2004) Petrologic, geochemical and isotopic characteristics of potassic and ultrapotassic magmatism in Central-Southern Italy: inferences on its genesis and on the nature of mantle sources. Periodico Di Mineralogia 73:135–164

Conticelli S, Peccerillo A (1992) Petrology and geochemistry of potassic and ultrapotassic volanism in central Italy: petrogenesis and inferences on the evolution of the mantle sources. Lithos 28:221–240

Cooke DL, Moorhouse WW (1969) Timiskaming volcanism in the Kirkland Lake area, Ontario, Canada. Can J Earth Sci 6:117–132

Cordiani UG (1970) Idade do vulcanismo no oceano Atlantico. Bol Inst Astron Geofis Univers Sao Paulo 1:9–75

Cordwell L (1982) Extension in the Rio Grande rift. J Geophys Res 87:8561–8569

Coulon C, Maluki H, Bollinger C, Wang S (1986) Mesozoic and Cenozoic volcanic rocks from central and southern Tibet: ^{39}Ar-^{40}Ar dating, petrological characteristics and geodynamical significance. Earth Planet Sci Lett 79:281–302

Courteny RG, White RS (1986) Anomalous heat flow and geoid across the Cape Verde Rise: evidence for dynamic support from a thermal plume in the mantle. Geophys J Royal Astron Soc 87:815–867

Courtillot V, Feraud G, Maluski H, Vandamme D, Moreau MG, Bose J (1988) Deccan flood basalts and the Cretaceous-Tertiary boundary. Nature 333:843–846

Cox KG (1989) The role of mantle plumes in the develompent of continental drainage patterns. Nature 342:873–876

Cox KG, Hawkesworth CJ, O'Nions RK, Appleton JD (1976) Isotopic evidence for the derivation of some Roman regions volcanic rocks from anomalously enriched mantle. Contrib Mineral Petrol 56:173–180

Cross W (1897) The igneous rocks of the Leucite Hills and Pilot Butte, Wyoming. Am J Sci 4:115–141

Cross W, Iddings JP, Pirsson LV, Washington HS (1902) A quantitative chemico-mineralogical classification and nomenclature of igneous rocks. J Geol 10:555–690

Cundari A (1973) Petrology of the leucite-bearing lavas in New South Wales. J Geol Soc Aust 20:465–492

Cundari A (1975) Mineral chemistry and petrogenetic aspects of the Vico lavas, Roman Volcanic region Italy. Contrib Mineral Petrol 53:129–144

Cundari A (1979) Petrogenesis of leucite-bearing lavas in the Roman volcanic region, Italy. Contrib Mineral Petrol 70:9–21

Cundari A (1980) Role of subduction in the genesis of leucite-bearing rocks: facts or fiction? Contrib Mineral Petrol 73:432–434

Cundari A, Le Maitre RW (1970) On the petrogeny of leucite-bearing rocks of the Roman regions. J Petrol 11:33–47

Cundari A, Mattias PP (1974) Evolution of the Vico lavas, Roman volcanic region, Italy. Bull Volcan 38:98–114

Cundari A, Salviulo G (1987) Clinopyroxenes from Somma-Vesuvius: implications of crystal chemistry and site configuration parameters for studies of magma genesis. J Petrol 28:727–736

Daly RA (1910) Origin of alkaline rocks. Bull Geol Soc Am 21:87–118

Daly RA (1933) Igneous rocks and the depths of the earth. McGraw-Hill, New York, p 598

Danchin RV, Boyd FR (1976) Ultramafic nodules from the premier kimberlite pipe. S Afr Carnegie Inst Wash Yb 75:531–538

Darlymple GB (1963) Potassium-argon dates of some Cenozoic volcanic rocks of Sierra Nevada, California. Bull Geol Soc Am 74:379–390

Darlymple GB (1964) Cenozoic Chemistry of the Sierra Nevada, California. Univ Calif Publ Geol Sci 47:41

Dauphas N, Marty B (1999) Heavy nitrogen in carbonatites of the Kola Peninsula: a possible signature of the mantle. Science 286:2488–2490

Davies G (1998) A channelled plume under Africa. Nature 395:743–745

Davies JH (1999) The role of hydraulic fractures in generatng intermediate-depth earthquakes. Nature 398:142–144

Dawson, J.B. (1966): Oldoinyolengai: an active volcano with sodium carbonatite lava flows. In: O. F. Tuttle and J. Gittins (eds.), Carbonatite, Jhon Wiley and Sons, London - New York - Sydney - Toronto. 155–168

Dawson JB, Smith M (1988) Veined and metasomatised peridotites from Pello Hill, Tanzania; evidence for anomalously light mantle beneath the Tanzania sector of the eastern rift valley. Contrib Mineral Petrol 94:510–527

De Fino M, La Vole L, Peccerillo A, Piccarreta G, Poli G (1986) Petrogenesis of Mont. Vulture volcano (Italy), inferences from mineral chemistry, major and trace element data. Contrib Mineral Petrol 92:135–145

De Fino, M. La. Vole, L. and Piccarreta, G. (1982): Magma evolution at Mt. Vulture (Southern Italy) Bul. Volcanol. 45, 115-126

De Lorenzo G (1900) Studio Geologico del Monte Vulture. Atti Acc Sci Fis Mat Napoli 10:1–207

De Mulder, M. (1985): The Karisimbi volcano (Virunga). Mus. Afr. Cont. Tervuren Belg. Ann. Ser. Octavo. Sci. Geol. (90). 101 pp

De Mulder M, Hetogen J, Deutsch S, Andre L (1986) The role of crustal contamination in the potassic suite of the Karsimbi Volcano (Virunga, african Rift Valley). Chem Geol 57:117–136

Deer, W.A., Howie, R.A. and Zussman, J. (1963): Rock-forming minerals. 4, Longmans, London. p. 435

Deer, W.A., Howie, R.A.and Zussman, J. (1997): *Rock-Forming Silicates.* 2A. *Single Chain Silicates*. The Geological Society, London, U.K

Deer, W.A., Howie, R.A., Wise, W.S. and Zussman, J. (2004): *Rock-Forming Silicates.* 4B. *Framework Silicates. Silica Minerals, Feldspathoids, and the Zeolites.* The Geological Society, London, U.K

Dewey, J.F. and Windley, B.F. (1988): Palaceocene-Oligocene tectonics of N.W. Europe. In: A.C. Morton and L.M. Person (eds.). Early Tertiery volcanism and the opening of the N.E. Atlantic. Geol. Soc. London. Spec. Pub., 39, 25–31

Di Battistini G, Montanini A, Vernia L, Venturelli G, Tonarini S (2001) Petrology of melilite-bearing rocks from the Montefiascone Volcanic Complex (Roman Magmatic Province): New Insights Into The Ultrapotassic Volcanism of Central Italy. Lithos 59:1–24

Di Girolamo P (1977) Geotectonic setting of Miocene-Quaternary volcanism in the around the eastern Tyrrhenian Sea border (Italy) as deduced from major elemental geochemistry. Bull Volcanol 4(3):1–22

Di Girolamo P, Ghiara MR, Lirer L, Munno R, Rolandi G, Stanzione D (1984) Vulcanologia e petrologia, dei campi flegrei. Boll Geol Soc Italy 103:349–413

Di Girolamo P, Melluso L, Morra V, Secchi FAG (1995) Evidence of interaction between mafic and differentiated magmas in the youngest phase of activity at Ischia Island (Italy). Per Mineral 64:393–411

Di Muro A, Bonaccorsi E, Principe C (2004) Complex colour and chemical zoning of sodalite-group phases in a hauynophyre lava from Mt. Vulture, Italy. Mineral Mag 68:591–614

Radicati Di, Brozolo F, Hunecke JC, Papanastassiou DA, Wassurburg GJ (1981) $^{40}Ar/^{39}Ar$ and Rb/Sr age determination of quarternary volcanic. Earth Planet Sci Lett 51:445–456

Dickinson WR (1979) Cenozoic plate setting of the Cordilleran region in the United States. In: Armentrout JM, Cole MR, Terbest H (eds) Cenozoic Paleogeography of the Western United States. Society Economic Paleontolagists and Mineralogists, Pacific sections, Los, Angeles, Calif, pp 1–13

Dimitriev EA (1976) Cainozoic potasic alkaline rocks of East Pamir. Donish, Dushanbe 170 pp

Dodge, F.C.W. and Moore, J.G. (1981): Late Cenozoic volcanic rocks of southern Sierra Nevada, California: II. Geochemistry. Bull. Geol. Soc. Am. 92, 912–914 and 1670–1761

Dollase WA, Thomas WM (1978) The crystal chemistry of silica-rich alkali-deficient nepheline. Contrib Mineral Petrol 69:139–149

Downes H (1984) Sr and Nd isotope geochemistry of coexisting alkaline magma series of Massif Central. France Earth Planet Sci Lett 69:321–334

Downes, H. (1987): Ternary and Quaternary volcanism in the Massif Central France. In: J.G. Fitton and B.G. Upton (eds.) Alkaline Igneous rocks. Geol. Soc. Spec. Pub. 30, 517–530

Dubean MI, Edgar AD (1985) The stability of picrite up to 30 kb: inference on its potential as a reservoir for K, Ba and Ti in mantle source for lamproitic magmas. Geol Soc Canad Abstr 10: A15

Duda A, Schmincke HU (1985) Polybaric differention of alkali basaltic magmas evidence from green-core clinopyroxene (Eifel, FRG). Contrib Mineral Petrol 91:340–353

Duda A, Schminke HU (1978) Petrology and Chemistry of Potassic rocks from the Laacher Sec area. Neues Jahrb Mineral Abh 132:1–33

Dudas FO (1991) Geochemistry of igneous rocks from the Cragy Mountains, Montana and tectonic models for the Montana alkalic province. J Geophys Res 96:13261–13277

Dudas FO, Carlson RW, Eggler DH (1987) Regional middle Proterozoic enrichment of the subcontinental mantle source of igneous rocks from central Montana. Geology 15:22–25

Duncan RA, Pyle DG (1988) Rapid eruption of the Deccan flood basalts at the Createaceous. Tertiary boundary. Nature 333:841–843

Dupuy C, Dostal J, Girod M, Liotard M (1981) Origin of volcanic rocks from Stromboli (Italy). J Volcanol Geotherm Res 10:49–65

Dwivedi MM, Gupta AK, Arima M (2007) An Experimental Study of the Join Leucite-Albite-Akermanite and Leucite-Akermanite- $Albite_{50}$ $Anorthite_{50}$ at atmospheric pressure and 1 GPa in presence of excess H_2O: Its significance in Leucite Sodic plagioclase and Melilite-Plagioclase incompatbility. J Mineral Petrol Sci Stage 1:1–11

Ebinger CJ, Sleep NH (1998) Cenozoic magmatism throughout east Africa resulting from impact of a single plume. Nature 395:788–791

Eby GN, Woolley AR, Din V, Platt RG (1998) Geochemistry and petrogenesis of nepheline syenites: Kasungu, Chipala, Ilomba and Ulinidi nepheline syenite intrusions, North Nyasa Alkaline Province, Malawi. J Petrol 39:1405–1424

Edel JB, Fuchs K, Gelbke C, Prodehl C (1975) Deep Strutures of the southern Rhine graben area. J Geophys Res 11:333–356

Edgar AD, Pizzolato LA (1995) Distribution of F in the model mantle systems K-richterite-apatite and phlogopite-K-richterite at 20 kbar: Consequence for F as a compatible element in ultrapotasssic magma genesis. Contrib Mineral Petrol 121:247–257

Edgar AD (1964) Phase equilibria relations in the system $CaMgSi_2O_6$ (diopside)-$NaAlSiO_4$ (nepheline)-$NaAlSi_3O_8$ (albite)-H_2O at 1000 kg/cm^2 water vapour pressure. Am Mineral 49:573–585

Edgar AD (1980) Role of subduction on the genesis of leucite-bearing rocks. Contrib Mineral Petrol 73:429–431

Edgar AD (1989) Barium- and strontium-enriched apatites in lamproites from West Kimberly. Western Australia. Am Mineral 74:889–895

Edgar AD (1992) Barium-rich phlogopite and biotite from some Quaternary alkaline mafic lavas from West Eifel, Germany. Eur J Mineral 4:321–330

Edgar AD, Charbonneau HE, Mitchell RH (1992) Phase relations of an armalcolite-phlogopite lamproite from Smokey Butte, Montana: applications to lamproite genesis. J Petrol 33:505–520

Edgar AD Condelifee, Barnett RL, Shirran RJ (1980) An experimental study of an olivine ugandite magma and mechanism for the forming of its enriched derivations. J Petrol 21(3):475–491

Edgar AD, Green DH, Hibberson WO (1976) Experimental Petroogy of a highly potassic magma. J Petrol 17:339–356

Edgar AD, Lloyd FE, Vukadinovic (1994) The role of fluorine in the evolution of ultrapotassic magmas. Mineral Petrol 51:173–193

Edgar AD, Pizzolato LA, Sheen J (1996) Fluorine in igneous rocks and minerals with emphasis on ultrapotassic mafic and ultramafic magmas and their manle source regions. Mineral Mag 60:243–257

Edgar AD, Vukadinovic D (1993) Potassium-rich clinopyroxenes in the mantle: An experimental investigation of a K-rich lamproite up to 60 kbar. Geochim Cosmochim Acta 57:5061–5072

Edwards CMH, Merizies MA, Thirlwall MF (1991) Evidence from Muria, Indonesia for the interplay of suprasubduction zone and intraplate process in the genesis of potassic alkaline magma. J Petrol 32(3):555–592

Edwards CMH, Morris JD, Thirlwall MF (1993) Separating mantle from slab signatures in arc lavas using B/Be and radiogenic isotope systematics. Nature 362:530–533

Edwards CMH, Merizies MA, Thirlwall MF, Moriss JD, Leman WP, Harmon R (1994) The transition to potassic alkaline volcanism in island areas. The Ringit-Beser complx, East Java, Indonesia. J Petrol 35:1557–1595

Eggler, D.H. (1974): Effect of CO_2 on the melting of peridotite: Carnegie Inst. Wash.Yb. 73, 215–224

Eggler, D.H. (1987): Solubility of major and trace elements in matnle metasomatic fluids: experimental constraints. In: M.A. Menzies, and C.J. Hawkesworth (eds.), Mantle Metasomatism, Academic Press. 21–41

Eggler, D.H., Meen. Ja, Welt, F., Dudas, F.O., Furlong, K.P. McCallum, M.E. and Carlson, R.W. (1988): Tectonism of the Wyoming Province. Colorado School Mines. 83, 25–40

Eggler DH, Wendlant RF (1979) Experimental studies on the relationship between kimberlite magmas and partial melting of peridotite. In: Boyd FR, Meyer HOA (eds) Kimberlites, diatremes, and diamonds: Their Geology. Washington D.C. Am. Geophys. Union, Petrology and Geochemistry, pp 330–338

Ehrenberg SN (1982) Petrogenesis of garnet lherzolite and megacrystalline nodules from the Thumb. Navajo, volcanic field. J Petrol 23:507–547

Elkins-Tanton, L., T., Draper, D., S., Agee, C., B., Jewell, J., Thorpe, A., Hess, P., C. (2007): The last lavas erupted during the main phase of the Siberian Flood Volcanic Province: Results from Experimental Petrology. Contrib. Mineral. Petrol. 153(2), 191–209

Elkins-Tanton, L., T., Grove, TL. (2003): Evidence for deep melting of hydrous metasomatized mantle: Pliocene high-potassium magmas from the Sierra Nevadas. Journal Geophysic. Res.- Solid Earth, 108, article 2350, 29, 2003

Ellam, R.M. (1986): The transition from calc-alkaline to potassic volcanism in the Aeolian islands, southern Italy. Ph.D. thesis, the Open University, Milton, Keynes, U.K

Ellam, R.M., Hawkesworth, C.J., Menzies, M.A., and Rogers, N.W. (1989): The volcanism of southern Italy: Role of subduction and the relationship between potassic and sodic alkaline magnatism. J. Geophys. Res. 94(B-4), 4589-4601

Emerick CM, Duncun R (1982) Age of progressive volcanism in the Comores Archipelago, western Indian Ocan and implication for small plate motion. Earth Planet Sci Lett 60:415–428

Eremeyev NV (1984) Volcano plutonic complexes of potassic alkaline rocks. Nauka, Moscow 134 pp

Eremeyev NV (1987) Potassic alkaline magmatism Verhne-Yakokutsy graben (central Aldan). Izvestiya. Akadem. Nauk S.S.S.R. Seriya Geologiya 1:126–130

Erlank AJ, Waters FG, Hawkesworth CJ, Haggerty SE, Allsopp HL, Rickard RS, Menzes MA (1987) Evidence for mantle metasomatism in peridotite nodules from the kimberley Pipes, South Africa. In: Manzies MA, Hawkesworth CJ (eds) Mantle metasomatism. Academic Press, London, pp 291–311

Ernst WG, Seki Y, Onuki H, Gilbert MC (1970) Comparative study of low grade metamorphism in the California Coast Range and the outer metamorphic belt of Japan. Geol Soc Am Mem 124:1–276

Esperanca S, Holloway JR (1987) On the origin of some mica lamprophyres: experimental evidence from a mafic minette. Contrib Mineral Petrol 95:207–216

Eugster HP, Wones DR (1962) Stability relations of the ferruginous biotite and annite. J Petrol 3:82–125

Evensen NM, Hamilton PJ, O'Nions RK (1978) Rare Earth abundances in chondritic meteorites. Geochem Cosmochim Acta 42:1199–1212

Fairley KA (1995) Rapid cycling of subducting sediments into the Samoan mantle plume. Geology 23(6):531–534

Fan QC, Hooper PR (1991) The Cenozoic basaltic rocks of eastern China: petrology and chemical composition. J Petrol 32:765–810

Fasshauer DW, Wunder B, Chatterjee ND, Hohne GWH (1998) Heat capacity of wadeite-type $K_2Si_4O_9$ and the pressure-induced stable decomposition of K-feldspar. Contrib Mineral Petrol 131:210–218

Federico M, Peccerillo A (2002) Mineral chemistry and petrogenesis of granular ejecta from the Alban Hills Volcano (Central Italy). Mineral Petrol 74:223–252

Faure, G. (1977): Principles of Isotope Geology, Wily, New York, 589 pp.S

Feldstein SN, Lange RA (1999) Pliocene Potassic Magmas from the Kings River region, Sierra Nevada, California: Evidence for Melting of a subduction modified mantle. J Petrol 40 (8):1301–1320

Ferguson AK, Cundari A (1975) Petrological aspects of evolution of the leucite-bearing lavas from Bufumbira, Southwest Ugand. Contrib Mineral Petrol 52:25–46

Ferguson JB, Buddington AF (1920) The binary system akermanite-gehlenite. Am J Sci 50 (4):131–140

Ferguson JB, Merwin HE (1919) The ternary system $CaO\text{-}MgO\text{-}SiO_2$. Am J Sci 48:81–123

Ferrara, G., Laurenzi M.A., Taylor, H.P. (Jr), Tonarini, S. and Turi, B. (1985): Oxygen and strontium isotope studies of K-rich volcanic rocks from Alban Hills, Italy. Earth Planet. Sci. Lett. 75, 13–28

Ferrara, G., Preite-Martinez, M., Taylor, H.P. (Jr), Tonarini, S. and Turi, B. (1986): Evidence for crustal assimilation mixing of magmas, and a 87Sr-rich upper mantle: an oxygen and strontium isotope study of the M. Vulsini volcanic area, Central Italy. Contrib Mineral Petrol 92, 269–80

Finckh L (1912) Dei jung-vulkanischen gesteine des Kiwusee gebietes. Wiss Ergeb Dtsch Zentral-Africa Expedition 1907–1908(1):44–45

Fitch TJ (1970) Earthquake mechanisms and island arc tectonics in the Indonesian-Philippine region. Bull Seism Soc Am 60:565–591

Fitch TJ, Molnar P (1970) Focal Mechanism along inclined earthquake zone in the Indinesia - Philippine reion. J Geophys Res 75:1431–1444

Flerov GV, Koloskov AV (1976) Alkali basaltic magmatism of central Kamchatka. Nauka, Moscow 147 pp

Foden JD, Varne R (1980) The petrology and tectonic setting of Quarternary-Recent volcanic centre of Lombok and Sumbawa, Sunda. Chem Geol 30:201–226

Foley SF (1985) The oxidation state of lamproite magmas Tschermaks' mineralogy. Petrol Mitt 34:217–238

Foley, S.F. (1989a): The genesis of lamproitic magmas in the reduceed, fluorine-rich mantle: In: A. L. Jaques, Ferguson and D.H. Green (eds.) Kimberlite and relted rocks, their composition occurrence, origin and emplacement. Black Wells, Melbourne. 616–632

Foley SF (1989) Experimental constrainsts on petrology and Chemistry in lamproites: 1. The effect of water activity and oxygen fugacity. Eur J Mineral 1:411–426

Foley SF (1990) A review and assessment of experiments on kimberlites, lamproites and lamprophyres as a guide to their origin. Proc Ind Acad Sci 99:57–80

Foley SF (1992) Petrological characterization of the source components of potassic magmas: Geochemical and experimental constrainst. Lithos 28:187–204

Foley SF (1992) Vein plus wall rock melting mechanisms in the lithosphere and the origin of potassic alkaline magmas. Lithos 28:435–454

Foley SF (1993) An experimental study of olivine lamproite: First results from the diamond stability field. Geochim Cosmochim Acta 57:483–489

Foley SF, Hofer H, Brey G (1994) High pressure synthesis of priderite and members of the lindsleyiite-mathiasite and hawthorneite-yimengite series. Contrib Mineral Petrol 117:164–174

Foley SF, Taylor WR, Green DH (1986) The effect of fluorine on phase relationship in the system Mg_2SiO_4-$KAlSiO_4$-SiO_2 at 28 kbar and the solution mechanism of fluorine in silicate melts. Contrib Mineral Petrol 93:46–55

Foley SF, Taylor WR, Green DH (1986) The role of fluorine and oxygen fugacity in the genesis of the ultrapotassic rocks. Contrib Mineral Petrol 94:183–192

Foley SF, Venturelli G, Green DH, Toscani L (1987) The ultrapotassic Rocks: Characteristics, classification, and constraints for petrogenetic models. Earth Sc Rev 24:81–134

Forbes WC, Flower MFJ (1974) Phase relations of titanphlogopite, $K_2Mg_4TiAl_2Si_6O_{20}(OH)_4$: A refractory phase in the upper mantle? Earth Planet Sci Lett 22:60–66

Fornaseri M (1985) Geochronology of volcanic rocks from Latium (Italy). Rend Soc Ital Mineral Petrol 40:73–106

Fornaseri M, Scherillo A, Ventriglia U (1963) La Regione Vulcanica Dei Colli Albani. C.N.R, Rome 394 pp

Francalanci L, Peccerillo A, Poli G (1987) Partition coefficients for minerals in potassium-alkaline rock: data from Central Italy. Geochem J 21:1–10

Fraser KJ, Hawkesworth CJ, Erlank AJ, Mitchell RH, Scott-Smith BH (1985) Sr, Nd and Pb isotope and minor element geochemistry of lamproites and kimberlites. Earth Planet Sci Lett 76:57–70

Freda C, Gaeta M, Misiti V, Mollo S, Dolfi D, Scarlato P (2008) Magma-carbonate interaction: An experimental study on ultrapotassic rocks from Alban Hills (Central Italy). Lithos 101:397–415

Frey FA, Green DH, Roy SD (1978) Integrated models of basalt petrogenesis: a study of quartz tholeiites to olivine melilites from south-easten Australia utilizing geochemical and expeirmental petrological data. J Petrol 19:463–513

Fudali RF (1963) Experimental study on the origin of pseudoleucite and associated problems of alkali rock systems. Bull Geol Soc Am 74:1101–1126

Fuster, J.M., Gastesi, P., Sagredo, J., and Fermoso, M.L. (1967): Las rocas lamproiticas del S.E. de Espana Studious Geologicos. 23, 35-49

Fyfe WS (1973) Dehydration reactions. Bull Am Assoc Petrogr Geo 57:190–197

Fyfe, W.S. (1992): Geosphere interactions on a convecting planet mixing and Separation. In: O. Hutzinger (ed.), The Handbook of Environmental Chemistry, Vol. I. Part F., Springer Verlag

Fyfe WS (1997) Deep fluids and volatile recycling: crust to mantle. Tectonophysics 275:243–251

Fyfe, W.S. and McBirney, A.R. (1975): Subduction and the structure of andesitic volcanic basalt. Am. J. Sci. 275(A), 285–297

Gaeta, M., Freda, C., Christensen, J., N., Dallai, L., Marra, F., Karner, D., B., Scarlato, P. (2006): Time-dependent geochemistry of clinopyroxene from the Alban Hills (Central Italy): Clues to the source and evolution of ultrapotassic magmas. Lithos, 86, 330–346

Gallo, F., Giammetti, F., Venturelli, G. and Vernia, L. (1984): The kamafugitic rocks of Sanvenanzo and Cupaello, Central. Italy. Neues. Jahrb. Mineral. Monatosh 198–210

Gans PB, Mahood GA, Schermer E (1989) Syn extensional magmatism in the Basin and Range Province. Jemez Volcanic field, New Mexico. J Geophys Res 91:1763–1778

Gatta, G., D., Rotiroti, N., Ballaran, T., B., Pavese, A. (2008): Leucite at high pressure: Elastic behavior, phase stability, and petrological implications. American Mineral. 93 (10), 1588–1596

Gee LL, Sock RO (1988) Experimental Petrology of melilite nephelinites. J Petrol 29 (6):1233–1255

Genik G (1992) Regional framework, structure, and petroleum aspects of rift basins in Niger, Chad and Central African Republic. Tectonophysics 213:169–185

Ghiorso MS, Carmichael ISE (1981) A Fortran IV computer programme for evaluating temperatures and oxygen fugacities from the composition of coexisting iron-titanium oxides. Comp Geosci 7:123–129

Ghosh C (1949) Petrochemical study of the lamprophyres and associated intrusive rocks of the Jharia coal field. Q J Min Met Soc Ind 21:133–147

Ghosh SK, Mukhopadhyay A (1985) Tectonic history of the Jharia basin in interacratonic Gondwana basin in eastern India. Q J Geol Min Met Soc Ind 57(1):33–58

Giannetti B (1979) The geology of Roccamonfina caldera (Campanian Province, Italy). Giornale Geologia 43(2):187–206

Gibson SA, Morrison MA, Hendry GL, Dickin AP (1991) The Flat Tops volcanic field NW Colorado, I. Lower Miocene open-system multi-source magmatism at Flander, trappers Lake. J Geophys Res 96(13):609–627

Gibson SA, Thompson RN, Leat PT, Dickin AP, Morrison MA, Hendry, GL, Mitchell JG (1992) Asthenosphere-derived magmatism in the Rio Grande Rift, Western U.S.A.: implications for continental break-up. In: Story BC, Alabaster T, Pankhurst RJ (eds.). Magmatism and the causes of contenental break-up, vol 68. Geological Society of London. Spececial Pubications, pp 61–89

Gibson SA, Thompson RN, Leat PT, Morrison MA, Hendry GL, Dickin AP, Mitchell JG (1993) Ultrapotassic magmas along the flanks of the Oligo-Miocene Rio Grande Rift, U.S.A.: Monitors of the zone of lithospheric mantle extension and thinning beneath a continental rift. J. Petrol. 34(1):187–228

Gibson SA, Thompson RN, Lenardos OH, Dickin AP, Mitchell JG (1995) The late Cretaceous impact of the Trinidade Mantle Plume: Evidence from large volcanic, mafic potassic magmatism in S.E. Brazil. J. Petrol. 36(1):189–229

Gibson SA, Thompson RN, Leonardos OH, Turner S, Mitechell JG, Dickin AP (1994) The Serra do Bueno potassic diatreme: A possible hypabyssal equivalent of the ultramafic potassic volcanics in the Cretaceous Alto Paranaiba Igneous Province, S.E. Brazil. Mineral. Mag. 58:357–373

Gilbert, M.C. and Brieggs, D.F. (1974): Comparison of the stabilities of OH and F potassic richterites. A Preliminary report. Trans. Am. Geophys. Union. 480–481

Gilluly J (1971) Plate tectonics and magmatic evolution. Bull. Geol. Soc. Am. 82:2387–2396

Gittins J, Fawcett JJ, Rucklidge JC (1982) An occurrence of the spinel end-member Mg_2TiO_4 and related spinel solid solutions. Mineral. Mag. 45:135–137

Gittins J, Fawcett JJ, Brooks CK, Rucklidge JC (1980) Intergrowth of nepheline-potassium feldspar and kalsilite-potassium feldspar: a re-examination of the pseudo-leucite problem. Contrib Mineral Petrol 73:119–226

Gough ST, Morgan WJ, Hargraves RB (1980) Kimberlites: Their relations to mantle hot spots. Earth Planet. Sci. Lett. 50:560–274

Green DH (1973) Experimental melting studies on a model upper mantle compostion at high pressure under water-saturated and water undersaturated conditions. Earth Planet. Sci. Lett. 19:37–53

Green DH, Nicholls IA, Viljoen MJ, Viljoen RP (1975) Experimental demonstration of the existence of peridotite liquidus in earliest Archean magmatism. Geol. 3:11–14

Green, D.H., and Taylor, W.R. (1987): Mantle-derived magmas: Role of variable source peridotite and variable C-H-O fluid compositions. In: B.O. Mysen (ed.), Magmatic Process: Physico-chemical Principles. Geochem. Soc. Spec. Pub. 139–154

Green V, Achauer U, Meyer R (1991) A three dimensional image of the crust and upper mantle beneath the Kenya rift. Nature 354:199–203

Greig JW, Barth TFW (1938) The system Na_2O-Al_2O_3-$2SiO_2$ (nepheline, carnegieite)-Na_2O. Al_2O_3.$6SiO_2$ (albite). Am. J. Sci. 35A(5):93–112

Griffiths R, Cambell I (1991) Interaction of mantle pulme heads with Earth's surface at the onset of small scale convection. J. Geophys. Res. 96:18295–18310

Gummer WK, Burr SV (1946) Nephelinized paragenesis in the Bancroft area. Ontario. J. Geol. 54:137–168

Gupta AK (1972) The system forsterite-diopside-akermanite-leucite and its significance in the origin of potassium-rich mafic and ultramafic rocks. Am. Mineral. 57:1242–1259

Gupta AK, Onuma K, Yagi K, Lidiak EG (1973) Effect of silica concentration on the diopsidic pyroxenes in the system Diopside-$CaTiAl_2O_6$-SiO_2. Contrib Mineral Petrol 41:333–344

Gupta, A.K., Chattopadhyay, B., Fyfe, W.S. and Powell, M. (2002): Experimental studies on the three ultramafic lamproitic rocks from Damodar Valley East India. Mineral. Petrol. 74, 343–360 (Edgar Volume)

Gupta AK, Chattopadhyay S, Chattopadhyay B, Arima M (2006) Experimental Study of the system diopside-nepheline-sanidine at 0.1, 1 and 2 GPa [P(H_2O)=P (Total)]: Its significance in the genesis of alkali-rich basic and ultrabasic rocks. Lithos 86:91–109

Gupta AK, Dwivedi MM, Bhattachariya H, Dasgupta S (2010) Silica-undersaturated portion of the system nepheline-kalsilite-SiO_2 at 2 GPa [P(H_2O)=P (Total)]. Canad'Mineral. 48:297–213

Gupta AK, Edgar AD (1975) Leucite-Na-feldspar incompatibility: An experimental study. Mineral. Mag. 40:337–384

Gupta AK, Fyfe WS (1975) Leucite survival: The alteration to analcime. Canad. Mineral. 13:361–363

Gupta AK, Fyfe WS (2003) The young potassic Rocks- Theophrastus Pub. S.A 33 J. Theologou Str Zographou, Athens 15773 Greece, p 370

Gupta VK, Gupta AK (1985) Study of the system leucite-akermanite-SiO_2 under one atmospheric pressure. Ind. J. Earth Sci. 12(2):125–139

Gupta VK, Gupta AK (1997) Phase relations in the system Leucite-Akermanite-Albite-SiO_2 under one atmospheric pressure. Synthetic and Natural Rock System (Yagi volume), Allied Publishers Ltd. 48-67

Gupta AK, Green DH (1988) Experimental study of the system forsterite-kalsilite-SiO_2 in presence or absence of volatiles (CO_2 or H_2O) under 28 kb. Mineral. Petrol. 39:163–174

Gupta AK, Green DH, Taylor WR (1987) The liquidus surface of the system forsterite-nepheline-silica at 28 kb. Am. J. Sci. 287(6):560–565

Gupta, A.K., Le Maitre, R.W., Haukka, M.L. and Yagi, K. (1983): Geochemical studies on the carbonated apatite glimmerite, from Damodar valley India, Proc. Japan Acad. Sci. 59 Ser. B (No. 5), 113-116

Gupta AK, Lidiak EG (1973) The system diopside-nepheline-leucite. Contrib Mineral Petrol. 41 (3):231–239

Gupta AK, Venkatesh TL (1993) Experimental studies on synthetic richterite and its implications. Ind. J. Earth Sci. 20:67–70

Gupta AK, Venkateshwaran GP, Lidiak EG, Edgar AD (1973) The system diopside-nepheline-akermanite-leucite and its bearing on the genesis of alkali-rich mafic and ultramafic volcanic rocks. J. Geol. 81:209–218

Gupta AK, Yagi K (1979) Experimental study of some assimilative reactions related to the genesis of the leucite-bearing rocks. Bull. Volcan. Soc. Japan 22:65–74

Gupta AK, Yagi K (1980) Petrology and Genesis of Leucite-bearing rocks. Springer-Verlag, Berlin-Heidelbergh, New York 252 pp

Gupta AK, Yagi K, Hariya Yu, Onuma K (1976) Experimental investigations of some synthetic leucite-rocks under water vapour pressures. Proc. Japan Acad. Sci. 52:469–472

Hacksworth CJ, Vollmer R (1979) Crustal contamination versus enriched mantle. $^{143}Nd/^{144}Nd$ and $^{87}Sr/^{86}Sr$ evidence from the Italian volcanism. Contrib Mineral Petrol. 69:151–165

Haggerty SE (1983) The mineralogy and Chemistry of new titanaties from the Jagersforntein kimberlite, South Africa: Implication of metasomatism in the upper mantle. Geochim. Cosmochim. Acta. 47:1833–1854

Haggerty SE, Smyth JR, Erlank AJ, Rickard RS, Danchin RV (1983) Lindsleyite (Ba) and Mathiasite (K): two new chromium titanates as the crichtonite series from the upper mantle. Am. Mineral. 68:494–505

Hall A (1999) Ammonium in granites and its petrogenetic significance. Earth Sci. Rev. 45:145–165

Hamilton DL (1961) Nephelines as crystallization temperatures indicators. J. Geol. 69:321–329

Hamilton DL (1972) The system $NaAlSi_3O_8$ (Ab)-$NaAlSiO_4$ (Ne)-H_2O at 8 and 12 kilobars. N.E. R.C. Prog Exp Petrol 2:24–27

Hamilton DL, MacKenzie WS (1960) Nepheline solid solutions in the system $NaAlSiO_4$-$KAlSiO_4$-SiO_2. J Petrol 1:56–72

Hamilton DL, MacKenzie WS (1965) Phase equilibria studies in the system $NaAlSiO_4$ (nepheline)-$KAlSiO_4$ (Kalsilite)-SiO_2-H_2O. Mineral Mag 34:214–231

Hamilton W (1974) Bathymetric map of the Indonesian region. US Geol Surv 1:5,000,000, Map I-875-A

Hamilton W (1979) Tectonic model of the Indonesian region. US Geol Surv Prof Pap 1078:345

Harker A (1909) The natural history of Igneous rocks. Methuen, London

Harlow GE, Veblen DR (1991) Potassium in clinopyroxene inclusions from diamonds. Science 251:652–655

Harris PG (1957) Zone refining and origin of potassic melts. Geochim Cosmochim Acta 12:195–208

Hart SR (1984) A large scale isotope anomaly in the southern hamisphere mantle. Nature 309:753–757

Hatherton T, Dickinson WR (1969) The relationship between andesitic volcanism and seismicity in Indonesia, the Lesser Antilles and other island arc. J Geophys Res 74:5301–5310

Heller PH, Tabor HRW, Suezek CA (1987) Palaeogeographic evolution of the United States, Pacific northwest during Paleocene time. Cand J Earth Sci 24:1652–1667

Helmstead H, Doig P (1975) Eclogite nodules from the kimberlite pipes of the Colorado Plateau: samples of Franciscan type oceanic lithosphere. Phys Chem Earth 9:95–112

Henshaw DE (1955) The structure of Wadeite. Mineral Mag 30:585–595

Hentschel G, Abraham K, Schreyer W (1980) First terrestrial occurrence of roedderite in volcanic ejecta of the Eifel, Germany. Contrib Mineral Petrol 73:127–130

Hergt JM, Chapell BW, McCulloch MT, McDougall I, Chivas AR (1989) Geochemical and isotopic constraints on the origin of the Jourassic dolerits of Tasmania. J Petrol 30:841–883

Herz N (1977) Timing and spreading in the south Atlantic from Brazilia alkaline rocks. Bull Geol Soc Am 88:101–112

Higazy A (1954) Trace elements of volcanic and ultrabasic potassic rocks of south western Uganda and adjoining part of Belgain Congo. Bull Geol Soc Am 65:37–70

Hill RI (1991) Starting plumes and Continental breakup. Earth Planet Sci Lett. 104:398–416

Hilton DR, Barlings J, Wheller GE (1995) Effect of shallow level contamination on the helium isotope systematics of ocean island lavas. Nature 373:326–330

Hirose K, Kushiro I (1993) Partial meltings of dry peridotites at high pressures: Determination of compositions of melts segregated from peridotite using aggregates of diamonds. Earth Planet Sci Lett 114:477–489

Hoder HS (1969) Anorthite-akermanite and albite-sodamelilite reaction relations. Carnegie Inst Wash Yb 67:105–108

Hoernle K, Zhang YS, Graham D (1995) Seismic and geochemical evidence for large-scale mantle upwelling beneath the eastern Atlantic and western and central Europe. Nature 374:34–39

Hogarth DD (1997) Mineralogy of leucite-bearing dykes from Napoleon Bay, Baffin Island: multistage Proterozoic lamproites. Can Mineral 35:53–78

Holm PM, Munksgaard NC (1982) Evidence for mantle metasomatism: an oxygen and Sr isotopic study of the vulsinian district, central Italy. Earth Planet Sci Lett 60:376–388

Holm PM, Lou S, Nielsen A (1982) The geochemistry and petrogenesis of the lava of the Vulsinian district, Roman Province, central Italy. Contrib Mineral Petrol 80:367–378

Holmes A (1932) The origin of igneous rocks. Geol Mag 69:543–558

Holmes A (1937) A contribution to petrology of kimberlite and its inclusion. Trans Geol Soc S Africa 39:379–428

Holmes A (1942) A suit of volcanic rocks from southwest Uganda containing kaliophyllite (a polymorph of $KAlSiO_4$). Mineral Mag 26:167–217

Holmes A (1950) Petrogenesis of katungite and its associates. Am Mineral 35:772–792

Holmes A (1964) Principles of Physical Geology. Nelson and Sons. Ltd., London 1054 pp

Holmes A (1965) Principles of Physical Geology. Nelson and Sons. Ltd., London 1288 pp

Holmes A, Harwood HF (1932) Petrology of the volcainc fields east and southeast of Ruwenzori. Uganda Q J Geol Soc 88:370–442

Holmes A, Harwood HF (1937) The petrology of the volcanic rocks of Bufumbira. Mem Geol Sur Ugand 3:300

Holmes A, Paneth FA (1936) Helium ratios of rocks and minerals from diamond pipes of south Africa. Proc Royal Soc 154A:385–413

Hovis GL, Roux J, Rodrigues E (2002) Thermodynamic and structural behavior of analcime-leucite analogue systems. Am Mineral 87(4):523–532

Huang WL, Wyllie PJ (1975) Melting reactions in the system $NaAlSi_3O_8$-$KAlSi_3O_8$-SiO_2 to 35 kbars, dry and with excess water. J Geol 83:737–748

Huckenholz HG (1973) The origin of fassaitic augite in the alkali basalt suit of the Hoch Eifel area, Western Germany. Contrib Mineral Petrol 40:315–326

Hussak E (1890) Uber Leucit-Pseudokrystalle in Phonolith (tinguait) der Serre de tingea Estado Rio de Janeiro, Brazil. Neues Jahrb Mineral Abh 1:166–169

Hytonen K, Schairer JF (1961) Crystallization in forsterite-anorthite-diopside-silica. Carnegie Inst Wash Yb 60:139–141

Iddings JP (1913) Igneous rocks. Wiley, London 685

Iddings JP, Morely EW (1915) Contribution to the petrography of Java and Celebes. J Geol 23:231–245

Imbo G (1965) Catalogue of active volcanoes of the world including solfatra fields. Int Assoc Volcanol 18:1–71

Ionov DA, Griffin WL, O'Reilly SY (1997) Volatile-bearing minerals and lithophile trace elements in the upper mantle. Chem Geol 141:153–184

Itihara Y, Honma H (1979) Ammonium in biotite from metamorphic and granitic rocks of Japan. Geochim Cosmochim Acta 43:503–509

Itihara Y, Tainosho Y (1989) Ammonium and insoluble nitrogen in Precambrian rocks from the Gawler Craton, Australia: inference of life activity. J Geol Soc Japan 95:439–445

Itihara Y, Honma H (1983) Content and origin of ammonium in biotites of granitic and metamorphic rocks. In: Augustithis SS (ed) The Significance of trace elements in solving petrogenetic problems and controversies. Theophrastus Publications, Athens, pp 431–444

Iwao S (1953) Albitite and associated jadeite rock from Kotaki district, Japan; a study in ceramic raw material. Geol Surv Japan Rep 153:25 (Mineral Abstr 12–381)

Jamtveit B, Putnis C, Malthe-Sorenssen A (2009) Reaction induced fracturing during replacement processes. Contrib Mineral Petrol 157:127–133

Jaques AL, Foley SF (1985) The origin of Al-rich spinel inclusions in leucite from the leucite lamproites of Western Australia. Am Mineral 70:1143–1150

Jaques AL, Boxer G, Lucas H, Haggerty SF (1986) Mineralogy and petrology of the Argyle lamproite pipe, Western Australia. Extended Abstract 4th International Kimberlite Confererence, Perth, Australia, 11–15 Aug pp 48–50

Jaques AL, Lewis JD, Smith CB, Gregory GP, Ferguson J, Chappel BW, McCuleoch MT (1984b) The diamond-bearing ultrapotassic (lamproitic) rocks of the West Kimberley region, Western Austraial. In: Kornprobst J (ed) Kimberlites 1: Kimberlites and related rocks. Elsevier, Amsterdam, pp 225–254

Jaques AL, Webb AW, Fanning CM, Black LP, Pidgeon RT, Ferguson J, Smith CB, Gregory GP (1984) The age of the diamond-bearing pies and associated leucite lamproite of the West Kimberley region, Western Australia. BMRJ Aust Geol Geophys 9:1–8

Javoy M, Pineau F, Demaiffe D (1984) Nitrogen and carbon isotopic composition in the diamonds of Mbuji Mayi (Zaire). Earth Planet Sci Lett 68 (1984):399–412

Jener GA, Longerich HP, Jacson SE, Fryer BJ (1990) ICPMS—A powerful tool for trace element analysis in earth sciences: evidence from analysis of selected USGS reference samples. Chem Geol 83:133–148

Jia Y, Kerrich R, Gupta AK, Fyfe WS (2003) ^{15}N-enriched Gondwana lamproites, eastern India: crustal N in mantle source. Earth Planet Sci Lett 215:43–56

Jones AP, Larsen LM (1985) Geochemistry and REE minerals of nepheline syenites from the Motzfeldt centre, South Greenland. Am Mineral 70:1087–1100

Jones AP, Smith JV (1983) Petrological significance of mineral chemistry in the Agathila Peak and the Thumb Minettes, Navajo Volcanic Field. J Geol 91:643–656

Jones RH, MacKenzie WS (1991) Liquidus phase relationships in the system $KAlSi_3O_8$-$CaAl_2Si_2O_8$-$KAlSiO_4$ at $P(H_2O) = 5$ kbar. Am Mineral 76:200–204

Judd JW (1887) On the discovery of leucite in Australia. Mineral Mag 7:194–195

Kanukov BV, Makhotkin LL, Golovanova TI (1991) Petrology of the potassic volcanic series of the Yakokut volcanic-plutonic complex of the Centrl Aldan. Izvestia Akadem. Nauk SSSR Seriva, Geologiya 12:83–101

Karlsson HR, Clayton RW (1991) Analcime phenocrysts in igneous rocks: primary or secondary? Am Mineral 76:189–199

Katili JA (1973) Geochronology of West Indonesia and its implication on plate tectonics. Tectonophysics 19:195–212

Kemp JF, Knight WC (1903) Leucite Hills of Wyoming. Bull Geol Soc Am 14:305–336

Kempton PD (1987) Mineralogic and geochemical evidence for differing styles of metasomatism in spinel lherzolite xenoliths : enriched mantle source regions of basalts. In: Menzies MA, Hawkesworth CJ (eds) Mantle Metasomatism. Academic Press, London, pp 21–41

Kennedy GC (1955) Some aspects of the roll of water in rock melts. Geol Soc Am Spec Pap 62:489–503

King BC (1965) Petrogenesis of the alkaline igneous rock suites of the volcanic and intrusive centers of eastern Uganda. J Petrol 6:67–100

King BC, Suther land DS (1966) The carbonatites complexes of eastern Uganda. Carbonatites 73–126

Kogarko LN, Konova VA, Orlova MP, Wooley AR (1995) Alkaline rocks and Carbonatites of the world, part 2: former U.S.S.R. Champman and Hall, London, p 226

Konzett J, Sweeney RJ, Thompson AB, Ulmer P (1997) Potassium amphibole stability in the upper mantle: an experimental study in a peralkaline KNCMASH system to 8.5 Gpa. J Petrol 38(4):538–568

Koster Van Groos AF, Wyllie PJ (1973) Liquid immiscibility in the join $NaAlSi_3O_8$-$CaAl_2Si_2O_8$-Na_2CO_3-H_2O. Am J Sci 273:465–487

Kuehner SM (1980) Petrogenesis of ultrapotassic rocks, Leucite Hills, Wyoming. M.Sc. thesis, University of Western Ontario, London, Ontario

Kuehner SM, Edgar AD, Arima M (1981) Petrogenesis of the ultrapotassic rocks from Leucite Hills. Wyoming Am Mineral 66:663–667

Kushiro I (1964) Stability field of akermanite. Carnegie Inst Wash Yb 63:84–86

Kushiro I (1975) On the nature of silicate melt and its significance in magma genesis: regularities in the shift of liquidus boundaries involving olivine, pyroxene and silicate minerals. Am J Sci 275:411–431

Kushiro I, Erlank AJ (1970) Stability of potassic richterite. Carnegie Inst Wash Yb 68:231–233

Kushiro I, Yoder HS (1966) Anorthite-forsterite and anorthite-enstatite reaction and their bearing on basalt-eclogite transformation. J Petrol 7:337–362

Kushiro I, Syono Y, Akimoto S (1967) Stability of phlogopite at high pressures and possible presence of phlogopite in the earth's upper mantle. Earth Planet Sci Lett 3:197–203

La Volpe L, Patella D, Rapisardi L, Tramacere A (1984) The evolution of Montle Vulture volcano, southern Italy: inferences from volcanological geological and deep dipole electrical sounding data. J Volcanol Geotherm Res. 22:147–162

Lacroix A (1926) La systematique des roches leucitiques; les types de la famille syentique. Acad Sci CR 182:597–601

Lacroix A (1930) La jadéite de Birmanie: les roches qu'elle constitue ou qui l'accompagnent. Composition et origine. Bull Soc fr Minéral Cristallogr. 53:216–264

Lange RA, Carmichael ISE, Stebbins JF (1986) Phase transistions in leucite ($KAlSi_2O_6$), orthorhombic ($KAlSiO_4$), and their iron analogue ($KFeSi_2O_6$, $KFeSiO_4$). Am Mineral 71:937–945

Larimar WM (1964) The oxidation states of the elements and their potentials in aqueous solutions. Prentice Hall, Inc., Englewood Cliffs, p 392

Larsen ES, Buie BF (1938) Potash analcite and pseudoleucite from the Highwood Mountains of Montana. Am Mineral 23:837–849

Larsen ES, Hurlbut CS, Burges CH, Buie BF (1941) Igneous rocks of Highwood Mountains (Part 5). Bull Geol Soc Am 52:1829–1840

Latin D, Norry M, Tazery R (1993) Magmatism in the Gregory rift; evidence for melt generation by a plume. J Petrol 34:1007–1027

Le Bas MJ (1977) Carbonatite-Nephelinite Volcanism. John Wiley and Sons, New York, p 347

Le Bas MJ (1987) Nephelinites and carbonatites. In: Fitton JG, Upton BGJ (eds) Alkaline igneous rocks, vol 30. Geological Society Special Pubications, London, pp 53–83

Le Bas MJ (1989) Nephelinitic and basanitic rocks. J Petrol 20:1299–1312

Le Bas MJ, Le Maitre RW, Streckeisen A, Zanettin B (1986) A chemical classification of volcanic rocks based on the total alkali-silica diagram. J Petrol 27:745–750

Le Maitre RW (1969) Kaersutite-bearing plutonic xenoliths from Tristan da Cunha, South. Atlantic Mineral Mag 37:185–197

Le Roxe AP, Cliff RA, Adair BJI (1990) Triston da Cunha, South Atlantic: geochemistry and petrogenesis of a basanite-phonolite lava series. J Petrol 31:779–812

Lee DC, Hallidayk AN, Fitton JG, Pol G (1994) Chronology of volcanism along the Cameroon volcanic line. Earth Planet Sci Lett 123:119–138

Lefèvre R, Michard A (1965) La jadéite dans le métamorphisme alpin, à propos des gisements de type nouveau, de la bande d'Acceglio (Alpes cottiennes, Italie). Bull Soc fr Minéral Cristallogr 88:664–677

Leonardos OH, Carvalho JB, Tallarico FHB, Gibson SA, Thompson RN, Meyer HOA, Dickin AP (1993) Oxenolito de granada lherzolito de Treas Ranchoas 4: uma rocha matriz do diamante na provincia magmatica cretacea do Alto Paraniba. Goias, Annais de 1°, vol 2/93. Simposio de Geologia do Diamantes Cuiaba: Universidade Federal do Mato Grosso, Special Publications, pp 3–19

Limpan PW, Precora HJ, Christiansen RL (1971) Evolving subduction zone in the western United States as interpreted from igneous rocks. Science 174:822–825

Lindgrens W (1933) Differentiation and Ore deposition: Lindgren volume: Ore deposits of the Western Deposits of Western States. Am Inst Min Metall Eng 797:152–180

Lindsley DH (1966) P-T projection for part of the system kalsilite-silica. Carnegie Inst Wash Yb 65:244–247

Lipman PW (1980) Cenezoic volcanism in the Western United States: Implicatins for continental tectonics: In: continental tectonics. National Academic Sciences, Washington, pp 64–174

Lipman PW (1983) Tectonic setting of the mid to late Tertiary in the Rocky Mountain region: A review: Proceedings of society Denver region exploration geologists society symposium. The genesis of Rocky Mountain ore deposits: changes with time and tectonics. In: Babcock JW (ed) Denver regional exploration geologists society 1964, Wheat Ridge, Colorado, pp 125–131

Lippard SJ (1973) The petrology of phonolites from the Kenya Rift. Lithos 6:217–254

Lloyd FE (1981) Upper-mantle metasomatism beneath a continental rift clinopyroxenes in alkali mafic lavas and nodules from south-west Uganda. Mineral Mag 44:315–323

Lloyd FE (1987) Characterization of mantle metasomatic fluids in spinel lherzolites and alkali clinopyroxenites from the west Eifel and south west Uganda. In : Mantle Metasomatism Menzies MA, Hawkesworth CJ (eds) Academic Press, London, pp 21–41

Lloyd FE, Bailey DK (1969) Carbonatite in the Tuffs of the west Eifel, Germany. Contrib Mineral Petrol 23:130–139

Lloyd FE, Bailey DK (1975) Light element metasomatism of the continental mantle: the evidence and the consequences. Phys Chem Earth 9:389–416

Lloyd FE, Arima M, Edgar AD (1985) Partial melting of a phologpite clinopyroxenite from south-west Uganda: An experimental study bearing on the origin of highly potassic continental rift volcanism. Contrib Mineral Petrol 91:321–329

Lopez Ruiz J, Rodriguez Badoila E (1980) La region volcanica Neogena del sureste de Espana. Etudios Geol 36:5–63

Lucazeau F, Vasseur G, Bayer R (1984) Interpretation of heat flow data in the French massif Central. Tectonophysics 103:99–119

Luhr JF, Carmichael ISE (1981) The Colima Volcanic complex, Mexico: Part II Late-Quatenary cinder cones. Contrib Mineral Petrol 76:127–147

Luhr JF, Giannetti B (1987) The Brown Leucitic Tuff of Rocamonfina volcano (Roman Region, Italy). Contrib Mineral Petrol 95:420–436

Luth RW (1997) Experimental study of the system phlogopite-diopside from 3.5 to 17 GPa. Am Mineral 82:1198–1209

Luth RW, Boettcher AL (1986) Hydrous melting of silicates. Am Mineral 71:264–267

Luth WC (1967) Studies in the system $KAlSiO_4$-Mg_2SiO_4-SiO_2-H_2O: I. Inferred phase relations and applications. J Petrol 8:372–416

Macdonald R, Upton BG, Collerson KD, Hearn BC, James D (1992) Potassic lavas of the Bearpaw Mountains, Montana: mineralogy, chemistry and origin. J Petrol 33(2):305–346

MacKenzie WS, Rahman S (1968) The paragenesis of leucite-Na-feldspar. Contrib Mineral Petrol 19:339–342

Makhotkin II, Arakelyanz MM, Vladykin NV (1989) On the age of the lamproites of the Aldan Province. Doklad Akadem Nauk SSR 306:703–707

Maksunov Ye P (1973) Mesozoic annular magmatic complexes in the Aldan Shield. Int Geol Rev 15:46–56

Marty B (1995) Nitrogen content of the mantle inferred from N_2-Ar correlation from ocean basalts. Nature 377:326–329

Marty B, Humbert F (1997) Nitrogen and argon isotopes in ocean basalts. Earth Planet Sci Lett 152:101–112

Marvin RF, Hearn BC, Menhert HH, Naeser CW, Zartman BE, Lindsley DA (1980) Late Cretaceous-Paleocene-Eocene igneous activity in north-central Montana. Isochron 29:5–25

Mason B (1966) Principles of geochemistry. Wiley, New York 329

Massonne HJ (1992) Evidence for low-temperature ultrapotassic siliceous fluids in subsuction zone environments from experiments in the system K_2O-MgO-Al_2O_3-SiO_2-H_2O (KMASH), vol 28. Elsevier Science Publishing, pp 421–434

McBirney A, Aoki K-I, Bass MN (1967) Eclogites and jadeite from the Motagua fault zone. Guatemala Am Mineral 52:908–918

McCauley JW, Newnham RE, Gibbs GV (1973) Crystal structure analyses of synthetic fluorphlogopite. Am Mineral 58:249–254

McCulloch MT, Jaques AL, Neslon DR, Lewis JD (1983) Nd and Sr isotopes in kimberlites and lamproites from Western Australia, an enriched mantle origin. Nature 302:400–403

McDowell FW, Roden MF, Smith D (1986) Comments on tectonoc implications of the age, composition and orientation of lamprophyre dikes, Navajo volcanic field. In: Laughlin et al (eds) Earth Planet Sci Lett, vol 80. Arizona, pp 415–417

McElhinny MW, Haile NS, Crawford AR (1974) Palaeomagnetic evidence shows Malay Peninsula was not part of Gondwana land. Nature 252:641–645

McKenzie DP (1989) Some remarks on the movement of small melt fractions in the mantle. Earth Planet Sci Lett 95:53–72

Meen JK, Eggler DH (1987) Petrology and geochemistry of the Cretaceous Independence volcanic suit, Absoraka Mountains, Montana:clues to the composition of the Archean sub-Montana mantle. Bull Geol Soc Am 98:238–247

Melluso L, Conticelli S, D'Antonio M, Mirco NP, Saccani E (2003) Petrology and mineralogy of wollastonite- and melilite-bearing Paralavas from the Central Apennines, Italy. Am Mineral 88:1287–1299

Melluso L, Morra V, Girolama Di (1996) The Mt. Vulture volcanic complex (Italy): evidence for distinct parental magmas and for residual melts with melilite. Mineral Petrol 56:225–250

Melluso L, Morra V, Annamaria P, Clandio S, Mariarosaria A (1995) The equption of the Breccia Museo (Campi Flegrei, Italy): fractional crystallization processes in a shallow zoned magma chamber and implications for the eruptive dynamics. J Volcanol Geotherm Res 68:325–339

Melzer S, Foley S (1996) Phase relationship of phlogopite in the system $KAlSiO_4$-Mg_2SiO_4-SiO_2-F and the stability of F-phlogopite at atmospheric pressure, 10 and 18 kbar. Neues Jahrbuch fur Mineralogie Abbandlungen 7:1–31

Menzies M, Rogers N, Tindle A, Hawkesworth C (1987) Metasomatic and enrichment processes in lithospheric peridotites, an effect of asthenosphere-lithosphere interaction: Mentle metasomatism. Academic Press, London, pp 313–361

Menzies MA, Hawkesworth C (1987) Mantle Metasomatism. Academic Press, London 472 pp

Merill RB, Wyllie PJ (1975) Kaersutite and kaersutite eclogite melting from Kakanui, New Zealand, water-excess and water-deficient melting to 30 kilobars. Bull Geol Soc Am 86:555–570

Mertes H, Schminke HU (1983) Age distribution of volcanoes in the west Eifel. Neues Jahrb Geol Palaontol Abh 166(2):260–293

Meyer W, Alberts HJ, Bernes HP, Von Gehlen K, Glatthaar D, Lohnertz W, Peffer KH, Schutgen A, Wienecke K, Zakosek H (1983) Pre-quaternary uplift of the central part of the Rhenish Massif. In: Fuchs K, Von Gehlen K, Malzer H, Murawski H, Semmel A (eds) Plateau uplift. The Rhenish Shield—a case histroy. Springer, Berlin, pp 39–46

Middlemost EAK, Paul DK, Flectcher IR (1988) Geochemistry and mineralogy of the minette lamproite association from the Indian Gondwanas. Lithos 22:131–142

Mihajlovic T, Lengauer CL, Ntaflos T, Kolitsch U, Tillmanns E (2004) Two new minerals, rondorfite, $Ca_8Mg[Sio_4]_4Cl_2$, and almarudite [K□, $(Na)_2$ $(Mn,Fe,Mg)_2(Be,Al)_3[Si_{12}O_{30}]$ and a study of iron-rich wadalite $[Ca_{12}[(Al_8Si_4Fe_2)O_{32}]C_4$, from the Bellerberg (Bellberg) Volcano, Eifel, Germany. Neues Jahrbuch Fur Mineralogie-Abhandlungen 179:265–294

Mineyva IG (1972) Principal geochemical characteristics of potassic alkaline rocks. Geochem Int 9:173–179

Mirnejad H, Bell K (2006) Origin and source evolution of the Leucite Hills lamproites: evidence from Sr-Nd-Pb-O isotopic compositions. J Petrol. 47:2463–2489

Mitchell RH (1985) A review of the mineralogy of lamproites. Trans Geol Soc S Africa 88:411–434

Mitchell RH (1995) Melting experiments on a sanidine-phlogopite lamproite at 7 GPa and their bearing on the source of lamproitic magma. J Petrol 36:1455–1474

Mitchell RH, Bergman SO (1991) Petrology of lamproites. Plenum, New York 441

Mitchell RH, Edgar AD (2002) Melting experiments on SiO_2-rich lamproitest to 6.4 GPa and their bearing on the sources of lamproite magmas. Mineral Petrol 74:115–128

Mitchell RH, Platt RG, Downey M (1987) Petrology of lamproites from Smoky Bute, Montana. J Petrol 28:645–677

Miyashiro A (1960) Thermodynamics of reactions of rock-forming minerals with silica, Part-I. J Japan Geol Geogr 33:71–78

Modeski PJ, Boettcher AL (1972) The stability of phlogopite-enstatite at high pressure: A model for micas in the interior of the earth. Am J Sci 272:852–869

Molner P, Tapponier P (1975) Cenozoic tectonics of Asia: effect of continental collision. Science 189:414–426

Morey GW, Hesselgesser JM (1952) The system H_2O-Na_2O-SiO_2 at 400°C. Am J Sci Bowen 343–372

Morgan WJ (1983) Hot-spot tracks and the early rifting of the Atlantic. Tectonophysics 94:123–139

Morse SA (1969) Alkali feldspar-water at 5 kb. Carnegie Inst Wash Yb 67:120–126

Muelenbachs K, Kushiro I (1974) Oxygen isotope exchange and equilibrium of silicates with CO_2 or O_2. Carnegie Inst Wash Yb 73:232–236

Mysen BO (1983) The CO_2-clinopyroxene system. Neues Jahrb Mineral Abh 146:41–65

Nag K, Arima M, Gupta AK (2007) Experimental study of the join forsterite-diopside-leucite and forsterite-leucite-akermanite up to 2.3 GPa [P (H_2O)=P (total)] and variable temperatures; its petrological significance. Lithos 98(1-4):177–194

Nakamura Y, Yoder HS (1974) Analcite, hyalophane, and phillipsite from the Highwood Mountains, Montana. Carnegie Inst Wash Yb 73:354–358

Nappi G, Renzulli A, Santi P (1991) Evidence of incremental growth in the Vulsinian calderas (central Italy). J Volcanol Geotherm Res 47:13–31

Nash WP, Wilkinson JPG (1970) Shonkin Sag Laccolith, Montana, (part I). Mafic minerals and estimates of pressure, oxygen fugacity and silica activity. Contrib Mineral Petrol 33:162–170

Nash WP, Wilkinson JPG (1971) Shonkin Sag Laccolith, Montana, (part II). Bulk rock geochemistry. Contrib Mineral Petrol 33:162–170

Navon O, Stolper E (1987) Geochemical consequences of melt percolation: The upper mantle as a chromatographic column. J Geol 95:285–307

Nelson DR, McCulloch MT, Ringwood AE (1986b) Ultrapotassic magmas: end-porducts of subductions and mantle recycling of sediments. In: 4th international Kimberlite conference. Geological society of Australia. Abstract Series, vol 16. 11–15 Aug 1986

Nelson DR, McCulloch MT, Sun SS (1986) The origin of ultrapotassic rocks as inferred from Sr, Nd and Pb isotopes. Geochim Cosmochim Acta 50:231–245

Nelson IA, Whitford DJ (1983) Potassium-rich volcanic rocks of the Muriah Complex, Java, Indonesia products of multiple magma source. J Volcanol Geotherm Res 18:337–359

Nelson RN (1992) Isotopic characteristics of potassic rocks: evidence for the involvement of subduction sediments in magma genesis, vol 28. Elsevier Science Publishing, pp 403–420

Newton RC, Sharp WE (1975) Stability of forsterite + CO_2 and its bearing on the role of CO_2 in the mantle. Earth Planet Sci Lett 26:239–244

Nicolas A (1986) A melt extraction model based on structural model based on structural studies in mantle peridotites. J Petrol 27:999–1022

Nielson JE, Wilshire HG (1993) Magma transport and metasomatism in the mantle: a critical review of current geochemical models. Am Mineral 78:1117–1134

Niggli P (1923) Gesteins und Mineralprovinzen. Band I: Ein-fuhrung. Verlag von Gebruder Borntrager, Berlin

Ninkovich D, Hays JD (1972) Mediterranean island arcs and origin of potash volcanoes. Earth Planet Sci Lett 16:331–345

Nixon PH, Boyd FR (1973) Petrogenesis of the granular and sheared ultrapotassic nodule suite in kimberlite. In: Nixon PH (ed) Lesotho Kimberlites. Lesotho National Development Corporation, Maseru, pp 48–56

Nixon PH, Thirwall MF, Buckley F, Davies CJ (1984) Spenish and Western Australian lamproites: aspects of whole rock geochemistry. In: Kornprobst J (ed), Kimberlites and related rocks, vol 2A, pp 285–296

Nkoumbou C, Deruelle B, Velde D (1995) Petrology of Mt. Etinde nephelinite series. J Petrol 35 (2):373–395

Nockolds SR, Knox RW, Chinner GA (1978) Petrology for students. Cambridge University Press, Cambridge, pp 203–215

Norris K (1951) Priderite a new minerals from the leucite lamproites of the West Kimberley area. Mineral Mag 29:496–506

Nougier J (1972) Volcanic associations in Iles Kerguelen. In: Adie RJ (ed) Antarctic geology and geophysics. Oslo University Press, Oslo, pp 809–815

Nurlibayev AN (1993) Alkaline rocks of Kazakishtan and their ore deposits. Nouka, Alme-Ata, p 296

O'Brien HE, Thirlwall MF (1988) Characterization of source components of potassic mafic magmas of the Highwood Mountains Province, Montana. EOS. Trans Am Geophys Union 69:519

O'Brien HE, Irving AJ, McCallum IS (1988) Complex zoing and resorption of phenocrysts in mixed potassic mafic magmas of the Highwood mountains Province, Montana. Am Mineral 73:1007–1024

O'Brien HE, Irving AJ, McCallum IS (1991) Eocene potassic magmatism in the Highwood Mountains, Montana; Petrology, geochemistry and tectonic implications. J Geophys Res 96:13237–13260

O'Connor JM, Duncan RA (1990) Evolution of the Walvis Ridge-Rio Grande Rise. Hot-spot system imphications for African and south American Plate motions over plumes. J Geophys Res 95:17475–17502

O'Hara MJ, Yoder HS (1967) Formation and fractionation of basic magmas at high pressures. Scott J Geol 3:67–117

Okey AI (1997) Jadeite - K-feldspar rocks and jadeitites from northwest Turkey. Mineral. Mag. 61:835–843

Onuma K, Yagi K (1967) The system diopside-akermanite-nepheline. Am Mineral 52:227–243

Orlova MP (1990) Mesozoic stage of magmatism. In: Polakov GP, Depezinskas VV (eds) Potassium alkaline magmatism of the Baikal-Stanovoy rifting system. Siberian Branch of the Academy of Sciences, Novosibirsk, pp 65–123

Orlova MP, Ardontsev SN, Shchadenkov EM (1986) Alkaline magmatism of the Aldan shield and its specific mineralogical features. In: Frunmkin IM (ed) Geology and geochemistry of the ore-bearing magmatic and metasomatic associations from the Maly BAM area. Yakutia branch of the Siberian Division of the USSR Academy of Sciences, Yakutsk, pp 4–12

Orville PM (1967) Unit Cell parameters of the microcline-low albite and the sanidine-high albite solid solution series. Am Mineral 52:58–86

Osann A (1906) Uber einige Alkaligesteine aus Spanien. Festschr. Rosenbusch Stuttgart, pp 263–310

Osborn EF, Schairer JF (1941) The ternary system pseudowollastonite-akermanite-gehlinite. Am J Sci 239:715–763

Osborn EF, Tait DB (1952) The system diopside-forsterite-anorthite. Am J Sci (Bowen Vol):413–433

Osborne FF (1930) The nepheline-gneiss complex near Egan Chute, Dungannon Township, and its bearing on the origin of nepheline syenite. Am J Sci 200:33–60

Panomareva LG, Dobrestsov NL (1965) Jadeite-bearing and other amygdules in meta-extrusives of north-western Kamchatka Dokl. Acald. Sci. U.S.S.R. Earth Sci Sect 167:87–91

Patacca E, Scandone P (1987) Tectonic evolution of the outer margin of the Apenend and the foredeep system: the lithosphere in Italy. Adv Earth Sci Res Acad Naz Lincei 139–142

Paul DK, Potts PJ (1981) Rare earth abundunces of some Indian lamprophyres. Geol Mag 118:393–399

Paul DK, Sarkar A (1984) Petrogenesis of some Indian lamprophyres. Spec Pub Geol Surv Ind 12:45–54

Pearce TH (1970) The analcite-bearing volcanic rocks of the Crowsnest formation. Alberta Canad J Earth Sci 7:46–66

Peccerillo A (1985) Roman Comagmatic Province (central Italy): Evidence for subduction-related magma genesis. Geology 13:103–106

Peccerillo A (2001) Geochemical similarities between the Vesuvius, Phlegraean Fields and Stromboli Volcanoes: petrogenetic, geodynamic and volcanological implications. Mineral Petrol 73:93–105

Peccerillo A (2004) Carbonate-Rich pyroclastic rocks from central apennines: carbonatites or carbonated rocks? a commentary. Periodico Di Mineralogia 73:165–175

Peccerillo A, Taylor SR (1976) Geochemistry of Eocene calc-alkaline volcanic rocks from the Kastomonu area, northern Turkey. Contrib Mineral Petrol 58:63–81

Perchuk LL (1964) Physico-chemical petrology for granitoid and alkaline intrusives of the central Turkestan-Alay. Nauka, Moscow 159 pp

Perini G, Conticelli S (2002) Crystallization Conditions of leucite-bearing magmas and their implications on the magmatological evolution of ultrapotassic magmas: The Vico Volcano, Central Italy. Mineral Petrol 74:253–276

Perini G, Francalanci L, Davidson JP, Conticelli S (2004) Evolution and genesis of magmas from Vico volcano, Central Italy: Multiple differentiation pathways and variable parental magmas. J Petrol 45:139–182

Perini G, Tepley FJ, Davidson JP, Conticelli S (2003) The origin of K-feldspar megacrysts hosted in alkaline potassic rocks from central Italy: a track for low-pressure processes in mafic magmas. Lithos 66:223–240

Peters KE, Weeney, RE, Kaplan, IR (1978) Correlation of carbon and nitrogen stable ratios in sedimentary organic matter. Limnol Oceanogr 23:598–604.S

Phelps DW, Gust DA, Wooden JL (1983) Petrogenesis of the mafic feldspathoidal lavas of the Raton-Clayton volcanic field, New Mexico. Contrib Mineral Petrol 84:182–190

Phillips D, Harris JW (1995) Geothermobarometry of diamond inclusions from the DeBeers pool Mines, Kimberly, South Africa. In: Sixth international Kimberlite conference. Extended Abstract Novosibirsk. United Institute of Geology, Geophysics and Mineralogy, Siberian Branch of the Russian Academy of Sciences, pp 441–443

Pilot J, Werner C, Hulbrich F, Berkman N (1988) Paleozoic and Proterozoic, Zircon from the Mid-Atlantic ridge. Nature 393:676–679

Platt RG, Edgar AD (1971) The system nepheline-diopside-sanidine and its significance in the genesis of melilite and olivine-bearing alkaline rocks. J Geol 80:224–236

Poli G, Frey FA, Ferrara G (1984) Geochemical characteristics of south Tuscany (Italy) volcanic province: Constrainst on lava petrogenesis. Chem Geol 43:203–221

Pollack HN, Chapman DS (1977) On the regional variation of heat flow, geotherms and lithospheric thickness. Tectonophysics 38:279–296

Post JE, Von Dreele RB, Buseck PR (1982) Symmetry and cation displacements in hollandites: structure refinements of hollandite, cryptomelane and priderite. Acta Crystal B38:1056–1065

Prelevic D, Foley SF, Cvetkovic V, Romer RL (2004) The analcime problem and its impact on the geochemistry of ultrapotassic rocks from Serbia. Mineral Mag 68(4):633–648

Prider RT (1939) Some minerals from the leucite-rich rocks of West Kimberley area of Western Australia. Mineral Mag 25:373–383

Prider RT (1982) A glassy lamproite from the West Kimberley area, Western Australia. Mineral Mag 45:279–282

Prodehl C (1989) Central European Rift system—A review of crustal structure from seismic interpretations. Sonder-forschungsbereich 108, Spannung und Spannungsumwandlung in der Lithosphare. University of Karlsruhe, Representative 1987–1989, pp 277–322

Prodehl C, Mechin J (1991) Crustal thinning in relationship to the evolution of the Afro-Arabian rift system: a review of seismic refraction data. Tectonophysics 198:311–327

Purdy GM, Detrick R, Shor GG (1977) Crustal structure of the Banda sea and Weber Deep. EOS. Trans Am Geophys Union 58:509

Putnis CV, Geisler T, Schmid-Beurmann P, Stephan T, Giampaolo C (2007) An experimental study of the replacement of leucite by analcime. Am Mineral 92:19–26

Raja Rao CS (1987) Coalfields of India. Bull Geol Surv Ind Series A 45(4 Pt. I):336

Rice JM (1977) Progressive Metamorphism of impure dolomitic limestone in the Boulder aureole, Montana. Contrib Mineral Petrol 59:237–529

Richerds MA, Duncan RA, Courtillot BE (1989) Flood basalts and hot sopt tracks, plume heat and tails. Science 246:103–107

Richter FH, Rousley DB, De Paolo DJ (1992) Sr isotope evaluation of sea water. The role of tectonics. Earth Planet Sci Lett 109:11–23

Rittmann A (1933) Die geologisch bedingte evolution und differentiation des Somma-Vesuvius magma. Z Vulkanol 5:8–94

Rittmann A (1951) Magmatic character and tectonic position of the Indonesian volcanoes: Nomenclature of volcanic rocks. Bull Volcanol Ser II Tome 12:46–58

Robert JL (1976) Titanium solubility in synthetic phlogopite solid solution. Chem Geol 17:213–227

Rock NMR (1984) Nature and origin of calc-alkaline lamprophyres: minettes, vogesites, kersantites and spessartites. Trans Royal Soc Edinb Earth Sci 74:193–227

Rock NMR (1991) Lamprophyre. Blackie, Glasgow, p 285

Rock NMR (1987) The nature and origin of lamprophyres, an orerview. In: Fitton JG, Upton BGJ (eds) Alkaline igneous rocks, vol 30. Special Publications and Geological Society, London, pp 191–226

Rock NMS, Griffin BJ, Edgar AD, Paul DK, Hergt JM (1992) Olivine lamproites and minettes from Jharia coal field, Eastern India: A spectrum of petentially diamondiferous lmaporites and minettes from the Jharia coal field, eastern India. J Volcanol Geotherm Res 50:55–83

Roden, MF, Smith D (1979) Field geology and chemistry of Buell Park Minette, Diatreme, Apache country, Arizona. In: Boyd, FR, Meyer HOA (eds) Kimberlites, diatremes and diamond: their geology petrology and geochemistry. Proceedings of 2nd international conference, vol 1. A.G.U., Washington, DC, pp 364–381

Roden MP, Smith D, McDowell FW (1979) Age and extent of potassic volcanism on the Colorado Plateau. Earth Planet Sci Lett 43:279–284

Rogers NW (1992) Potassic magmatism as a key to trace element enrichment process in the upper mantle. J Volcanol Geotherm Res 50:85–99

Rogers NW, Hawksworth CJ, Parker PRJ, Marsh JS (1985) The geochemistry of potassic lavas from Vulsini, central Italy and implication form mantle enrichment process beneath the Roman regons. Contrib Mineral Petrol 90:244–257

Rogers NW, James D, Kelley SP, De Mulder M (1998) The generation of potassic lavas from the eastern Virunga Province, Rwanda. J Petrol 39(8):1223–1247

Rowell WF, Edgar AD (1983) Cenozoic potassium-rich mafic volcanism in the western USA: its relationship to deep subduction. J Geol 91:338–431

Ryabchikov ID, Green DH (1978) The role of CO_2 in the petrogenesis of highly potassic magmas. In: Problems of highly potassic Earth's crust and upper mantle. Tr Inst Geol Geofiz Akad Nauk S.S.S.R. No. 403:49–64 (in Russian)

Sack RC, Carmicheal ISE, Rivers ML, Ghiroso MS (1980) Ferric-Ferrous equilibrium in natural silicate liquids at 1 bar. Contrib Mineral Petrol 75:369–376

Sahama ThG (1973) Evolution of Nyiragongo magma. J Petrol 14:33–48

Sahama ThG (1974) Potassium-rich alkaline rocks. In: Sorensen H (ed) The Alkaline rocks. Wiley, London, pp 96–100

Sahama ThG, Meyer A (1958) Study of the volcano Nyiragongo, a progress report. Inst. Parcs. Nationaux Cango Belge, Mission d' Etudes Vulcanologiques fasc, vol 2. pp 1–18

Sahama ThG, Neuvonen KJ, Hytonen K (1956) Determination of the composition of kalsilites by an X-ray method. Mineral Mag 31:200–210

Sahama ThG, Wiik HB (1952) Leucite, potash nepheline and clinopyroxene from volcanic lavas from southwestern Uganda and adjoining Belgian Cango. Am J Sci (Bowen Vol):457–470

Sandiford M, Santosh M (1991) A granulite facies kalsilite-leucite-hibonite association from Punalur Southern India. Mineral Petrol 43:225–236

Sano Y, Takahata N, Nishio Y, Marty B (1998) Nitrogen recycling in subduction zones. Geophys Res Lett 25:2289–2292

Sano Y, Takahata N, Nishio Y, Fischer TP, Williams SN (2001) Volcanic flux of nitrogen from the Earth. Chem Geol 171:263–271

Santi P (1990) New geochronological data of the Vulsini volcanic district (central Italy). Plinius N. 4 Suppl. Euro J Mineral Porc Sci Ital Mineral Petrol Confress. Ischia, Italy, pp 15–18

Sato M (1972) Intrinsic oxygen fugacities for iron-bearing oxide and silicate minerals under low total pressure. Geol Soc Mem 135:289

Savelli C (1967) The problem of rock assimilation by Somma-Vesuvious magma. Contrib Mineral Petrol 16:328–353

Scarf CM, Luth WC, Tuttle OF (1966) An experimental study bearing on the absence of leucite in plutonic rocks. Am Mineral 51:726–735

Schairer JF (1948) Phase equilibrium relations in the quaternary system K_2O-MgO-Al_2O_3-SiO_2 (preliminary report) Bull. Geol Soc Am Abstr 59:1349

Schairer JF (1950) The alkali feldspar join in the system $NaAlSiO_4$-$KAlSiO_4$-SiO_2. J Geol 58:512–517

Schairer JF (1954) The system K_2O-MgO-Al_2O_3-SiO_2: I, results of quenching experiments on four joins in the tetrahedron cordierite-forsterite-leucite-silica and on the join cordierite-mullite-potash feldspar. J Am Ceram Soc 37:501–533

Schairer JF, Bowen NL (1935) Preliminary report on the equilibira relations between feldspathoids, alkali-feldspars and silica. Trans Am Geophys Union In: 16th Annual Meeting, 325–328

Schairer JF, Bowen NL (1938) The system leucite-diopside-silica. Am J Sci 35A(5):289–309

Schairer JF, Bowen NL (1947) The system anorthite-leucite-SiO_2. Geol Finl Bull 20:67–87

Schairer JF, Bowen NL (1955) The system K_2O-Al_2O_3-SiO_2. Am J Sci 253:681–746

Schairer JF, Yoder HS (1964) The join akermanite ($Ca_2MgSi_2O_7$)-sodamelilite ($NaCaAlSi_2O_7$). Carnegie Inst Wash Yb 63:89–90

Schairer JF, Yagi K, Yoder HS (1962) The system diopside nepheline. Carnegie Inst Wash Yb 61:96–98

Schairer JF, Yoder HS, Tilley CE (1965) Behaviour of melilite in the join gehlentite-soda-melilite-akermanite. Carnegie Inst Wash Yb 64:95–100

Schairer JF, Yoder HS, Tilley CE (1967) The high temperature behaviour of synthetic melilite in the join gehlenite-sodamelilite-akermanite. Carnegie Inst Wash Yb 65:217–226

Scharirer JF, Yoder HS (1970) Critical planes and flowsheet for a portion of the system CaO-MgO-Al_2O_3-SiO_3. Carnegie Inst Wash Yb 68:202–214

Schmidt MW (1996) Experimental constraints of recycling of potassium from subducted oceanic crust. Science 272:1927–1930

Schneider H (1965) Petrographie des Latera-vulkans und die magmenenwicklung der Monte Volsini (prov. Grosseto. Viterbo und Orvieto, Italian). Schweiz Min Petr Mitt 43:331–455

Scott Smith BH (1981) Kimberlite and lamproite dykes from Holsteinborg, West Greenland. Middelelser, Om Grandland. Geoscience 4:1–24

Scott Smith BH (1989) Lmaproites and kimberlites in India. Neues Jahrb Mineral Abh 161 (2):193–225

Scott Smith BH, Skinner EMW (1984) A new look at lamproitic rocks from Prairie Creek Arkansas. In: Kornprobst J (ed) Kimberlites I: kimberlites and related rocks, Elsevier, Amsterdam, pp 225–283

Scott Smith BH, Skinner EMW (1984) Diamondiferous lamproites. J Geol 92:433–438

Scott Smith BH, Skinner EMW, Loney PE (1987) The Kapamba lamproites of the Luangwa Valley, eastern Zambia. Geology A2-1G1(8):95–103

Seifert F, Schreyer W (1971) Synthesis and stability of micas in the system K_2O-MgO-SiO_2-H_2O and their relations to phlogopite. Contrib Mineral Petrol 30:196–215

Sekl Y (1961) Jadeite from Kanasake of the kanto Mountains, Central Japan. J Geol Soc Japan 67:101–104

Semenova VG, Soloveva IV, Vladimirov BM, Zavyalova LL, Barankevich VG (1987) Glass and quenching phases from the deep-seated inclusions from alkaline basaltoids of Tokinsky Stanovik. In: Zharikov VA, Grachev AF (eds) Deep-seated xenoliths and structure of the lithosphere. Nauka, Moscow, pp 73–95

Semenova VG, Soloveva LV, Vladimirov BM (1984) Deep-seated inclusions from alkaline basaltoids from Tokinsky Stanovik. Nauka, Siberian Division of the U.S.S.R. Academy of Sciences, Novosibirsk, p 119

Shand SJ (1910) On borolannite and its associates in Assynt. Trans Edin Geol Soc 2:202–210

Shand SJ (1931) The granite-syenite-limestone complex of Palabora, Transvaal. Trans Geol Soc S Africa 67:81–92

Shand SJ (1933) The lavas of Mauritius. Q J Geol Soc London 89:1–13

Shand SJ (1943) Eruptive rocks, vol 2. Wiley, New York, p 444

Sharda P, Moreira M, Staudacher T (1999) Argon-led isotopic correlation in Mid-Atlantic Ridge basalts. Science 283:666–668

Shea T, Larsen JF, Gurioli L, Hammer JE , Houghton BF, Cioni R (2009) Leucite crystals: Surviving witnesses of magmatic processes preceding the 79AD eruption at Vesuvius, Italy. EPSL 281(1–2):88–98

Sheraton JW, Cundari A (1980) Leucites from Gaussberg, Antarctica. Contrib Mineral Petrol 71:417–427

Sheraton JW, England RN (1980) Highly potassic dykes from Antarctica. J Geol Soc Aust 27:129–135

Singh P, Gupta AK (1992) Phase relations in the join diopside-nepehline under 10 kb in absence of volatiles. In: Abstracts 29th international geological congress, Kyoto, Japan

Singh P, Arima M, Gupta AK (2000) Phase equilibrium constraints on the petrogenesis of nephelinites with reference to the system diopside-nepheline under variable P-T conditions in the presence or absence of water. J Mineral Petrol Sci 95:113–124

Sittler C (1969) The sedimentary trough of the Rhine Graben. Tectonophysics 8:543–560

Skinner EMW, Scott BH (1979) Petrography, mineralogy and geochemistry of kimberlite and associated lamprophyre dykes near Swatruggens, Western Transwaal, R.S.A. In: De Beers kimberlite symposium, II

Sleep NH (1996) Leteral flow of hot plume material ponded at sub-lithospheric depths. J Geophys Res 101:28065–28083

Solovova IP, Girnis AV, Kogarko LN, Konorikova NN, Stoppa F, Rosatelli G (2005) Compositions of magmas and carbonate-silicate liquid immiscibility in the Vulture alkaline igneous complex, Italy. Lithos 85:113–128

Smith CB, Lorenz V (1986) Volcanology of the Ellendale diatremes. In: 4th international Kimberlite conference, vol 16. Geological Society of Australia (Abstract Series), pp 250–207

Smith JV, Tuttle OF (1957) The nepheline-kalsilite system: I. X-ray data for the crystalline phases. Am J Sci 255:282–305

Snavely PD Jr, MacLeod NS, Flanagan FJ, Berman S, Neiman HG, Bastron H (1976) Nepheline syenite, STM-1, from Table Mountain, Oregon. In: Flanagan FJ (ed) Descriptions and analyses of eight new USGS rock standards, vol 840. US Geological Survey Professional Paper, pp 7–10

Sobolev VS, Bazarova TJ, Yagi K (1975) Crystallization temperature of wyomigite from Leucite Hills. Contrib Mineral Petrol 49:301–308

Sparks RSJ (1975) Stratigraphy and geology of the ignimbrites of Vulsini volcano, central Italy. Geol Rundsch 64:497–523

Spencer AB (1969) Alkali igneous rocks of the Balcones Province, Texas. J Petrol 10:272–306

Stephensen AL (1972) Geochemistry of some Heard Island igneous rocks. In: Adei RJ (ed) Antarctic geology and geophysics. pp 793–801

Stewart K, Turner S, Kelley S, Hawkesworth C, Kirstein L, Montovani M (1996) ^{40}Ar-^{39}Ar geochronology in the Parana continental flood basalt province. Earth Planet Sci Lett 143:95–109

Stoppa F (1988) L'euremite di colle Fabbri (spoleto): Un Litipoad Affinita Carbonatitica in Italia. Boll Geol Soc Italy 107:239–248

Stoppa F, Woolley AR, Cundari A (2002) Extension of the melilite-carbonatite province in the Apennines of Italy: the kamafugite of Grotta Del Cervo, Abruzzo. Mineral Mag 66:555–574

Strong D (1972) Petrology of the Island of Moheli, Western Indian Ocean. Bull Geol Soc Am 83:389–406

Sudo A, Tatsumi Y (1990) Phlogopite and K-amphibole stability in the mantle: Implication for magma genesis in subduction zone. Geophys Res Lett 17:29–32

Sugimura A (1959) Geographische Verteilung der Charaketerstendenzen des magma in Japan. Bull Volcanol Soc Japan 4:77–103

Sugimura A (1960) Zonal arrangement of geophysical and petrological feature in Japan and its environs. J Tokyo Univ Fac Sci 12(2):133–153

Sugimura A (1968) Spatial relations of basaltic magmas in island area. In: Hess HH, Poldervaart A (eds) Basalts. Interscience, New York, pp 537–571

Sun SS, McDonough WF (1989) Chemical and isotopic systematics of oceanic basalts: implications for mantle compositions and process. In: Saunders AD, Norry MJ (eds) Magmatism in the Ocan Basins, vol 42. Geological Society, London, pp 131–345 (Special Publication)

Sweeney RJ, Thompson AB, Ulmer P (1993) Phase relation of a natural MRID compositon and implications for lithospheric melting and mantle metasomatism. Contrib Mineral Petrol 115:225–241

Takahashi E, Kushiro I (1983) Melting of a dry peridotite at high pressures and basalt magma genesis. Am Mineral 68:859–879

Tanton KM, McKenzie D (1994) The generation of kimberlites, lamproites and their source rocks. J Petrol 35(3):787–817

Tateiwa I (1976) The Korea-Japanese geotectonic zone. Tokyo University Press, Tokyo, p 654

Tatevosyan TSh (1966) Intrusive rocks of the Bargushatsk ridge. Geology for the Armenian SSR Perography. Instrusive Rocks 3:114–214

Taylor D, MacKenzie WS (1975) A contribution to the pseudoleucite problem. Contrib Mineral Petrol 49:321–333

Taylor SR, McLennan SM (1985) Continental crust: Its composition and evolution. Blackwell, London, p 312

Taylor WR (1985) The role of C-O-H fluid in the upper mantle process: theoretical experimental and spectroscopic study Ph.D. dissertation, Universtiy of Tasmania

Templeman-Kluit DJ (1969) Re-examination of pseudoleucite from the spotted Fawn Creek, West Central Yukon. Can J Earth Sci 6:55–62

Thibault Y, Holloway JR (1994) Solubility of CO_2 in a Ca-rich leucitite: effects of pressure, temperature, oxygen fugacity. Contrib Mineral Petrol 160:216–224

Thompson AB (1992) Water in the earth's upper mantle. Nature 358:295–302

Thompson RN (1977) Primary basalts and magma genesis (III), Alban hills, Roman Comagamatic Province, central Italy. Contrib Mineral Petrol 60:91–108

Thompson RN, Leat PT, Humphreys E (1989) What is the influence of Yellowshone mantle plume on pliocene-recent Western USA magmatism. Bull NM Bur Min mineral Res 131:269

Thompson RN, Leat PT, Dicin AP, Morrison MA, Hendry GL, Gibson SA (1990) Strongly potassic mafic magmas from lithospheric mantle sources during continental extension and heating: evidence from Miocene minettes of northwest colorado, USA. Earth Planet Sci Lett 98:139–153

Thompson RN, Morrison MA, Hendry GL, Parry SJ (1984) As assessment of the relative roles of crust and mantle in magma genesis: an elemental approach. Phil Trans Royal Soc London A 310:549–590

Tiepolo M, Zanetti A, Oberti R, Brumm R, Foley S, Vannucci R (2003) Trace-element partitioning between synthetic potassic-richterites and silicate melts, and contrasts with the partitioning behaviour of pargasites and kaersutites. Eur J Mineral 15:329–340

Tilley CE (1956) Nepheline associations. Kon Ned Geol Min Gen Geol Soc (Brower Vol):403–414

Tilley CE, Muir ID (1962) The Hebridean magma type, Edinburgh. Geol Soc Trans 19:208–215

Tilley CE, Muir ID (1964) Intermediate members of the oceanic basalt trachyte association. Geol Foeren Stockholn Foerh 85:436–444

Tilley CE, Yoder HS, Schairer JF (1965) Melting relations of volcanic tholeiite and alkali rock series. Carnegie Inst Wash Yb 64:69–82

Tomita T (1967) Volcanic geology of the Cenozoic alkaline petrographic province of eastern Asia. Geol Mineral Resour Far East 1:139–202

Tronnes RG (2002) Stability range and decomposition of potassic richterite and phlogopite end members at 5-15 GPa. Mineral Petrol, 74 (Edgar volume), 360−368

Tronnes RG, Edgar AD, Arima M (1985) A high pressure-high temperature study of TiO_2 solubility in the Mg-rich phlogopites: implications to phlogopite chemistry. Geochim Cosmochim Acta 49:2323–2329

Tsuboi S (1920) On a leucite rock vulsinite vicoite from Utsuryo Island in the sea of Japan. Geol Soc Japan 27:91–103

Turbeville B (1993) Petrology and petrogenesis of Latera Caldera, central Italy. J Petrol 34:77–123

Turi B, Taylor HP (1976) Oxygen isotope studies potassic volcanic rocks of the Roman Province, central Italy. Contrib Mineral Petrol 55:1–31

Turner F, Verhoogen J (1960) Igneous and metamorphic petrology. McGraw-Hill, New York, p 694

Turner S, Arnaud N, Liu J, Rogers N, Hawksworth G, Haris N, Kelley S, Van Galstern P, Deng W (1996) Post-collision, shoshonitic volcanicsm on the Tibetan Plateau: implications for convective thinning of the lithosphere and the source of Ocean Island Basalt. J Petrol 37 (1):45–71

Tuttle OF, Bowen NL (1958) Origin of granite in the light of experimental studies in the system $NaAlSi_3O_8$-$KAlSi_3O_8$-SiO_2-H_2O. Geol Soc Am Mem 74:153
Uffen RJ (1959) On the origin of rock magma. J Geophys Res 64:117–122
Uffen RJ, Jessop AM (1963) The stress release hypothesis of magma formation. Bull Volcanol 29:7–23
Urano H (1971) Geochemical and petrological study on the origins of metamorphic rocks and granitic rocks by determination of fixed ammoniacal nitrogen. J Earth Sci Nagoya Univ 19:1–24
Uyeda S, Kanamori K (1979) Back-arc opening and the mode of subduction. J Geophys Res 84:1049–1061
van Bemmelem RW (1949) The geology of Indonesia, 1A. The Government Printing office, The Hague, p 732
Van Kooten GK (1980) Mineralogy, petrology and geochemistry of an ultrapotassic basaltic suite, Central Sierra Nevada, California, USA. J Petrol 21:651–684
Varne R (1968) The petrology of Moroto Mountain, Eastern Uganda, and the origin of nephelinites. J Petrol 9:169–190
Varne R (1985) Ancient subcontinental mental: a source for K-rich orogenic volcanics. Geology 13:405–408
Veksler LV, Fedorchuk YM, Nielson TFD (1998) Phase equilibria in the silica-undersaturated part of the $KAlSiO_4$-Mg_2SiO_4-Ca_2SiO_4-F system at 1 atom and the larnite-normative trend of melt evolution. Contrib Mineral Petrol 131:347–363
Velde D (1975) Armalcolite-Ti-phlogopite-diopside-analcite-bearing lamproites from Smoky Butte, Garfield County. Montana Am Mineral 60:556–573
Velde D (1979) Trioctahedral mica in melilite-bearing eruptive rocks. Carnegie Inst Wash Yb 78:468–475
Velde D, Yoder HS (1977) Melilite and melilite-bearing igneous rocks. Carnegie Inst Wash Yb 76:478–485
Venturelli G, Thorpe RS, Dal Piaz GV, Del Moro A, Potts MJ (1984) Petrogenesis of calc-alkaline shoshonite and associated ultrapotassic Oligocene volcanic rocks from the northwestern Alps, Italy. Contrib Mineral Petrol 86:209–220
Vezzoli L (1988) Island of Ischia. In: Vezzoli L (ed) Quaderni de La ricerca Scientifica, vol 114, No 10. CNR, p 126
Vollmer R (1976) Sr and Pb isotope isotope studies of potassic volcanic rocks of the Roman province, Italy: Rb-Sr and U-Th-Pb systematic alkailc rocks from Italy. Geochim Cosmochim Acta 40:283–290
Vollmer R, Norry MJ (1983) Unusual isotopic variations in Nyiragongo nephelinites. Nature 301:141–143
Vollmer R, Norry MJ (1983) Petrology and geochemistry of a U and Th-enriched nephelinites from Mt. Nyiragongo, Zaire: Its bearing on ancient mantle metasomatism. Bull Geol Soc Finland 57:37–46
Vollmer R, Ogden P, Schulting JG, Kingsley RH, Waggoner DG (1984) Nd and Sr isotopes in ultrapotassic volcanic rocks from the Leucite Hills, Wyoming. Contrib Mineral Petrol 87:359–386
Von Rath G (1864) Skizzen aus dem vulkanischen gebiet des Niederrheins. II, Fortsetz Z.D. Geol Ges 16:82
Vukadinovic D, Edgar AD (1993) Phase relations in the phlogopite-apatite system at 20 kbar: implications for the role of fluorine in the mantle melting. Contrib Mineral Petrol 114:247–254
Wade A, Prider RT (1940) The leucite-bearing rocks in the west kimberley area, Western Australia. Q J Geol Soc London 96:39–98
Wagner C, Velde D (1986) The mineralogy of K-richterite-bearing lamproites. Am Mineral 71:17–37
Walker GPL (1984) Downsag calderas, ring faults, caldera sizes and incremental caldera growth. J Geophys Res 89(B-10):8407–8416

Wallace GM, Whalen JB, Martin RF (1990) Agpaitic and miaskitic nepheline syenites of the McGerrigle plutonic complex, Gaspé, Quebec: an unusual petrological association. Can Mineral 28:251–266

Washington HS (1906) The Roman comagmatic region. Carnegie Inst Wash Yb 57:140

Washington HS (1917) Chemical analyses of igneous rocks. US Geological Survey of Professional Paper, p 99

Washington HS (1923) Italite, a new rock. Am J Sci 51:33–42

Washington HS (1927) Italite locality of Vila Senni. Am J Sci 14:173–182

Waters FG, Earlank AJ (1988) Assessment of vertical section and distribution of mantle metasomatism below Kimberley, South Africa. J Petrol (Special Lithosphere Issue):185–204

Watkinson DH (1973) Pseudoleucite from plutonic alkali rock-carbonate complexes. Can Mineral 12:129–134

Watson EB, Brenan JM, Baker DR (1990) Distribution of fluids in the continental mantle. In: Menzies MA (ed) Continental mantle. Clarend Press, Oxford, pp 121–125

Wedepohl KH (1995) The composition of the continental crust. Geochim Cosmochim Acta 59:1217–1232

Weed WH, Pirsson LV (1896) Missourite, a new leucite rock from Highwood mountains of Montana. Am J Sci 2:315–323

Wellman P (1970) Potassium-argon ages for leucite-bearing rocks from New South Wales, Australia. Proc Royal Soc NSW 103:103–107

Wendlandt RF (1984) An experimental and theoretical analysis of partial melting in the system $KAlSiO_4$-CaO-MgO-SiO_2-$CO_2$and application to the genesis of potassic magmas, carbonatites and kimberlites. In: Kronprobst J (ed) Kimberlites and related rocks. Elsevier, Amsterdam, pp 359–369

Wendlandt RF, Eggler DH (1980) The origin of potassic magmas: I. Melting relations in the systems $KAlSiO_4$-Mg_2SiO_4-SiO_2 and $KAlSiO_4$-MgO-SiO_2-CO_2 to 30 Kbar. Am J Sci 280:385–420

Wendlandt RF, Eggler DH (1980) Stability of sanidine + forsterite and its bearing on the genesis of potassic magma and the distribution of potassium in the upper mantle. Earth Planet Sci Lett 51:215–220

Wendlandt RF, Eggler DH (1980) The origins of potassic magmas II: stability of phlogopite in natural spinel lherzolite and in the system $KAlSiO_4$-MgO-SiO_2-H_2O-CO_2 at high pressure and temperature. Am J Sci 280:421–458

White RS, McKenzie DP (1989) Magmatism at rift zones. J Geophys Res 94:11933–11944

White WH, Hofmann AW, Puchelt H (1987) Isotope geochemistry of Pacific Mid-ocean Ridge basalts. J Geophys Res 92:4881–4893

Whiteford DJ, Jezek PA (1979) Origin of Late Cenozoic lavas from the Banda Arc. Indonesia: trace element and Sr isotope evidence. Contrib Mineral Petrol 68:141–150

Whiteford DJ, Nicholls IA, Taylor SR (1979) Spatial variations in the geochemistry of Quaternary lavas across the Sunda arc in Java and Bali. Contrib Mineral Petrol 70:341–356

Wilkinson JFG (1977) Analcime phenocrysts in a vitriophyric analcimite primary or secondary? Contrib Mineral Petrol 64:1–10

Wilkinson JFG, Hensel HD (1994) Nephelines and analcimes in some alkaline igneous rocks. Contrib Mineral Petrol 118:79–91

Willemse J, Bensch JJ (1964) Inclusions of original carbonate rocks in gabbro and norite of the eastern part of the Bushveld Complex. Trans Geol Soc S Africa 67:1–87

Williams H (1936) Pliocene volcanoes of Navajo-Hopi County. Bull Geol Soc Am 47:111–172

Williams H, McBirney AR (1979) Volcanology. Freeman and Coopera, San Francisco

Williams LB, Ferrell RE, Chinn EW, Sassen R (1989) Fixed-ammonium in clays associated with crude oils. Appl Geochem 4:605–616

Wilson M (1992) Magmatism and continental rifting during the opening of the South Atlantic Ocean: a consequence of lower cretaceous super plume activity. In: Storey BC, Alabaster T, Pankhurst RJ (eds) Magmatism and the cause of continental break-up, vol 68. Geological Society, London, pp 241–255
Wilson M, Downes H (1991) Tertiary-Quaternary extension-related alkaline magmatism in Western and Central Europe. J Petrol 32(4):811–849
Wilson M, Giraud R (1993) Magmatism and rifting in western and central Africa from late Jurassic to recent times. Tectonophysics 213:203–225
Wilson NM (1989) Igneous petrogenesis. Unwin Hyman, London, p 466
Wimmenauer W (1974) The alkaline province of central Europe and France. In: Sorensen H (ed) The alkaline rocks. Wiley, London, pp 238–271
Wones DR (1967) A low pressure investigation of the stability of phlogopite. Geochim Cosmochim Acta 31:2248–2253
Woolley AR (1987) Alkaline rocks and carbonatites of the world. Part 1 north and south America. British museum (natural history), London, p 216
Worner G, Schminke HU (1984) Mineralogical and geochemical evolution of the Laacher Sea magma chamber. J Petrol 25:805–835
Worner G, Harmon RS, Hoefs J (1987) Stable isotope relations in an open magma system. Laacher See. Eifel (FRG). Contrib Mineral Petrol 95:343–349
Wright JB (1963) A note on possible differentiation trends in ternary to recent lavas of Kenya. Geol Mag 100:164–180
Wyllie PJ (1974) Limestone assimilation. In: Sorenson H (ed) The alkaline rocks. Wiley, London, pp 459–474
Wyllie PJ (1979) Kimberlite magmas from the system peridotite-CO_2-H_2O. In: Boyd FR, Meyer HOA (eds) Am Geophys Union, vol 1. Washington D.C., pp 319–329
Yagi K, Gupta AK (1977) Experimental study on assimilation reactions of some granitic rocks related to the genesis of leucite rocks. Bull Volcanol Soc Jpn 22:65–74
Yagi K, Matsumoto H (1966) Note on the leucite-bearing rocks of Leucite Hills, Wyoming, USA. J Fac Sci Hokkaido Univ 13(4):301–312
Yagi K, Haira Y, Onuma K, Fukushima N (1975) Stability relation of kaersutite. J Fac Sci Hokkaido Univ 22(16):331–342
Yoder HS (1952) Change of melting point of diopside with pressure. J Geol 60:364–374
Yoder HS (1969) Calc-alkalic andesites: experimental data bearing on the origin of their assumed characteristics. In: McBirney AR (ed) Proceedings of the Andesite conference on Oregon department of geology and mineral industrial bulletin vol 65, pp 77–89
Yoder HS (1973) Melilite stability and paragenesis. Forstschr Mineral 50:140–173
Yoder HS (1986) Potassium-rich rocks: phase analysis and heteromorphic reactions. J Petrol 27 (5):1215–1228
Yoder HS (1989) Igneous and metamorphic facies of potassium-rich rocks: coexisting assemblage in kalsilite-forsterite-larnite-quartz at 950 °C and 2 kbar with or without H_2O. Carnegie Inst Wash Yb 88:54–58
Yoder HS (1990) Igneous and Metamorphic facies of potassium-rich rocks-coexisting assemblages in kalsilite-forsterite-larnite-quartz at 950 °C and 2 kbar with or without H_2O. Carnegie Inst Wash Yb 89:54–59
Yoder HS, Kushiro I (1969) Melting of hydrous phase phlogopite. Am J Sci 267:558–582
Yoder HS, Kushiro I (1972) Composition of residual liquids in the nepheline-diopside system. Carnegie Inst Wash Yb 71:413–416
Yoder HS, Schairer JF (1969) The melilite-plagioclase incompatibility dilemma in igneous rocks. Carnegie Inst Wash Yb 67:101–105
Yoder HS, Tilley CE (1962) Origin of basalt magmas: an experimental study of natural and synthetic rock systems. J Petrol 3:342–532
Yoder HS, Upton BGJ (1971) Diopside-Sanidine-H_2O at 5 and 10 kb. Carnegie Inst Wash Yb 70:108–112

Zalutski VV, Chulkov NT (1971) Find of kalsilite syenites in areas of the Synnyr alkaline pluton. Zap Vses Mineral Obshchest 100:642–644

Zhang ZC, Feng CY, Li ZN, Li SC, Xin Y, Li ZM, Wang XZ (2002) Petrochemical study of the Jingpohu Holocene alkali basaltic rocks, northeastern China. Geochem J 36:133–153

Zhang M, Suddaby P, Thompson RN, Thirlwall MF, Menzies MA (1995) Potassic volcanic rocks in NE China: geochemical constrains on mantle source. Magma J Petrol 36(5):1275–1303

Zheng RS, MacKenzie WS (1984) Preliminary report on the system $NaAlSiO_4$-$KAlSiO_4$-H_2O at PH_2O=5 kbar. Bull Mineral 107:571–577

Zhidikov AY (1990) The Plaeozoic stage of magmatism. In: Polykov GV, Kepezinskar VV (eds) Potassic alkaline magmatism of the Baikal-Stanovoy, rift system. Nauka (Siberian Division), USSR Academy of Science, Novosibirsk, pp 32–64

Ziegler PA (1988) Evolution of the Arctic, North Atlantic and the Western Tethys. Am Assoc Pet Mem 43:198

Ziegler PA (1990) Geological Atlas of western and central Europe. In: Shell international petroleum mineral (2nd edn). Geological Society, London, p 239

Ziegler PA (1992) European Cenozoic rift System. Tectonophysics 208:91–111

Author Index

A.K. Gupta, *Origin of Potassium-rich Silica-deficient Igneous Rocks*,
Springer Geology, DOI 10.1007/978-81-322-2083-1

P

R

S

T

U

Subject Index

A.K. Gupta, *Origin of Potassium-rich Silica-deficient Igneous Rocks*, Springer Geology, DOI 10.1007/978-81-322-2083-1

X

Y

Z